# Metal Nanoparticles: Concepts and Applications

# Metal Nanoparticles: Concepts and Applications

Edited by **Peggy Rusk**

**WILLFORD PRESS**

New York

Published by Willford Press,
118-35 Queens Blvd., Suite 400,
Forest Hills, NY 11375, USA
www.willfordpress.com

**Metal Nanoparticles: Concepts and Applications**
Edited by Peggy Rusk

International Standard Book Number: 978-1-68285-154-8 (Hardback)

The publisher's policy is to use permanent paper from mills that operate a sustainable forestry policy. Furthermore, the publisher ensures that the text paper and cover boards used have met acceptable environmental accreditation standards.

**Trademark Notice:** Registered trademark of products or corporate names are used only for explanation and identification without intent to infringe.

Printed in the United States of America.

# Contents

# Preface

Extremely useful in the fields of engineering and biomedical sciences, metal nanotechnology is an upcoming field of science that has undergone rapid development over the past few years. This book brings forth some of the most innovative concepts and elucidates various applications of nanoparticles. This book will be ideal for the students and researches in the fields of engineering, physical metallurgy, biotechnology, nanotechnology, etc. Some significant syntheses of nanoparticles revolving around gold have been discussed in this book. It also presents studies related to combining of metal nanoparticles such as graphene, silver, etc.

The information contained in this book is the result of intensive hard work done by researchers in this field. All due efforts have been made to make this book serve as a complete guiding source for students and researchers. The topics in this book have been comprehensively explained to help readers understand the growing trends in the field.

I would like to thank the entire group of writers who made sincere efforts in this book and my family who supported me in my efforts of working on this book. I take this opportunity to thank all those who have been a guiding force throughout my life.

**Editor**

# Novel synthesis of Au nanoparticles using fluorescent carbon nitride dots as photocatalyst

Xiaoyun Qin · Wenbo Lu · Guohui Chang ·
Yonglan Luo · Abdullah M. Asiri ·
Abdulrahman O. Al-Youbi · Xuping Sun

**Abstract** The present paper reports on a novel synthesis for Au nanoparticles (AuNPs) with the use of fluorescent carbon nitride dots (CNDs) as a photocatalyst. It suggests that the resultant CND-protected AuNPs (CNDs/AuNPs) exhibit good catalytic activity toward 4-nitrophenol reduction and that the CNDs enhance the catalytic activity via a synergistic effect.

**Keywords** Au nanoparticle · Carbon nitride dot · Fluorescence · Photocatalyst · 4-Nitrophenol reduction · Synergistic effect

## Introduction

In the past few years, metal nanoparticles have been the focus of significant scientific research owing to their size-dependent physical and chemical properties, and substantial effort has been invested into their synthesis and characterization [1, 2]. Among them, Au nanoparticles (AuNPs) are the most stable noble nanoparticles, and as key materials and building blocks for the twenty-first century [3], they represent one of the most widely studied nanomaterials [4]. Although the use of soluble Au dates back to the fifth to fourth century BC [3], the first report of the formation of stable gold colloids appeared as recently as 1857 [5]. After that, many preparative methods have been developed, including chemical reduction, photo-chemistry, sonochemistry, radiolysis, and thermolysis [1–3]. Among these methods, the photochemical one is unique due to its excellent spatial and temporal control. It has no rigorous temperature requirements and does not require the use of harmful strong reducing agents [6]. Indeed, the photochemical route has drawn great attention in synthesizing metal NPs [7–10]. Photocatalysts like polyoxometalates, porphyrin, and semiconductors have been applied to the photochemical synthesis of metal NPs [11–14]. However, most of the photo-catalysts are either expensive or involve complex preparation. Accordingly, there is a strong desire to develop economical and simple photocatalysts for the synthesis of metal NPs.

Carbon nitride materials have been the focus of materials research because of their unique properties and wide applications in catalysis, sensors and corrosion protection, etc. [15–21]. More recently, photoluminescent carbon nitride dots (CNDs) have been easily prepared by the polymerization of $CCl_4$ and 1,2-ethylenediamine (EDA) under reflux, microwave, or solvothermal heating [22]. In this paper, we demonstrate a novel application of such CNDs as an effective photocatalyst to synthesize AuNPs for the first time, carried out by UV irradiation of an aqueous $HAuCl_4$ solution in the presence of ethanol and CNDs. It suggests that the resultant CND-protected AuNPs (CNDs/AuNPs) exhibit good activity in catalyzing the reduction of 4-nitrophenol

X. Qin · W. Lu · G. Chang · Y. Luo · X. Sun
Chemical Synthesis and Pollution Control Key Laboratory
of Sichuan Province, School of Chemistry and Chemical Industry,
China West Normal University,
Nanchong 637002 Sichuan, China

A. M. Asiri · A. O. Al-Youbi · X. Sun
Chemistry Department, Faculty of Science,
King Abdulaziz University,
Jeddah 21589, Saudi Arabia

A. M. Asiri · A. O. Al-Youbi · X. Sun (✉)
Center of Excellence for Advanced Materials Research,
King Abdulaziz University,
Jeddah 21589, Saudi Arabia
e-mail: sun.xuping@hotmail.com

(4-NP) by NaBH$_4$. In addition, CNDs play a synergistic role in enhancing the catalytic activity of Au catalysts.

## Experimental

The chemicals CCl$_4$, EDA, and HAuCl$_4$ were purchased from Aladin Ltd. (Shanghai, China). 4-NP was purchased from Sinopharm Chemical Reagent Co. Ltd, and sodium borohydride (NaHB$_4$) was purchased from Tianjin Fuchen Chemical Corp. Gold sol solution (average, 15 nm) was purchased from Shanghai Jieyi Biotechnology Co., Ltd. All the chemicals were used as received without further purification. The water used throughout all experiments was purified through a Millipore system.

The photoluminescent CNDs were prepared following our previous work [22] by the polymerization of CCl$_4$ and EDA under reflux heating. CCl$_4$ and EDA are used as carbon and nitrogen sources, respectively. In a typical synthesis, 0.69 mL of EDA was added to 1 mL of CCl$_4$ solution. After that, the mixture was heated to 80°C for 60 min. The excess precursors and the resulting small molecules were removed by dialyzing against water using a dialysis membrane for 1 day. The resultant CNDs were dispersed in water for further characterization and use.

The synthesis of AuNPs was carried out in a quartz cuvette using CNDs as a photocatalyst. Twenty microliters of CNDs and 40 μL of HAuCl$_4$ (24.3 mM) were injected into 3 mL 30 % (volume ratio) ethanol aqueous solution in the quartz cuvette. The mixture was placed under a 50-W high-pressure mercury lamp ($\lambda$=254 nm) for 1 h to obtain AuNPs.

The UV/Vis absorption spectra were recorded on a UV5800 spectrophotometer. The suspension was added into a quartz cuvette (1×1 cm, 4 mL) to obtain the UV/Vis absorbance data. Transmission electron microscopy (TEM) measurements were made on a Hitachi H-8100 EM (Hitachi, Tokyo, Japan) with an accelerating applied potential of 200 kV. The sample for TEM characterization was prepared by placing a drop of the dispersion on a carbon-coated copper grid and drying at room temperature. Fluorescence emission spectra were recorded on a Shimadzu RF-5301 spectrofluorometer (Shimadzu Corporation, Kyoto, Japan). A 75-fold diluted solution was introduced to a quartz cuvette (1×1 cm, 2 mL) to obtain fluorescence emission data (excitation wavelength, 360 nm). Zeta potential measurements were performed on a Nano-ZS Zetzsozer ZEN3600 (Malvern Instruments Ltd., UK). The dispersion was transferred to the test cell and measured by the device for three replicates. X-ray photoelectron spectroscopy (XPS) analysis was measured on an ESCALAB MK II X-ray photoelectron spectrometer using Mg Kα (1,253.6 eV) as the exciting source for approximately 14 h in a permanent working mode (12 kV, 20 mA). The sample for XPS characterization

was prepared by placing a drop of the dispersion on a bare indium tin oxide-coated glass substrate, which was air-dried at room temperature. All binding energies were determined after charge correction of the XPS data measured without the use of ion gun by setting, $C_{1s}$=285.0 eV.

## Results and discussion

Figure 1 shows the UV/Vis absorption spectra of the aqueous dispersions of CNDs, HAuCl$_4$, and CNDs–HAuCl$_4$ mixture before and after 1 h UV irradiation. It is seen that CND dispersion shows two peaks at 226 and 360 nm (curve a). The yellowish HAuCl$_4$ solution shows an intense absorption band at 320 nm (curve b) due to the metal-to-ligand charge transfer (MLCT) transition of the AuCl$_4^-$ complex [23]. The CNDs–HAuCl$_4$ mixture before irradiation shows two peaks at 230 and 320 nm (curve c). In contrast, the mixture shows a new intense surface plasmon resonance (SPR) absorption band at 527 nm with the disappearance of the MLCT band at the same time after 1 h irradiation (curve d), suggesting that AuCl$_4^-$ is reduced and AuNPs are generated [3]. Colloidal gold is known to have an intense SPR absorption in the visible region which generally occurs around an energy corresponding to 520 nm. The SPR is caused by light waves that are trapped on the surface due to the interaction with the free electrons of the metal, and the free electrons will oscillate in a collective fashion when the light wave meets the resonance conditions [24]. The color change of the mixture from yellow to purple before and after UV irradiation (inset) provides another piece of evidence to support the formation of AuNPs.

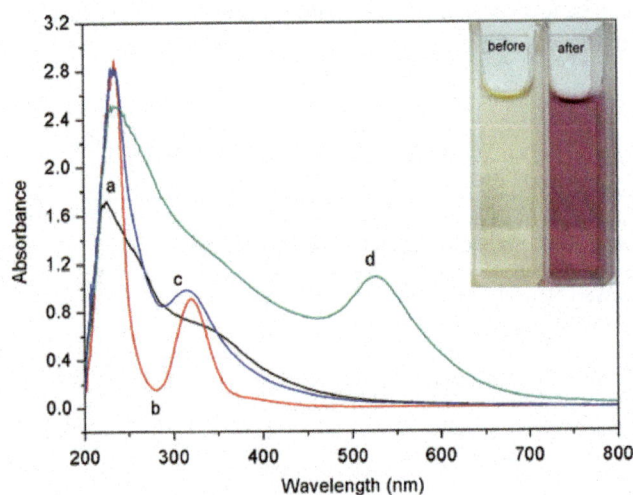

Fig. 1 UV/Vis absorption spectra of the aqueous dispersions of CNDs (*a*), HAuCl$_4$ (*b*), CNDs–HAuCl$_4$ mixture before (*c*) and after (*d*) 1 h of UV irradiation (*inset*, photographs of CNDs–HAuCl$_4$ mixture before and after UV irradiation)

Figure 2 shows the TEM images of the product thus formed, indicating the formation of nanoscale particles that are nearly spherical in shape. The corresponding particle size distribution histograms shown in Fig. 2c indicate that they have diameters ranging from 5 to 25 nm. The chemical composition of the product was further determined by the energy-dispersed spectrum, as shown in Electronic supplementary material (ESM) Fig. S1. The peaks of C, N, and Au elements are observed, indicating the existing of CNDs and AuNPs. The peaks of Si and Cu elements originate from the copper grid substrate used for TEM measurements, and the Cl element observed originates from $Cl^-$ ions released from $AuCl_4^-$. The high-resolution TEM (HRTEM) image taken from one nanoparticle (see the inset in Fig. 2b) reveals clear lattice fringes with an interplane distance of 0.236 nm corresponding to the (111) lattice space of metallic gold [25], further suggesting that these particles are AuNPs. It is worth mentioning that these AuNPs can be very stable for several weeks without observation of any floating or precipitated particles. The zeta potential of AuNPs was measured as 8.35 mV as a positive net charge, which can be attributed to the adsorption of positively charged CNDs. The discrepancy of the zeta potential between AuNPs and CNDs (27.2 mV) [22] is due to the simultaneous adsorption of negatively charged counterions, i.e., $Cl^-$ ions, on AuNPs [26]. The observation of zeta potential at charge stoichiometry being still positive of incomplete charge neutralization is due to the stabilized assembly charge caused by steric factors [27]. It is well established that AuNPs can quench fluorescence when fluorescent molecules and AuNPs are brought into close proximity with each other [28, 29]. Indeed, substantial fluorescence quenching is observed for the CNDs–$HAuCl_4$ mixture after UV irradiation (ESM Fig. S2), which can be attributed to the fact that some CNDs are associated with the AuNPs formed; subsequently, fluorescence quenching occurs with the fluorescence resonance energy transfer between CNDs and AuNPs

[30]. The residual fluorescence observed originates from free CNDs existing in the dispersion.

The surface composition and elemental analysis for the resultant nanoparticles were characterized by the XPS technique. The XPS spectra of the nanoparticles are shown in Fig. 3. The five peaks at 56.8, 85.2, 334.2, 353.0, and 546.3 eV in Fig. 3a are attributed to $Au_{5p3}$, $Au_{4f}$, $Au_{4d5}$, $Au_{4d3}$, and $Au_{4p3}$, respectively. Another three peaks at 285.0, 398.8, and 531.8 eV associated with $C_{1s}$, $N_{1s}$, and $O_{1s}$ are also observed. The XPS results show that the nanoparticles obtained are mainly composed of C, N, and Au and limited amounts of O element. O atoms may come from the $O_2$, $H_2O$, or $CO_2$ adsorbed on the surface of particles [22]. The $Au_{4f}$, $C_{1s}$, and $N_{1s}$ regions were analyzed using the peak deconvolution method. The spectrum of $Au_{4f}$ shows two peaks at 83.9 eV for $4f_{7/2}$ and 87.5 eV for $4f_{5/2}$ (Fig. 3b). The position and difference between the two peaks (3.8 eV) exactly match the value reported for $Au^0$ [31]. The absence of a binding energy at 84.9 eV, corresponding to $Au^{3+}$, further indicates that the gold atoms exist as $Au^0$ (AuNPs) [32]. Furthermore, the $C_{1s}$ spectrum of the nanoparticles (Fig. 3c) exhibits three peaks at 284.6, 285.8, and 288.1 eV, which are attributed to C–C, C–N, and C=N, respectively. The $N_{1s}$ spectrum of such nanoparticles (Fig. 3d) exhibits three peaks at 398.5, 399.4, and 400.6 eV, which are associated with the C–N–C, N–$(C)_3$, and N–H groups, respectively [33]. The XPS spectrum data are consistent with that of CNDs reported by our previous work [22] and prove the existence of CNDs in the product. The XPS data further confirm the formation of CNDs/AuNPs [34].

In light of the total absence of any specific reducing agent in the solution, the formation of the AuNPs is quite surprising and can be ascribed to a photoinduced reductive pathway with the use of CNDs as a photocatalyst. Scheme 1

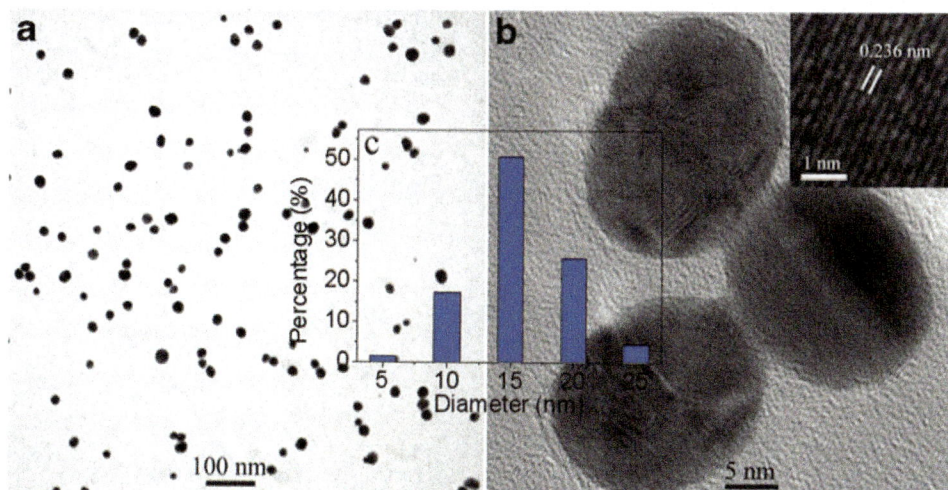

Fig. 2 a, b Typical TEM images at different magnifications of the AuNPs thus formed. c Corresponding particle size distribution histograms. *Inset* in (b) shows a HRTEM image of one single AuNP

**Fig. 3 a** XPS spectrum of AuNPs thus formed. **b** Au$_{4f}$ spectrum. **c** C$_{1s}$ spectrum. **d** N$_{1s}$ spectrum of AuNPs thus formed (*red line*, the sum of all peaks)

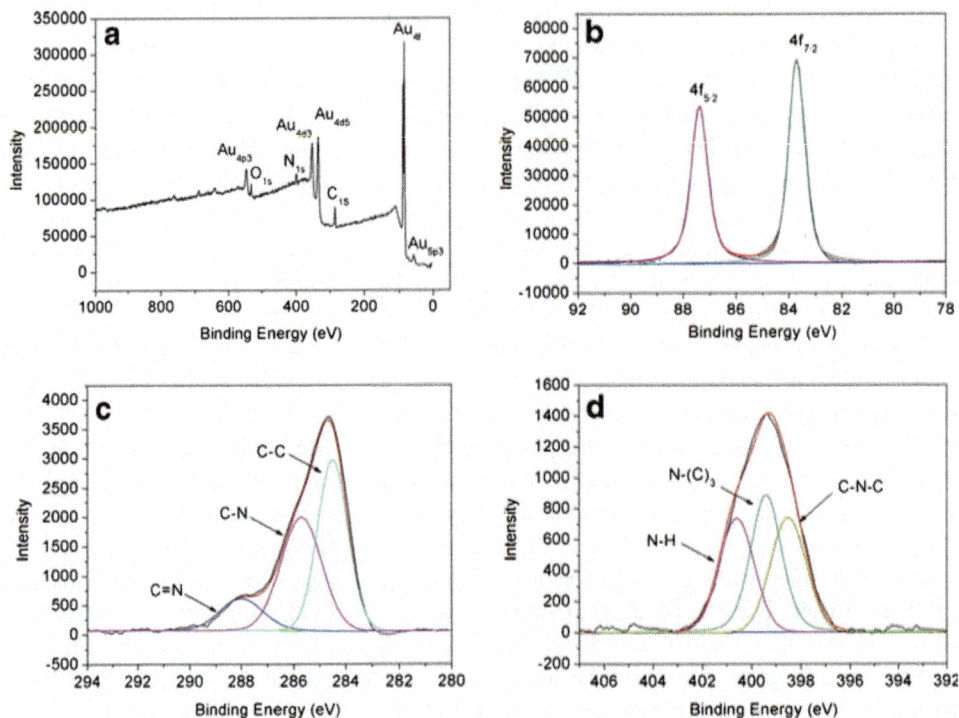

illustrates the proposed mechanism for the photochemical synthesis of AuNPs. The fluorescent CNDs have a strong UV absorption before 360 nm (a in Fig. 1). Electron and hole pairs are generated in the CNDs upon UV excitation (reaction 1) [35, 36]. The photogenerated electrons reduced the surrounding AuCl$_4^-$ (reaction 2), while the holes were scavenged by the alcohols (reactions 6) [37]. Au$^0$ atoms are generated after the fast disproportionation of Au$_2^+$ (reaction 3) and the disproportionation of Au$^+$ and Au$_2^+$ (reaction 4) [38]. After the start of the photocatalytic reaction, the concentration of Au$^0$ atoms steadily increases with elapsed time as AuCl$_4^-$ is reduced. When the concentration reaches the point of supersaturation, the Au$^0$ atoms start to form small clusters, i.e., nuclei by diffusion-limited aggregation (reaction 5), and then finally grow into CND-protected nanocrystals [39]. As reported previously, AuCl$_4^-$ itself can be

photochemically reduced to Au$^0$ atoms [40]. To confirm the catalytic effect of CNDs in synthesizing AuNPs, control experiments were conducted by the UV irradiation of an equal amount of HAuCl$_4$ in aqueous solution of ethanol for 1 h, in the absence of CNDs. The solution color changed from clear light yellow to colorless due to the photoinduced reduction of Au$^{3+}$ to Au$^+$ [41]. However, no AuNPs were obtained. Also, keeping the CNDs–HAuCl$_4$ mixture with the presence of ethanol in the dark for several days or under visable light irradiation for several hours failed to produce AuNPs. All these observations indicate that CNDs indeed serve as an effective photocatalyst for the photoreduction of AuCl$_4^-$.

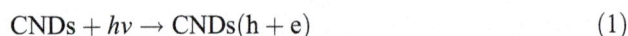

$$\text{CNDs} + h\nu \rightarrow \text{CNDs}(h + e) \tag{1}$$

**Scheme 1** Scheme (not to scale) illustrating the proposed mechanism for the photochemical synthesis of AuNPs

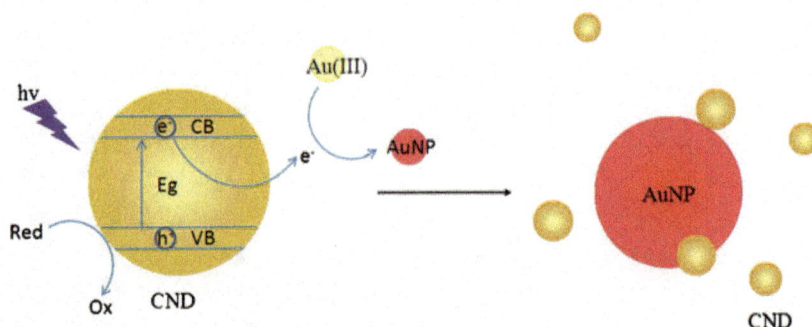

**Fig. 4** **a** UV/Vis absorption spectra of 4-NP and time-dependent absorption spectra for the catalytic reduction of 4-NP by NaBH$_4$ in the presence of CNDs/AuNPs (**b**) and AuNPs (**c**). **d** Plot of ln($C_t/C_0$) of 4-NP against time for the catalysts. Conditions: [4-NP]= 3.5 mM; [Catalyst]=30 mg/L; [NaBH$_4$]=80 mM

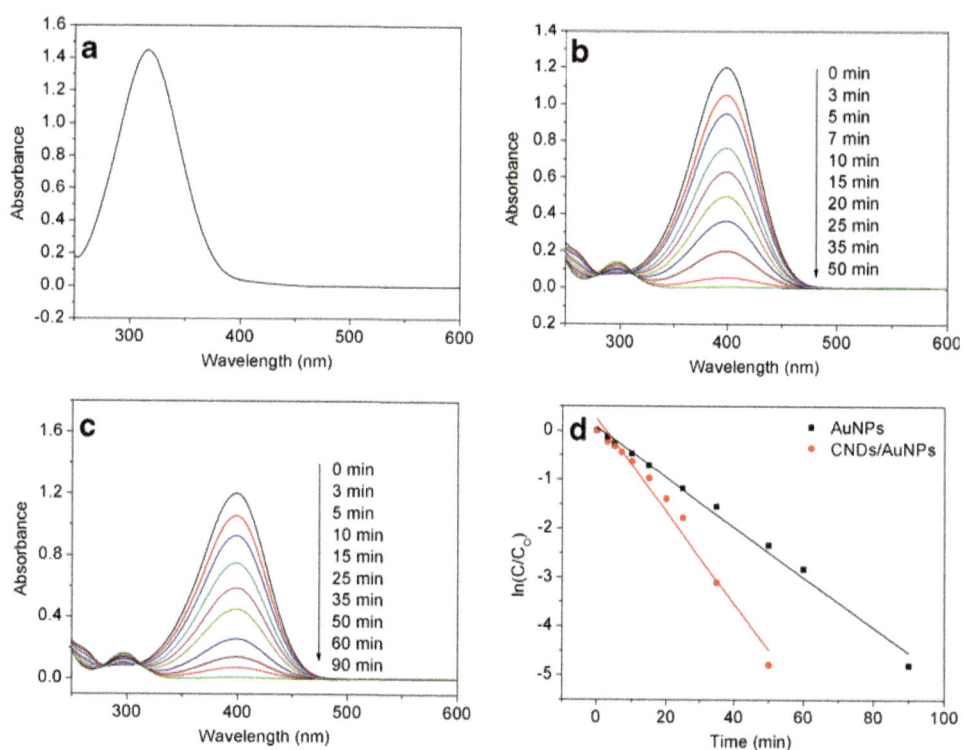

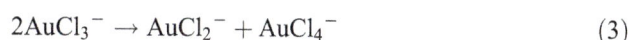

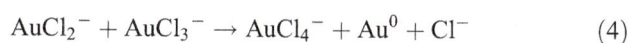

$$CNDs(e) + AuCl_4^- \rightarrow CNDs + AuCl_3^- + Cl^- \qquad (2)$$

$$2AuCl_3^- \rightarrow AuCl_2^- + AuCl_4^- \qquad (3)$$

$$AuCl_2^- + AuCl_3^- \rightarrow AuCl_4^- + Au^0 + Cl^- \qquad (4)$$

$$nAu^0 \rightarrow Au_n \qquad (5)$$

$$CNDs(2h) + CH_3CH_2OH \rightarrow CNDs(e) + 2H^+ + CH_3CHO \qquad (6)$$

It is well known that 4-aminophenol (4-AP) is very useful and important in many applications that include analgesic and antipyretic drugs, photographic developer, corrosion inhibitor, anti-corrosion lubricant, and so on [42, 43]. The reduction of 4-NP by NaBH$_4$ in the presence of noble metal nanoparticles as catalysts has been intensively investigated for the efficient production of 4-AP [44, 45]. Therefore, the reduction of 4-NP to 4-AP with an excess amount of NaBH$_4$ was used as a model system to quantitatively evaluate the catalytic activities of CNDs/AuNPs. As indicated in Fig. 4a, pure 4-NP shows a distinct spectral profile with an absorption maximum at 316 nm. When a NaBH$_4$ solution was added into the 4-NP, a new absorption peak was observed at 400 nm, corresponding to the 4-NP ions in alkaline conditions [46]. The absorption intensity of 4-NP ions at 400 nm decreased quickly with time, accompanied with the appearance of the peak of 4-AP at 300 nm, indicating successful conversion of 4-NP to 4-AP.

**Scheme 2** Scheme (not to scale) illustrating the proposed mechanism of synergistic enhancement in CNDs/AuNPs

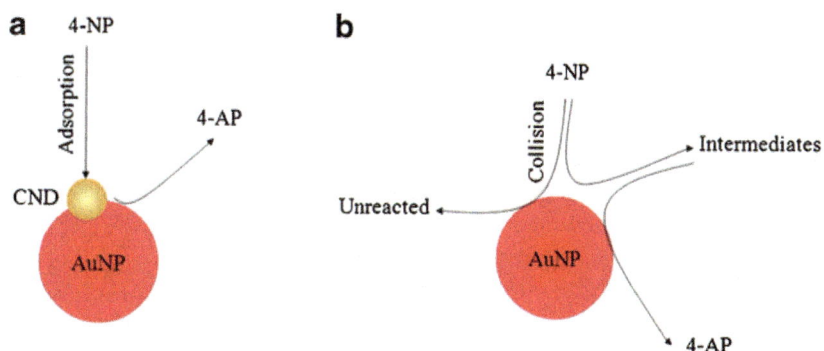

The total reduction reaction is summarized by the following equation:

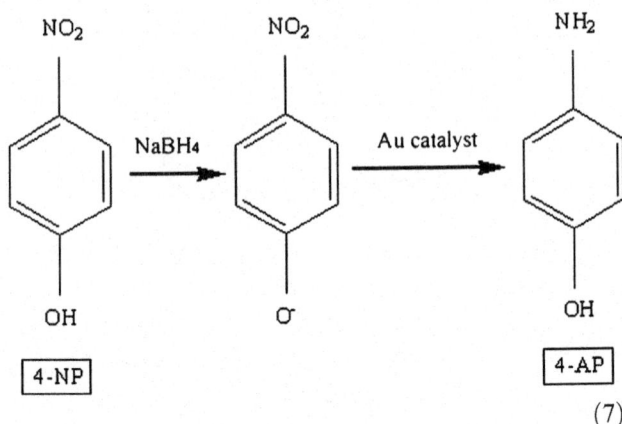

$$\tag{7}$$

It should be noted that the reduction of 4-NP by NaBH₄ was finished within 50 min upon the addition of CNDs/AuNPs (Fig. 4b), with the observation of a fading and ultimate bleaching of the yellow green color of the reaction mixture in aqueous solution. However, a longer reaction time of 90 min was required to achieve the full reduction using gold sol solution (AuNPs stabilized by sodium citrate, about 15 nm in diameter) as a catalyst which contained equivalent Au atoms (Fig. 4c). The reproducibility of the result was tested for three replicates to confirm the enhanced catalytic activity of CNDs/AuNPs than AuNPs. Without the addition of a catalyst, the reduction did not proceed, and the absorption peak centered at 400 nm remains unaltered over time. However, due to the presence of a large excess of NaBH₄ compared to 4-NP, the rate of reduction is independent of the concentration of NaBH₄. ESM Fig. S3 presents the analysis of the experimental data using different kinetic models, indicating that the first-order reaction model is the most precise for the 4-NP catalytic reduction system [47]. Hence, $\ln(C_t/C_0)$ versus time can be obtained based on the absorbance as a function of time, and good linear correlations are observed, as shown in Fig. 4d, suggesting that the reactions follow first-order kinetics. Then, the kinetic reaction rate constants (defined as $k_{app}$) are estimated from the slopes of the linear relationship to be $5.25 \times 10^{-2}$ min$^{-1}$ (AuNPs) and $9.43 \times 10^{-2}$ min$^{-1}$ (CNDs/AuNP), respectively.

These results clearly indicate the CNDs/AuNPs have a higher catalytic activity for the reduction of 4-NP than the AuNPs only. The π-rich nature of CNDs absorbed on the surface of AuNP may play an active part in the catalysis, yielding a synergistic effect. The synergistically enhanced catalytic activity may be explained as follows: it is expected that 4-NP can be adsorbed onto CNDs via π–π stacking interactions. Such adsorption provides a high concentration of 4-NP near to the CNDs/AuNPs, leading to a highly efficient contact between them (Scheme 2a). In contrast, without the presence of CNDs on the nanoparticle surface, 4-NP must collide with AuNPs by chance and remains in contact for the catalysis to proceed. When this is not achieved, 4-NP will pass back into the solution and can only react further when it collides with AuNPs again (Scheme 2b) [48, 49].

## Conclusions

In summary, AuNPs were synthesized using fluorescent CNDs both as an effective photocatalyst and a stabilizer to protect the synthesized AuNPs from aggregation. Such CNDs/AuNPs are found to exhibit good catalytic performance toward 4-NP reduction, and the synergistic effect of adsorbed CNDs enhances the catalytic activity of Au catalysts. We believe that these studies will provide a general CND-based synthesis strategy and lead to the development and design of simple chemical synthesis of metal nanoparticles methodologies.

## References

1. Cushing BL, Kolesnichenko VL, O'Connor CJ (2004) Recent advances in the liquid-phase syntheses of inorganic nanoparticles. Chem Rev 104:3893–3946
2. Watanabe K, Menzel D, Nilius N, Freund HJ (2006) Photochemistry on metal nanoparticles. Chem Rev 106:4301–4320
3. Daniel M-C, Astruc A (2004) Gold nanoparticles: assembly, supramolecular chemistry, quantum-size-related properties, and applications toward biology, catalysis, and nanotechnology. Chem Rev 104:293–346
4. Hayat MA (ed) (1989) Colloidal gold: principles, methods and applications. Academic, San Diego
5. Faraday M (1857) The Bakerian Lecture: experimental relations of gold (and other metals) to light. Philos Trans R Soc London 147:145–181
6. Stamplecoskie KG, Scaiano JC (2010) Light emitting diode irradiation can control the morphology and optical properties of silver nanoparticles. J Am Chem Soc 132:1825–1827
7. Kim F, Song JH, Yang P (2002) Photochemical synthesis of gold nanorods. J Am Chem Soc 124:14316–14317
8. Dahl JA, Maddux BS, Hutchison JE (2007) Toward greener nanosynthesis. Chem Rev 107:2228–2269
9. Zhu JM, Shen YH, Xie AJ, Qiu LG, Zhang Q, Zhang SY (2007) Photoinduced synthesis of anisotropic gold nanoparticles in room-temperature ionic liquid. J Phys Chem C 111:7629–7633
10. Huang X, Zhou X, Wu S, We Y, Qi X, Zhang J, Boey F, Zhang H (2010) Reduced graphene oxide-templated photochemical synthesis and in situ assembly of Au nanodots to orderly patterned Au nanodot chains. Small 6:513–516

11. Sakamoto M, Fujistuka M, Majima T (2009) Light as a construction tool of metal nanoparticles: synthesis and mechanism. J Photochem Photobiol C 10:33–56

12. Troupis A, Hiskia A, Papaconstantinou E (2002) Synthesis of metal nanoparticles by using polyoxometalates as photocatalysts and stabilizers. Angew Chem Int Ed 41:1911–1914

13. Quaresma P, Soares L, Contar L, Miranda A, Osório I, Carvalho PA, Franco R, Pereira E (2009) Green photocatalytic synthesis of stable Au and Ag nanoparticles. Green Chem 11:1889–1893

14. Huang X, Qi X, Huang Y, Li S, Xue C, Gan CL, Boey F, Zhang H (2010) Photochemically controlled synthesis of anisotropic Au nanostructures: platelet-like Au nanorods and six-star Au nanoparticles. ACS Nano 4:6196–6202

15. Liu L, Ma D, Zheng H, Li X, Cheng M, Bao X (2008) Synthesis and characterization of microporous carbon nitride. Microporous Mesoporous Mater 110:216–222

16. Datta KKP, Reddy BVS, Ariga K, Vinu A (2010) Gold nanoparticles embedded in a mesoporous carbon nitride stabilizer for highly efficient three-component coupling reaction. Angew Chem Int Ed 49:5961–5965

17. Qiu Y, Gao L (2003) Chemical synthesis of turbostratic carbon nitride, containing C–N crystallites, at atmospheric pressure. Chem Commun 18:2378–2379

18. Liu A, Cohen ML (1989) Prediction of new low compressibility solids. Science 245:84–842

19. Yang GW, Wang JB (2000) Carbon nitride nanocrystals having cubic structure using pulsed laser induced liquid–solid interfacial reaction. Appl Phys A 71:343–344

20. Bai Y-J, Lu B, Liu Z, Li L, Cui D, Xu X, Wang Q (2003) Solvothermal preparation of graphite-like $C_3N_4$ nanocrystals. J Cryst Growth 247:505–508

21. Yang L, May PW, Huang YZ, Yin L (2007) Hierarchical architecture of self-assembled carbon nitride nanocrystals. J Mater Chem 17:1255–1257

22. Liu S, Tian J, Wang L, Luo Y, Zhai J, Sun X (2011) Preparation of photoluminescent carbon nitrate dots form $CCl_4$ and 1,2-ethylenediamine: a heat treatment-based strategy. J Mater Chem 21:11726–11729

23. Gangopadhayay AK, Chakravorty A (1961) Charge transfer spectra of some gold(III) complexes. J Chem Phys 35:2206–2209

24. Yao H, Mo D, Duan J, Chen Y, Liu J, Sun Y, Hou M, Schäpers T (2011) Investigation of the surface properties of gold nanowire arrays. Appl Surf Sci 258:147–150

25. Lu L, Ai K, Ozaki Y (2008) Environmentally friendly synthesis of highly monodisperse biocompatible gold nanoparticles with urchin-like shape. Langmuir 24:1058–1063

26. Sau TK, Murphy CJ (2005) Self-assembly patterns formed upon solvent evaporation of aqueous cetyltrimethylammonium bromide-coated gold nanoparticles of various shapes. Langmuir 21:2923–2929

27. Willerich I, Li Y, Gröhn F (2010) Influencing particle size and stability of ionic dendrimer–dye assemblies. J Phys Chem B 114:15466–15476

28. Dubertret B, Calame M, Libchaber AJ (2001) Single-mismatch detection using gold-quenched fluorescent oligonucleotides. Nat Biotechnol 19:365–370

29. Maxwell DJ, Taylor JR, Nie S (2002) Self-assembled nanoparticle probes for recognition and detection of biomolecules. J Am Chem Soc 124:9606–9612

30. Oh E, Hong M-Y, Lee D, Nam S-H, Yoon HC, Kim H-S (2005) Inhibition assay of biomolecules based on fluorescence resonance energy transfer (FRET) between quantum dots and gold nanoparticles. J Am Chem Soc 127:3270–3271

31. Maye MM, Luo J, Lin Y, Engelhard MH, Hepel M, Zhong CJ (2003) X-ray photoelectron spectroscopic study of the activation of molecularly-linked gold nanoparticle catalysts. Langmuir 19:125–131

32. Kannan P, John SA (2003) Fabrication of conducting polymer-gold nanoparticles film on electrodes using monolayer protected gold nanoparticles and its electrocatalytic application. Electrochim Acta 56:7029–7037

33. Zou XX, Li GD, Wang YN, Zhao J, Yan C, Guo MY, Li L, Chen JS (2011) Direct conversion of urea into graphitic carbon nitride over mesoporous $TiO_2$ spheres under mild condition. Chem Commun 47:1066–1068

34. Kannan P, John SA (2011) Fabrication of conducting polymer-gold nanoparticles film on electrodes using monolayer protected gold nanoparticles and its electrocatalytic application. Electrochim Acta 56:7029–7037

35. Willianms G, Seger B, Kamat PV (2008) $TiO_2$–graphene nanocomposites. UV-assisted photocatalytic reduction of graphene oxide. ACS Nano 2:1487–1491

36. Kamat PV, Bedja I, Hotchandani S (1994) Photoinduced charge transfer between carbon and semiconductor clusters. One-electron reduction of $C_{60}$ in colloidal $TiO_2$ semiconductor suspensions. J Phys Chem 98:9137–9142

37. Wang C-Y, Liu C-Y, Zheng X, Chen J, Shen T (1998) The surface chemistry of hybrid nanometer-sized particles I. Photochemical deposition of gold on ultrafine $TiO_2$ particles. Colloid Surf A 131:271–280

38. Shukla S, Vidal X, Furlani EP, Swihart MT, Kim K-T, Yoon Y-K, Urbas A, Prasad PN (2011) Subwavelength direct laser patterning of conductive gold nanostructures by simultaneous photopolymerization and photoreduction. ACS Nano 5:1947–1957

39. Fang J, Du S, Lebedkin S, Li Z, Kruk R, Kappes M, Hahn H (2010) Gold mesostructures with tailored surface topography and their self-assembly arrays for surface-enhanced raman spectroscopy. Nano Lett 10:5006–5013

40. Eustis S, Hsu HY, El-Sayed MA (2005) Gold nanoparticle formation from photochemical reduction of $Au^{3+}$ by continuous excitation in colloidal solutions. A proposed molecular mechanism. J Phys Chem B 109:4811–4815

41. Kurihara K, Kizling J, Stenius P, Fendler JH (1983) Laser and pulse radiolytically induced colloidal gold formation in water and in water-in-oil microemulsions. J Am Chem Soc 105:2574–2579

42. Du Y, Chen H, Chen R, Xu N (2004) Synthesis of $p$-aminophenol from $p$-nitrophenol over nano-sized nickel catalysts. Appl Catal A 277:259–264

43. Zhang Z, Shao C, Zou P, Zhang P, Zhang M, Mu J, Guo Z, Li X, Wang C, Liu Y (2011) In situ assembly of well-dispersed gold nanoparticles on electrospun silica nanotubes for catalytic reduction of 4-nitrophenol. Chem Commun 47:3906–3908

44. Deng Y, Cai Y, Sun Z, Liu J, Liu C, Wei J, Li W, Liu C, Wang Y, Zhao D (2010) Multifunctional mesoporous composite microspheres with well-designed nanostructure: a highly integrated catalyst system. J Am Chem Soc 132:8466–8473

45. Jiang H-L, Akita T, Ishida T, Haruta M, Xu Q (2011) Synergistic catalysis of Au@Ag core–shell nanoparticles stabilized on metal-organic framework. J Am Chem Soc 133:1304–1306

46. Praharaj S, Nath S, Ghosh SK, Kundu S, Pal T (2004) Immobilization and recovery of Au nanoparticles from anion exchange resin: resin-bound nanoparticle matrix as a catalyst for the reduction of 4-nitrophenol. Langmuir 20:9889–9892

47. Huang J, Vongehr S, Tang S, Lu H, Meng X (2010) Highly catalytic Pd–Ag bimetallic dendrites. J Phys Chem C 114:15005–15010

48. Leary R, Westwood A (2011) Carbonaceous nanomaterials for the enhancement of $TiO_2$ photocatalysis. Carbon 49:741–772

49. Zhang Y, Liu S, Lu W, Wang L, Tian J, Sun X (2011) In situ green synthesis of Au nanostructures on graphene oxide and their application for catalytic reduction of 4-nitrophenol. Catal Sci Technol 1:1142–1144

# 2

# Sensing behavior of silica-coated Au nanoparticles towards nitrobenzene

Suman Singh · Pooja Devi · Deepak Singh ·
D. V. S. Jain · M. L. Singla

**Abstract** In the present work, we report silica-stabilized gold nanoparticles (SiO$_2$/Au NPs) as a wide-range sensitive sensing material towards nitrobenzene (NB). Surface hydroxyl groups of silica selectively form Meisenheimer complex with electron-deficient aromatic ring of NB and facilitate its immobilization and subsequent catalytic reduction by Au cores. Silica-coated Au NPs were synthesized and characterized for their chemical, morphological, structural, and optical properties. SiO$_2$/Au NPs-modified electrodes were characterized with impedometric and cyclic voltammetric electrochemical techniques. SiO$_2$/Au NPs are found to have a higher optical detection window of range, 0.1 M to 1 µM and a lower electrochemical detection window of range, $10^{-4}$ to $2.5 \times 10^{-2}$ mM with a detection limit of 12.3 ppb. A significant enhancement in cathodic peak current, $C_1$, and sensitivity (102 µA/mM) was observed with modified electrode relative to bare and silica-modified electrodes. The $I_P$ was found to be linearly correlated to NB concentration ($R^2 = 0.985$). The interference of cationic and anionic species on sensor sensitivity was also studied. Selectivity in the present sensing system may be further improved by modifying silica with specific functional moieties.

S. Singh · P. Devi · M. L. Singla (✉)
Central Scientific Instruments Organization,
Chandigarh, India
e-mail: singla_csio@yahoo.com

D. Singh
Indian Institute of Pulses Research, (IIPR),
Kanpur, India

D. V. S. Jain
Panjab University,
Chandigarh, India

**Keywords** Silica-coated Au NPs · Impedance · Sensitivity · NB

## Introduction

Nitrobenzene (NB) is a widely used solvent in manufacturing processes and is often discharged by industries as a waste. Increased industrial development and increased discharge of waste is resulting in surface and ground water pollution. NB is a high priority pollutant in the nitroaromatics (NACs) declared by the Environment Protection Agency on the basis of its known carcinogenicity, mutagenicity, and acute toxicity [1, 2]. Conventional analytical methods used for the detection of NB include liquid–liquid extraction–gas chromatography, solid phase microextraction, nuclear quadrupole resonance, and electron capture techniques [3–7]. These techniques are highly selective and sensitive but are very expensive, laborious, and time consuming. Nanomaterials of size few to 100 nm have enabled new types of sensors that are capable of detecting extremely small amounts of analytes in lower limit range [8, 9]. However, nanoparticles (NPs) suffer from the aggregation problems, which revert them back to the bulk materials. To prevent agglomeration, metal NPs are often coated with ligands, polymers, organic surfactants, mesoporous/nanoporous supports, which not only delimit the particle size but also help in immobilization of the resulting NPs [10–12]. Owing to the large internal surface area and small pore size, mesoporous material finds applications in catalysis, chromatography supports, optics, photonics, semiconductor devices, and chemical sensors [13]. In the present work, polyvinylpyrollidone (PVP)-coated gold nanoparticles (Au NPs) have been synthesized and then functionalized with silica for NB detection. The formation of PVP and the silica layer on the particle surface helps in increasing the aggregation stability of the NPs by decreasing the inter phase tension via strengthening interactions between the dispersed phase and the dispersion medium. This in turn increases the entropy component of the system due to the involvement of molecules and ions of the surface layer in thermal motion together with particles of the dispersed phase [14]. Additionally, porous silica helps in the adsorption of analytes. Silica minerals are reported as one of the most efficient adsorbents for NACs contaminants [15].

Herein, we are reporting $SiO_2$/Au NPs as a dynamic range sensor towards NB.

## Experimental details

### Materials

All chemicals were of analytical grade and used as received without further purification: $HAuCl_4$ (Spectrochem Pvt. Ltd., Mumbai, India), PVP (Sisco Research Lab, Mumbai, India), tetraethyl orthosilicate (TEOS, Merk Specialties Pvt. Ltd., Darmstadt, Germany), ammonium hydroxide (S.D. Fine Chem. Ltd., Mumbai, India), ethanol (Changshu Yangyuan Chemical China, Changshu, China), ethylene glycol (Loba Chemie, Mumbai, India), and NB (Spectrochem Pvt. Ltd.). De-ionized water obtained from Millipore was used for all synthesis and experimental studies.

### Preparation of $SiO_2$/Au NPs

Au NPs were synthesized using hydrazine hydrate reducing agent and PVP capping agent. Briefly, PVP (1 g) was added into distilled water (50 ml) under continuous stirring with following additions of 1 % $HAuCl_4$ solution (5 ml). To this mixture, hydrazine hydrate (1 ml) was added and stirred until the appearance of wine red color which confirms the formation of Au NPs. Synthesized Au NPs were functionalized by silica using Lu et al. approach [11]. Briefly, Au NPs colloidal solution (4 ml) was added to ethanol (20 ml) under constant stirring followed by addition of ammonium hydroxide and TEOS (5–10 μl). This solution was stirred at room temperature for about 1 h and then centrifuged at 8,000 rpm for 30 min to collect coated NPs. Figure 1 shows

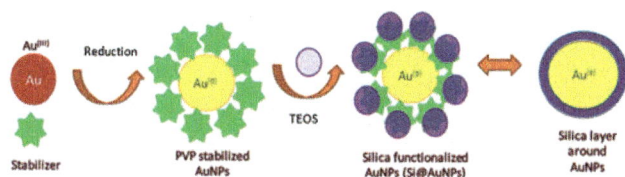

**Fig. 1** Schematic for synthesis of $SiO_2$/Au NPs

the pictorial presentation of steps used in the synthesis and silica coating of Au NPs.

For electrochemical sensing, NB stock solution was prepared in acetonitrile and NaCl (0.1 M) was used as an electrolytic medium. Varying amounts (1–300 μl) of NB (0.01 mM) was added to the electrolyte solution and mixed properly prior to each voltammetric measurement.

### Instrumentation

Optical studies were done using PerkinElmer® Lamda 35 UV–visible spectrophotometer. Scanning electron microscope (SEM) and energy-dispersive x-ray analysis (EDXA) analysis was carried out on FE-SEM (Oxford Company) equipped with an X-ray analyzer for morphological and elemental information. Glassy carbon (GC) electrode spin coated with $SiO_2$/Au NPs was used as working electrode for voltammetric and impedance studies in 0.1 M$K_3[Fe(CN)_6]$/KCl solution (used as a redox probe) using the CHI-660 Instrument.

## Results and discussion

### Characterization of $SiO_2$/Au NPs

Figure 2 shows EDXA spectrum of $SiO_2$/Au NPs indicating presence of both Au (AuM, 2.2; AuL, 10.0 keV) and silica (SiK, 1.5–1.8 keV) along with a prominent peak for copper (CuK, 8.2 KeV; CuL, 0.9 KeV) which is possibly due to X-ray emission from the copper substrate. EDXA spectra were collected over random areas to monitor the homogeneity of elemental composition. The presence of Au and silica peaks is consistent with the formation of silica shell on Au NPs. Inset shows the transmission electron microscopy (TEM) image of $SiO_2$/Au NPs at high resolution confirming coating of Au NPs with silica. It can be clearly seen that the size of Au NP is around ~38 nm and shell is of thickness ~18 nm.

Figure 3 shows UV–visible spectrum of Au (a) and $SiO_2$/Au NPs (b and c). The surface plasmon resonance (SPR) peak position at 528 nm confirms the formation of spherical Au nanoparticles [12]. A shift of ~3 nm was observed on coating synthesized Au NPs with silica and is attributed to the change in the refractive index of the surrounding medium from 1.36 (ethanol) to 1.45 (silica) [16]. On further increasing TEOS amount (5 to 10 μl) for silica coating, no change in SPR peak position is observed, however, the intensity of SPR band increased (Fig. 3). This might be because of the increased scattering from thicker shells. The broadening of the SPR band after coating can be attributed to the roughness of the shell surface or the presence of a few particles with incomplete shells as earlier reported by Lu et al. [13, 17].

**Fig. 2** EDXA spectrum
of SiO$_2$/Au NPs (*inset*: (*a*)
SEM image of SiO$_2$/Au NPs
corresponding to EDXA
spectrum and (*b*) TEM image
of SiO$_2$/Au NP at higher
resolution)

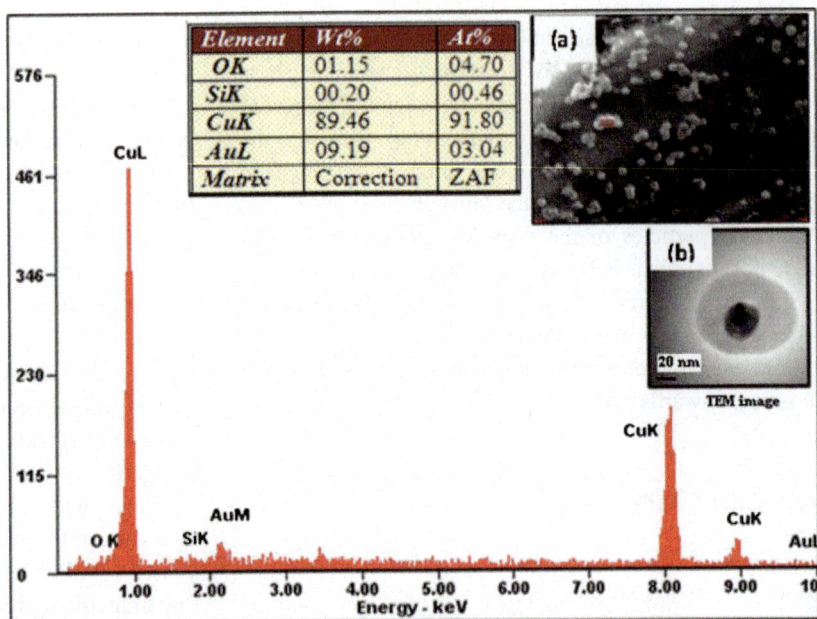

| Element | Wt% | At% |
|---------|-----|-----|
| OK | 01.15 | 04.70 |
| SiK | 00.20 | 00.46 |
| CuK | 89.46 | 91.80 |
| AuL | 09.19 | 03.04 |
| Matrix | Correction | ZAF |

Characterization of SiO$_2$/Au NPs-modified electrode

*Impedance and voltammetric studies*

Figure 4 shows interfacial features of bare (GC) and modified (GC/SiO$_2$/Au NPs) electrodes in 0.1 MK$_3$[Fe(CN)$_6$]/KCl solution. Nyquist plot (real ($Z'$) vs. imaginary parts ($Z''$) of the impedance) of SiO$_2$/Au NPs-modified electrode (GC/SiO$_2$/Au NPs) shows somewhat flattened semicircle at higher frequencies and a straight line forming an angle of 45° to the real axis at lower frequencies. The flattened circle is a consequence of roughness due to the material (SiO$_2$/Au NPs) deposited on the electrode surface. A perfect semicircle corresponds to perfectly smooth surface and this circularity decreases with an increase in surface roughness

[18, 19]. The straight line at lower frequencies can be attributed to the Warburg impedance which becomes dominant at lower frequencies for diffusion limited processes. Its appearance for modified electrode (GC/SiO$_2$/Au NPs) may be due to the presence of insulating silica coating around Au NPs, which may partially limit the transport of ions to the electrode surface as indicated by increase in charge transfer resistance ($R_{ct}$) of the bare electrode from $1.88 \times 10^2$ to $2.51 \times 10^2$ after modification. This increase in $R_{ct}$ value is possibly due to the steric hindrance and electrostatic repulsion between surface SiO$_2^-$ groups and negatively charged redox couple [20, 21].

Figure 5 shows cyclic voltamograms of bare (GC) and modified (GC/SiO$_2$/Au NPs) electrodes in 0.1 MK$_4$(Fe (CN)$_6$)+KCl solution. Cyclic voltammogram (CV) of

**Fig. 3** UV–visible spectra of
(*a*) Au NPs and (*b*, *c*) SiO$_2$/
Au NPs synthesized with
varying TEOS amount, 5 μl
and 10 μl, respectively

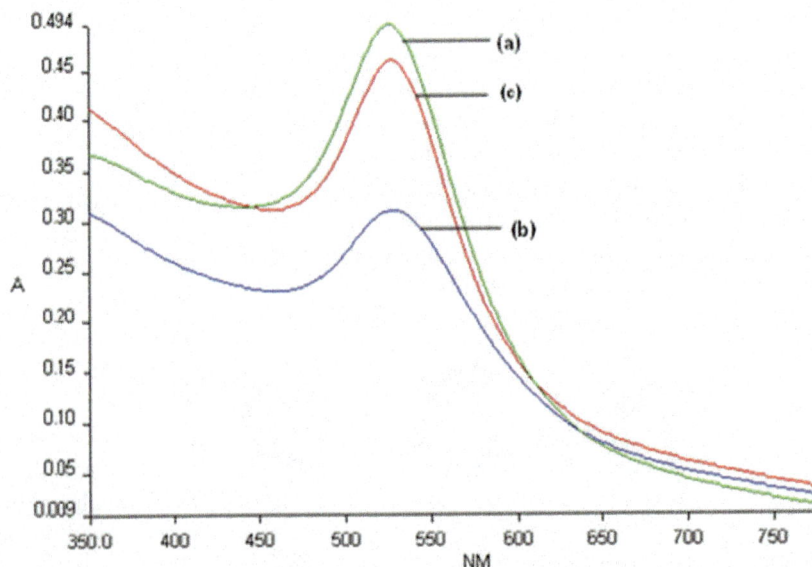

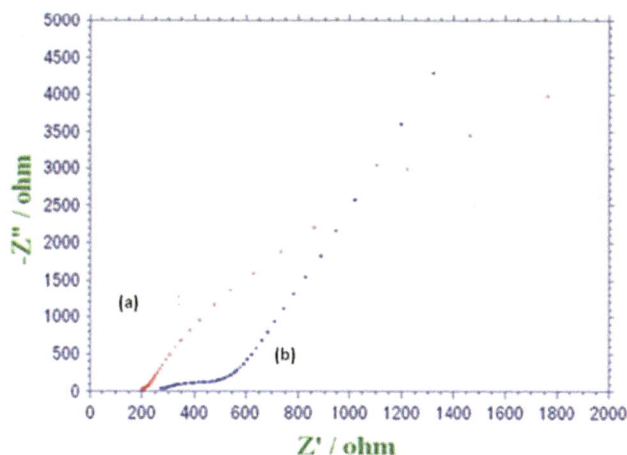

**Fig. 4** Nyquist Plot of (*a*) bare GC and (*b*) SiO$_2$/Au NPs-modified GC electrode in 0.1 M [K$_4$Fe(CN)$_6$+KCl] solution

modified electrode in ferricyanide solution is a valuable tool to monitor the barrier of the modified electrode since the electron transfer between the solution species and the electrode must occur either through the barrier itself by tunneling or through the defects in the material/barrier. It is well proved that when the bare electrode surface is modified by somematerials,theelectron-transferkineticsofFe(CN)$_6$$^{4-/}$$^{3-}$ is perturbed.

It can be clearly seen from Fig. 5a, b that the CV response of the bare GC electrode shows only one oxidation peak (Fe$^{+2}$ to Fe$^{+3}$) whereas modified (GC/SiO$_2$/Au NPs) electrode shows the presence of a proper redox couple for Fe$^{2+}$/Fe$^{3+}$ ions with an enhanced current sensitivity (102 μA/mM) relative to the bare GC (inset of Fig. 5b). This appearance of a redox couple and increase in the current signal for modified electrode could be explained by the increase in the

effective electrode surface area because of modification with NPs [22]. Peak current (*I*$_P$) increases linearly with the square root of scan rate which indicates the dominance of diffusion in this electrochemical process. This behavior can be explained by Randles Sevick equation [23].

$$i_p = 2.69 \times 10^5 n^{\frac{3}{2}} A D^{\frac{1}{2}} C v^{\frac{1}{2}} \tag{1}$$

From the above equation, the diffusion coefficient (*D*) for modified electrode is found to be $1.93 \times 10^{-8}$ cm$^2$/s.

Detection of NB

*Optical detection*

The optical detection of NB using SiO$_2$/Au NPs is based on the change in the intensity of the SPR band of NPs after incubation with NB and is basically based on adsorption chemistry. Because of high mesoporosity and surface roughness resulting in a large surface-to-volume ratio, SiO$_2$/Au NPs offer a vast surface area for efficient interaction with the analyte. In the presence of NB, intensity of the SPR band of SiO$_2$/Au NPs decreases with no effect on band position (Fig. 6).

In our case, this sensing might be due to the interaction between –NO$_2$ groups of NB and surface hydroxyls present on the silica surface. This interaction results in the formation of Meisenheimer complex which is a σ-complex formed by covalent addition of nucleophile to an arene carrying electron deficient aromatic compound. Since NB is electron deficient due to the strong electron withdrawing effect of NO$_2$ group, NB is able to form Meisenheimer complex. From available literature, it is evident that such an interaction has been

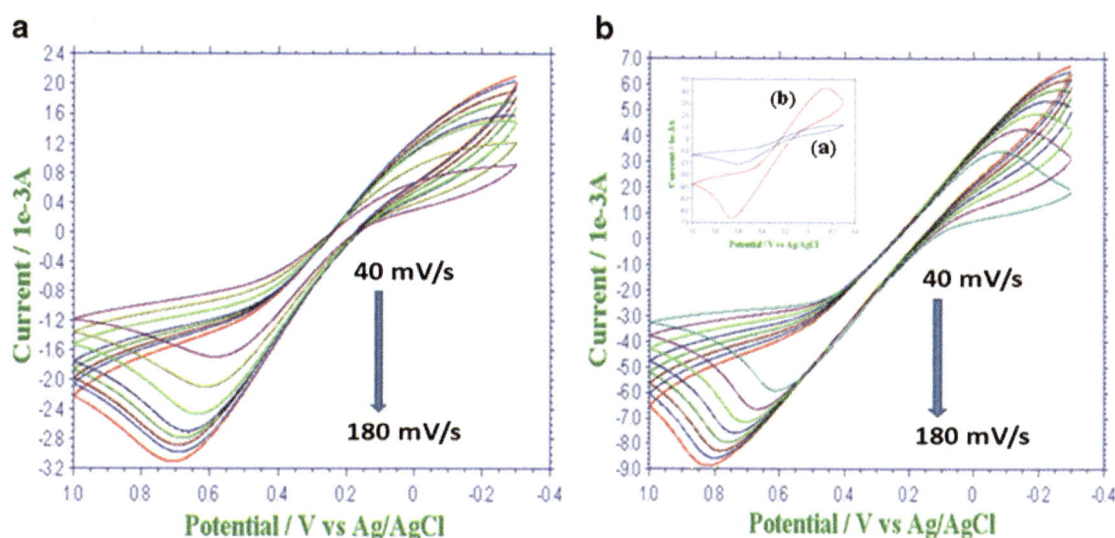

**Fig. 5** CV of (*a*) bare GC and (*b*) SiO$_2$/Au NPs-modified GC electrode in 0.1 M [K$_4$Fe(CN)$_6$+KCl] solution as a function of scan rate (40 to 180 mV/s), *inset*: an overlay of bare GC (*blue curve*) and Si@Au NPs/GC (*red curve*) response in K$_4$Fe(CN)$_6$ electrochemical solution

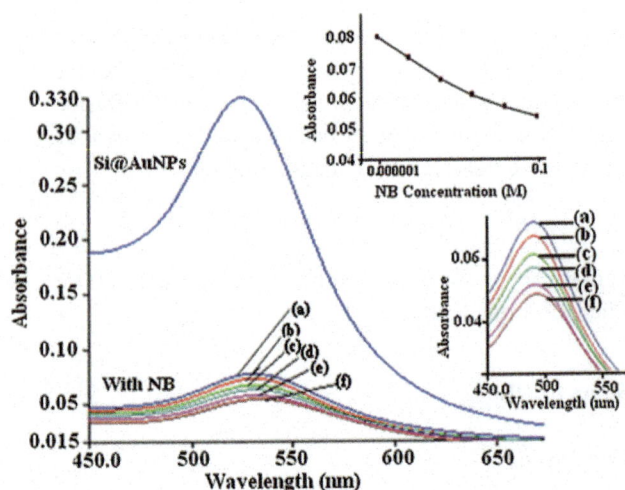

**Fig. 6** UV–visible spectra of SiO$_2$/Au NPs in the absence (*blue curve*) and presence of (*a*) 10$^{-6}$ M, (*b*) 10$^{-5}$ M, (*c*) 10$^{-4}$ M, (*d*) 10$^{-3}$ M, (*e*) 10$^{-2}$ M, (*f*) 10$^{-1}$ M NB

employed for designing the surface chemistry of nanostructures and electrodes to achieve the selectivity and sensitivity for the detection of TNT using fluorescence and electrochemical techniques [24–26]. This complex formation is believed to cause the damping of surface plasmon resonance due to gold core. The exact mechanism by which this occurs is not known, but the literature suggests the possibility of an increase in the imaginary part of the dielectric constant of gold [27–29]. The adsorption of NB introduces a thin layer of with the modified electron density on gold cores which produces a damping effect on SPR.

The sensing ability of SiO$_2$/Au NPs towards NB has been determined as a function of NB concentration as shown in

Fig. 6. As the concentration of NB increased from 10$^{-6}$ to 10$^{-1}$ M, intensity of SPR of SiO$_2$/Au NPs at 530 nm decreased linearly (inset, Fig. 6).

*Electrochemical detection*

It is evident from the literature that metallic NPs can catalyze oxidation/reduction of various organic compounds [30] but they suffer the aggregation problem. To explore the capability of SiO$_2$/Au NPs to enhance the sensitivity of the electrode towards NB via surface absorption by silica and reduction at Au cores [30–32], CV studies were performed by incubating SiO$_2$/Au NPs-modified electrode in 0.5 M NaCl electrolyte solution containing different amounts of 0.01 mM NB.

In the absence of NB, no reduction and oxidation peaks were observed for SiO$_2$/Au NPs-modified electrode (Fig. 7, inset, curve c) incubated in 0.1 M NaCl. With the addition of NB in electrolytic solution, two cathodic reduction peaks, $C_1$ (−0.74 V) and $C_2$ (−0.45), and two anodic oxidation peaks, $A_1$ (−0.62 V) and $A_2$ (0.03 V), were observed as shown in Fig. 7 (inset, curve d). $A_1/C_1$ can be attributed to the four-electron irreversible reduction of the nitro group (−NO$_2$) to the hydroxylamine derivative (−NHOH) and $A_2/C_2$ is assigned to the two-electron reversible oxidation of the hydroxylamine group (−NHOH) to a nitroso group (−NO) as shown below in Eqs. 2 and 3 [33, 34].

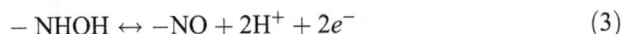

$$-NO_2 + 4e^- + 4H^+ \rightarrow -NHOH + H_2O \qquad (2)$$

$$-NHOH \leftrightarrow -NO + 2H^+ + 2e^- \qquad (3)$$

**Fig. 7** CV of (*a*) bare GC, (*b*) SiO$_2$/Au NPs/GC electrode in 0.1 M NaCl solution containing 0.01 mM NB, *inset*: CV of (*c*) SiO$_2$/Au NPs/GC NaCl without NB and (*d*) with NB (0.01 mM)

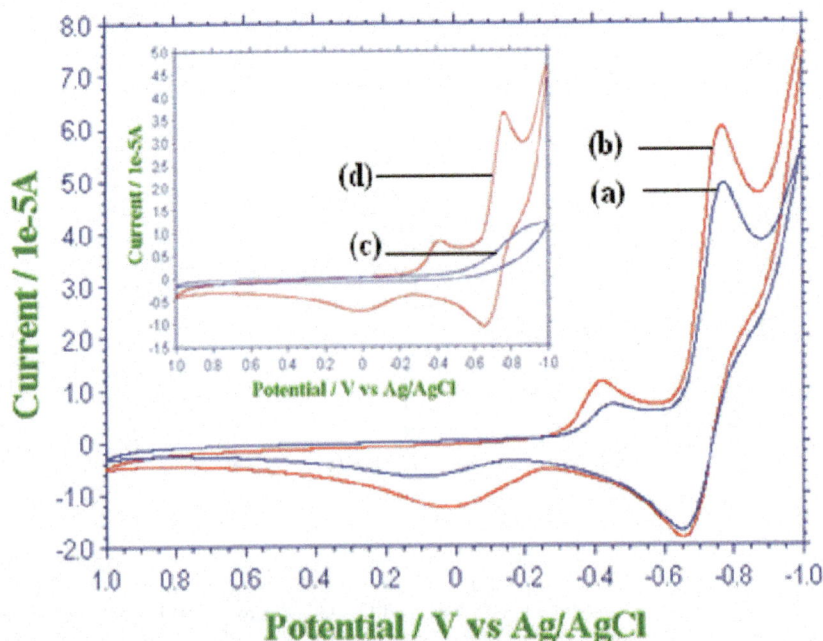

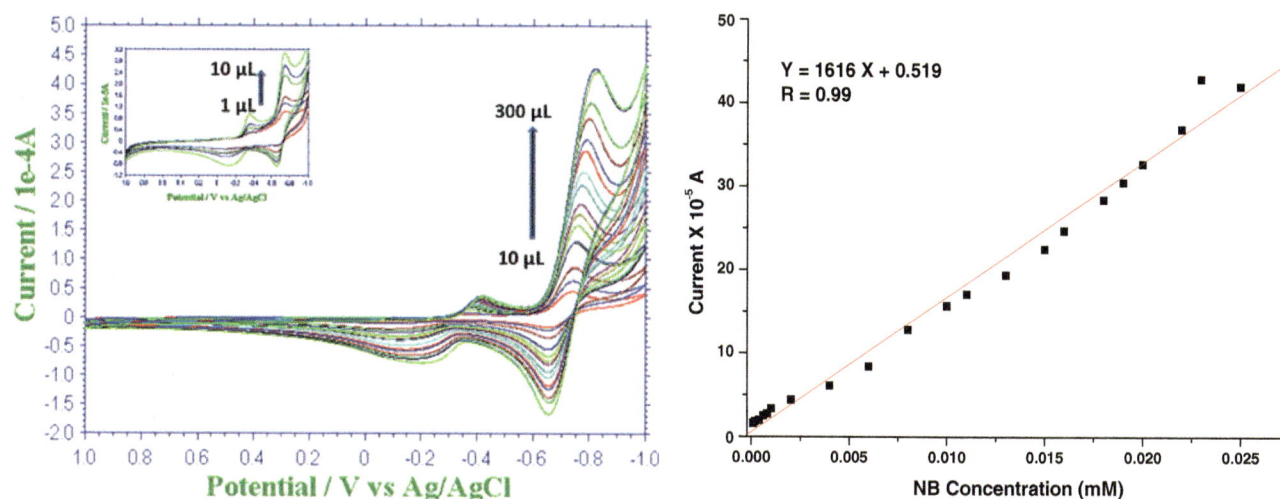

**Fig. 8** (*a*) CV of SiO$_2$/Au NPs/GC electrode in 0.1 M NaCl solution contacting varying amount of NB (10 to 300 μl), *inset*: for 1 to 10 μl of NB and (*b*) a relation curve between reduction peak current, $C_1$, and NB concentration ($1 \times 10^{-4}$ to $2.5 \times 10^{-2}$ mM)

A significant enhancement (~30 %) in reduction peak current ($C_1$) is observed at modified electrode (GC/SiO$_2$/Au NPs) relative to that at bare electrode (Fig. 7, curves a and b) and is attributed to the synergic effect of surface –OH groups selectivity of mesoporous silica towards NB and its subsequent catalytic reduction by metallic Au cores [31, 32].

Figure 8 shows the CV response of the modified electrode as a function of NB concentration ($1 \times 10^{-4}$ to $2.5 \times 10^{-2}$ mM). A monotonic increase in peak current with an increase of NB concentration is observed. The detection limit calculated from linear response curve (Fig. 8 (b)) is found to be 12.3 ppb and the current sensitivity is ~102 μA/mM ($R^2 = 0.985$). It can be seen in Fig. 8 (b) that the curve is having linearity up to $2.3 \times 10^{-2}$ mM NB concentration and a kink after that. The level at which the kink is observed may be assigned as the threshold point of the sensing platform as the graph is curvilinear after this value.

Interference studies with ions NO$_3^-$, Ni$^{2+}$, and Zn$^{2+}$ and phenol (0.1 M) were done under the same experimental conditions and no interference of these species on sensor performance was observed. The selectivity of this sensing platform towards specific nitro-compound can be further introduced by modifying the silica surface with specific organic moieties [35]. The effect of other parameters such as film thickness, pH, and temperature on sensitivity is under study.

## Conclusion

Silica-coated Au NPs (SiO$_2$/Au NPs) have been explored as a wide-range sensitive sensing platform towards NB. Acid–base chemistry (Meisenheimer complex) between hydroxyl groups presented on the mesoporus silica surface and

electron-deficient aromatic ring of NB was used for NB immobilization and subsequent catalytic reduction by metal (Au) core. Damping in SPR intensity of Au NPs was observed on NB immobilization, which may be explained by an increase in the imaginary part of the dielectric constant on NB adsorption. The optical detection range is observed in window $10^{-1}$ to $10^{-6}$ M while electrochemical detection range is found to be in the narrow window ($10^{-4}$ to $2.5 \times 10^{-2}$ mM) with a detection limit of 12.3 ppb and sensitivity of 102 μA/mM relative to bare and silica-modified electrode.

**Acknowledgments** Authors are thankful to Dr. Pawan Kapur, director, Central Scientific Instruments Organization, Chandigarh, for his kind permission to carry out this work.

## References

1. Zhang H-K, Liang S-X, Liu S-J (2007) Determination of nitrobenzene by differential pulse voltammetry and its application in wastewater analysis. Anal Chem 387(4):1511–1516
2. Davies L (2003) Nitrobenzene, W.H.O.T.G.o.E.H.C.f
3. Chen M, Yin Y, Tai C, Zhang Q, Liu J, Hu J, Jiang G (2006) Analyses of nitrobenzene, benzene and aniline in environmental water samples by headspace solid phase microextraction coupled with gas chromatography-mass spectrometry. Chin Sci Bull 51:1648–1651
4. Håkånsson K, Coorey RV, Zubarev RA, Talrose VL, Hakansson PJ (2000) Low-mass ions observed in plasma desorption mass spectrometry of high explosives. Mass Spectrom 35:337–346
5. Anferov VP, Mozjoukhine GV, Fisher R (2000) Pulsed spectrometer for nuclear quadrupole resonance for remote detection of nitrogen in explosives. Rev Sci Instrum 71:1656–1659

6. Luggar RD, Farquharson MJ, Horrocks JA, Lacey RJ (1998) Multivariate analysis of statistically poor edxrd spectra for the detection of concealed explosives. J X-ray Spectrom 27:87–94

7. Rouhi AM (1997) Seeking drugs in natural products. Chem Eng News 75:14–20

8. Fujihara H, Nakai H (2001) Fullerenethiolate-functionalized gold nanoparticles: A new class of surface-confined metal−c60 nanocomposites. Langmuir 17:6393–6395

9. Nooney RI, Dhanasekaran T, Chen Y, Josephs R, Ostafin AE (2002) Self-assembled highly ordered spherical mesoporous silica/gold nanocomposites. Adv Mater 14:529–532

10. Zhang HX, Cao AM, Hu JS, Wan LJ, Lee ST (2006) Electrochemical sensor for detecting ultratrace nitroaromatic compounds using mesoporous sio2-modified electrode. Anal Chem 78:1967–1971

11. Lu Y, Yin Y, Li YZ, Xia Y (2002) Synthesis and selfassembly of au@sio2 core−shell colloids. Nano Lett 2:785–788

12. Shimizu T, Teranishi T, Hasegawa S, Miyake M (2003) Size evolution of alkanethiol-protected gold nanoparticles by heat treatment in the solid state. J Phys Chem B 107:2719–2724

13. Graf C, Blaaderen AV (2002) Metallodielectric colloidal core-shell particles for photonic applications. Langmuir 18:524–534

14. Bard AJ, Faulkner LR (1980) Techniques based on concepts of impedance In: Electrochemical methods: fundamentals and applications. Wiley, New York, pp 316–330

15. Caschera D, Federici F, Zane D, Focanti F, Curulli A, Padeletti G (2009) Gold nanoparticles modified gc electrodes: Electrochemical behaviour dependence of different neurotransmitters and molecules of biological interest on the particles size and shape. J Nanopart Res 11:1925–1936

16. Kobayashi Y, Katakami H, Mine E, Nagao D, Konno M, Liz MLM (2005) Silica coating of silver nanoparticles using a modified Stöber method. J Colloid Interface Sci 283(2):392–396

17. Oldenburg SJ, Westcott SL, Averitt RD, Halas NJ (1999) Surface enhanced raman scattering in the near infrared using metal nano-shell substrates. J Chem Phys 111:4729–4735

18. Cachet C, Stroder U, Wiart R (1982) The kinetics of zinc electrode in alkaline zincate electrolytes. Electrochim Acta 27:903–908

19. Williams DE, Asher J (1984) Measurement of low corrosion rates: Comparison of a.C. Impedance and thin layer activation methods. Corrosion Sci 24:185–196

20. Bonanni A, Pumera M, Miyahara Y (2011) Influence of gold nanoparticle size (2-50 nm) upon its electrochemical behavior: An electrochemical impedance spectroscopic and voltammetric study. Phys Chem Chem Phys.

21. Katz E, Willner I (2003) Probing biomolecular interactions at conductive and semiconductive surfaces by impedance spectroscopy: Routes to impedimetric immunosensors, DNA-sensors, and enzyme biosensors. Electroanalysis 15:913–947

22. Curulli A, Valentini F, Viticoli M, Caschera D, Palleschi G (2005) Gold nanotubules arrays as new materials for sensing and biosensing: Synthesis and characterization. Sens Actuators B 111–112:526–531

23. Zanello P (2003) Inorganic electrochemistry: Theory, practice and application, royal society of chemistry ISBN 0-85404-661-5

24. Xie C, Zhang Z, Wang D, Guan G, Gao D, Liu J (2006) Surface molecular self-assembly strategy for tnt imprinting of polymer nanowire/nanotube arrays. Anal Chem 78:8339–83346

25. Gao D, Zhang Z, Wu M, Xie C, Guan G, Wang D (2007) A surface functional monomer-directing strategy for highly dense imprinting of tnt at surface of silica nanoparticles. J Am Chem Soc 129:7859–7866

26. Guan G, Zhang Z, Wang Z, Liu B, Guo D, Xie C (2007) Single-hole hollow polymer microspheres toward specific high-capacity uptake of target species. Adv Mater 19:2370–2374

27. Mirkin CA, Letsinger RL, Mucic RC, Storhoff JJ (1996) A DNA-based method for rationally assembling nanoparticles into macroscopic materials. Nature 382:607–609

28. Linnert T, Mulvaney P, Henglein A (1993) Surface chemistry of colloidal silver: surface plasmon damping by chemisorbed iodide, hydrosulfide (SH-), and phenylthiolate. J Phys Chem 97:679–682

29. Henglein A, Meisel D (1998) Spectrophotometric Observations of the Adsorption of Organosulfur Compounds on Colloidal Silver Nanoparticles. J Phys Chem B 102:8364–8366

30. Zhu H, Ke X, Yang X, Sarina S, Liu H (2010) Reduction of nitroaromatic compounds on supported gold nanoparticles by visible and ultraviolet light. Angew Chem Int Ed 49:9657–9661

31. Turner MB, Golovko V, Vaughan PHO, Abdulkin P, Berenguer-Murcia A, Tikhov MS, Johnson BFG, Lambert RM (2008) Selective oxidation with dioxygen by gold nanoparticle catalysts derived from 55-atom clusters. Nature 454(7207):981–983

32. Grirrane A, Corma A, GarcÃ-a H (2008) Gold-Catalyzed Synthesis of Aromatic Azo Compounds from Anilines and Nitroaromatics. Science 322(5908):1661–1664

33. Núñez-Vergara LJ, Bonta M, Navarrete-Encina PA, Squella JA (2001) Electrochemical characterization of *ortho* and *meta*-nitrotoluene derivatives in different electrolytic media. Free radical formation Electrochim Acta 46:4289–4300

34. Cavalheiro ETG, Brajter-Toth A (1999) Amperometric determination of xanthine and hypoxanthine at carbon electrodes. Effect of surface activity and the instrumental parameters on the sensitivity and the limit of detection J Pharm Biomed Anal 19:217–220

35. Engel Y, Elnathan R, Pevzner A, Davidi G, Flaxer E, Patolsky F (2010) Supersensitive Detection of Explosives by Silicon Nanowire Arrays. Angew Chem Int Ed 38:6830–6835

# 3

# Facile synthesis of hollow urchin-like gold nanoparticles and their catalytic activity

Wei Wang · Yujia Pang · Jian Yan · Guibao Wang · Hui Suo · Chun Zhao · Shuangxi Xing

**Abstract** The galvanic replacement reaction between Ag nanoparticles (NPs) and $HAuCl_4$ followed by addition of ascorbic acid led to the formation of AuNPs sharing both urchin-like and hollow structures. The AgNPs took as sacrificial templates to guide the hollow structure and the intermediates provided rough surface and active sites for the further deposition of AuNPs, which originated from the reduction of excess $HAuCl_4$ by ascorbic acid. These unique structured AuNPs presented excellent optical properties and great advantages in catalysis applications.

**Keywords** Hollow structure · Urchin-like structure · Gold nanoparticles · Catalytic activity

## Introduction

Nanostructured gold nanoparticles (AuNPs) have revealed great potential applications in many fields owing to their unique optical and catalytic properties [1–6]. Among them, urchin-like or branched gold particles with a rough surface are of great significance. The high surface area endows them excellent performance in catalysis, surface plasmon resonance (SPR) [7], surface-enhanced Raman spectra (SERS) [8], electronic devices, and biological applications [1, 2,

W. Wang · Y. Pang · J. Yan · S. Xing (✉)
Institute of Colloid and Interface Chemistry, Faculty of Chemistry, Northeast Normal University,
Changchun 130024, People's Republic of China
e-mail: xingsx737@nenu.edu.cn

W. Wang · G. Wang · H. Suo · C. Zhao
State Key Laboratory on Integrated Optoelectronics,
College of Electronic Science and Engineering, Jilin University,
Changchun 130012, People's Republic of China

9].In particular, owing to the existence of a large electromagnetic field enhancement at the tips of branched particles, a strong SERS activity can be detected with intensity over $10^7$ and a relatively high reproducibility [8]. In addition, the urchin-like structure leads to large SPR shift from 500 to 800 nm and tunable SPR enhancement light scattering and absorption, making them novel and highly effective contrast agents for in vivo cancer diagnosis and therapy [10]. Many strategies emerge for the construction of size and branch-controlled urchin-like gold particles [11], mainly through seeding growth approach [12–14], where the preferential deposition of atoms on certain facets is dominant [12]. Besides, various molecules are adapted as stabilizer or capping agent to guide the formation of urchin-like structures, such as thiol-terminated molecules [15], cetyltrimethylammonium bromide [8], and sodium dodecyl sulfate [16].

On the other hand, the hollow metal nanostructures present many potential applications in catalysts, drug delivery, optical imaging, and nano-reactors [17, 18]. In the hollow nanosphere system, SPR peak locations shift over a region of more than 100 nm due to changes of shell thickness [19], rendering them great potential application in optical sensors [17]. As for the generation of NPs with hollow interiors, the sacrificial template approach is the most adaptable one [18–21]. Generally, AgNPs have always been taken as sacrificial templates and their reaction with $HAuCl_4$ or other metal salts leads to the formation of metal NPs with voids [17, 22–28], in which the morphology of the AgNP seeds directly controls that of the resulting metal shells. These hollow-structured materials have been explored for biomedical and catalytic applications. An improved performance has already been achieved for their use as contrast enhancement agents for both optical coherence tomography and photoacoustic tomography [22, 26].

In this paper, we provided a facile method to prepare gold particles sharing both hollow interior and urchin-like shell.

Such a unique nanostructure facilitates the access of species to the Au surface and improves their catalytic performance [29].

## Experimental

### Chemicals

Deionized water was used in all the experimental processes (18.0 MΩ cm$^{-1}$). Hydrogen tetrachloroaurate (III) (Au≥47.8 % on metals basis), silver nitrate, polyvinylpyrrolidone (PVP), sodium borohydride were bought from Sinopharm Chemical Reagent Co. Ltd., ascorbic acid, sodium citrate, and glucose were from Beijing Chemical Works and all these above chemicals were used as received. 4-Nitrophenol (4-NP, Shanghai Chemical Reagent Co. Ltd.) was purified before use.

### Synthesis of silver seeds

Typically, 4.5 mg of AgNO$_3$ was dissolved in 25 mL of water and preheated at 70°C for 10 min. After that, the temperature was increased to 110°C followed by addition of 1.5 mL of sodium citrate (1 % in mass) to initiate the reduction of AgNO$_3$. After 1.5 h, the colloidal Ag solution was cooled to room temperature.

The AgNPs in (ethylene) glycol (EG) system were prepared based on the previous report [30]: 0.025 g of AgNO$_3$ and 0.10 g of PVP were dissolved in 10 mL of EG at room temperature. After that, the solution was heated at 160°C in an oil bath for 3 h under stirring.

### Synthesis of hollow urchin-like gold nanoparticles

Four milliliters of Ag colloid was centrifuged at 4,000 rpm for 5 min and the concentrated Ag deposites were dropped into 1 mL of HAuCl$_4$ aqueous solution (2.94 mM). Right after that, 1 mL of ascorbic acid (10 mM) was added. The solution turned dark blue in a short period of time. Five minutes later, 400 μL of PVP (50 mM in monomer concentration) was mixed with the product solution to avoid the possible aggregation in the further purification process. A control experiment was conducted with the same conditions in the absence of Ag seeds. The morphology of the products was investigated by a high-resolution transmission electron microscopy (HRTEM, JEOL-2100F) operated at 200 kV and scanning electron microscopy (SEM, XL-30 ESEM FEG). The optical spectra of the samples were recorded on a UV2400PC UV–vis spectrometer.

### Catalytic reaction for degradation of 4-NP

Three hundred microliters of as-prepared solution was concentrated to a total of 10 μL by centrifugation at 6,000 rpm for 4 min. After removal of the supernatant, the isolated AuNPs were immediately added to a freshly prepared reaction mixture (3 mL) containing 4-NP (0.15 mM) and NaBH$_4$ (5.5 mM). The optical property of the reaction system was analyzed by using a UV–vis spectroscopy (UV-2550) at every 5-min interval.

### Electrocatalytic oxidation of glucose

A piece of indium–tin–oxide (ITO)-coated glass (1×5 cm) was firstly pre-cleaned with acetone and immersed into a solution of H$_2$O$_2$: NH$_3$·H$_2$O/H$_2$O (1:1:5 $v/v$) for 30 min. After that, the ITO glass was rinsed with deionized water and dried at 40°C.

The cleaned ITO glass was treated with aqueous 3-aminopropyltriethoxysilane (0.25 wt%) for 15 min. After washing, the glass was immersed in a purified hollow urchin-like AuNPs solution for 30 min, washed with water, and dried in air.

The electrocatalytic oxidation of glucose was performed on a CHI 852C electrochemical workstation (CH Instruments, Chenhua Co., Shanghai, China). A conventional three-electrode cell was utilized with a saturated Ag/AgCl electrode as the reference electrode, a platinum plate as the counter electrode, and the Au-loaded ITO glass as the working electrode. A potential scan in the range of −0.6–0.85 V with a scan rate of 50 mV s$^{-1}$ was implemented to explore the electrochemical behavior of glucose in stirred 0.1 M NaOH aqueous solutions containing 1.33 mM glucose.

The estimation of the electrochemically active surface area of the hollow urchin-like and spherical gold electrodes were carried out using cyclic voltammetry (CV) and the Randles-Sevcik equation for a reversible redox couple, which at 25°C is [31]

$$I_p = \left(2.69 \times 10^5\right) n^{3/2} A D^{1/2} \nu^{1/2} C_\infty$$

where $I_p$ is the peak current (ampere), $n$ is the number of electrons transferred, $A$ is the electrode area (square centimeter), $D$ is the diffusion coefficient of the electroactive species (square centimeter per second), $\nu$ is the scan rate (volt per second), and $C_\infty$ is the bulk concentration of the same electroactive species (moles per cubic centimeter). Here, we used 5 mM K$_3$Fe(CN)$_6$ in 0.1 M KCl aqueous solution for this purpose, and K$_3$Fe(CN)$_6$ has a diffusion coefficient of 1.0×10$^{-5}$ cm$^2$ s$^{-1}$ at 25°C. The scan rate kept at 50 mV s$^{-1}$ in the measurement.

## Results and discussion

Figure 1a, b present a typical TEM image of the resulting AuNPs. All observed particles are urchin-like with hollow interiors. The average particle size is 104±11 nm and the interior size ranges from 23 to 45 nm. The interior turns smaller compared to the AgNP seeds ($d_{av}$=65 nm, Fig. 1c), which may originate from the balance between the diffusion of the Ag$^+$ ions (or Ag atoms) and the shrinkage of Au

Fig. 1 a, b TEM images of hollow urchin-like AuNPs at low and high magnification; c TEM image of Ag seeds obtained with centrifugation speed of 4000 rpm for 5 min; d EDS of the hollow urchin-like AuNPs; e TEM image of spherical AuNPs obtained in the absence of Ag seeds; f SEM image of hollow urchin-like AuNPs

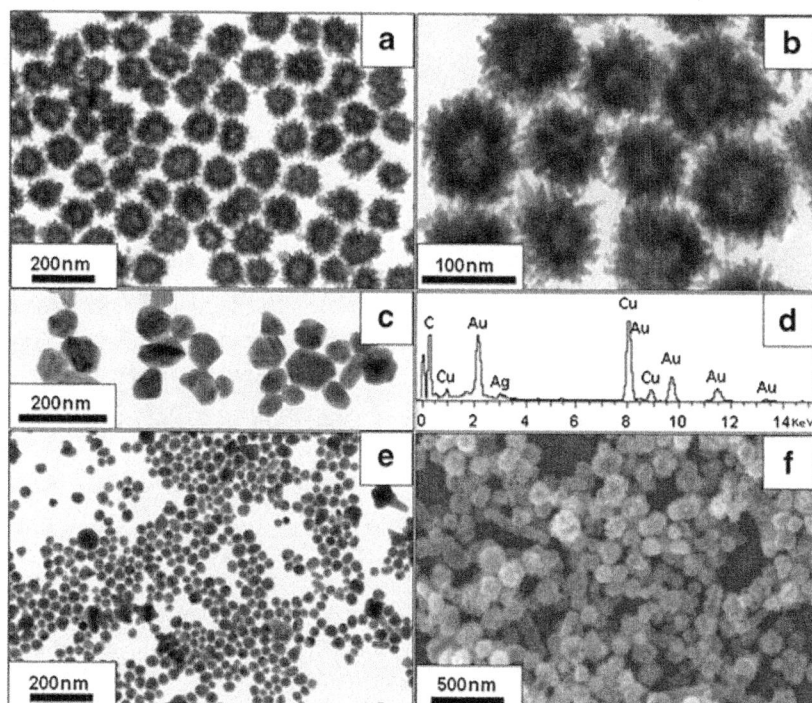

atoms. The galvanic replacement reaction between Ag and $HAuCl_4$ leads to the dissolution of Ag atoms and their following migration to the outside [32]. Meanwhile, $HAuCl_4$ is reduced to Au atoms and they deposit on the original Ag surface. Along with the disappearance of the AgNPs, AuNPs tend to shrink in order to minimize the surface energy. The diffusion rate of silver may be faster than that of the shrinkage of gold atoms; therefore, after the complete dissolution of Ag, the shrinkage process may stop, leaving the center part unoccupied. In comparison, only spherical AuNPs ($d_{av}$=33 nm) are obtained in the absence of Ag seeds (Fig. 1e). The SEM image of the products confirms that all the particles present rough surface with sharp tips (Fig. 1f).

The UV–vis spectra verify the structure evolution process (Fig. 2). Initially, the SPR peak of Ag colloid is observable at 420 nm (curve 1). Upon mixing the AgNPs with $HAuCl_4$ solution, the original Ag peak red-shifts to 550 nm (curve 2). This indicates the beginning of the galvanic reaction between Ag seeds and $HAuCl_4$ and partial formation of the nanoshells. After addition of ascorbic acid into the above system, the peak continues red-shifting to nearly 830 nm (curve 3), which derives from the combination of formation of the nanoshells and the growth of the Au branches onto their surface. In this sense, this structured NP can be found in application in optics and clinical diagnostics [22].

The ratio of $HAuCl_4$ to AgNP seeds plays an important role in the formation of uniform hollow urchin-like AuNPs. Decreasing the amount of AgNP seeds by half results in the free growth of spherical AuNPs in bulk solution (Fig. 3a). Onto the AgNPs surface, only limited $HAuCl_4$ is consumed

to generate the aimed gold structures. The excessive $HAuCl_4$ relative to the typical usage will be reduced into spherical AuNPs coexisting with the hollow ones. On the other hand, if we decrease the $HAuCl_4$ content to half, less uniform structure is achieved (Fig. 3b), which should result from the incomplete dissolution of Ag or AgCl on the surface (vide post).

The selection of reductant appears significant for the generation of well-structured hollow urchin-like AuNPs. If $NaBH_4$ was applied instead of ascorbic acid, no uniform urchin-like products were observed (Fig. 3c). Only small NPs with average diameter of 7 nm are distributed around

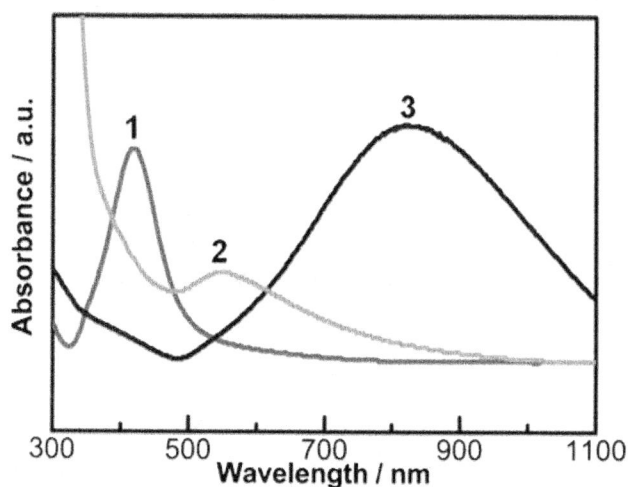

Fig. 2 UV–vis spectra of synthetic mixture. Curve 1: Ag colloid; curve 2: addition Ag seeds into $HAuCl_4$ aqueous solution; curve 3: addition of ascorbic acid into the above solution

**Fig. 3** TEM images of AuNPs obtained from a system **a** with less Ag seeds; **b** with lower HAuCl$_4$ concentration; **c** with NaBH$_4$ as reductant instead of ascorbic acid; **d** operated at 100°C

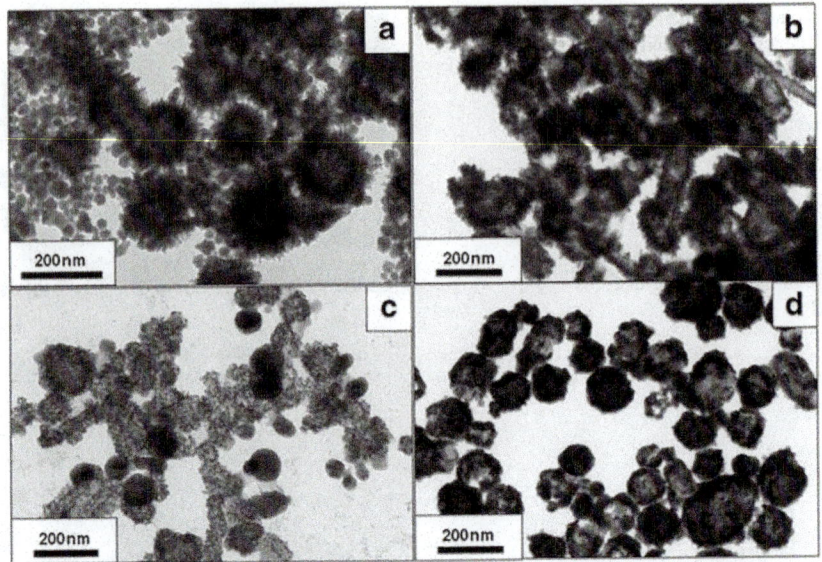

the as-obtained hollow gold structures. This indicates that a very fast reaction (NaBH$_4$ is a stronger reductant than ascorbic acid) does not benefit the further deposition of gold on the preexisting hollow gold surface.

The formation of the hollow urchin-like gold nanostructure can be induced by two separated steps (Fig. 4): (1) upon introducing concentrated AgNPs into the HAuCl$_4$ aqueous solution, a galvanic replacement reaction occurs on the original AgNPs surface. After that, the Au atoms are generated and AgNPs simultaneously transfer into Ag$^+$ ions (or Ag atoms) followed by migration to the bulk solution (vide ante). Attention should be paid that no heat treatment is conducted, which allows for the fast diffusion of the newly formed Au, forming alloy with the coexisting Ag atoms and constructing to seamless shell [24]. (2) Owing to the rough surface provided via the presence of undissolved AgCl [27], an urchin-like structure is readily produced when ascorbic acid is added to reduce the excessive HAuCl$_4$. This occurs because the reduced gold has great tendency to nucleate on the surface of preexisting NPs for lowering the energy cost. In the case of NaBH$_4$ system, a kinetic process may dominate in the second step. In addition, AgCl taking as archoring sites can be coordinated or dissolved via the excess addition of HAuCl$_4$, and no Cl signal is detected in energy-dispersive spectrometer (EDS, Fig. 1d, the Cu peaks come from the TEM grid). Finally, such interesting structured AuNPs are achieved. A control experiment was done by carrying out the reaction at 100°C. As mentioned in the previous reports, such condition helps for the generation of

seamless and smooth metal shell without the presence of AgCl. As a result, introducing ascorbic acid does not lead to

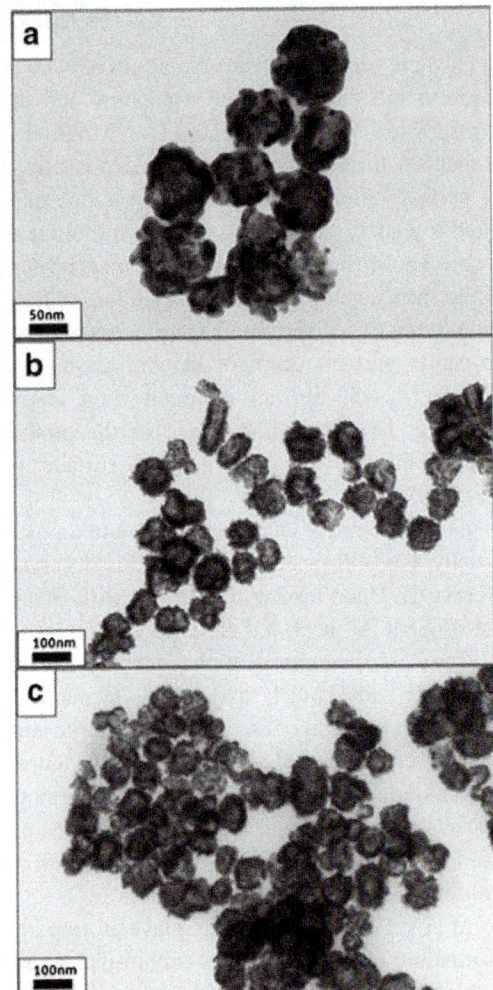

**Fig. 5** TEM images of the hollow urchin-like AuNPs at different intervals. **a** 10 s; **b** 30 s; **c** 1 min

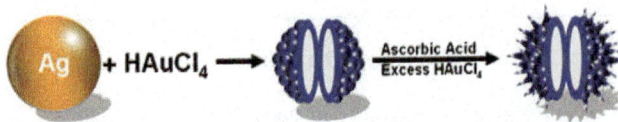

**Fig. 4** Scheme for the fabrication of hollow urchin-like AuNPs

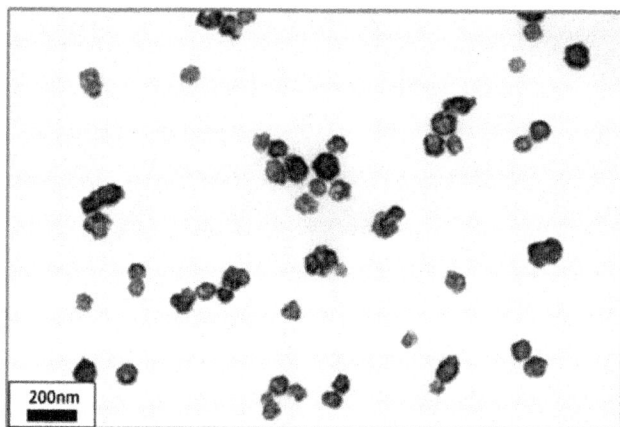

**Fig. 6** TEM image of the hollow urchin-like AuNPs via adding HAuCl$_4$ to concentrated AgNPs

the urchin-like structure (Fig. 3d). Temporal evolution of the hollow urchin-like structure was studied and it showed that the formation of these unique structured NPs is a fast process. Upon addition of all the ingredients, the galvanic replacement reaction occurs in 10 s, resulting hollow-structured NPs with rough surface. After 30 s, the tips grow around the hollow NPs via the reduction of HAuCl$_4$ by ascorbic acid. No apparent difference is observed in the later incubation time except a structure optimization from the diffusion process and Oswald ripening (Fig. 5). Attention should be paid that the feeding order of the reactants slightly makes effects on the formation of the hollow urchin-like AuNPs. If HAuCl$_4$ is added to the concentrated Ag NPs, similar structure can be obtained; however, the NPs show less uniform distribution (Fig. 6), which may originate from the uneven galvanic replacement reaction between the seeds and HAuCl$_4$ because of the limited HAuCl$_4$ available at the initial stage.

The morphology of the AgNP seeds controls the shape and size of the resulting AuNPs. Changing the AgNP size can effectively tune the diameter of the corresponding gold structures. After centrifugation at lower speed, the supernatant of the Ag colloid is further centrifuged at higher speed, thereby AgNPs with smaller size were roughly obtained (inset in Fig. 7a).

Using these AgNP seeds ($d_{av}$=46 nm), we synthesized hollow urchin-like AuNPs with average diameter of 95±15 nm

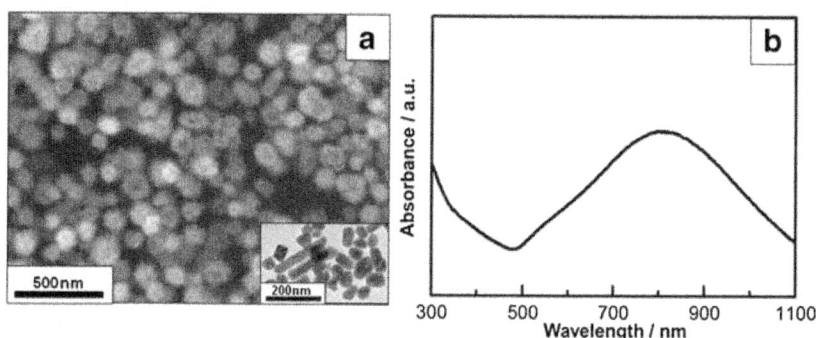

**Fig. 8** TEM images of **a** AgNPs obtained from (ethylene) glycol system and **b** hollow urchin-like AuNPs via using the above AgNPs as seeds. *Inset*: A larger view for the AuNPs in **b**

(Fig. 7a). Their SPR peak can thus be regulated to 800 nm owing to the morphology change (Fig. 7b), and the shoulder peak at around 560 nm (also observable in curve 3, Fig. 2) is related to the gold tips, which is always detectable in previous reports [33, 34]. Additionally, in Figs. 1f and 7a, some fiber-like structured particles are also observable, which should originate from the similar morphological AgNPs (inset in Fig. 7a). To exclude the presence of Ag nanofibers and the broad distribution of the AgNPs, we applied AgNPs obtained from EG system as seeds for the formation of hollow urchin-like NPs. As expected, similar structures were generated except for a more compact tips configuration (Fig. 8), which may result from the different surface properties of the Ag seeds.

It is anticipated that such structured AuNPs behave excellent catalytic performance. We selected the reduction of 4-NP to 4-aminophenol by NaBH$_4$ in aqueous media as a model reaction. The catalytic properties of the hollow

**Fig. 7 a** SEM image and **b** UV–vis spectrum of hollow urchin-like AuNPs obtained by using smaller AgNPs as seeds. *Inset*: TEM image of AgNPs with smaller average diameter

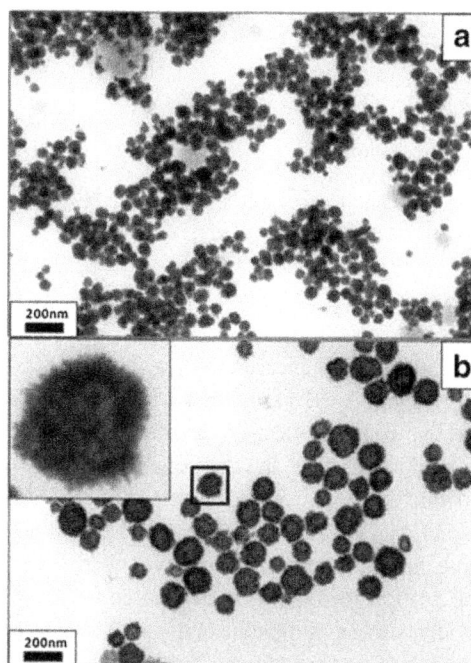

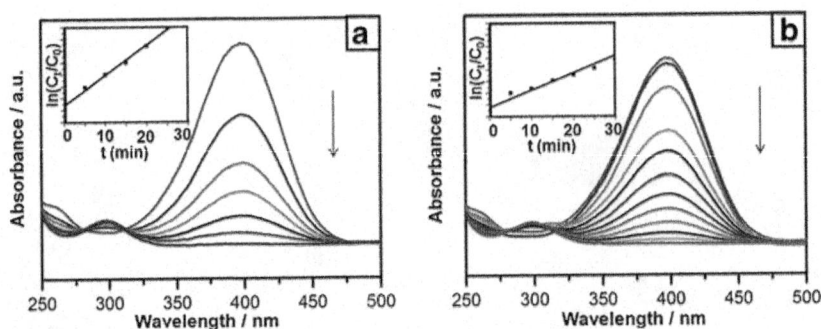

urchin-like AuNPs and spherical AuNPs synthesized in the similar system without AgNPs as seeds were compared based on their respective degrade effects. When hollow urchin-like AuNPs were used as catalysts, the complete degradation was detected in 30 min; however, using equivalent spherical AuNPs as catalysts accomplished the degradation in 55 min (Fig. 9a, b). Furthermore, in Fig. 9b, the first two curves nearly overlap, indicating that a longer period of time is required for the 4-NP to adsorb onto the surface of the spherical AuNPs due to their lower activity [24].

Considering this reaction a first-order reaction, the rate constant is determined by the slope of the linear fit of $-\ln(C_t/C_0)$ versus time, where $C_t/C_0$ represents the ratio of 4-NP concentration at time $t$ and 0 as calculated based on their corresponding absorbance intensity in the kinetic UV–vis spectra. The rate constants are 0.124 and 0.073 $min^{-1}$ for the reaction by using hollow urchin-like and spherical AuNPs as catalysts, respectively (insets in Fig. 9a, b). Following the previous discussion on the catalyst, a relatively large surface permits the oxidation and reduction reaction occurring together while a small particle size is essential to keep its high activity [24]. The hollow urchin-like AuNPs fit this point well. The sharp tips and thin shell not only render them high activity, but keep linked to avoid separating the two half

reactions. Therefore, a higher catalytic activity was achieved for the hollow urchin-like AuNPs.

The catalytic property of the hollow urchin-like AuNPs was further evaluated by exploiting the NPs in the electrocatalytic oxidation of glucose. The CV curves of the glucose oxidation using gold as catalyst are documented in previous reports [33, 35, 36]. Typically, using the spherical AuNPs electrode (Fig. 10), an oxidation potential located at 0.21 V is corresponding to the electrosorption of glucose to form adsorbed intermediate, which releases one proton per glucose molecule. These intermediates accumulated and occupied on the active sites of the electrode surface, inhibiting the direct oxidation of glucose. As a more positive potential is attained (0.48 V), the intermediates are catalytically oxidized by the formed adsorbed OH, leaving free Au active sites to directly oxidize the glucose. However, the formation of the gold oxide at a higher potential blocks the glucose to adsorb onto the active sites. As the negative potential is carrying on, the peak at $-0.07$ V is observed, relating to the formation of intermediates on the electrode surface again. Compared to the result by using spherical AuNPs, a much higher current density is achieved for the direct oxidation of glucose around 0.32 V for the hollow urchin-like gold system, indicating a much faster electron transfer on the electrode surface and thus higher catalytic performance [37]. During the negative potential sweep, glucose can be oxidized again when the Au surface is free from oxides, leading to the cathodic reoxidation peak at 0.66 V. This peak along with the peak located at 0.76 V are rarely detected for the spherical AuNPs, which probably results from the special hollow urchin-like structure and needs further detailed investigation.

In summary, we successfully synthesized hollow urchin-like gold nanoparticles by using silver nanoparticles as sacrificial seeds followed by reduction of excessive $HAuCl_4$ with ascorbic acid. This unique structure makes them excellent application in optics and catalysis fields. Similar strategies will be adopted for generation of other hollow urchin-like metal nanoparticles to dig their wider application.

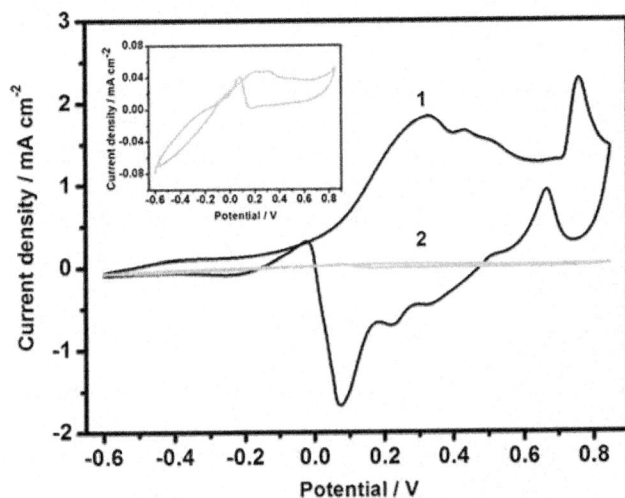

**Fig. 10** CVs for the oxidation of glucose (1.33 mM) at hollow urchin-like (curve *1*) and spherical (curve *2*) AuNPs electrodes in 0.1 M NaOH at 50 mV $s^{-1}$. *Inset*: A magnified view for curve *2*

**Acknowledgments** This work was supported by National Natural Science Foundation of China (grant no. 21103018) and Jilin Provincial Science and Technology Development Foundation (grant no. 201101010).

## References

1. Daniel M-C, Astruc D (2003) Gold nanoparticles: assembly, supramolecular chemistry, quantum-size-related properties, and applications toward biology, catalysis, and nanotechnology. Chem Rev 104:293–346.

2. Hu M, Chen J, Li Z-Y, Au L, Hartland GV, Li X, Marquez M, Xia Y (2006) Gold nanostructures: engineering their plasmonic properties for biomedical applications. Chem Soc Rev 35:1084–1094.

3. Yam VW-W, Cheng EC-C (2008) Highlights on the recent advances in gold chemistry-a photophysical perspective. Chem Soc Rev 37:1806–1813.

4. Soulé J-F, Miyamura H, Kobayashi S (2011) Powerful amide synthesis from alcohols and amines under aerobic conditions catalyzed by gold or gold/iron, -nickel or -cobalt nanoparticles. J Am Chem Soc 133:18550–18553.

5. Zhang Y, Cui X, Shi F, Deng Y (2011) Nano-gold catalysis in fine chemical synthesis. Chem Rev.

6. Haruta M (2004) Gold as a novel catalyst in the 21st century: preparation, working mechanism and applications. Gold Bull 37:27–36.

7. Nehl CL, Hafner JH (2008) Shape-dependent plasmon resonances of gold nanoparticles. J Mater Chem 18:2415–2419.

8. You H, Ji Y, Wang L, Yang S, Yang Z, Fang J, Song X, Ding B (2012) Interface synthesis of gold mesocrystals with highly roughened surfaces for surface-enhanced Raman spectroscopy. J Mater Chem.

9. Sperling RA, Rivera Gil P, Zhang F, Zanella M, Parak WJ (2008) Biological applications of gold nanoparticles. Chem Soc Rev 37:1896–1908.

10. Lu L, Ai K, Ozaki Y (2008) Environmentally friendly synthesis of highly monodisperse biocompatible gold nanoparticles with urchin-like shape. Langmuir 24:1058–1063.

11. Lim B, Xia Y (2011) Metal nanocrystals with highly branched morphologies. Angew Chem In Ed 50:76–85.

12. Li J, Wu J, Zhang X, Liu Y, Zhou D, Sun H, Zhang H, Yang B (2011) Controllable synthesis of stable urchin-like gold nanoparticles using hydroquinone to tune the reactivity of gold chloride. J Phys Chem C 115:3630–3637.

13. Broek BVd, Frederix F, Bonroy K, Jans H, Jans K, Borghs G, Maes G (2011) Shape-controlled synthesis of NIR absorbing branched gold nanoparticles and morphology stabilization with alkanethiols. Nanotechnology 22:015601.

14. Waqqar A, Kooij ES, Arend van S, Bene P (2010) Controlling the morphology of multi-branched gold nanoparticles. Nanotechnology 21:125605.

15. Bakr OM, Wunsch BH, Stellacci F (2006) High-yield synthesis of multi-branched urchin-like gold nanoparticles. Chem Mater 18:3297–3301.

16. Kuo C-H, Huang MH (2005) Synthesis of branched gold nanocrystals by a seeding growth approach. Langmuir 21:2012–2016.

17. Sun Y, Mayers B, Xia Y (2003) Metal nanostructures with hollow interiors. Adv Mater 15:641–646.

18. Hu J, Chen M, Fang X, Wu L (2011) Fabrication and application of inorganic hollow spheres. Chem Soc Rev 40:5472–5491.

19. Liang H-P, Wan L-J, Bai C-L, Jiang L (2005) Gold hollow nanospheres: tunable surface plasmon resonance controlled by interior-cavity sizes. J Phys Chem B 109:7795–7800.

20. Liu J, Liu F, Gao K, Wu J, Xue D (2009) Recent developments in the chemical synthesis of inorganic porous capsules. J Mater Chem 19:6073–6084.

21. Ortiz N, Skrabalak SE (2011) Controlling the growth kinetics of nanocrystals via galvanic replacement: synthesis of au tetrapods and star-shaped decahedra. Crys Growth Des 11:3545–3550.

22. Xia Y, Li W, Cobley CM, Chen J, Xia X, Zhang Q, Yang M, Cho EC, Brown PK (2011) Gold nanocages: from synthesis to theranostic applications. Acc Chem Res 44:914–924.

23. Chen J, Yang M, Zhang Q, Cho EC, Cobley CM, Kim C, Glaus C, Wang LV, Welch MJ, Xia Y (2010) Gold nanocages: a novel class of multifunctional nanomaterials for theranostic applications. Adv Funct Mater 20:3684–3694.

24. Zeng J, Zhang Q, Chen J, Xia Y (2010) A comparison study of the catalytic properties of Au-based nanocages, nanoboxes, and nanoparticles. Nano Lett 10:30–35.

25. Personick ML, Langille MR, Zhang J, Mirkin CA (2011) Shape control of gold nanoparticles by silver underpotential deposition. Nano Lett 11:3394–3398.

26. Skrabalak SE, Chen J, Sun Y, Lu X, Au L, Cobley CM, Xia Y (2008) Gold nanocages: synthesis, properties, and applications. Acc Chem Res 41:1587–1595.

27. Sun Y, Xia Y (2004) Mechanistic study on the replacement reaction between silver nanostructures and chloroauric acid in aqueous medium. J Am Chem Soc 126:3892–3901.

28. Lu X, Tuan HY, Chen J, Li ZY, Korgel BA, Xia Y (2007) Mechanistic studies on the galvanic replacement reaction between multiply twinned particles of Ag and HAuCl$_4$ in an organic medium. J Am Chem Soc 129:1733–1742.

29. Ataee-Esfahani H, Nemoto Y, Wang L, Yamauchi Y (2011) Rational synthesis of Pt spheres with hollow interior and nanosponge shell using silica particles as template. Chem Commun 47:3885–3887.

30. Sun Y, Wiley B, Li Z-Y, Xia Y (2004) Synthesis and optical properties of nanorattles and multiple-walled nanoshells/nanotubes made of metal alloys. J Am Chem Soc 126:9399–9406.

31. Gooding JJ, Praig VG, Hall EAH (1998) Platinum-catalyzed enzyme electrodes immobilized on gold using self-assembled layers. Anal Chem 70:2396–2402.

32. Liu Y, Hight Walker AR (2011) Preferential outward diffusion of Cu during unconventional galvanic replacement reactions between HAuCl$_4$ and surface-limited Cu nanocrystals. ACS Nano 5:6843–6854.

33. Xu F, Cui K, Sun Y, Guo C, Liu Z, Zhang Y, Shi Y, Li Z (2010) Facile synthesis of urchin-like gold submicrostructures for nonenzymatic glucose sensing. Talanta 82:1845–1852.

34. Pandian Senthil K, Isabel P-S, Benito R-G, Abajo FJGd, Luis ML-M (2008) High-yield synthesis and optical response of gold nanostars. Nanotechnology 19:015606.

35. Zhang H, Xu J-J, Chen H-Y (2008) Shape-controlled gold nano-architectures: synthesis, superhydrophobicity, and electrocatalytic properties. J Phys Chem C 112:13886–13892.

36. Li Y, Song Y-Y, Yang C, Xia X-H (2007) Hydrogen bubble dynamic template synthesis of porous gold for nonenzymatic electrochemical detection of glucose. Electrochem Commun 9:981–988.

37. Zhao J, Kong X, Shi W, Shao M, Han J, Wei M, Evans DG, Duan X (2011) Self-assembly of layered double hydroxide nanosheets/Au nanoparticles ultrathin films for enzyme-free electrocatalysis of glucose. J Mater Chem 21:13926–13933.

# 4

# Synthesis of DNA-templated fluorescent gold nanoclusters

Guiying Liu · Yong Shao · Kun Ma · Qinghua Cui ·
Fei Wu · Shujuan Xu

**Abstract** Water-soluble and red-emitting gold nanoclusters (Au NCs) were synthesized with single-stranded DNA as a promising biotemplate and dimethylamine borane as a mild reductant. The fluorescent Au NCs can be formed in a weakly acidic aqueous solution that is free from the simultaneous formation of large nanoparticles. The cluster feature of the formed Au species has been revealed by fluorescence spectra, absorption spectra, and transmission electron microscopy. Additionally, DNA sequences could be used to tune the Au NCs' emissions. The as-prepared Au NCs display high stability at physiological pH condition, and thus, wide potential applications are anticipated for the biocompatible fluorescent Au NCs serving as nanoprobes in bioimaging and related fields.

**Keywords** Gold nanoclusters · Fluorescence · DNA · Dimethylamine borane · Template

## Introduction

Decreasing the size of noble metal nanostructures (mainly Au and Ag) down to less than 2 nm will produce nanoclusters (NCs) and restrict the motion of their free electrons in a very confined space that results in discrete electronic band structures. When the discrete band energies become larger than thermal energies, the NCs will behave like molecules in respect of optical properties such as light absorption and emission. Au NCs have emerged as novel fluorescent nanomaterials because of their better performance in many aspects like biocompatibility, photostability, and non-toxicity relative to organic dyes and semiconductor quantum dots [1–4].

Fluorescent Au NCs have been prepared mainly in a bottom–up manner by the reduction of gold precursors in the presence of various templates such as macromolecules (dendrimers [5–9], proteins [10–24], poly-butadiene [25]), small molecules (histidine [26], carbohydrate [27], thiols [28–33], N, N-dimethylformamide [34, 35], penicillamine [36]), and even solid functional organisms (eggshell membrane [37]). Alternatively, top–down etching of preformed large nanoparticles down to desired NC sizes has received much attention due to many available synthetic strategies for the large nanoparticles. In this aspect, polyethylenimine [38], dihydrolipoic acid [39, 40], thiols [41–43], Good's buffers [44], cyclodextrins [45], and even hydrochloric acid [46] have been employed as effective etchants. Recently, large nanoparticles have been reported to be even fluorescent after being sensitized by thiols [47, 48].

It is widely accepted that the formation of stable Au NCs is controlled by a slow thermodynamic process for a narrow size distribution following their relative rapid formation [49] or by a cyclic process of growth and etching reactions around the most stable cluster species to form nearly monodisperse product distributions [46, 50]. In addition, the optical properties of the Au NCs are related to the ligands that protect them from aggregations [51] and redox state of the gold core [52, 53], or the gold core geometry tuned by the oxidation states [53]. On the basis of this mechanism understanding, many applications, for example, detections of $Hg^{2+}$ [12–15, 37, 47], $Cu^{2+}$ [19, 20, 32], $CN^{-}$ [17], $H_2O_2$

G. Liu · Y. Shao (✉) · K. Ma · Q. Cui · F. Wu · S. Xu
Zhejiang Key Laboratory for Reactive Chemistry
on Solid Surfaces, Institute of Physical Chemistry,
Zhejiang Normal University,
Jinhua 321004 Zhejiang, People's Republic of China
e-mail: yshao@zjnu.cn

[18], glucose [24], and dopamine [21], and cell labeling or imaging [13, 23, 40, 48], have been achieved with Au NCs as reporters by direct or indirect reaction of the NCs' protecting ligands or gold cores with the species of interest. However, in comparison to the fruitful strategies for the DNA-templated synthesis and optical tunability of silver nanoclusters (Ag NCs) [54, 55], there are fewer reports for the successful synthesis of Au NCs with DNA as template. Recently, atomically monodisperse fluorescent Au NCs were obtained by etching gold particles (either spheres or rods) with the assistance of DNA under sonication in water [56]. Due to the photosensitivity of Ag species, the most prominent advantage of DNA-templated Au NCs over Ag NCs in biocompatible applications would be the Au NCs' favorable stability. In this work, single-stranded DNA was first employed as an alternative template during reduction of Au precursor to produce Au NCs (see Fig. 1).

## Experimental

### Synthesis of fluorescent Au NCs

Twenty-three-mer DNAs with the sequences of 5'-GAGGCGCTGCCYCCACCATGAGC-3' (named 23-Ys, Y = C, A, G, and T) were synthesized by TaKaRa Biotechnology Co., Ltd. (Dalian, China). All the DNA samples were HPLC purified by the manufacturer. Other reagents were of analytical grade and used without further purification. Nanopure water (18.2 m$\Omega$; Millipore Co., USA) was used in all experiments. In a typical experiment, chloroauric acid (HAuCl$_4$, Sigma Chemical Co., St. Louis, USA) solution was added to the single-stranded DNA solution in 20 mM phosphate containing 1 mM magnesium acetate (PBS) by an appropriate HAuCl$_4$/DNA concentration ratio. After being thoroughly mixed, the solution was aged at room temperature for 10 h to allow for the completion of the interaction of HAuCl$_4$ with DNA. Then, the freshly prepared dimethylamine borane (DMAB, Sigma Chemical Co., St. Louis, USA) solution was added to the aged HAuCl$_4$/DNA solution, which was followed by another 36-h reaction at room temperature in the dark to produce fluorescent Au NCs. The resulting solutions were examined at room temperature (22±1 °C). For control experiments, sodium borohydride was used as the reductant to replace DMAB.

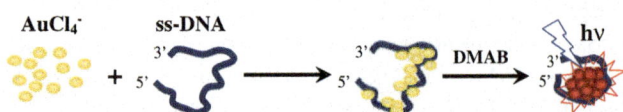

Fig. 1 Schematic illustration for the formation of Au NCs templated by DNA

### Characterization of fluorescent Au NCs

Fluorescence spectra were acquired with a FLSP920 spectrofluorometer (Edinburgh Instruments Ltd., UK) at 22±1 °C, equipped with a temperature-controlled circulator (Julabo, Germany). UV/vis absorption spectra were determined with a UV2550 spectrophotometer (Shimadzu Corp., Japan). Transmission electron microscopy (TEM) images were acquired on a JEOL 2010F transmission electron microscope at the acceleration voltage of 200 kV. The TEM samples were prepared by dropping a dispersion of the as-prepared Au NCs onto a Cu grid covered by a holey carbon film.

## Results and discussion

Fluorescent Au NCs have been widely synthesized in a bottom–up manner by reduction of gold precursors that are associated with various biotemplates [10] such as bovine serum albumin [11], horseradish peroxidase [18], lysozyme [12], and transferrin protein [20]. Nevertheless, synthesis of Au NCs templated by DNA has rarely been reported maybe because of the weak association between the negatively charged DNA and commonly used precursor AuCl$_4^-$. Here, we tried to qualify the right conditions to synthesize fluorescent Au NCs in the presence of DNA (Fig. 1). Twenty-three-mer single-stranded DNAs with the sequences of 5'-GAGGCGCTGCCYCCACCATGAGC-3' (named 23-Ys, Y = C, A, G, and T) were employed in this work. These sequences are stable in aqueous solution free from any secondary structure at room temperature. In an optimized experiment, the concentration ratio 1:15:75 of DNA/HAuCl$_4$/DMAB was used to produce fluorescent Au NCs in PBS at pH 4.4 (Fig. S1 and S2 in the Supporting information). As shown in Fig. 2, the red fluorescent Au NCs can be prepared in aqueous solution by reducing the gold salt with DMAB using single-stranded 23-C as the template. The DNA-

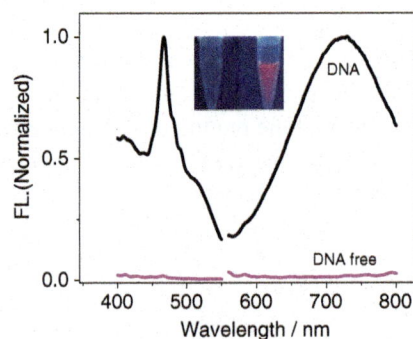

Fig. 2 Fluorescence excitation (measured at 725 nm) and emission (excited at 467 nm) spectra of 20 mM PBS (pH 4.4) containing 75 μM HAuCl$_4$ and 375 μM DMAB in the absence and presence of 5 μM 23-C. *Inset*: photographs of the solutions in the absence and presence of DNA (from *left* to *right*) under UV illumination

**Fig. 3** TEM images of Au nanomaterials prepared in PBS at pH 4.4 for NCs (**a**) and pH 7.0 for larger nanoparticles (**b**)

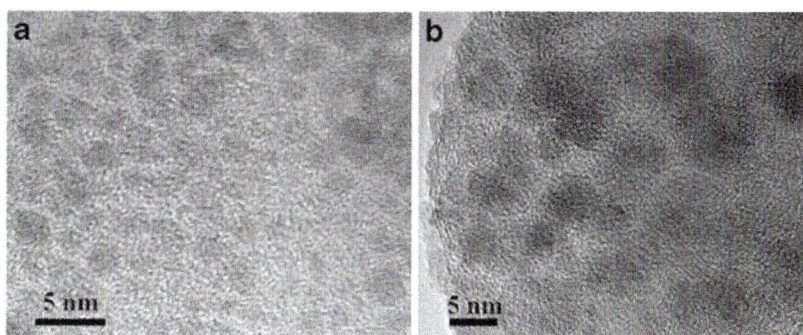

templated Au NCs display excitation and emission bands at 467 and 725 nm, respectively. However, reducing the HAuCl$_4$ solution by DMAB in the absence of 23-C induces a light pink sample without any noticeable emission, confirming the crucial role of DNA for the formation of fluorescent Au NCs. Under UV illumination, a bright red emission from the as-prepared Au NC solution can be clearly distinguished from that of the solution without 23-C by the naked eye, indicating that highly fluorescent Au NCs are formed in the presence of DNA. Previously, Dickson et al. [5] have explained their experimental results with the spherical Jellium model for predicting the size of Au NCs by fitting the Au NCs' emission energy with the scaling relation of $E_{Fermi}/N^{1/3}$, where $E_{Fermi}$ is the Fermi level of gold element and $N$ is the number of Au atoms composed of Au NCs. From the observed emission energy of 1.71 eV in our experiment for the DNA-templated Au NCs, we roughly estimate that the number of gold atoms composed of the Au NCs is about 21. As revealed by the TEM analysis (Fig. 3a), it is difficult to accurately determine the diameter of the fluorescent Au NCs due to the low TEM contrast with the background for such small-sized materials, which is in good agreement with the cluster dimension predicted by the Jellium model. However, the cluster profile can be easily seen from the TEM image.

We found that many factors strongly affected the formation of fluorescent Au NCs. As shown in Fig. 4a, the solution pH plays a key role in modulating the emissions of Au NCs templated by 23-C. By comparison to the emission from the solution prepared in PBS at pH 4.4, the resulting solutions prepared in PBS at pH 5.0 and 6.0 exhibit

1.2- and 3.1-fold decreases in the fluorescence intensities, respectively. However, there is almost unnoticeable fluorescence emission for those prepared in PBS at pH 7.0 and 8.0. Therefore, acidic solution conditions seem to facilitate the creation of fluorescent Au NCs. Absorption spectra were then followed to further confirm the influence of the solution pH on the formation of fluorescent Au NCs. As shown in Fig. 4b, the solutions prepared at pH above 6.0 accordingly exhibit clear absorption peaks located at about 525 nm, which suggests the formation of larger gold nanoparticles with characteristic surface plasmon resonance absorption. As an example, the production of such gold nanoparticles at pH 7.0 is thus evidenced by TEM analysis with diameter larger than 5 nm (Fig. 3b). By contrast, featureless absorption spectra are observed for the solutions prepared at lower pH values. This fact indicates that the fluorescent Au NCs produced at the weakly acidic conditions should be smaller than 2 nm in diameter [57, 58], which is in agreement with the TEM results and the Jellium model-based prediction. Previously, the similar absorption spectra with such featureless behaviors were observed for fluorescent Au NCs by their intensities decaying roughly exponentially toward the visible region from the UV region [14, 36, 38, 39]. Therefore, the production of fluorescent Au NCs is free from the simultaneous formation of large gold nanoparticles at the acidic conditions. On the basis of these observations, our method would expand the potential applications of fluorescent Au NCs with DNA as the biotemplate because the previously reported protein-based synthesis of fluorescent Au NCs was mostly carried out at strong alkaline conditions (pH≥12) [10].

**Fig. 4** Effects of solution pH on the formation of fluorescent Au NCs: **a** fluorescence spectra, **b** absorption spectra

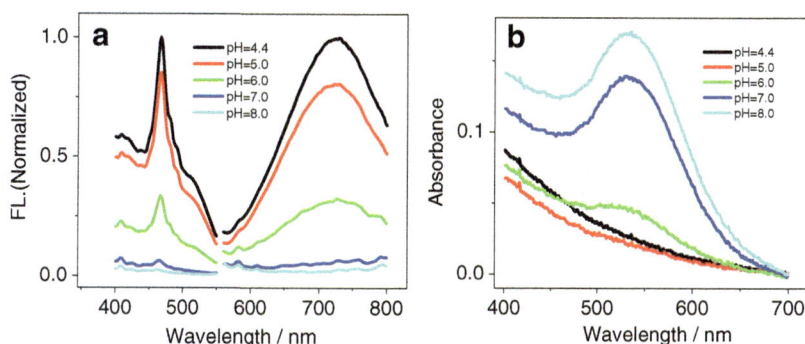

However, the immediate addition of DMAB into the freshly mixed DNA–HAuCl$_4$ solution prepared at whatever pH results in the prompt formation of pink samples without any fluorescence response to be observed. Consequently, we speculate that the first crucial step for the creation of fluorescent Au NCs is the formation of an Au(III)–DNA complex, which occurs by replacing the Cl$^-$ ligands in AuCl$_4^-$ with DNA bases before the reduction. To follow the reaction between AuCl$_4^-$ and DNA, we monitored the time evolution of the DNA absorption spectra at 260 nm after the addition of HAuCl$_4$. As shown in Fig. 5, an abrupt decrease in the absorption is evidenced after aging the sample prepared at pH 4.4 for 10 h, whereas there is no distinct change in the absorption for the sample prepared at pH 7.0 even with the reaction time extending up to 50 h. Although the exact interaction mechanism of the DNA base with HAuCl$_4$ is not yet clear, the coordination and chelation between gold and both the ring and amino nitrogens of the nucleic acid bases [59] should contribute to this process. At an acidic solution, cytosine in DNA should be partially protonated [60] to facilitate its interaction with the negatively charged AuCl$_4^-$, while at neutral and alkaline conditions, there is a relative large repulsion force to prevent the negatively charged DNA from approaching the negatively charged AuCl$_4^-$. The possible protonation of cytosine in DNA would induce less base stacking, which is reflected by the higher absorption at pH 4.4 than that at pH 7.0 as observed at the initial stage of AuCl$_4^-$ addition (Fig. 5). Nevertheless, the at-least 10-h aging time for the production of fluorescent Au NCs at the weakly acidic condition shows that the specific interaction between AuCl$_4^-$ and DNA is still a slow process. Thus, without the aging step prior to reduction, Au(III) mainly in the form of AuCl$_4^-$ free in water can be directly reduced by DMAB into large gold nanoparticles. Further works will be expected in this laboratory to identify the interaction mode of DNA bases with the fluorescent Au NCs by, for example, infrared and circular dichroism spectra.

It is well known that different DNA sequences and lengths can be used to modulate the emissions of DNA-

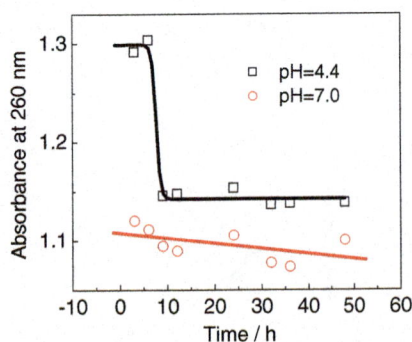

**Fig. 6** Dependences of DNA sequences on the fluorescence spectra of the as-prepared Au NCs

templated silver nanoclusters [54]. Thus, it is expected that the DNA sequences could be also used to tune the Au NCs' emissions. To examine the impact of DNA sequences, we only changed the central base in 23-C from cytosine to adenine (23-A), guanine (23-G), and thymine (23-T) and kept the other reaction conditions unchanged. As shown in Fig. 6, the Au NCs' emissions are dependent on the DNA sequences with the intensities decreasing in the order of 23-C>23-A>23-T>23-G. The emission maxima are also blue shifted in the same order. Due to the one-base alteration for all the used DNAs at the same length, small changes in Au NCs' emissions could be imaged as observed here. Therefore, we believe that it is feasible to synthesize Au NCs with different emission behaviors by DNA sequence alterations.

We found that the used reductant had a profound effect on the formation of fluorescent Au NCs. For example, replacement of DMAB with NaBH$_4$, a common reductant in the synthesis of noble metal nanoclusters [54], mainly resulted in prompt production of large nanoparticles with barely faint fluorescence even though the aging procedure was still carried out, which is in agreement with the previous observation that NaBH$_4$ was an ineffective reductant for the production of fluorescent Au NCs [36]. By comparison to

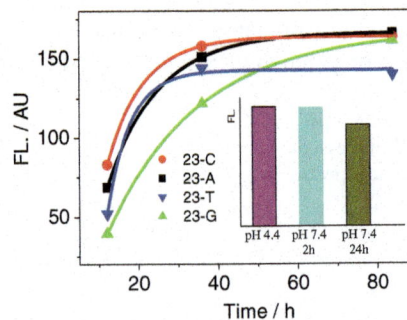

**Fig. 5** Time evolutions of the corresponding absorbances at 260 nm for the HAuCl$_4$–DNA solutions at pH 4.4 and 7.0 before DMAB addition

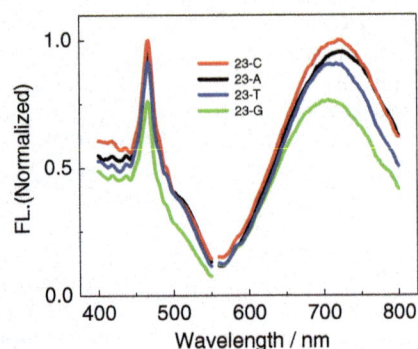

**Fig. 7** Time evolutions of the Au NCs' fluorescence emissions after the addition of DMAB. *Inset*: the emission intensities of the preformed Au NCs templated by 23-C at pH 4.4 and then after 2 and 24 h of adjusting the solution pH to 7.4

the strong reduction capacity related to NaBH$_4$, DMAB was a weak reductant [61] and proved to be a fine candidate to reduce the DNA-bound gold species to fluorescent Au NCs. As shown in Fig. 7, an incubation time of 36 h after the addition of DMAB is needed to get the stable emissions on account of the weak reducing capacity of DMAB at the weakly acidic condition. The formed Au NCs are stable enough to keep their emissions for more than 2 days. Thus, we reasonably conclude that a slow reduction process of the DNA-bound gold species is crucial to prevent the preformed Au NCs from aggregating into large nanoparticles.

Lastly, we tested the stability of fluorescent Au NCs at the solution with different pH from that for their preparation. As shown in the inset of Fig. 7, the fluorescence intensities of the preformed Au NCs at pH 4.4 decrease only 1 and 15.6 % after 2 and 24 h of adjusting the solution pH value to 7.4, indicating that the preformed Au NCs' emission is not seriously affected by electrolyte's pH. Accordingly, we expect that although the fluorescent Au NCs can be formed only at the weakly acidic conditions, the high stability of the preformed Au NCs at the physiological pH condition would greatly facilitate their potential use in bioimaging applications due to biocompatibility of the used DNA template.

## Conclusion

In summary, we presented a new approach for the synthesis of water-soluble, red fluorescent Au NCs templated by DNA. Investigations by fluorescence, TEM, and absorption spectra convince that the fluorescent Au NCs can be formed by reducing the Au precursor with DMAB at weakly acidic pH conditions. During this process, the aging time for completing the interaction of DNA with HAuCl$_4$ before reduction is critical to form the fluorescent Au NCs. In addition, the Au NCs' emissions could be tuned by DNA sequences. The high stability of the preformed Au NCs at the physiological pH condition and the biocompatibility of the used DNA template would support their wide applications as novel nanoprobes.

**Acknowledgments** This study was supported by the National Natural Science Foundation of China (grant no. 21075112), the Zhejiang Provincial Natural Science Foundation of China for Distinguished Young Scholars (grant no. R12B050001), the Foundation of State Key Laboratory of Electroanalytical Chemistry, Changchun Institute of Applied Chemistry (grant no. SKLEAC2010001), and the Scientific Research Foundation for Returning Overseas Chinese Scholars, State Education Ministry.

## References

1. Zheng J, Nicovich PR, Dickson RM (2007) Highly fluorescent noble metal quantum dots. Annu Rev Phys Chem 58:409–431
2. Lin CAJ, Lee CH, Hsieh JT, Wang HH, Li JK, Shen JL, Chan WH, Yeh HI, Chang WH (2009) Synthesis of fluorescent metallic nanoclusters toward biomedical application: recent progress and present challenges. J Med Biol Eng 29:276–283
3. Yang QF, Liu JY, Chen HP, Wang XX, Huang QM, Shan Z (2011) Preparation of noble metallic nanoclusters and its application in biological detection. Prog Chem 23:880–892
4. Shang L, Dong SJ, Nienhaus GU (2011) Ultra-small fluorescent metal nanoclusters: synthesis and biological applications. Nano Today 6:401–4184
5. Zheng J, Zhang CW, Dickson RM (2004) Highly fluorescent, water soluble, size-tunable, gold quantum dots. Phys Rev Lett 93:077402
6. Zheng J, Petty JT, Dickson RM (2003) High quantum yield blue emission from water-soluble Au$_8$ nanodots. J Am Chem Soc 125:7780–7781
7. Shi X, Ganser TR, Sun K, Balogh LP, Baker JR Jr (2006) Characterization of crystalline dendrimer-stabilized gold nanoparticles. Nanotechnology 17:1072–1078
8. Bao Y, Zhong C, Vu DM, Temirov JP, Dyer RB, Martinez JS (2007) Nanoparticle free synthesis of fluorescent gold nanoclusters at physiological temperature. J Phys Chem C 111:12194–12198
9. Jao YC, Chen MK, Lin SY (2010) Enhanced quantum yield of dendrimer-entrapped gold nanodots by a specific ion-pair association and microwave irradiation for bioimaging. Chem Commun 46:2626–2628
10. Xavier PL, Chaudhari K, Baksi A, Pradeep T (2012) Protein-protected luminescent noble metal quantum clusters: an emerging trend in atomic cluster nanoscience. Nano Rev 3:14767
11. Xie J, Zheng Y, Ying JY (2009) Protein-directed synthesis of highly fluorescent gold nanoclusters. J Am Chem Soc 131:888–889
12. Wei H, Wang Z, Yang L, Tian S, Hou C, Lu Y (2010) Lysozyme-stabilized gold fluorescent cluster: synthesis and application as Hg$^{2+}$ sensor. Analyst 135:1406–1410
13. Hu D, Sheng Z, Gong P, Zhang P, Cai L (2010) Highly selective fluorescent sensors for Hg$^{2+}$ based on bovine serum albumin-capped gold nanoclusters. Analyst 135:1411–1416
14. Kawasaki H, Yoshimura K, Hamaguchi K, Arakawa R (2011) Trypsin-stabilized fluorescent gold nanocluster for sensitive and selective Hg$^{2+}$ detection. Anal Sci 27:591–596
15. Pu KY, Luo Z, Li K, Xie J, Liu B (2011) Energy transfer between conjugated-oligoelectrolyte-substituted poss and gold nanocluster for multicolor intracellular detection of mercury ion. J Phys Chem C 115:13069–13075
16. Retnakumari A, Setua S, Menon D, Ravindran P, Muhammed H, Pradeep T, Nair S, Koyakutty M (2010) Molecular-receptor-specific, non-toxic, near-infrared-emitting Au cluster-protein nanoconjugates for targeted cancer imaging. Nanotechnology 21:055103
17. Liu Y, Ai K, Cheng X, Huo L, Lu L (2010) Gold-nanocluster-based fluorescent sensors for highly sensitive and selective detection of cyanide in water. Adv Funct Mater 20:951–956
18. Wen F, Dong Y, Feng L, Wang S, Zhang S, Zhang X (2011) Horseradish peroxidase functionalized fluorescent gold nanoclusters for hydrogen peroxide sensing. Anal Chem 83:1193–1196
19. Durgadas CV, Sharma CP, Sreenivasan K (2011) Fluorescent gold clusters as nanosensors for copper ions in live cells. Analyst 136:933–940
20. Xavier PL, Chaudhari K, Verma PK, Pal SK, Pradeep T (2010) Luminescent quantum clusters of gold in transferrin family protein, lactoferrin exhibiting FRET. Nanoscale 2:2769–2776

21. Li L, Liu H, Shen Y, Zhang J, Zhu JJ (2011) Electrogenerated chemiluminescence of Au nanoclusters for the detection of dopamine. Anal Chem 83:661–665

22. Guével XL, Daum N, Schneider M (2011) Synthesis and characterization of human transferrin-stabilized gold nanoclusters. Nanotechnology 22:275103

23. Retnakumari A, Jayasimhan J, Chandran P, Menon D, Nair S, Mony U, Koyakutty M (2011) CD$_{33}$ monoclonal antibody conjugated Au cluster nano-bioprobe for targeted flow-cytometric detection of acute myeloid leukaemia. Nanotechnology 22:285102

24. Jin L, Shang L, Guo S, Fang Y, Wen D, Wang L, Yin J, Dong S (2011) Biomolecule-stabilized Au nanoclusters as a fluorescence probe for sensitive detection of glucose. Biosens Bioelectron 26:1965–1969

25. Yabu H (2011) One-pot synthesis of blue light-emitting Au nanoclusters and formation of photo-patternable composite films. Chem Commun 47:1196–1197

26. Yang X, Shi M, Zhou R, Chen X, Chen H (2011) Blending of HAuCl$_4$ and histidine in aqueous solution: a simple approach to the Au$_{10}$ cluster. Nanoscale 3:2596–2601

27. Barrientos AG, de la Puente JM, Rojas TC, Fernandez A, Penades S (2003) Gold glyconanoparticles: synthetic polyvalent ligands mimicking glycocalyx-like surfaces as tools for glycobiological studies. Chem Eur J 9:1909–1921

28. Link S, Beeby A, FitzGerald S, El-Sayed MA, Schaaff TG, Whetten RL (2002) Visible to infrared luminescence from a 28-atom gold cluster. J Phys Chem B 106:3410–3415

29. Huang T, Murray RW (2001) Visible luminescence of water-soluble monolayer-protected gold clusters. J Phys Chem B 105:12498–12502

30. Shibu ES, Radha B, Verma PK, Bhyrappa P, Kulkarni GU, Pal SK, Pradeep T (2009) Functionalized Au$_{22}$ clusters: synthesis, characterization, and patterning. ACS Appl Mater Interfaces 1:2199–2210

31. Polavarapu L, Manna M, Xu QH (2011) Biocompatible glutathione capped gold clusters as one- and two-photon excitation fluorescence contrast agents for live cells imaging. Nanoscale 3:429–434

32. Tu X, Chen W, Guo X (2011) Facile one-pot synthesis of near-infrared luminescent gold nanoparticles for sensing copper (II). Nanotechnology 22:095701

33. Yu M, Zhou C, Liu J, Hankins JD, Zheng J (2011) Luminescent gold nanoparticles with pH-dependent membrane adsorption. J Am Chem Soc 133:11014–11017

34. Liu X, Li C, Xu J, Lv J, Zhu M, Guo Y, Cui S, Liu H, Wang S, Li Y (2008) Surfactant-free synthesis and functionalization of highly fluorescent gold quantum dots. J Phys Chem C 112:10778–10783

35. Kawasaki H, Yamamoto H, Fujimori H, Arakawa R, Iwasaki Y, Inada M (2010) Stability of the DMF-protected Au nanoclusters: photochemical, dispersion, and thermal properties. Langmuir 26:5926–5933

36. Shang L, Dörlich RM, Brandholt S, Schneider R, Trouillet V, Bruns M, Gerthse D, Nienhaus GU (2011) Facile preparation of water-soluble fluorescent gold nanoclusters for cellular imaging applications. Nanoscale 3:2009–2014

37. Shao CY, Yuan B, Wang HQ, Zhou Q, Li Y, Guan Y, Deng Z (2011) Eggshell membrane as a multimodal solid state platform for generating fluorescent metal nanoclusters. J Mater Chem 21:2863–2866

38. Duan HW, Nie SM (2007) Etching colloidal gold nanocrystals with hyperbranched and multivalent polymers: a new route to fluorescent and water-soluble atomic clusters. J Am Chem Soc 129:2412–2413

39. Lin CAJ, Yang TY, Lee CH, Huang SH, Sperling RA, Zanella M, Li JK, Shen JL, Wang HH, Yeh HI, Parak WJ, Chang WH (2009) Synthesis, characterization, and bioconjugation of fluorescent gold nanoclusters toward biological labeling applications. ACS Nano 3:395–401

40. Wang HH, Lin CAJ, Lee CH, Lin YC, Tseng YM, Hsieh CL, Chen CH, Tsai CH, Hsieh CT, Shen JL, Chan WH, Chang WH, Yeh HI (2011) Fluorescent gold nanoclusters as a biocompatible marker for in vitro and in vivo tracking of endothelial cells. ACS Nano 5:4337–4344

41. Muhammed MAH, Verma PK, Pal SK, Kumar RCA, Paul S, Omkumar RV, Pradeep T (2009) Bright, NIR-emitting Au$_{23}$ from Au$_{25}$: characterization and applications including biolabeling. Chem Eur J 15:10110–10120

42. Qian H, Zhu M, Lanni E, Zhu Y, Bier ME, Jin R (2009) Conversion of polydisperse Au nanoparticles into monodisperse Au$_{25}$ nanorods and nanospheres. J Phys Chem C 113:17599–17603

43. Li X, Jiang P, Ge G (2011) Synthesis of small water-soluble gold nanoparticles and their chemical modification into hollow structures and luminescent nanoclusters. Colloid Surf A 384:62–67

44. Bao Y, Yeh HC, Zhong C, Ivanov SA, Sharma JK, Neidig ML, Vu DM, Shreve AP, Dyer RB, Werner JH, Martinez JS (2010) Formation and stabilization of fluorescent gold nanoclusters using small molecules. J Phys Chem C 114:15879–15882

45. Shibu ES, Pradeep T (2011) Quantum clusters in cavities: trapped Au$_{15}$ in cyclodextrins. Chem Mater 23:989–999

46. Shichibu Y, Konishi K (2010) HCl-induced nuclearity convergence in diphosphine-protected ultrasmall gold clusters: a novel synthetic route to "magic-number" Au$_{13}$ clusters. Small 6:1216–1220

47. Huang CC, Yang Z, Lee KH, Chang HT (2007) Synthesis of highly fluorescent gold nanoparticles for sensing mercury(II). Angew Chem Int Ed 46:6824–6828

48. Huang CC, Chen CT, Shiang YC, Lin ZH, Chang HT (2009) Synthesis of fluorescent carbohydrate-protected Au nanodots for detection of concanavalin A and *Escherichia coli*. Anal Chem 81:875–882

49. Wu Z, MacDonald MA, Chen J, Zhang P, Jin R (2011) Kinetic control and thermodynamic selection in the synthesis of atomically precise gold nanoclusters. J Am Chem Soc 133:9670–9673

50. Pettibone JM, Hudgens JW (2011) Gold cluster formation with phosphine ligands: etching as a size-selective synthetic pathway for small clusters. ACS Nano 5:2989–3002

51. Zhu ZK, Jin RC (2010) On the ligand's role in the fluorescence of gold nanoclusters. Nano Lett 10:2568–2573

52. Zhou C, Sun C, Yu M, Qin Y, Wang J, Kim M, Zheng J (2010) Luminescent gold nanoparticles with mixed valence states generated from dissociation of polymeric Au (I) thiolates. J Phys Chem C 114:7727–7732

53. Kamei Y, Shichibu Y, Konishi K (2011) Generation of small gold clusters with unique geometries through cluster-to-cluster transformations: octanuclear clusters with edge-sharing gold tetrahedron motifs. Angew Chem Int Ed 50:7442–7445

54. Díez I, Ras RHA (2011) Fluorescent silver nanoclusters. Nanoscale 3:1963–1970, and references therein

55. Xu H, Suslick KS (2010) Water-soluble fluorescent silver nanoclusters. Adv Mater 22:1078–1082, and references therein

56. Zhou R, Shi M, Chen X, Wang M, Chen H (2009) Atomically monodispersed and fluorescent sub-nanometer gold clusters created by biomolecule-assisted etching of nanometer-sized gold particles and rods. Chem Eur J 15:4944–4951

57. Templeton A, Wuelfing W, Murray R (2000) Monolayer protected cluster molecules. Acc Chem Res 33:27–36

58. Liu C, Ho M, Chen Y, Hsieh C, Lin Y, Wang Y, Yang M, Duan H, Chen B, Lee J (2009) Thiol-functionalized gold nanodots: two-photon absorption property and imaging in vitro. J Phys Chem C 113:21082–21089

59. Gibson DW, Beer M, Barrnett RJ (1971) Gold (III) complexes of adenine nucleotides. Biochemistry 10:3669–3678

60. Nakamoto K, Tsuboi M, Strahan GD (2008) Drug-DNA interactions: structures and spectra. Wiley, USA

61. Watanabe H, Abe S, Honma H (1998) Gold wire bondability of electroless gold plating using disulfiteaurate complex. J Appl Electrochem 28:525–530

# 5

# Effect of high gold salt concentrations on the size and polydispersity of gold nanoparticles prepared by an extended Turkevich–Frens method

**Kara Zabetakis · William E. Ghann · Sanjeev Kumar · Marie-Christine Daniel**

**Abstract** The Turkevich–Frens synthesis starting conditions are expanded, ranging the gold salt concentrations up to 2 mM and citrate/gold(III) molar ratios up to 18:1. For each concentration of the initial gold salt solution, the citrate/gold(III) molar ratios are systematically varied from 2:1 to 18:1 and both the size and size distribution of the resulting gold nanoparticles are compared. This study reveals a different nanoparticle size evolution for gold salt solutions ranging below 0.8 mM compared to the case of gold salt solutions above 0.8 mM. In the case of $[Au^{3+}]<0.8$ mM, both the size and size distribution vary substantially with the citrate/gold(III) ratio, both displaying plateaux that evolve inversely to $[Au^{3+}]$ at larger ratios. Conversely, for $[Au^{3+}]\geq0.8$ mM, the size and size distribution of the synthesized gold nanoparticles continuously rise as the citrate/gold(III) ratio is increased. A starting gold salt concentration of 0.6 mM leads to the formation of the most monodisperse gold nanoparticles (polydispersity index < 0.1) for a wide range of citrate/gold(III) molar ratios (from 4:1 to 18:1). Via a model for the formation of gold nanoparticles by the citrate method, the experimental trends in size could be qualitatively predicted: the simulations showed that the destabilizing effect of increased electrolyte concentration at high initial $[Au^{3+}]$ is compensated by a slight increase in zeta potential of gold nanoparticles to produce concentrated dispersion of gold nanoparticles of small sizes.

**Keywords** Gold nanoparticles · Synthesis · Citrate reduction · High concentration

Kara Zabetakis and William Ghann contributed equally to this work.

K. Zabetakis · W. E. Ghann · M.-C. Daniel (✉)
Department of Chemistry and Biochemistry,
University of Maryland, Baltimore County,
Baltimore, MD 21250, USA
e-mail: mdaniel@umbc.edu

K. Zabetakis
e-mail: kzab@umd.edu

W. E. Ghann
e-mail: wghann1@umbc.edu

S. Kumar
Department of Chemical Engineering, Indian Institute of Science, Bangalore 560012, India

*Present Address:*
K. Zabetakis
School of Civil and Environmental Engineering,
University of Maryland, College Park,
Baltimore, MD 20742, USA

## Introduction

The synthesis of colloidal gold via citrate reduction was first introduced by Turkevich et al. in 1951 [1], and was later refined by Frens in the 1970s [2, 3]. This citrate reduction method is also reviewed in Hayat's book on colloidal gold, along with the other well-known techniques for preparing gold nanoparticles of different sizes (using reducing agents such as white phosphorus, sodium borohydride, ascorbic acid) [4]. The citrate reduction process involves hot gold chloride and sodium citrate as reactants. In this reaction, the citrate molecules act as both reducing and stabilizing agents, allowing for the formation of the colloidal gold [5]. The synthesis process typically creates gold nanoparticles (GNPs) in the 10–150-nm size range with two major disadvantages: this technique produces very dilute GNP solutions ($[Au^{3+}]\leq0.25$ mM) [6] and the size distribution broadens with increase in particle size, leading to polydisperse GNPs for sizes over 50 nm.

Gold nanoparticles display a variety of properties [7] and have important applications as diverse as cosmetics [8], electronics [9], therapeutics [10–12], imaging [13, 14], drug delivery [15, 16], and pollution remediation [17, 18]. The mentioned applications often call for monodisperse nanoparticles of a particular size in large quantities and/or at high concentrations [19, 20]. Therefore, it is important to understand how changes in the synthesis conditions can affect the nanoparticle characteristics.

Variations of the Turkevich–Frens method have been investigated in the past years. Ji et al. [5] have demonstrated that molar ratios of citrate to gold(III) salt (Ct/Au) higher than 3:1 raise the pH of the reaction mixture (pH>6.5), a condition that favors the formation of monodisperse GNPs. They also have observed that the diameter of the obtained GNPs levels off at very high citrate to gold(III) molar ratios (Ct/Au≥14:1) [5]. Puntes group [21] and Sivaraman et al. [22] have shown that if the sequence of the addition of citrate and gold chloride is reversed, the GNPs formed are smaller and narrower in size distribution. Moreover, gold salt concentrations much higher than 0.25 mM have been studied by other groups. Kimling et al. [6] have noticed a significant increase in size (as well as in polydispersity) of the GNPs obtained from gold salt solutions with concentrations higher than 0.8 mM and Ct/Au ratios<2. Li et al. [23] have reported the formation of concentrated GNP solutions ($[Au^{3+}]$=2.5 mM) with narrower size distributions by adjusting the pH and temperature of the reaction mixtures. Along with the decrease in polydispersity, a decrease of the diameter of the formed GNPs has also been observed [23].

Herein, by expanding the conditions of the Turkevich–Frens synthesis, we report a systematic study regarding the evolution of both size and size distribution of the GNPs formed, when the concentration of initial gold salt solutions and their respective citrate/gold(III) ratios are increased. To examine these trends, GNP solutions have been made under different starting conditions: a series of seven gold salt concentrations (ranging from 0.3 to 2 mM) have been investigated, and Ct/Au molar ratios from 2:1 to 18:1 have been studied. The sizes and polydispersity indices (PDI) of resulting gold nanoparticles have been measured via dynamic light scattering (DLS) spectroscopy and analyzed to assess for trends in the particle sizes and size distributions. The concentration of the starting gold solution is found to have a significant effect on the size and PDI of the formed nanoparticles. Simulations have also been performed in order to better understand these observations.

## Experimental

### Materials/chemicals

Tetrachloroauric acid monohydrate ($HAuCl_4 H_2O$, 99.9 % assay) and trisodium citrate dihydrate (99.9 % assay) were purchased from Electron Microscopy Services (Fort Washington, PA). These chemicals were used without further purification. The TEM carbon-coated 200-mesh copper grids were also purchased from Electron Microscopy

Services. All glassware used for GNPs syntheses was cleaned with freshly prepared aqua regia solution (three parts HCl, one part $HNO_3$) and rinsed with Milli-Q water (18 MΩ cm resistivity). The same ultrapure water was used in the preparation of GNP solution and other aqueous solutions.

### Gold nanoparticle synthesis

The GNP solutions were synthesized using a modified Frens method [5]. First, a gold salt stock solution was prepared using the entire content of a commercial vial with an average of 0.1 g of chloroauric acid. The content of the vial was accurately weighed and dissolved in 10.0 mL of pure water (Millipore Milli-Q). Next, a 5 % citrate aqueous stock solution (0.17 M) was prepared using trisodium citrate. The accurate molarities of the gold salt and sodium citrate solutions were calculated. An accurately measured quantity of the gold salt solution (calculated based on the desired concentration of gold) was then transferred to a 100-mL volumetric flask with a micropipette (Gilson). The concentration of gold chloride solution was varied from 0.3 to 2 mM. All 100 mL of the chloroauric acid solution was then transferred to a three-neck round bottom flask equipped with a stir bar. Next, a micropipette was used to accurately transfer a portion of the 5 % (0.17 M) sodium citrate solution (that was calculated from the desired molar ratio of sodium citrate to gold salt) to a separate vial. The molar ratios of citrate to gold(III) ranged from 2:1 to 18:1. The vial was then set aside for later use. The three-neck round bottom flask containing both the gold salt solution and a stir bar was equipped with a condenser and placed in a hot oil bath. The stirring speed was increased until a vortex was noticed. When the reflux started, the citrate solution from the vial was quickly poured into the round bottom flask. Then, the reaction was allowed to run for no less than 20 min, but no more than 30 min. In other words, the reaction was allowed to run until a deep ruby red color was observed; otherwise, the reaction was stopped after 30 min. The reaction was observed to go to completion by one of two visual pathways: the gold chloride solution went either (1) from yellow to black to ruby (for low Ct/Au ratios) or (2) from yellow to pale pink to ruby (for high Ct/Au ratios). The hue and intensity of the final ruby color of the particle solution depended on the initial gold salt concentration and on the Ct/Au ratio.

### DLS measurements

Samples from the obtained GNP solutions were analyzed by DLS spectroscopy to quantify the average hydrodynamic diameter (Zave) and PDI. DLS measurements were performed with a Malvern Zetasizer Nano ZS (Malvern, Southborough, MA) equipped with a 633-nm He–Ne laser and operating at an angle of 173°. The software used to collect and analyze the data was the Dispersion Technology Software version 5.02

from Malvern. The GNP solutions were stored securely in a hood. These stored solutions have been monitored by DLS over a period of up to 3 months, with an average of four DLS measurements for each stored solution. All the DLS measurements have been then averaged for each individual GNP solution, and the resulting Zave and PDI have been recorded.

## Model

The model used for predictions of the GNPs sizes was the model of Kumar et al. [24], taking into consideration the role of coagulation during the synthesis of citrate-stabilized GNPs. For the lower range of $[Au^{3+}]$ studied here (0.3 to 0.6 mM), the original model (published in 2007) was first used to obtain predictions. Then, in order to better correlate with the experimental data, the expression of the surface potential in the original model was modified by changing the prefactor in this expression: the value 90 was replaced by 95 and 100, respectively and two new sets of predictions were obtained for the GNPs sizes.

## Results and discussion

To date, the previous studies on the formation of GNPs based on the Turkevich–Frens method have mostly involved lower initial concentrations of gold salts (≤0.25 mM). The use of higher concentrations of gold salts has received little attention and their effect on the size of the formed GNPs has not been systematically investigated. Herein, we report the study on the formation of GNPs, starting with a series of gold salt concentrations ranging from 0.3 to 2 mM and using Ct/Au molar ratios ranging from 2:1 to 18:1 for each of these concentrations. DLS spectroscopy has been used to

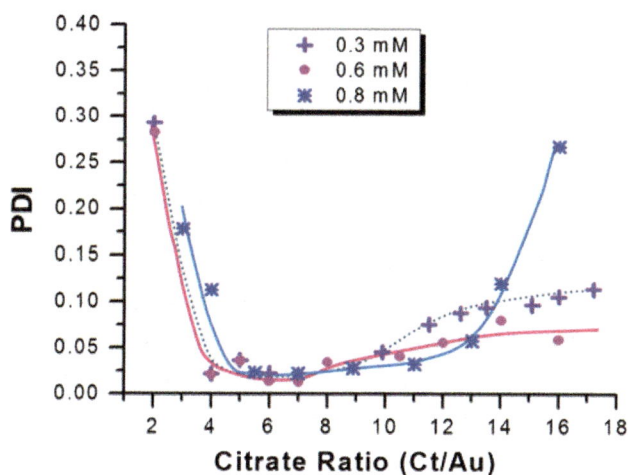

**Fig. 2** GNP polydispersity vs citrate ratio for gold chloride concentrations>0.8 mM. The *lines* are guides for the eye

measure the hydrodynamic diameter (HD) and the PDI of each prepared GNP solution.

### Effect of the initial gold salt concentration on the GNP size distribution

Figures 1 and 2 present the trends of the PDI of the obtained GNPs with respect to the Ct/Au ratio for different concentrations of gold salt. In the case of starting gold chloride solutions of 0.3 and 0.6 mM, the PDI of the formed GNPs decreases as Ct/Au is increased from 2:1 to 4:1 and remains low (between 0.02 and 0.1) for higher Ct/Au ratios (up to 17:1) (Fig. 1). However, the 0.8 mM gold salt solutions lead to GNPs with increasing PDI for Ct/Au ratios higher than 13:1 (Fig. 1). Also, when using more concentrated gold chloride solutions (up to 2 mM), the resulting GNPs present a low PDI for a range of citrate/gold(III) ratios that shrinks as the concentration of gold salt increases (Fig. 2). For instance, the 0.8 mM gold chloride solution forms monodisperse GNPs (PDI <0.1) [25, 26] for a range of Ct/Au

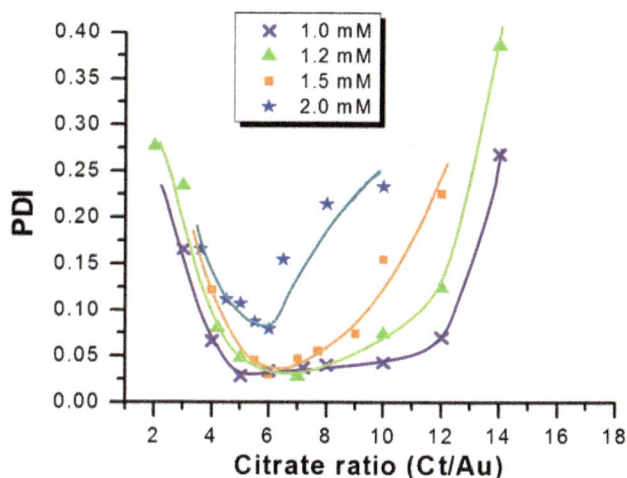

**Fig. 1** GNP polydispersity vs citrate ratio for gold chloride concentrations≤0.8 mM. The *lines* are guides for the eye

**Table 1** Ranges of Ct/Au ratios that can be used to form monodisperse GNPs at different $[Au^{3+}]$

| Initial $[Au^{3+}]$ (mM) | Range of Ct/Au molar ratios leading to monodisperse GNPs (PDI<0.1) | Range of Ct/Au molar ratios leading to formation of GNPs (PDI<0.2) |
|---|---|---|
| 0.3 | 3.2–13.5 | 3–>17 |
| 0.6 | 3.5–>17 | 3–>17 |
| 0.8 | 4.5–13.5 | 3.5–14.5 |
| 1 | 4–12.5 | 3.5–13 |
| 1.2 | 4.5–11 | 4–12 |
| 1.5 | 4.5–9 | 4–10.5 |
| 2 | 5–6 | 4–7.5 |

ratios from 4:1 to 13:1, whereas only a very narrow range of Ct/Au ratios (5:1 to 6:1) produces monodisperse GNPs when starting with 2.0 mM gold chloride solutions (Table 1). From these data, we can conclude that, depending on the starting concentration of gold salt, a wider or narrower range of Ct/Au ratios allows for the formation of monodisperse GNPs. This way, by choosing the appropriate Ct/Au ratios, it is possible to generate concentrated solutions of monodisperse GNPs. Also, it can be noted that an initial gold chloride concentration of 0.6 mM results in the formation of the most monodisperse GNPs (with PDIs down to 0.03) for the widest range of citrate/gold(III) ratios (from 3.5:1 and up) (Table 1).

### Effect of the initial gold salt concentration on the GNP diameter

When varying the initial gold chloride concentration, we find that the size evolution of the formed GNPs follows different trends (as a function of the Ct/Au ratios) for gold salt concentrations in the range below 0.8 mM (Fig. 3) and in the range above 0.8 mM (Fig. 4).

The GNPs synthesized from 0.3 and 0.6 mM gold salt solutions display large hydrodynamic diameters (HD was around 32 and 27 nm, respectively) when a Ct/Au ratio of 2:1 is used (Fig. 3). The GNPs then rapidly decrease in size and reach a minimum diameter for Ct/Au ratios of 4:1 and 5:1, for 0.3 and 0.6 mM gold salt solution, respectively, and slowly increase back in size for higher ratios of citrate to gold, finally leveling off at high Ct/Au ratios (>14:1). A similar behavior of the size evolution as a function of Ct/Au ratio has already been reported by Ji et al. for gold chloride solutions of 0.25 mM [5]. Although the evolution of the GNP sizes follows similar trends for 0.3 and 0.6 mM gold

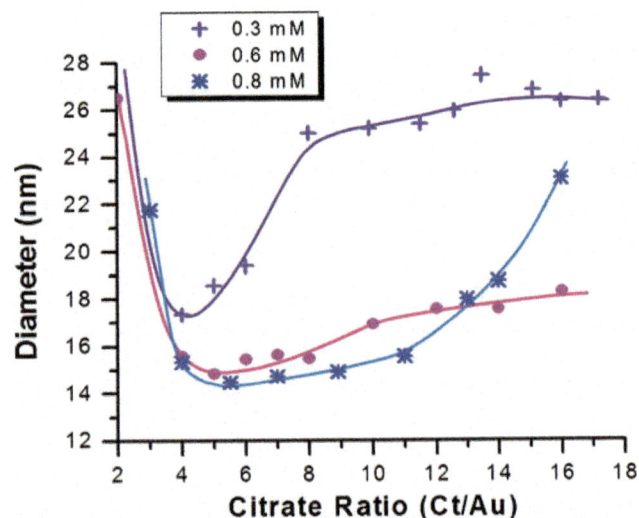

**Fig. 4** GNP size vs citrate ratio for gold chloride concentrations> 0.8 mM. The *lines* are guides for the eye

salt solutions, the GNPs obtained from 0.6 mM gold salt solutions present smaller diameters than when using 0.3 mM gold salt solutions, for the same Ct/Au ratios. This size difference is even more pronounced when reaching the plateau at higher Ct/Au ratios. For instance, the HD trend of GNPs prepared from 0.3 mM gold salt solutions levels off at diameters of around 26 nm, while it levels off at diameters of around 18 nm when starting with 0.6 mM gold chloride solutions. Likewise, 0.3 mM gold salt solutions lead to GNPs with sizes reaching a plateau of around 26 nm, which is lower than the plateau reported for GNPs from 0.25 mM gold chloride solutions (HD>30 nm, Peng's data [5]). These results indicate that, for the gold salt concentration range below 0.8 mM, higher concentrations of gold salt lead to a plateau with smaller GNP sizes. The size plateaus at large Ct/Au ratios for low [Au$^{+3}$] because, for low gold salt concentrations (below 0.8 mM), the total electrolyte concentration is still low enough at high Ct/Au ratios, so that there is no occurrence of coagulation which would lead to size increase. Also, for [Au$^{3+}$]<0.8 mM, the pH has a major effect on the GNP formation (vide supra), and Ji et al. [5] have shown that the stabilization of pH at high Ct/Au ratios is associated to the plateau observed at these same ratios.

The size evolution found for GNPs prepared from 0.8 mM gold salt solutions is intermediate between the trends observed with 0.3 and 0.6 mM gold(III) solutions and the trends displayed with Au$^{3+}$ solutions of 1 mM and over. Indeed, 0.8 mM gold salt solutions form smaller GNPs than with those of 0.6 mM but only for 4<Ct/Au ratio<13. Also, the minimum size is reached for a Ct/Au ratio of around 5:1–6:1, then increases very slowly until Ct/Au ratios of around 11:1, as it was going to reach a plateau, but instead a sudden increase in HD is observed for Ct/Au ratios>13:1 (Fig. 3).

Figure 4 displays the HD of the GNPs formed from 1–2 mM gold chloride solutions as a function of the Ct/Au

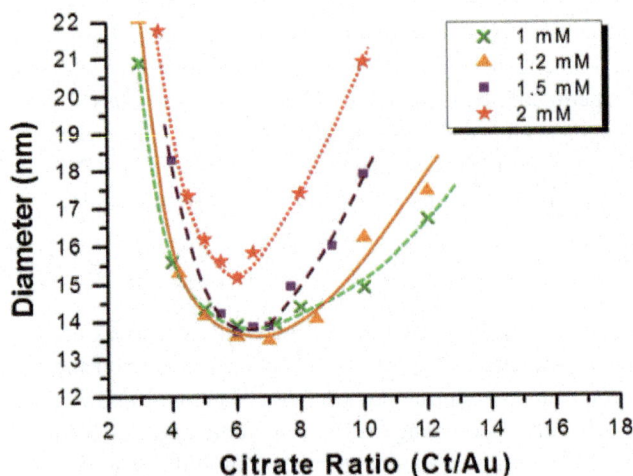

**Fig. 3** GNP size vs citrate ratio for gold chloride concentrations≤ 0.8 mM. The *lines* are guides for the eye

ratios used. Similar to the case of lower $[Au^{3+}]$, the sizes of GNPs formed from 1–2 mM gold salt solutions first decrease as the Ct/Au ratio is increased, reaching a minimum size at a ratio of around 6:1. However, the diameter range of the GNPs produced from gold salt solutions of 1 mM up to 2 mM does not level off at high Ct/Au ratios, but instead shows a continuous increase (Fig. 4). Furthermore, the plot slopes of the decrease and then the increase in the GNP size (as Ct/Au ratio increases) are becoming steeper as the concentrations of gold chloride increases, leading to larger GNPs from concentrated gold salt solutions at Ct/Au ratios of 4:1 or 10:1. Also, the minimum HD obtained from 1, 1.2, and 1.5 mM $HAuCl_4$ solutions remained around 13.5–14 nm (as well as from 0.8 mM $HAuCl_4$), but it slightly increases to about 15 nm when 2 mM gold salt solution is used.

In order to assess the reproducibility of the size trends observed when varying Ct/Au ratios, we used initial gold salt solutions of 1.5 mM as a representative example. We performed three gold nanoparticles syntheses in triplicates, using Ct/Au ratios of 4:1, 6:1, and 10:1. The resulting sizes were plotted in Figure S1 of the Electronic supplementary material (ESM), along with their respective error bars. Good reproducibility was obtained, even at this high gold salt concentration. Although the presence of few aggregates could be detected by DLS for some of the experiments, filtration with 0.22 μm filters eliminated any trace of aggregation.

To help visualize the difference in the GNP sizes obtained using same Ct/Au ratios but different gold(III) concentrations, Figs. 5 and 6 present the HD of GNPs formed using different Ct/Au ratios as a function of $[Au^{3+}]$. As shown in Fig. 5, when the Ct/Au ratio is increased from 2:1 up to 5:1, a general decrease of the GNP size is observed across the range of gold chloride concentrations studied here (except at 0.3 mM $HAuCl_4$, for which GNP sizes start increasing back for a Ct/Au ratio of 5:1, since the minimum size is observed for a ratio

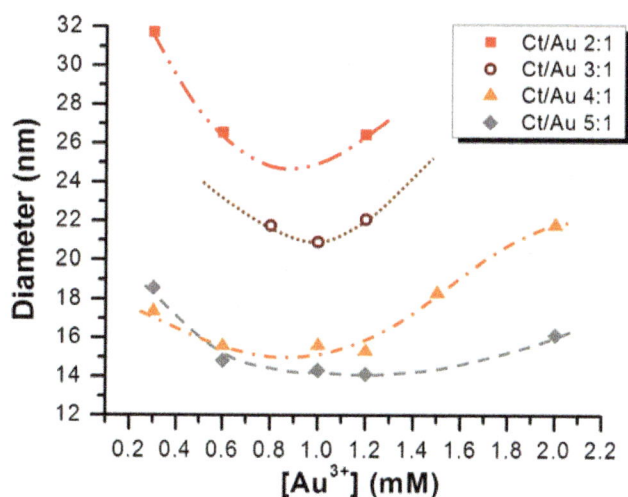

**Fig. 6** GNP size vs gold(III) concentration for Ct/Au≥6. The *lines* are guides for the eye

of 4:1 at this concentration). For further increase of the Ct/Au ratio from 6:1 to 14:1, the GNP size increases back for most of the concentrations, although this increase is minimal for gold concentrations below 0.8 mM (Fig. 6).

On the other hand, for the same Ct/Au ratio, different initial gold chloride concentrations give GNPs of different sizes (Figs. 5 and 6). For instance, for a Ct/Au ratio of 10:1, 0.3 mM gold salt solutions produce GNPs of around 25 nm, while the sizes of GNPs formed with 0.6, 1.2, and 2 mM $HAuCl_4$ solutions are about 17, 16, and 20 nm, respectively. These results correlate to some of the data reported by Kimling et al. [6] in which they find that gold salt concentrations above 0.8 mM lead to a size increase of the formed GNPs (in their case, when using Ct/Au ratios mostly up to 2:1). We also observe this tendency of size increase at higher Ct/Au ratios when the gold salt concentrations increase over 0.8 mM, but this increase in GNP size is minimal at Ct/Au ratios around 6:1 and is enhanced as the Ct/Au ratios move away from 6:1.

### Effect of pH

The pH of the reacting mixture (gold salt + citrate) has an important role in the formation of the gold nanoparticles, as discovered by Ji et al. [5]. They reported that initial solutions with different Ct/Au ratios displayed different pH values, which affected the reaction mechanism of the nanoparticles formation. It was found that pH≥6.5 led to more monodisperse gold nanoparticles through a reaction mechanism which did not involve aggregates as intermediates (unlike those found in Frens' method). These studies were done with $[Au^{3+}]$=0.25 mM using Ct/Au ratios of 0.7:1 to 28:1, corresponding to a pH range of 3.7 to 7.7, respectively. By comparison, our studies were performed with Ct/Au ratios of 2:1 to 14:1 which corresponds to a pH range of 4.1 to 6.8, respectively (different concentrations led to small variations in

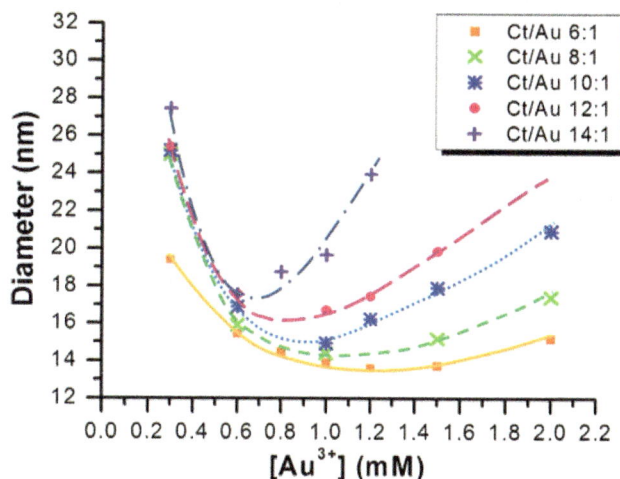

**Fig. 5** GNP size vs gold(III) concentration for 2≤Ct/Au<6. The *lines* are guides for the eye

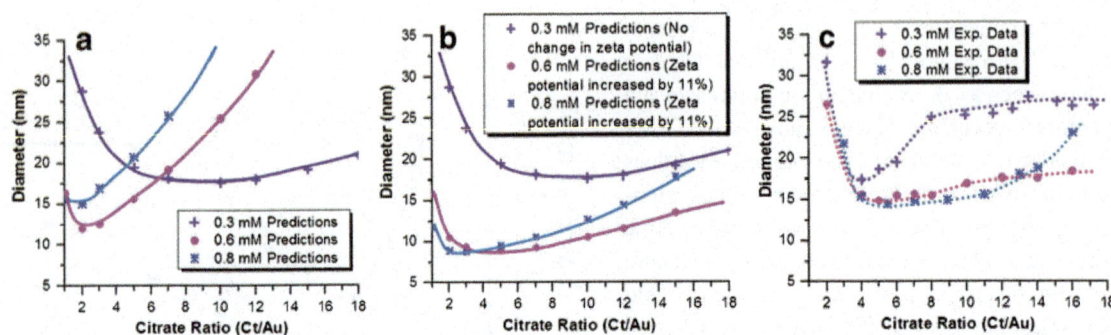

**Fig. 7** Comparisons of experimental data and simulations for the sizes of GNPs from 0.3, 0.6, and 0.8 mM initial [Au³⁺]. **a** Predictions obtained using the model of Kumar et al. [24] without modifications;

**b** Predictions obtained using the model of Kumar et al. [24] in which the zeta potential was increased by 11 % for GNPs from 0.6 and 0.8 mM [Au³⁺]; **c** Experimental data. The *lines* are guides for the eye

pH, Table S10 of the ESM). Our results with [Au³⁺]=0.3 mM (Fig. 3) show a comparable size trend vs Ct/Au than observed by Ji et al. [5] (i.e., size decrease followed by size increase and level off). This similarity is expected since the gold(III) concentration of 0.3 mM is very close to Ji et al. conditions (0.25 mM) [5]. The size trend we obtained for [Au³⁺]=0.6 mM also presents a size decrease followed by a size increase (Fig. 3), but the increase is minimal and the size quickly levels off at high Ct/Au ratios. With regards to the polydispersity, the low PDI values that we observe at high Ct/Au ratios (i.e., high pH) for [Au³⁺]=0.3 and 0.6 mM (Fig. 1) are in accordance with the findings of Ji et al. [5], which showed that higher pH (>6.5) led to more monodisperse nano-particles. However, the GNPs that we obtained for gold(III) concentrations over 0.6 mM did not display anymore low PDI at high Ct/Au ratios (Figs. 1 and 2), even though the pH of the solutions at these high ratios were over 6.5 (Table S10 of the ESM). This seems to indicate that, at high gold(III) concentrations and high Ct/Au ratios, the total electrolyte concentration counterbalances the effect of the pH and favors agglomeration. Further studies are needed to investigate this hypothesis.

Modeling

The model of Kumar et al. [24] was used to predict the sizes of GNPs obtained from initial gold chloride solutions of 0.3

to 2 mM, respectively. One should note that the complex nature of the GNP synthesis process, as brought out in the recent experimental findings [24], including those of the present work, and poor quantitative understanding of elec-trical interactions at high electrolyte concentration brings limitations to the model. In view of these complexities, the model is used only to obtain a physical insight into the synthetic process for the expanded set of experimental con-ditions. When the original model is used with no further modifications, only the predicted sizes of GNPs from 0.3 mM initial gold(III) solutions show some qualitative agreement with the experimental data, although displaying a lower asymptotic behavior than in the experimental trend observed. In order to better predict the sizes of GNPs from initial [Au³⁺]>0.3 mM, the expression of the surface poten-tial in the original model was altered. Indeed, Zukoski's data [27] have shown that when citrate concentrations are in-creased, the measured electrophoretic mobilities suggest larger zeta potentials.

As shown in Fig. 7, an increase in zeta potential (with respect to the original value) by 11 % in the model of Kumar et al. [24] leads to an improvement in the predictions of the GNP sizes. Indeed, the residuals (Fig. S2 of the ESM) are overall smaller for the predictions using a zeta potential increase of 11 % than for the ones using a zeta potential increase of only 5.5 %, especially in the case of [Au³⁺]=0.6

**Fig. 8** Comparisons of experimental data and simulations for the sizes of GNPs from 1, 1.2, 1.5, and 2 mM initial [Au³⁺]. **a** Predictions obtained using the model of Kumar et al. [24] in which the zeta potential was increased by 11 % for all initial [Au³⁺]; **b** Experimental data. The *lines* are guides for the eye

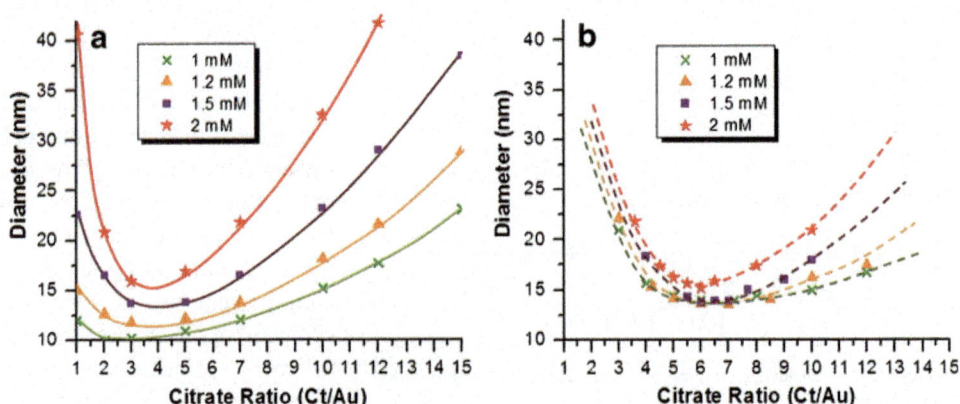

and 0.8 mM. The small differences of 2 to 5 nm can actually be explained by the fact that the model predicts core sizes of the GNPs while the experimental data correspond to HDs, which are always few nanometers larger than the core size. For the 0.6 mM data set, both the slow increase in size at high Ct/Au ratio and the smaller diameters (compared to the 0.3 mM data set) are well captured. The increase in size of GNPs from 0.8 mM over the ones from 0.6 mM at high Ct/Au ratios is also correctly predicted. The predictions for GNP sizes from 0.3 mM initial [$Au^{3+}$] could not be improved by modifying the surface potential: this can be explained by the fact that, at this low electrolyte concentration, particles are stabilized against coagulation through double layer repulsion and an increase in zeta potential does not promote further stabilization.

Figure 8 shows a comparison of model predictions (for surface potential increased by 11 %) with the experimental data for initial [$Au^{3+}$] ranging from 1 to 2 mM. The figure shows that the model accurately describes the trends found experimentally: the slope of the data set increases as the initial [$Au^{3+}$] increases, and all the trends display a minimum approximately around the same Ct/Au ratio. The minimum size of GNPs from 2 mM gold(III) solutions being larger than the minima observed for GNPs from 1, 1.2, and 1.5 mM gold salt solutions is also well captured. The large differences in predicted sizes vs experimental data at high Ct/Au ratios are attributed to the limitations of the model at high electrolyte concentration. Briefly, in our model, the calculation of stability factor for DLVO theory at high electrolyte concentration and in presence of a host of ionic

**Fig. 9** TEM images of GNPs obtained from starting [$Au^{3+}$] of 0.3, 0.6, 1.2, and 2 mM and using Ct/Au ratios of 4:1, 6:1, and 10:1. The *scale bars* represent 20 nm

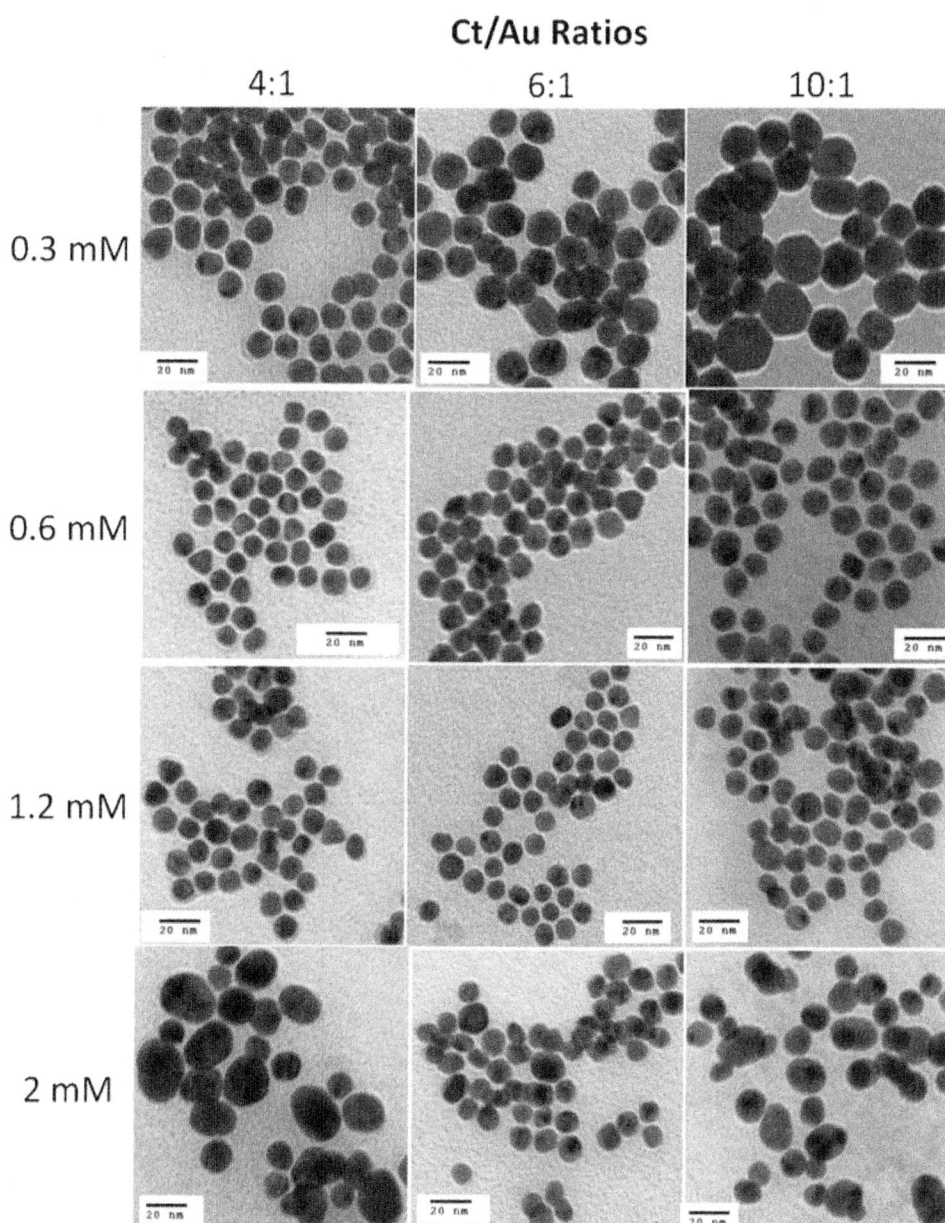

species in the system is based on a semi-empirical approach validated for the low concentrations used in the original Turkevich protocol. The model is therefore used here to qualitatively understand the experimental observations made in the present work under expanded set of experimental conditions.

## Relationship between the evolution of GNP sizes and their respective PDI

Interestingly, for each initial concentration of gold chloride solution, the evolution of the GNP diameter as a function of Ct/Au ratio follows a similar trend as to the evolution of its PDI. Consequently, for gold(III) solutions below 0.8 mM, the largest monodisperse GNPs (PDI<0.1) [25, 26, 28] have a polydispersity curve that reaches a plateau at high Ct/Au ratios. For instance, the largest HD that can be obtained from 0.3 and 0.6 mM gold(III) solutions in a monodisperse manner are about 27 and 18 nm, respectively. However, for initial HAuCl$_4$ concentrations above 0.8 mM, the increase in the particle size at higher Ct/Au ratios is associated with an increase in PDI. When drawing an imaginary line at PDI of 0.1 in Fig. 2, the PDI curve is intercepted twice for a given initial gold(III) concentration: the corresponding values of Ct/Au ratios can be used in Fig. 4 to determine the maximum monodisperse sizes from this concentration. The increase in PDI at higher gold salt concentrations and higher Ct/Au ratios is most probably due to the increase in ionic strength which contributes to a higher propensity of the nanoparticles for agglomeration.

The sizes and size distributions of GNPs were also measured by TEM to verify the trends observed by DLS. Samples of GNPs with initial gold(III) concentrations of 0.3, 0.6, 1.2, and 2 mM were selected. For each of these concentrations, TEM images were taken for Ct/Au ratios of 4:1, 6:1, and 10:1. The mean sizes and size distributions were measured for each of the 12 samples (Fig. 9) and compared with the values observed by DLS. As expected, the diameters measured by TEM were consistently smaller than the HD measured by DLS (in solution). However, the general trends in sizes and size distributions found from the TEM data correlated well with those observed by DLS. For instance, samples with Ct/Au ratios of 6:1 gave the narrowest standard deviations for gold(III) concentrations of 0.3 and 0.6 mM (6.8 and 7.4 %, respectively). Also, a larger mean size was measured for GNPs from 0.3 mM gold salt solutions (15.4 nm) than for GNPs from 0.6 mM gold salt solutions (12.0 nm). Furthermore, the GNPs from 2 mM gold salt solutions and Ct/Au ratio of 10:1 showed the largest standard deviation (24.7 %) out of all the samples studied by TEM, which matches with the highest PDI observed by DLS for this sample.

## Conclusion

By extending the conditions of the Turkevich–Frens method, it has been found that two groups of gold chloride concentrations present two different behaviors as a function of the Ct/Au ratios. Gold salt solutions below 0.8 mM lead to the formation of highly monodisperse GNPs for Ct/Au ratio over 4:1. The size of the formed GNPs presents a minimum at Ct/Au ratios around 4:1–5:1 and saturates at high ratios (>10:1). However, gold salt solutions over 0.8 mM lead to the formation of monodisperse GNPs (PDI<0.1) [25, 26, 28] only for a certain range of Ct/Au ratios, which becomes narrower as the concentration of gold(III) increases. Also, the size of these GNPs presents a minimum at specific Ct/Au ratios, but increases unboundedly at high ratios, accompanied by an increase of the corresponding PDI. It can be concluded that the increase in ionic strength induced by the high concentration of gold salt has a significant impact on both the size and size distribution of the formed GNPs, especially for HAuCl$_4$ solutions over 0.8 mM. The model of Kumar et al. [24], initially tested against the experimental data obtained for the original Turkevich protocol using 0.3 mM initial gold chloride concentration, is tested for the expanded protocol conditions. The model captures the experimental observations for the expanded operational range (concentration ranging from 0.3 to 2.0 mM) quite well. The model also brings out the sensitive dependence of particle synthesis process to small variations in zeta potential of particles.

**Acknowledgments**  The authors are grateful to Dr. Tiberiu-Dan Onuta for his very helpful suggestions and comments on this work, as well as for the creation of the residual data in Figure S2 of the ESM. We also acknowledge the financial support from the Department of Defense (Prostate Cancer Research Program, CDRMP, grant no. PC081299).

## References

1. Turkevich J, Stevenson PC, Hillier J (1951) Discuss Faraday Soc 11:55–75
2. Frens G (1972) Kolloid-Z Z Polym 250:736–741
3. Frens G (1973) Nature (London) Phys Sci 241:20–22
4. Hayat MA (eds) (1989) Colloidal Gold: principles, methods, and applications, Vol. 1. Academic
5. Ji X, Song X, Li J, Bai Y, Yang W, Peng XJ (2007) Am Chem Soc 129:13939–13948
6. Kimling J, Maier M, Okenve B, Kotaidis V, Ballot H, Plech AJ (2006) Phys Chem B 110:15700–15707
7. Daniel M-C, Astruc D (2004) Chem Rev (Washington, DC, U S) 104:293–346

8. Fathi-Azarbayjani A, Qun L, Chan YW, Chan SY (2010) AAPS PharmSciTech 11:1164–1170

9. Lee J-S, Cho J, Lee C, Kim I, Park J, Kim Y-M, Shin H, Lee J, Caruso F (2007) Nat Nano 2:790–795

10. Bowman M-C, Ballard TE, Ackerson CJ, Feldheim DL, Margolis DM, Melander C (2008) J Am Chem Soc 130:6896–6897

11. Daniel M-C, Grow ME, Pan H-M, Bednarek M, Ghann WE, Zabetakis K, Cornish J (2011) New J Chem 35:2366–2374

12. Bresee J, Maier KE, Boncella AE, Melander C, Feldheim DL (2011) Small 7:2027–2031

13. Hainfeld JF, Slatkin DN, Focella TM, Smilowitz HM (2006) Br J Radiol 79:248–253

14. Ghann WE, Aras O, Fleiter T, Daniel MC (2012) Langmuir 28:10398–10408

15. Boisselier E, Astruc D (2009) Chem Soc Rev 38:1759–1782

16. Bresee J, Maier KE, Melander C, Feldheim DL (2010) Chem Commun (Cambridge, U K) 46:7516–7518

17. Xia T, Kovochich M, Brant J, Hotze M, Sempf J, Oberley T, Sioutas C, Yeh JI, Wiesner MR, Nel AE (2006) Nano Lett 6:1794–1807

18. Dreher KL (2004) Toxicol Sci 77:3–5

19. Bergen JM, von Recum HA, Goodman TT, Massey AP, Pun SH (2006) Macromol Biosci 6:506–516

20. Gaumet M, Vargas A, Gurny R, Delie F (2008) Eur J Pharm Biopharm 69:1–9

21. Ojea-Jimenez I, Bastus NG, Puntes VJ (2011) Phys Chem C 115:15752–15757

22. Sivaraman SK, Kumar S, Santhanam VJ (2011) Colloid Interface Sci 361:543–547

23. Li C, Li D, Wan G, Xu J, Hou W (2011) Nanoscale Res Lett 6:440

24. Kumar S, Gandhi KS, Kumar R (2007) Ind Eng Chem Res 46:3128–3136

25. Nobbmann U, Connah M, Fish B, Varley P, Gee C, Mulot S, Chen J, Zhou L, Lu Y, Shen F, Yi J, Harding SE (2007) Biotechnol Genet Eng Rev 24:117–128

26. Zhang C, Chung JW, Priestley RD (2012) Macromol Rapid Commun 33:1798–1803

27. Chow MK, Zukoski CFJ (1994) Colloid Interface Sci 165:97–109

28. http://www.malvern.com; FAQ: What does polydispersity mean?

# 6

# Dynamic light scattering for gold nanorod size characterization and study of nanorod–protein interactions

Helin Liu · Nickisha Pierre-Pierre · Qun Huo

**Abstract** In recent years, there has been considerable interest and research activity in using gold nanoparticle materials for biomedical applications including biomolecular detection, bioimaging, drug delivery, and photothermal therapy. In order to apply gold nanoparticles in the real biological world, we need to have a better understanding of the potential interactions between gold nanoparticle materials and biomolecules in vivo and in vitro. Here, we report the use of dynamic light scattering (DLS) for gold nanorods characterization and nanorod–protein interaction study. In the size distribution diagram, gold nanorods with certain aspect ratios exhibit two size distribution peaks, one with an average hydrodynamic diameter at 5–7 nm, and one at 70–80 nm. The small size peak is attributed to the rotational diffusion of the nanorods instead of an actual dimension of the nanorods. When proteins are adsorbed to the gold nanorods, the average particle size of the nanorods increases and the rotational diffusion-related size distribution peak also changes dramatically. We examined the interaction between four different proteins, bovine serum albumin, human serum albumin, immunoglobulin G, and immunoglobulin A (IgA) with four gold nanorods that have the same diameter but different aspect ratios. From this study, we found that protein adsorption to gold nanorods is strongly dependent on the aspect ratio of the nanorods, and varies significantly from protein to protein. The two serum albumin proteins caused nanorod aggregation upon interaction with the nanorods, while the two immunoglobulin proteins formed a stable protein corona on the nanorod surface without causing significant nanorod aggregation. This study demonstrates that DLS is a valuable tool for nanorod characterization. It reveals information complementary to molecular spectroscopic techniques on gold nanorod–protein interactions.

**Keywords** Gold nanoparticle · Gold nanorod · Protein interaction · Dynamic light scattering

Helin Liu and Nickisha Pierre-Pierre made equal contributions to this work.

H. Liu · N. Pierre-Pierre · Q. Huo (✉)
NanoScience Technology Center, Department of Chemistry and Burnett School of Biomedical Science,
University of Central Florida,
Orlando, FL 32826, USA
e-mail: qun.huo@ucf.edu

## Introduction

Gold nanoparticle materials have attracted considerable attention due to their unique properties and potential applications as optical probes. Within their surface plasmon resonance (SPR) wavelength region, gold nanoparticles absorb and/or scatter light intensely, and such properties make them excellent optical materials for bioimaging, biomolecular detection, and photothermal therapy [1–4]. Gold nanoparticles can be made in a wide range of shapes and geometries such as spherical particles, nanorods, nanoshells, nanostars, and nanocages [5–10]. The optical properties of gold nanoparticles are strongly dependent on the shape and size of the particles [11–13]. For in vivo biomedical applications, gold nanoparticles with SPR band in the near-infrared (IR) region (700–900 nm) are preferred because light within this spectrum window can penetrate tissue more deeply than the visible light, and also is not substantially absorbed by the aqueous environment. Gold nanorods (GNRs) exhibit two SPR bands, the transverse band around 520–600 nm, and the longitudinal band in the near IR region, with the exact wavelength tunable by controlling the aspect ratio of the nanorods [13]. Because of their near IR SPR band, gold nanorods are considered as more promising than solid spherical nanoparticles for in vivo biomedical applications.

The potential interactions between nanoparticle materials and various biomolecules, particularly proteins, are a major research topic [14, 15]. These interactions can play a significant role on the biological activity, stability, outcome, and toxicity of the nanoparticle bioconjugate materials in vitro and in vivo. Dobrovalskaia et al. conducted a systematic study on blood plasma protein adsorption to citrate-protected gold nanoparticles and identified about 60 different proteins in the "protein corona" that is formed on the gold nanoparticle surface [16]. De Paoli Lacerda et al. reported an interaction study of common human blood proteins with spherical gold nanoparticles and determined their different binding affinities [17]. Our group recently discovered and developed a simple serum–gold nanoparticle adsorption assay for cancer detection based on the serum protein–gold nanoparticle interactions [18, 19]. We found that serum proteins adsorbed to gold nanoparticles from cancer patients differ from the normal healthy donors, and this difference can be used to predict the aggressiveness of cancer. More recently, Arvizo et al. reported a serum-adsorbed gold nanoparticle system to identify potential therapeutic targets for cancer [20].

There have been many methods and techniques reported for nanoparticle–protein interaction studies. From a literature survey, we found that the most commonly used method is fluorescence spectroscopy [16, 21, 22]. Almost all proteins have fluorescent amino acids, tryptophan, and tyrosine. When the fluorescent tryptophan or tyrosine interacts with gold nanoparticles, the fluorescence properties of tryptophan or tyrosine will change. These changes can occur as emission wavelength red-shift or blue-shift, fluorescence quenching or enhancement. Such changes have been used to determine the binding affinity and binding constant of proteins with gold nanoparticles quantitatively. However, there is a general concern on the fluorescence technique: a protein can contain multiple tryptophan and tyrosine residues. Depending on their actual distance to the gold metal core, the fluorescence of these amino acid residues can be quenched or enhanced [23, 24]. This problem could cause uncertainty in the quantitative analysis. In addition to fluorescence, circular dichroism [22, 25, 26] and FT-IR spectroscopy [22] are often used to monitor protein conformation change upon binding with gold nanoparticles. Calzolai et al. identified ubiquitin–gold nanoparticle interaction site using NMR spectroscopy [27]. All in all, multiple analytical techniques are needed to probe the complex interactions between proteins and gold nanoparticle materials.

Dynamic light scattering (DLS) is a technique that is used routinely for nanoparticle size analysis. Proteins are macromolecules. The hydrodynamic diameters of typical proteins are in the nanometer range (1–10 nm). When proteins are adsorbed to gold nanoparticles, the size of the nanoparticles will increase. We and many other groups have previously demonstrated that DLS can be used as a very convenient and powerful tool to monitor specific binding and non-specific adsorption of proteins to spherical gold nanoparticles [28–31]. Based on the nanoparticle size change, we and others have developed a novel platform technology, nanoparticle-enabled dynamic light scattering assay (Nano-DLSay™) for biological and chemical detection and analysis with high to ultrahigh sensitivity and excellent reproducibility [32–38]. In this work, we applied the DLS technique to study the gold nanorods and nanorod–protein interactions. Compared to citrate-protected spherical gold nanoparticles, the understanding of nanorod–protein interactions is substantially less. Nanorod–protein interactions are more complicated than that of spherical particles, due to their non-spherical geometry and different surface ligand layer than the citrate-protected spherical nanoparticles. The goal of this study is to find better conditions for making gold nanorods–protein bioconjugates for in vivo and in vitro applications.

## Experimental

*Reagents* CTAB (cetyl trimethylammonium bromide)-protected Gold nanorods (CTAB-GNRs, A-12-25-550, A-12-25-600, A-12-25-650, and A-12-25-700) were purchased from Nanopartz Inc. (Loveland, Co). Spherical, citrate-protected gold nanoparticles with an average diameter of 100 nm (GNP100, catalog number 15708–9) were purchased from Ted Pella Inc. (Redding, CA). The physical properties of the four gold nanorods and nanoparticles used in this study are summarized in Table 1. The four gold nanorods are denoted as GNR1.4, GNR1.9, GNR2.4, and GNR3.0, according to their aspect ratio. All four gold nanorods have the same diameter of 25 nm. Bovine serum albumin (BSA, A7888), human serum albumin (HSA, A9511) were purchased from Sigma (Saint Louis, MO). Human immunoglobulin G (IgG) (ab91102) and IgA (ab91025) proteins were purchased from Abcam (Cambridge, MA).

*UV–vis absorption spectroscopy measurement* The UV–vis absorption spectra were obtained with an Agilent 8453 spectrometer using a 1-cm path length quartz cuvette. The reference samples were deionized water.

*DLS measurements* The hydrodynamic diameters of the nanoparticles under investigation were measured using a Zetasizer Nano ZS90 DLS system equipped with a red laser (532 nm) and an Avalanche photodiode detector (quantum efficiency >50 % at 532 nm; Malvern Instruments Ltd., England). A Hellma cuvette QS 3 mm was used as a sample container. DTS applications 5.10 software was used to

**Table 1** Physical properties of the GNRs and GNP100 used in the current study

| Name | GNR1.4 | GNR1.9 | GNR2.4 | GNR3.0 | GNP100 |
|---|---|---|---|---|---|
| SPR peak wavelength (nm) | 538 | 580 | 650 | 698 | 560 |
| Aspect ratio | 1.4 | 1.9 | 2.4 | 3 | 1 |
| Concentration (pM) | 874 | 874 | 437 | 218 | 9.29 |

analyze the data. All sizes reported here were based on intensity average. The intensity average particle size was obtained using a non-negative least squares analysis method. For each sample, one measurement was conducted with a fixed run time of 10 s. A detection angle of 90° was used for the size measurement.

*Gold nanoparticle–protein adsorption study* To 100 μL of a GNR solution was added 2 μL of a protein solution. All protein solutions had a concentration of 1 mg/ml in 10 mM phosphate buffer. The average particle size of the mixed solution was measured after different incubation times.

## Results and discussions

### DLS analysis of CTAB-GNRs

We first used DLS to analyze the pure CTAB-GNRs. It should be noted that the DLS instrument used in the present study has a fixed detection angle of 90°. Particles with non-spherical shapes such as nanorods, can be more precisely characterized by multiple angles or depolarized DLS measurements [39–42]. However, because most users have fixed angle DLS instruments in their laboratories, therefore, we focused on the use of fixed angle DLS for nanorod characterization in this study.

The four GNRs have the same diameter of 25 nm, with different aspect ratios, 1.4, 1.9, 2.4, and 3.0, respectively (the four GNRs are denoted as GNR1.4, GNR1.9, GNR2.4, and GNR3.0, respectively). The longitudinal SPR peak of the four nanorods are 538, 580, 650, 698 nm, and the SPR peak of GNP100 is 560 nm (Fig. 1a). The size distribution curves of the four GNRs and GNP100 are shown in Fig. 1b. From the comparison, we noticed a number of interesting features that are absent from spherical gold nanoparticles: (1) all four nanorods showed a major particle size distribution peak at an average diameter of 75 nm; (2) with the exception of GNR1.4, other three GNRs exhibit a small particle size peak around 5–6 nm; and (3) the intensity distribution (expressed as the percentage of the total scattered light intensity) of the small size peak increases with increased aspect ratio. Figure 1c is a plot of the intensity distribution of the small size peak versus the aspect ratio of the nanorods. For comparison purposes, we also included spherical gold nanoparticles with an average diameter

around 100 nm in the study. Only one major peak with an average diameter of 105 nm was observed from the spherical nanoparticle.

The size distribution peak at 75 nm is worthy for further discussion. First, this number does not correspond to either the longitudinal or the transverse dimension of the nanorods. The diameter is 25 nm for all four nanorods and the length is

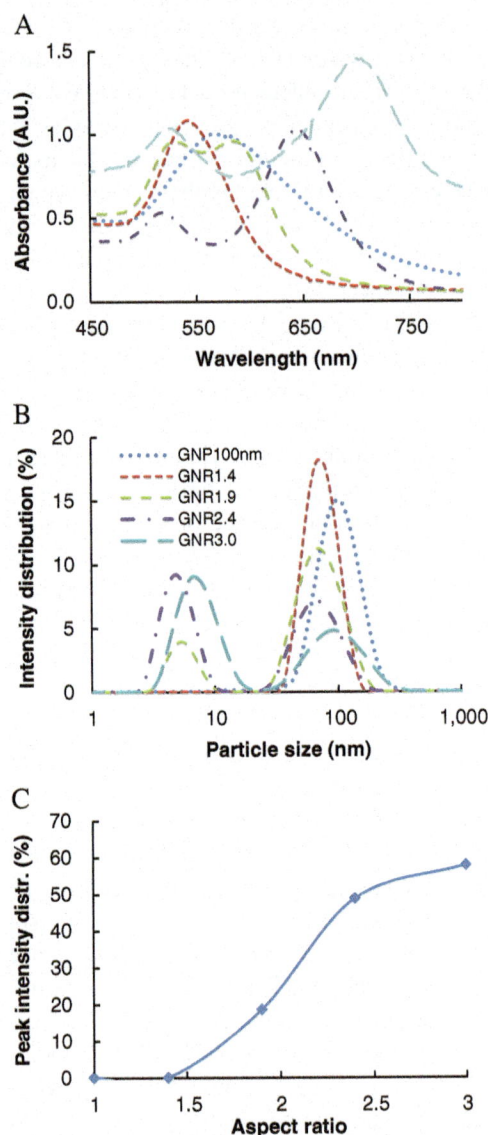

**Fig. 1** UV–vis absorption spectra (**a**) and size distribution curves of CTAB-GNRs and GNP100 (**b**). **c** Plot of the intensity distribution (percent) of the small size peak versus the aspect ratio of the nanorods. **a** and **b** have the same legends

34, 47, 60, and 73 nm, respectively. Second, the particle size value of this distribution peak is the same for all four nanorods, despite their obviously different mass. DLS obtains the nanoparticle size information by measuring the diffusion coefficient of the particle [43, 44]. This is the reason why the particle size obtained from DLS analysis is called the "hydrodynamic" dimension. The diffusion coefficient of a particle is not only dependent on the mass of the particle, but also the shape and the surface chemistry of the particles because these parameters affect the particle–solvent interactions, and therefore, the Brownian motion of the particles. The observed results simply mean that the four nanorods have the same diffusion coefficient as a spherical gold nanoparticle with a hydrodynamic diameter of 75 nm. In the case of nanorods, the shape and the surface chemistry of the nanorod perhaps play a more dominant role than the mass in its translational diffusion coefficient.

The small size peak around 5–7 nm is sometimes mistaken as the presence of small particle impurities. Several reports recently proposed that this small size peak is actually a representation of the rotational diffusion of the non-spherical nanorods [45, 46]. It is not an actual particle size distribution peak. It corresponds to neither the longitudinal and transverse dimension of the nanorods. This peak signifies that the rotational diffusion coefficient of the nanorods is equivalent to the translation diffusion coefficient of a spherical particle with an average diameter of 5–7 nm. The rotational diffusion appears to be strongly dependent on the aspect ratio of the nanorods: the intensity distribution (percent) of this peak increases significantly with increased aspect ratio (Fig. 1c). For GNR with an aspect ratio of 1.4, only the large particle size peak was observed, similar to the spherical nanoparticles. The findings from this study suggest that when using fixed angle DLS to analyze and interpret the size of gold nanorods, it is important to understand the true meaning of the average particle size data and size distribution peaks. The relative intensity of the size peak related to the rotational diffusion coefficient of the nanorods can provide additional size information of the nanorods.

Nanorods–protein interaction analysis

For spherical gold nanoparticles, proteins are believed to adsorb to the nanoparticle surface through a combined suite of chemical interactions including electrostatic interactions, van der Waals interactions, Au–S and Au–N bonding. Gold nanorods have a different surface chemistry from the citrate-protected spherical gold nanoparticles: the citrate ligands that protect the spherical gold nanoparticles are negatively charged, while the CTAB (cetyl trimethylammonium bromide) ligands protecting the nanorods are positively charged. It is expected that proteins interact with spherical particles and nanorods differently. Because DLS can be used

to monitor the nanoparticle size change continuously, we conducted the gold nanorod–protein interaction study under kinetic conditions.

Four proteins are investigated in this study: BSA, HSA, human IgG, and human IgA. In all of these protein–nanorod interaction studies, a pure phosphate buffer (PB) solution that is used to prepare the protein solutions was used as a negative control. Figure 2 is the DLS data of the GNR1.9 (A and B) and GNR2.4 (C and D) upon mixing with BSA or PB control solution. Three nanorods, GNR1.4, GNR1.9, and GNR3.0 showed no size and size distribution change at the presence of BSA (the data of GNR1.4 and GNR3.0 is not shown here, but very similar to GNR1.9). That is to say, the average particle size of the two peaks, 5–6 and 75 nm, remains unchanged (Fig. 2a) compared to phosphate buffer control solution, and the relative intensity of these two peaks remain unchanged as well (Fig. 2b), suggesting that there is no interaction between BSA and these three GNRs. Interestingly, there is a dramatic response from GNR2.4: First, the average particle size of the large size peak increases steadily and quickly from about 75 nm to more than 200 nm within 10 min of incubation time, and the average particle size of the small size peak also increases from 5–7 nm to about 20–30 nm (Fig. 2c). Second, the relative intensity of the small particle size peak decreases from the original 50 % to less than 10 % (Fig. 2d), and the relative intensity of the large particle size peak increases from 50 % to more than 90 %.

The interaction between HSA and nanorods are very similar (Fig. 3a, b: GNR1.9; Fig. 3c, d: GNR2.4). Again, three nanorods, GNR1.4, GNR1.9, and GNR3.0, showed no response to HSA at all (only the data for GNR1.9 is presented here), while GNR2.4 exhibited substantial particle size increase upon interaction with HSA. However, there are some slight differences between HSA and BSA: the particle size increase caused by HSA appears to be even larger than that is caused by BSA, and the small size peak that is indicative of rod-shaped particle disappeared completely over incubation time.

IgA and IgG are immunoglobulin proteins. IgA is a dimer of IgG, linked together through the Fc region, with the Fab region exposed outwards. Again, we did not observe any size change from GNR1.4, GNR1.9, and GNR3.0 (data not shown here). Only GNR2.4 showed response to IgA and IgG, however, in a very different way from the two serum albumin proteins (Fig. 4). The interaction of IgA or IgG caused the large particle size peak increase from 75 nm to about 100 nm for IgA (Fig. 4a), and to 85 nm for IgG (Fig. 4b). So the net increase is 15 nm for IgA and 10 nm for IgG. The small size peak, increased from about 5 to 15 nm for IgA (Fig. 4c), and from 5 nm to about 9 nm for IgG (Fig. 4d). Most noticeable is that the intensity distribution of the small size peak changed only very slightly: it decreased from 50 % to about 45–47 % (Fig. 4e, f) for both IgA and IgG.

**Fig. 2** Particle size analysis results of GNR1.9 (**a** and **b**) and GNR2.4 (**c** and **d**) upon incubation with bovine serum albumin (BSA) or 10 mMPB solution (control). All samples were prepared by adding 2 μL of 1 mg/mL protein solution or PB control solution into 100 μL GNR or GNP solution. All data points are average values of three experiments. **a** and **c** are the plots of the average particle size of the mixed sample solution versus the incubation time. **b** and **d** are the plots of the peak intensity distribution (percent) versus the incubation time. Legends for all four figures are the same as **a**

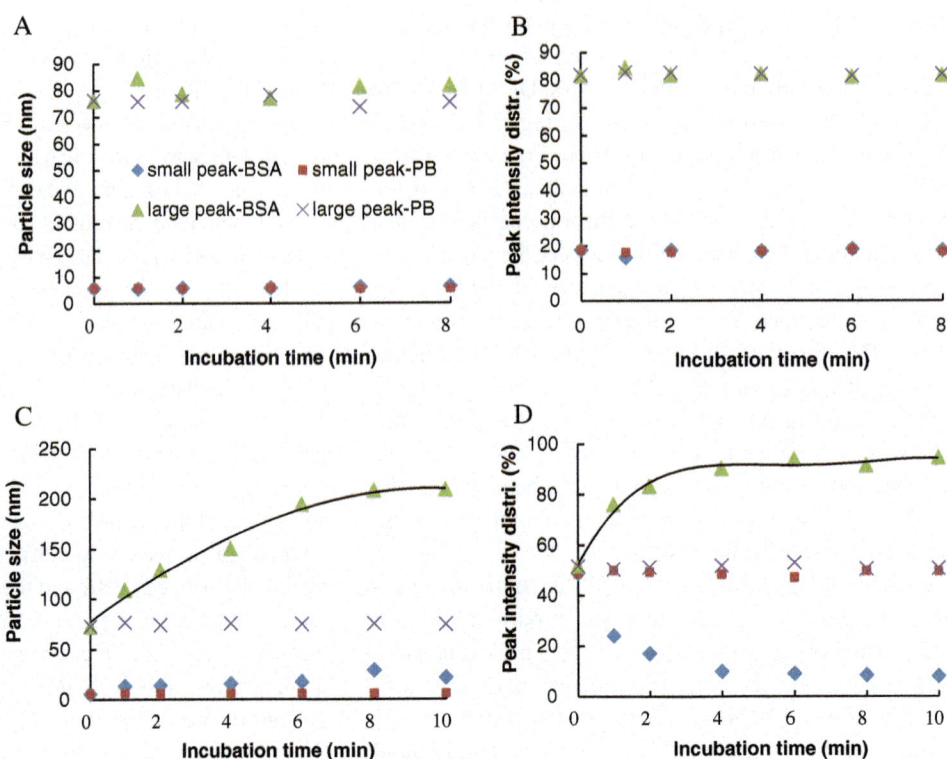

As a comparison to GNRs, the interaction between a spherical gold nanoparticle, GNP100 with a hydrodynamic diameter of 105 nm was also studied. Upon mixing with BSA and HSA at a concentration of 1 mg/mL, no obvious nanoparticle size increase was observed. Upon mixing with IgA and IgG, the nanoparticle size increased by about 20 nm, from 105 to 125 nm.

The particle size analysis revealed several unexpected results regarding the nanorod–protein interactions. The first noticeable result is that among the four nanorods studied

**Fig. 3** Particle size analysis results of GNR1.9 (**a** and **b**) and GNR2.4 (**c** and **d**) upon incubation with human serum albumin (HSA) or 10 mMPB solution (control). All samples were prepared by adding 2 μL of 1 mg/mL protein solution or PB control solution into 100 μL GNR or GNP solution. All data points are average values of three experiments. **a** and **c** are the plots of the average particle size of the mixed sample solution versus the incubation time. **b** and **d** are the plots of the peak intensity distribution (percent) versus the incubation time. Legends for all four figures are the same as **a**

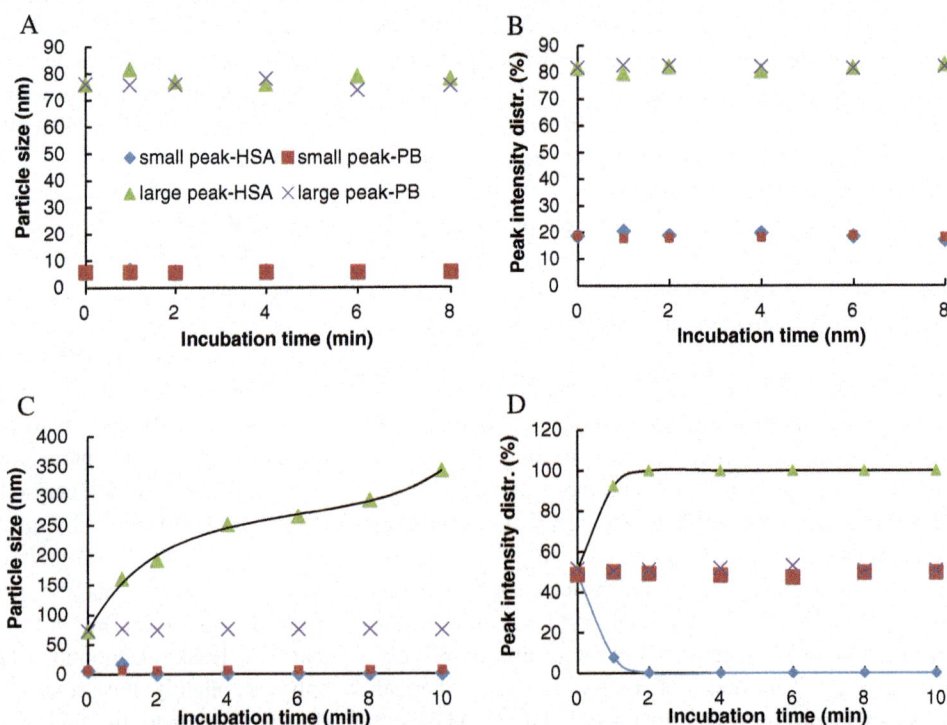

**Fig. 4** Particle size analysis results of GNR2.4 upon incubation with immunoglobulin A (IgA) or immunoglobulin G (IgG) solution. All samples were prepared by adding 2 μL of 1 mg/mL protein solution or PB control solution into 100 μL GNR solution. **a** and **b** are the plots of the average particle size of the large size peak versus incubation time; **c** and **d** are the plots of the average particle size of the small size peak versus incubation time. **e** and **f** are the plots of the small peak intensity distribution (percent) versus incubation time

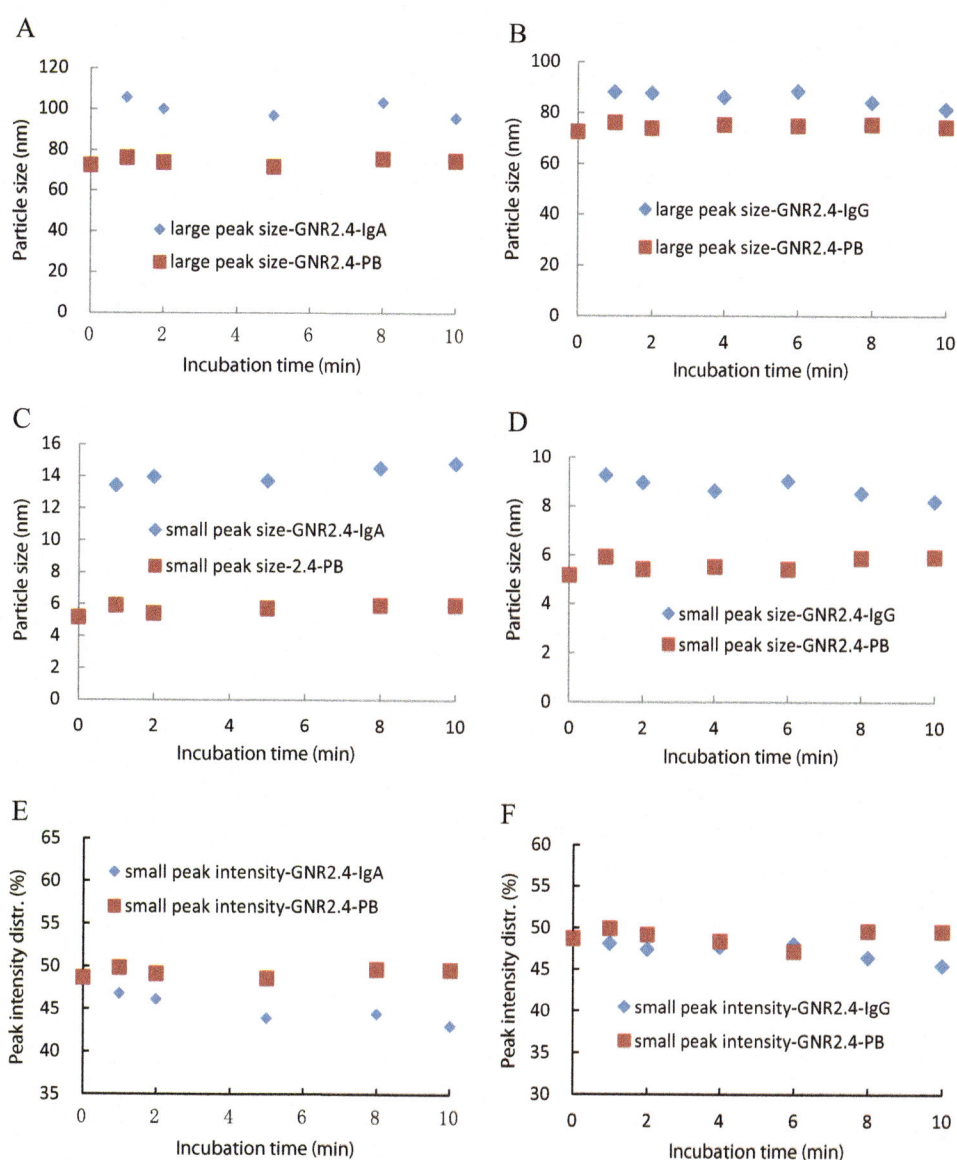

here, only nanorods with an aspect ratio of 2.4, GNR2.4, appears to interact with the two serum albumins and the two immunoglobulins. We have not noticed similar findings from previous reports on such a phenomenon. In one study reported by Pan et al., it was mentioned that BSA adsorbs more strongly to gold nanorods with an aspect ratio of 8.0 than nanorods with an aspect ratio of 3.0 [21]. The surface chemistry of the four nanorods investigated in this study is the same, because they are all CTAB-protected, and positively charged on the surface. At this moment, the exact reason behind this difference is unclear. Since only the aspect ratio is different, we can only hypothesize that this may be related to the different stability of the CTAB bilayer structure on the nanorods. It is possible that GNR with an aspect ratio of 2.4 may have the lowest stability, therefore, more easily displaced by proteins.

A second noticeable result is that it appears that the two serum albumin proteins (BSA and HSA) caused nanorod

aggregation (Fig. 5a) while the two immunoglobulin proteins formed a stable protein corona on the nanorod surface without causing nanorod aggregation (Fig. 5b). Serum albumin is a protein with a molecular weight around 60 KDa. The hydrodynamic diameter of a serum albumin is around 5 nm. If a complete layer of BSA molecule is adsorbed to the nanoparticle, the net increase of the particle size should not exceed about 10 nm. The dramatic size increase from 75 to more than 200 nm can only be interpreted as nanorod aggregate formation. Furthermore, the aggregate formation is most likely a random aggregate. With its non-spherical geometry, GNRs are known to form end-to-end or side-by-side assemblies under various conditions [38]. From the analysis of pure GNRs, we know that with higher aspect ratio, the intensity of the small size peak increases. If BSA interacts with GNR primarily through end-to-end mode, we should see intensity increase of the small particle size peak,

**Fig. 5** Schematic illustrations of two different interaction models between GNRs and proteins. Serum albumins (BSA and HSA) caused gold nanorods aggregate formation (**a**), while IgG and IgA formed a stable protein corona on the nanorod surface (**b**). Also in **a**, some CTAB ligands are intentionally eliminated to illustrate the potential disruption of the CTAB double-layer structure

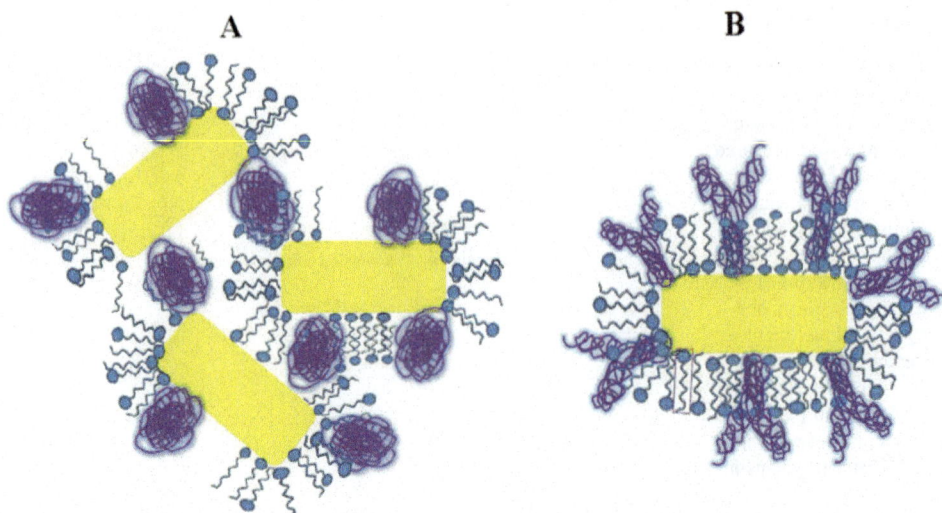

not decrease. As random aggregates are formed, the aggregates "look" more like a spherical particle, and the signature peak indicative of the rotational diffusion of a rod-like particle will disappear.

On the other hand, the interaction of the two immunoglobulin proteins (IgG and IgA) with the nanorod GNR2.4 appears to lead to the formation of a stable protein corona on the nanorod surface. The average particle size of the nanorods increased only about 10–15 nm, which is on par with the hydrodynamic diameter of the two proteins. The relative intensity of the small particle size peak, signature of the nanorod rotational diffusion, remains almost the same. This suggests that the nanorods, after their interaction with the two immunoglobulin proteins, maintained their rod-like shape.

Currently, we do not know the exact reasons behind these differences. We offer the following explanations as possible mechanisms. The isoelectric point of serum albumin is around 4.7; and the isoelectric points of immunoglobulins are higher, typically in the range of 6.1–8.5 (refer to the product information, Sigma-Aldrich). At neutral PB buffer solution (pH 7.4), serum albumins are negatively charged. The CTAB-GNRs are positive charged. It could be that the electrostatic interaction between negatively charged serum albumin and the nanorods caused the disruption of the CTAB biolayer. GNR2.4 may have the weakest CTAB bilayer structure and is easily perturbed by protein adsorption interaction. The CTAB bilayer structure may be disrupted too quickly to allow the serum albumin proteins to form a stable protecting layer on the nanorod surface. Subsequently, the exposed nanorods quickly aggregated together due to strong van der Waals interactions between the exposed metal cores. As to the immunoglobulin proteins, these proteins are membrane proteins that are produced by the B cells and located in the membrane of B cells. The immunoglobulins may bind to the gold nanorods by first inserting the Fc region of the antibody into the CTAB bilayer, and then further interaction with the gold nanorods.

The surface protection layer of the nanorods is not disrupted during the protein adsorption process. In the end, a stable protein corona is formed on the nanorods. These hypotheses certainly require further extensive studies.

## Conclusion

In this study, we demonstrate that fixed angle DLS may also be used for gold nanorods characterization and study. However, one should be aware that when using a fixed angle DLS instrument as most users have, the particle size data of nanorods obtained from DLS does not represent the true physical dimension of the nanorods. The actual information determined by DLS is the diffusion coefficient of the particle. If the nanoparticle is spherical, the hydrodynamic diameter can be revealed by Stokes–Einstein equation. If the particle has a rod shape, the diffusion coefficient determined by DLS is still accurate, but the hydrodynamic diameter cannot be deduced from the Stokes–Einstein equation. Despite this fact, we can still use single, fixed angle DLS to monitor the nanorod size change upon protein interaction, as demonstrated in this study. The small particle size peak that is a signature of the rotational diffusion of the nanorods is important for determining the nanorod aggregate formation. Complementary to various molecular spectroscopic techniques, DLS can provide additional information on the complex protein–nanorod interactions. Although many previous studies have demonstrated the use of antibody-conjugated gold nanorods for imaging and biomolecular assay development [47–51], it is also acknowledged by these previous studies that there are more difficulties involved in the preparation of gold nanorod–protein bioconjugates than the spherical nanoparticle bioconjugates. More often than not, the CTAB-protected gold nanorods need to be first modified with other ligands or functional groups to facilitate bioconjugation. Our study suggests that CTAB-protected gold nanorods with certain aspect ratios can be directly conjugated to antibodies through a simple adsorption process. The aspect ratio of the GNRs should be considered in selecting the most suitable nanorods for future application development

**Acknowledgment** This work is supported by a State of Florida Boost Scholar Award to Q.H. The authors also want to thank the reviewers of this manuscript for their highly constructive comments and suggestions, which have led to significant improvements of the paper.

# References

1. Jans H, Huo Q (2012) Gold nanoparticle-enabled biological and chemical detection and analysis. Chem Soc Rev 41:2849–2866
2. Dykman L, Khlebtsov N (2012) Gold nanoparticles in biomedical applications: recent advances and perspectives. Chem Soc Rev 41:2256–2282
3. Dreaden EC, Alkilany AK, Huang X, Murphy CJ, El-Sayed MA (2012) The golden age: gold nanoparticles for biomedicine. Chem Soc Rev 41:2740–2779
4. Boisselier E, Astruc D (2009) Gold nanoparticles in nanomedicine: preparations, imaging, diagnostics, therapies and toxicity. Chem Soc Rev 38:1759–1782
5. Newhouse RJ, Zhang JZ (2012) Optical properties and applications of shape-controlled metal nanostructures. Rev Plasmonics 2010:205–238
6. Chen J, Saeki F, Wiley BJ, Cang H, Cobb MJ, Li Z-Y, Au L, Zhang H, Kimmey MB, Li X, Xia Y (2005) Gold nanocages: bioconjugation and their potential use as optical imaging contrast agents. Nano Lett 5(3):473–477
7. Hu M, Chen J, Li Z, Au L, Hartland GV, Li X, Marquez M, Xia Y (2006) Gold nanostructures: engineering their plasmonic properties for biomedical applications. Chem Soc Rev 35:1084–1094
8. Sun Y, Xia Y (2003) Gold and silver nanoparticles: a class of chromophores with colors tunable in the range from 400 to 750 nm. Analyst 128:686–691
9. Halas NJ, Lal S, Chang W, Link S, Nordlander P (2011) Plasmons in strongly coupled nanostructures. Chem Rev 111:3913–3963
10. Wang H, Brandl D, Nordlander P, Halas NJ (2006) Plamonic nanostructures: artificial molecules. Acc Chem Res 40:53–62
11. Jain PK, Lee KS, El-Sayed IH, El-Sayed MA (2006) Calculated absorption and scattering properties of gold nanoparticles of different size, shape, and composition: applications in biological imaging and biomedicine. J Phys Chem B 110:7238–7248
12. Nehl CL, Liao H, Hafner JH (2006) Optical properties of star-shaped gold nanoparticles. Nano Lett 6:683–688
13. Link S, Mohamed MB, El-Sayed MA (1999) Simulation of the optical absorption spectra of gold nanorods as a function of their aspect ratio and the effect of the medium dielectric constant. J Phys Chem B 103:3073–3077
14. Mout R, Moyano DF, Rana S, Rotello VM (2012) Surface functionalization of nanoparticles for nanomedicine. Chem Soc Rev 41:2539–2544
15. Walkey CD, Chan WCW (2012) Understanding and controlling the interaction of nanomaterials with proteins in a physiological environment. Chem Soc Rev 41:2780–2799
16. Dobrovolskaia MA, Patri AK, Zheng J, Clogston JD, Ayub N, Aggarwal P, Neun BW (2009) Interaction of colloidal gold nanoparticles with human blood: effects on particle size and analysis of plasma protein binding profiles. Nanomedicine (Nanotechnol Biol Med) 5:106–117
17. Lacerda SHDP, Park JJ, Meuse C, Pristinski D, Becker ML, Karim A, Douglas JF (2010) Interaction of gold nanoparticles with common human blood proteins. ACS Nano 4:365–379
18. Huo Q, Colon J, Codero A, Bogdanovic J, Baker CH, Goodison S, Pensky MY (2011) A facile nanoparticle immunoassay for cancer biomarker discovery. J Nanobiotechnol 9:20, open access
19. Huo Q, Litherland SA, Sullivan S, Hallquist H, Decker DA, Rivera-Ramirez I (2012) Developing a nanoparticle test for prostate cancer scoring. J Transl Med 10:44, open access
20. Arvizo RR, Giri K, Moyano D, Miranda OR, Madden B, McCormick DJ, Bhattacharya R, Rotello VM, Kocher JP, Mukherjee P (2012) Identifying new therapeutic targets via modulation of protein corona formation by engineered nanoparticles. PLoS One 7:e33650, open access
21. Pan B, Cui D, Xu P, Li Q, Huang T, He R, Gao F (2007) Study on interaction between gold nanorod and bovine serum albumin. Colloids Surf A 295:217–222
22. Shang L, Wang Y, Jiang J, Dong S (2007) pH-dependent protein conformational changes in albumin:gold nanoparticle bioconjugates: a spectroscopic study. Langmuir 23:2714–2721
23. Iosin M, Toderas F, Baldeck PL, Astilean S (2009) Study of protein-gold nanoparticle conjugates by fluorescence and surface-enhanced Raman scattering. J Mol Struct 924–926:196–200
24. Matveeva EG, Shtoyko T, Gryczynski I, Zkopova I, Gryczynski Z (2009) Fluorescence quenching/ enhancement surface assays: signal manipulation using silver-coated gold nanoparticles. Chem Phys Let 454:85–90
25. Fischer NO, McIntosh CM, Simard JM, Rotello VM (2002) Inhibition of chymotrypsin through surface binding using nanoparticle-based receptor. Proc Natl Acad Sci 99:5018–5023
26. Hong R, Fischer NO, Verma A, Goodman CM, Emrick T, Rotello VM (2004) Control of protein structure and function through surface recognition by tailored nanoparticle scaffolds. J Am Chem Soc 126:739–743
27. Calzolai L, Franchini F, Gilliland D, Rossi F (2010) Protein–nanoparticle interaction: identification of the ubiquitin–gold nanoparticle interaction site. Nano Lett 10:3101–3105
28. Jans H, Liu X, Austin L, Maes G, Huo Q (2009) Dynamic light scattering as a powerful tool for gold nanoparticle bioconjugation and biomolecular binding study. Anal Chem 81:9425–9432
29. Austin L, Liu X, Huo Q (2010) An immunoassay for monoclonal antibody isotyping and quality analysis using gold nanoparticles and dynamic light scattering. Am Biotechnol Lab 28(8):10–12
30. Khlebtsov NG, Bogatyrev VA, Khlebtsov BN, Dykman LA, Englebienne P (2003) A multilayer model for gold nanoparticle bioconjugates: application to study of gelatin and human IgG adsorption using extinction and light scattering spectra and the dynamic light scattering method. Colloid J 65:679–693
31. Tsai DH, Delrio FW, Keene AM, Tyner KM, MacCuspie RI, Cho TJ, Zachariah MR, Hackley VA (2011) Adsorption and conformation of serum albumin protein on gold nanoparticles investigated using dimensional measurements and in situ spectroscopic methods. Langmuir 27:2464–2477
32. Liu X, Dai Q, Austin L, Coutts J, Knowles G, Zou J, Chen H, Huo Q (2008) A one-step homogeneous immunoassay for cancer biomarker detection using gold nanoparticle probes coupled with dynamic light scattering. J Am Chem Soc 130:2780–2782
33. Dai Q, Liu X, Coutts J, Austin L, Huo Q (2008) A one-step highly sensitive method for DNA detection using dynamic light scattering. J Am Chem Soc 130:8138–8139
34. Kalluri JR, Arbneshi T, Khan SA, Nelly A, Candice P, Varisli B, Washington M, McAfee S, Robinson B, Banerjee S, Singh AK, Senapati D, Ray PC (2009) Use of gold nanoparticles in a simple

colorimetric and ultrasensitive dynamic light scattering assay: selective detection of arsenic in groundwater. Angew Chem Int Ed 48:9668–9671

35. Gao D, Sheng Z, Han H (2011) An ultrasensitive method for the detection of gene fragment from transgenics using label-free gold nanoparticle probe and dynamic light scattering. Anal Chim Acta 696:1–5

36. Driskell JD, Jones CA, Tompkins SM, Tripp RA (2011) One-step assay for detecting influenza virus using dynamic light scattering and gold nanoparticles. Analyst 136:3083–3090

37. Wang X, Ramström O, Yan M (2011) Dynamic light scattering as an efficient tool to study glyconanoparticle-lectin interactions. Analyst 136:4174–4178

38. Wang L, Zhu Y, Xu L, Chen W, Kuang H, Liu L, Agarwal A, Xu C, Kotov NA (2010) Side-by-side and end-to-end gold nanorod assemblies for environmental toxin sensing. Angew Chem Int Ed 49:5472–5475

39. Pecora R (1968) Spectrum of light scattered from optically anisotropic macromolecules. J Chem Phys 49:1036–1043

40. Zero K, Pecora R (1985) Dynamic depolarized light scattering. In: Pecora R (ed) Dynamic light scattering applications of photon correlation spectroscopy. Plenum, New York, pp 83–99

41. Van der Zande BMI, Dhont JKG, Bohmer MR, Philipse AP (2000) Collidal dispersions of gold rods characterized by dynamic light scattering and electrophoresis. Langmuir 16:459–464

42. Lehner D, Lindner H, Glatter O (2000) Determination of the translational and rotational diffusion coefficients of rodlike particles using depolarized dynamic light scattering. Langmuir 16:1689–1695

43. Berne BJ, Pecora R (1976) Dynamic light scattering: with applications to chemistry, biology, and physics. Wiley, New York

44. A Technical Note from Malvern Instruments: http://www.malvern.com/common/downloads/campaign/MRK656-01.pdf

45. Rodrígues-Fernández J, Pérez-Juste J, Liz-Marzán LM, Lang PR (2007) Dynamic light scattering of short Au rods with low aspect ratios. J Phys Chem C 111:5020–5025

46. Khlebtsov BN, Khlebtsov NG (2011) On the measurement of gold nanoparticle sizes by the dynamic light scattering method. Colloid J 73:118–127

47. Wang C, Chen Y, Wang T, Ma Z, Su Z (2007) Biorecognition-driven self-assembly of gold nanorods: a rapid and sensitive approach toward antibody sensing. Chem Mater 19:5809–5811

48. Tong L, Wei Q, Wei A, Cheng J (2009) Gold nanorods as contrast agents for biological imaging: optical properties, surface conjugation and photothermal effects. Photochem Photobiol 85:21–32

49. Pissuwan D, Valenzuela SM, Killingsworth MC, Xu X, Cortie MB (2007) Targeted destruction of murine macrophase cells with bioconjugated gold nanorods. J Nanoparticle Res 9:1109–1124

50. Pissuwan D, Valenzuela SM, Miller CM, Cortie MB (2007) A golden bullet? Selective targeting of *Toxoplasma gondii* tachyzoites using antibody-functionalized gold nanorods. Nano Lett 7:3808–3812

51. Yu C, Irudayaraj J (2007) Multiplex biosensor using gold nanorods. Anal Chem 79:572–579

# 7

# Experimental and theoretical study on the synthesis of gold nanoparticles using ceftriaxone as a stabilizing reagent for and its catalysis for dopamine

Yuan-zhi Song · An-feng Zhu · Yang Song ·
Zhi-peng Cheng · Jian Xu · Jian-feng Zhou

**Abstract** Electrochemical synthesis of gold nanoparticles on the surface of glassy carbon electrode and preparation of GNPs in aqueous solution using ceftriaxone as an innocuous stabilizing reagent were proposed. The gold nanoparticles were characterized by scanning electron microscopy, transmission electron microscopy, infrared spectrometry, UV spectrophotometry, powder X-ray diffraction, and cyclic voltammetry. The catalysis of gold nanoparticles on the glassy carbon electrode for dopamine was demonstrated. The results indicate that the modified electrode has an excellent repeatability and reproducibility. The relationship between the molecular structure and the dispersion of GNPs on the surface of GCE as well as the catalysis of GNPs for dopamine was discussed.

**Keywords** Gold nanoparticles · Synthesis · Ceftriaxone · Stabilizing reagent

## Introduction

Gold nanoparticles (GNPs) have been widely used in the fields of physics, chemistry, biology, medicine, and material science [1]. To maximize the efficiency of GNPs in their applications, well-controlled particle size, efficient particle dispersion, and excellent reproducibility are necessary. The strategies for immobilization of GNPs layers onto the surfaces include electrostatic links and covalent bonding. The surface of functional groups (COOH, OH, SH, and $NH_2$) are suitable substrate for the deposition of GNPs [2–5]. The reduction of $HAuCl_4$ is the most used methods for the preparation of GNPs in aqueous solution; reductants such as ascorbic acid [6], citrate [7–9], and borohydride [10, 11] have been used in this reaction. Electrochemical deposition of metal nanoparticles has been found a better alternative because of their flexibility in controlling the size and coverage of the metal nanoparticles [12, 13].

However, it is difficult to control the size of GNPs in the aqueous solution because the size of the GNPs in aqueous solution is commonly controlled by changing the reaction parameters such as molar ratio of the reductant, gold precursor, pH, temperature, the stabilizing reagent, and the prepared process is time consuming. The stabilizing reagent of GNPs is most important factor; the size and dispersion are controlled mainly by the molecule of stabilizing reagent. Therefore, the relationship between the size and dispersion and the molecular is an interesting topic.

Cefoperazone, formerly known as cefoperazone sodium, is a semisynthetic, broad-spectrum cephalosporin antibiotic. Chemically, cefoperazone sodium is sodium of (6R,7R)-7-[(R)-2-(4-ethyl-2,3-dioxo-1-piperazinecarboxamido)-2-(p-hydroxyphenyl)-acetamido-3-[[(1-methyl-1H-tetrazol-5-yl)thio]methyl]-8-oxo-5-thia-1-azabicyclo[4.2.0]oct-2-ene-2-carboxylate. Its molecular formula is $C_{25}H_{26}N_9NaO_8S_2$ with a molecular weight of 667.65 [14]. The goals of this paper demonstrate ceftriaxone's versatile role in stabilizing GNPs prepared by electrochemical deposition and wet-chemistry reduction.

Our previous works have indicated that density-functional theory is a powerful method for predicting the geometry and harmonic vibration of organic compounds [15–19]. Therefore, the DFT-B3LYP/6-31G (d, p) was carried out to study the molecular structure of ceftriaxone.

Y.-z. Song (✉) · A.-f. Zhu · Z.-p. Cheng · J.-f. Zhou
Jiangsu Province Key Laboratory for Chemistry of
Low-Dimensional Materials, School of Chemistry
and Chemical Engineering, Huaiyin Normal University,
Huai An 223300, People's Republic of China
e-mail: singyuanzhi@126.com

Y. Song (✉) · J. Xu
College of Materials Science and Engineering,
Beijing University of Chemical Technology,
Peking 100029, People's Republic of China
e-mail: songyang@mail.buct.edu.cn

In present work, ceftriaxone as an innocuous stabilizing reagent was used for electrochemical synthesis of GNPs on the surface of glassy carbon electrode (GCE) and preparation for GNPs in aqueous solution for a stable sensor, DFT-B3LYP/6-31G (d, p) was carried out to study the molecular structure and the properties of ceftriaxone. The catalysis of GNPs for dopamine at ceftriaxone@GNP/GCE and GNP/GCE was demonstrated. The relationships between the molecular structure of ceftriaxone and the dispersion of GNPs on the surface of GCE as well as the catalysis of GNPs for dopamine were discussed.

## Experimental

### Materials

All reagents used herein were of analytical grade. Doubly distilled water was used throughout; 0.1 M phosphate-buffered solution (PBS) was prepared by dissolving 0.1 mol NaCl and 0.1 mol $Na_2HPO_4$ in double-distilled water of 1,000 mL and adjusted desired pH values with 6 M HCl or 1 M NaOH.

### Preparation of GNPs

The GNPs were deposited at a voltage of $-0.2$ V for 30 s on the surface of GCE that was immersed in the mixture of $HAuCl_4$ (2.0 mg $mL^{-1}$), $H_2SO_4$ (0.5 M), and ceftriaxone sodium (0.4 mg $mL^{-1}$), and then washed in doubly distilled water. In the typical synthetic process of GNPs in aqueous solution, 0.150 g of $NaBH_4$ were dissolved into the mixture of 2.0 mg $mL^{-1}$ $HAuCl_4$, 0.5 mol $L^{-1}$ $H_2SO_4$, and 0.4 mg $mL^{-1}$ ceftriaxone sodium. The solution was stirred with a magnetic stirrer for 10 min to ensure that the $NaBH_4$ completely dissolved; the black GNPs were soon produced, and followed by centrifugal separation, washing with absolute alcohol and drying in vacuum at 20 °C for 6 h.

### Characterization

For all electrochemical experiments, a CHI660B Electrochemical Analyzer (CHI, USA) was employed. The GNP-modified GCE was used as working electrode, a platinum wire served as the counter electrode, and a saturated calomel electrode was used as the reference electrode. The GNPs were characterized by scanning electron microscope (SEM; S-4800, HITACHI, Japan), transmission electron microscope (TEM, JEM 2100, JEOL, Japan), UV spectrophotometer (UV-1750, Shimadzu, Japan). Powder X-ray diffraction (XRD) spectra were recorded on a Switzerland ARL/X'TRA X-ray diffractometer rotating anode with Cu-Kα radiation source ($\lambda = 1.54056$ Å). IR spectra were measured by a NICOLET NEXUS470 spectrometer in the frequency range 4,000–400 $cm^{-1}$.

### Calculation

The calculations of ceftriaxone were performed with the Gaussian03 package [20]. The molecule in the vacuum was fully optimized using DFT-B3LYP with the 6–31G (d, p) basis sets.

## Results and discussion

### SEM and TEM images of GNPs

SEM images confirm the formation of a layer of GNPs on the GCE surface, several GNPs on the surface of GCE in presence of ceftriaxone were observed in Fig. 1a and b, and those on the surface of GCE in absence of ceftriaxone are shown in Fig. 1c, indicating that the well-dispersed GNPs on the surface of GCE in the presence of ceftriaxone were obtained. The size of GNPs deposited at $-0.2$ V for 30 s (Fig. 1a) are larger than those obtained at same potential for 10 s (Fig. 1b), indicating that the size of GNPs on the surface of GCE were controlled by the reduction time of $HAuCl_4$. The TEM images of GNPs obtained from aqueous solution were shown in Fig. 1b, the size of GNPs is about 25–100 nm. The GNPs on the GCE surface is similar to those in aqueous due to the reduction of $HAuCl_4$. From the molecular structure of ceftriaxone it can be seen that the gold atoms of surface of GNPs adsorb the negative ions (ceftriaxone) with the polar groups such as carbonyl, carboxyl, and amidocyanogen. Therefore, the GNPs on the surface of GCE are well dispersed.

### XRD of GNPs

The powder XRD pattern of the GNPs from aqueous solution is shown in Fig. 2. The major diffraction peaks can be indexed as the gold face-centered cubic (fcc) phase based on the data of the JCPDS file (JCPDS no. 04-0784) [21]. The diffraction peaks of GNPs appeared at 38.7°, 44.6°, 64.4°, and 78.7°, which can be assigned to (111), (200), (220), and (311) crystalline plane diffraction peaks of gold, respectively. The diffraction peaks at 16.0° and 27.9° may be assigned to the ceftriaxone. On the nanometer scale metals (most of them are fcc) tend to nucleate and grow into twinned and multiple twinned particles with their surfaces bounded by the lowest-energy (111) facets [22]. Other morphologies with less stable facets have only been kinetically achieved by adding chemical capping reagents to the synthetic systems [23–26].

**Fig. 1** SEM images of GNPs on the surface of GCE deposited at −0.2 V for 30 (**a**), 10 (**b**) in presence of ceftriaxone, and 30 s in absence of ceftriaxone (**c**), and TEM images of GNPs prepared by sodium borohydride reduction (**d**)

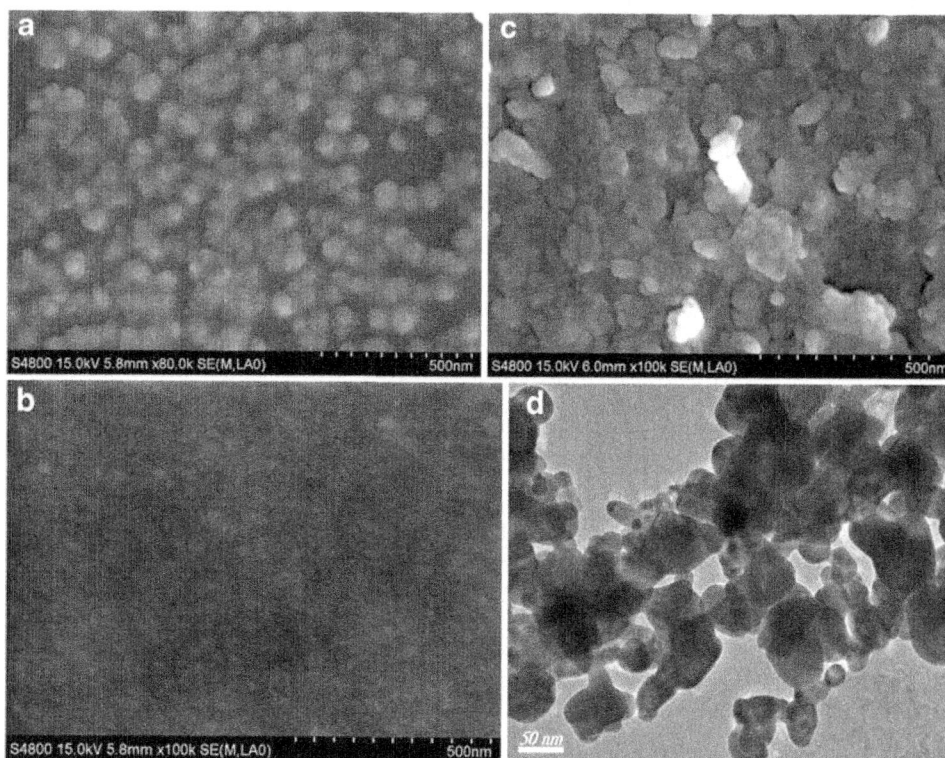

IR spectra of GNPs

The IR spectra of GNPs obtained from GCE and ceftriaxone are shown in Fig. 3. From the IR spectra of ceftriaxone, the bands at 3,436 and 3,244 cm$^{-1}$ could be assigned to the stretching vibrations of NH and OH, the bands of stretching vibrations of CH were found at 2,930 and 2,816 cm$^{-1}$, the band of stretching vibrations of C=O appeared at 1,743, 1,647, and 1,613 cm$^{-1}$, and the peaks at 1,535 and 1,499 cm$^{-1}$ were associated with the torsional vibrations of aromatic ring. The bands at 1,395 and 1,359 cm$^{-1}$ could be assigned to stretching vibrations of CN, and band of breath vibration of aromatic ring was observed at 1,037 cm$^{-1}$ [27]. However, the IR spectra of GNPs obtained from GCE, the bands of stretching vibrations of NH and OH were found at 3,402 and 3,130 cm$^{-1}$, the band at 1,621 cm$^{-1}$ could be assigned to the stretching vibrations of C=O, the bands of stretching vibrations of CN and the breath vibration of aromatic ring were found at 1,403 and 1,081 cm$^{-1}$, respectively, indicating ceftriaxone was assembled on the surface of GNPs.

UV spectra of GNPs

Figure 4 shows the UV absorption spectrum of the GNPs obtained from aqueous solution. A broad band centered at ca. 557 nm appears, characteristic of surface plasmon absorption on the GNPs.

Cyclic voltammograms of GNPs

The cyclic voltammograms (CVs) of GNPs on the surface of GCE in 0.1 M PBS of pH 7.3 are shown in Fig. 5. The oxidation peak of ceftriaxone@GNPs was found at 1.118 V, and the reduction peak was observed at 0.459 V. To remove ceftriaxone on the surface of GNPs, the ceftriaxone@GNPs-modified GCE was rinsed with a magnetic stirrer in 5 mol L$^{-1}$ H$_2$SO$_4$ aqueous solution and doubly distilled

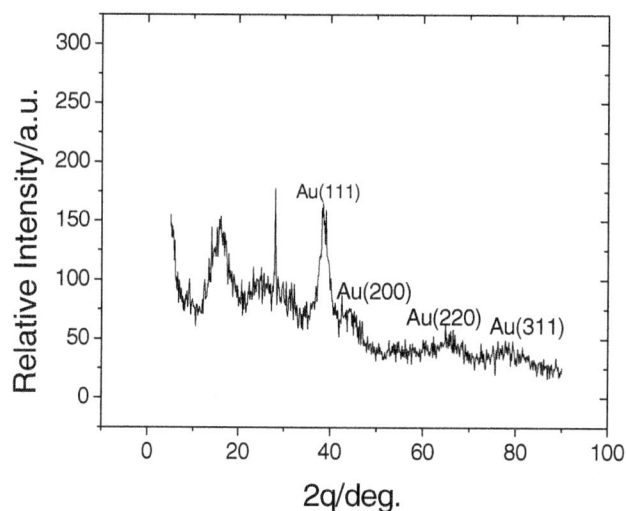

**Fig. 2** XRD pattern of GNPs obtained from aqueous solution

**Fig. 3** IR spectra of GNPs obtained from the surface of GCE (**a**) and ceftriaxone (**b**)

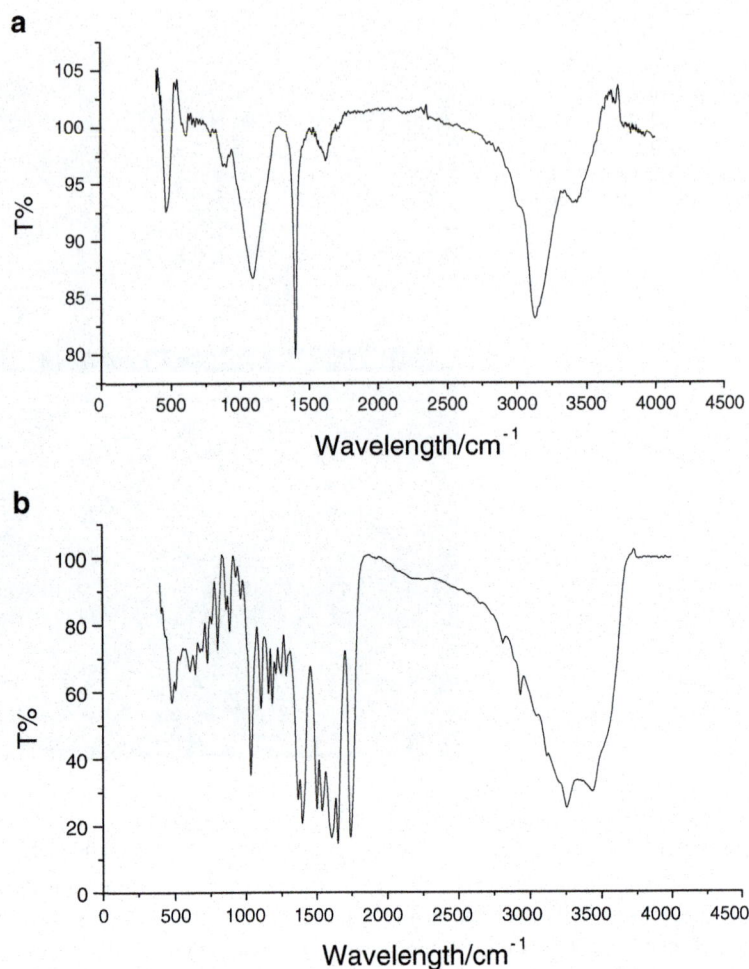

**a**

**b**

water for 10 min, sequentially. The peak potential of the rinsed GNPs shifted to positive direction, and the reduction current increases, indicating the ceftriaxone on the surface of GNPs was removed. The CVs of ceftriaxone at GCE are

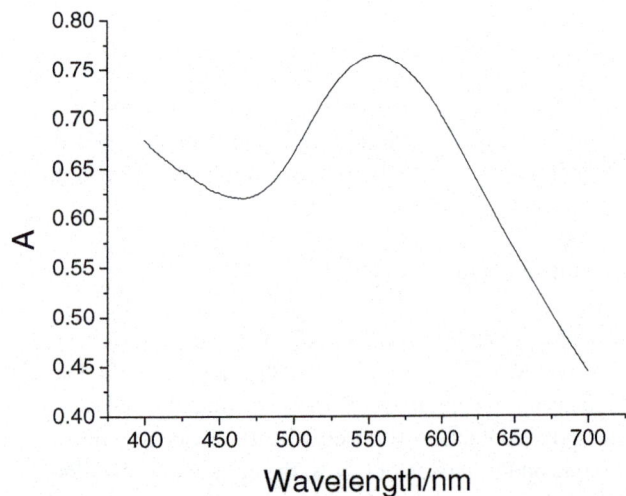

shown in Fig. 5 (inset), an oxidation peak was observed at 1.153 V, and a reduction peak at 0.900 V was observed, which was assigned to the sulfur atoms in ceftriaxone,

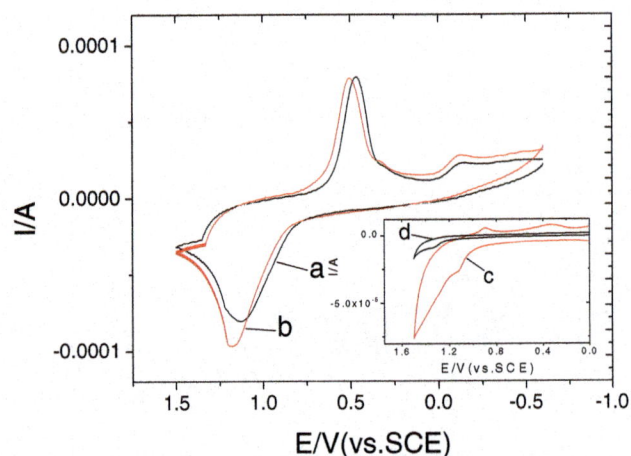

**Fig. 4** UV spectra of GNPs in absolute alcohol

**Fig. 5** CVs of the ceftriaxone@GNPs (*a*) and rinsed GNP/GCE (*b*). Supporting electrolyte: 0.1 mol $L^{-1}$ PBS of pH 7.3. *Inset* CVs of 0.4 mg/mL ceftriaxone sodium at GCE (*c*) and bare GCE (*d*), supporting electrolyte, 0.5 mol $L^{-1}$ $H_2SO_4$

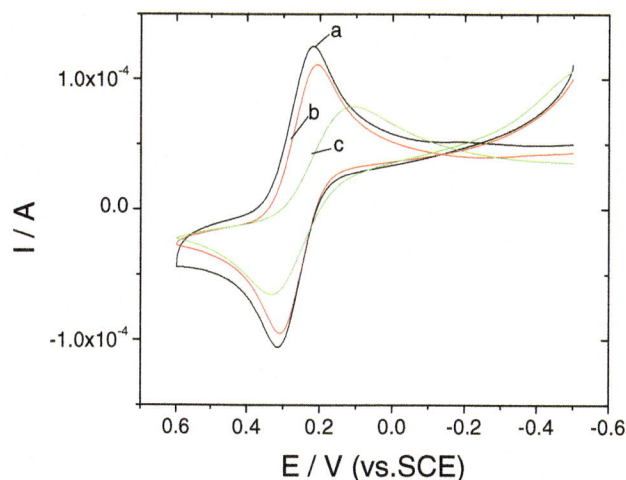

**Fig. 6** CVs of 5 mM $K_4[Fe(CN)_6]$ at ceftriaxone@GNP/GCE (*a*), rinsed GNP/GCE (*b*), and bare GCE (*c*). Supporting electrolyte, 0.1 mol $L^{-1}$ KCl

**Table 1** Mülliken charge of atoms in ceftriaxone

| Atom | Mülliken charge |
|------|-----------------|
| $S_1$ | 0.225548 |
| $N_4$ | −0.52594 |
| $N_7$ | −0.20545 |
| $O_9$ | −0.48835 |
| $N_{10}$ | −0.21175 |
| $N_{13}$ | −0.50661 |
| $S_{18}$ | 0.141123 |
| $S_{20}$ | 0.134278 |
| $N_{22}$ | −0.51138 |
| $N_{23}$ | −0.32606 |
| $N_{25}$ | −0.09105 |
| $O_{28}$ | −0.45712 |
| $O_{29}$ | −0.14326 |
| $O_{30}$ | −0.46103 |

indicating that ceftriaxone is stable under electrochemical synthesis of ceftriaxone@GNPs at −0.2 V.

## CVs of GNP-modified GCE in the $K_3Fe(CN)_6$–$K_4Fe(CN)_6$ system

The CVs of GNP modified GCE in the $K_3Fe(CN)_6$–$K_4Fe(CN)_6$ system were shown in Fig. 6. The real active surface area will be estimated. In a reversible process, the following Randles-Sevcik formula [28] at 298 K has been used:

$$i_p = 2.69 \times 10^5 n^{3/2} A\, C_o D_o^{1/2} v^{1/2}.$$

Where $i_p$ is the peak current (amperes), $n$ the number of electrons, $A$ the electrode area (in square centimeters), $C$ the concentration (in moles per cubic centimeter), $D$ the diffusion coefficient (in square centimeters per second), and $v$ the scan rate (in volts per second).

From the slope of the plot of oxidation current ($i_p$) versus $v^{1/2}$, the electrode surface area of the ceftriaxone@GNP/GCE, GNP/GCE, and the bare GCE is 0.115, 0.104, and 0.071 cm², respectively, indicating that the microscopic area of the GNP/GCE increased significantly and was about 1.46 times larger than the microscopic area of the bare GCE.

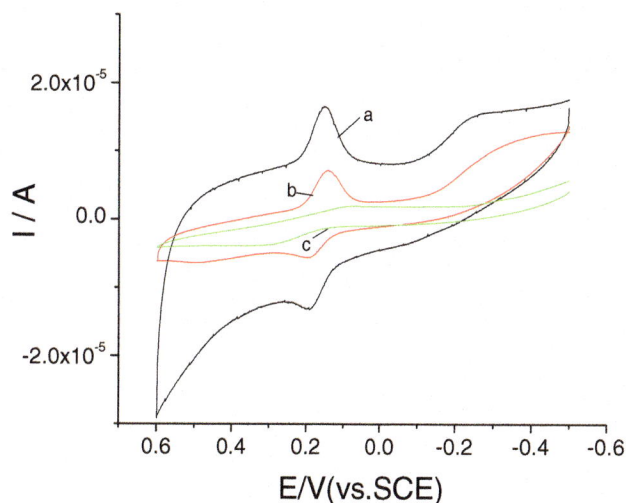

**Fig. 7** Geometry and numbering atoms of ceftriaxone

**Fig.8** CVs of 80.0 mg $L^{-1}$ dopamine at ceftriaxone@GNP/GCE (*a*), the rinsed GNP/GCE (*b*), and bare GCE (*c*). *Scan rate* 100.0 mV s$^{-1}$. Supporting electrolyte: 0.1 mol $L^{-1}$ PBS of pH 7.3

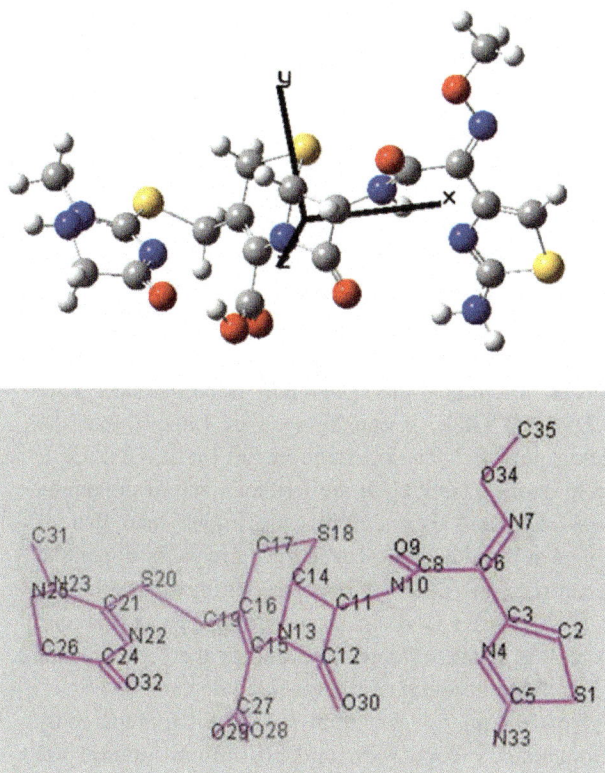

**Table 2** Peak and current of dopamine

| Electrode | Oxidation peak (V) | Oxidation current (μA), unit current (μA cm$^{-2}$/mg L$^{-1}$) | Reduction peak (V) | Reduction current (μA), unit current (μA cm$^{-2}$/mg L$^{-1}$) |
|---|---|---|---|---|
| Bare GCE | 0.276 | 1.14, 1.61 | 0.090 | 0.88, 1.24 |
| GNP/GCE | 0.194 | 3.64, 3.50 | 0.140 | 5.21, 5.01 |
| Ceftriaxone@GNP/GCE | 0.191 | 6.80, 5.91 | 0.151 | 7.11, 6.18 |
| GNP/DWCNT/GCE[12] | 0.247 | 108.5, 14.37 | 0.148 | 69.39, 9.19 |

## Optimized geometry and molecular properties of ceftriaxone

The molecular property is controlled by the structure. Ceftriaxone is an organic acid, the degree of ionization for ceftriaxone are controlled by sulfuric acid, thus the optimized geometry of protonated ceftriaxone at DFT-B3LYP/6–31G (d, p) level is shown in Fig. 7. It can be seen in Fig. 7 that the hydrophilic carbonyl and carboxyl locate at a side of ceftriaxone, while hydrophobic methylene and methyl appear at another side. The Mülliken charges with hydrogen summed into heavy atoms for ceftriaxone are listed in Table 1, in Table 1 the negative atoms are oxygen and nitrogen. The total dipole moment (vector $X=-7.0061$, $Y=3.9098$, $Z=-1.7552$) of ceftriaxone is 8.2130 Debye, indicating that the negative electron cloud in ceftriaxone shifted to $X$ direction. Therefore, the coordination bonds form between the gold atoms of surface of GNPs and the oxygen ($O_{28}$, $O_{30}$, and $O_{34}$) and nitrogen ($N_4$) in ceftriaxone, the hydrophobic ceftriaxone@GNP on the surface of GCE is well dispersed due to the absorbed ceftriaxone on the surface of the GNPs.

## Electrochemical catalysis of GNP

Dopamine is important in the regulation of sodium balance and blood pressure via renal mechanisms [29, 30]. The affinity of dopamine for its receptors is in the nanomolar range; higher concentrations occupy other G-protein-coupled receptor [29, 30]. Circulating dopamine concentrations (picomolar range) are not sufficiently high to activate dopamine receptors, but high nanomolar concentrations can be attained in dopamine-producing tissues (e.g., renal proximal tubule, jejunum). The concentration of dopamine is controlled by not only the taking drugs but also the human emotion. Therefore, the determination of dopamine in blood is important.

The CVs of dopamine at bare GCE, ceftriaxone@GNP/GCE, and rinsed GNP/GCE are shown in Fig. 8, respectively. The peak potentials, currents, and unit currents (the currents per square centimeter of electrode area for 1 mg L$^{-1}$ dopamine, μA/cm$^{-2}$/mg L$^{-1}$) of dopamine at bare GCE, ceftriaxone@GNP/GCE, and rinsed GNP/GCE are summarized in Table 2. For comparison with the gold nanoparticle/double-walled carbon nanotube-modified glassy carbon electrode (GNP/DWCNT/GCE), the currents and unit currents for 50 mg L$^{-1}$ dopamine hydrochloride at GNP/DWCNT/GCE are also shown in Table 2 [12]. The sensitivity of GNP/GCE for dopamine is lower than that GNP/DWCNT/GCE. However, when stirring the solution for renewing the modified electrode, the GNP/DWCNT on the surface of GCE removed easily due to adsorption of DWCNTs. Therefore, the GNP/GCE is more stable than GNP/DWCNT/GCE. It can be seen in Table 2 that the oxidation potential for dopamine at the rinsed GNP/GCE and ceftriaxone@GNP/GCE are less than that of dopamine at bare GCE, and their currents are higher than that of dopamine at bare GCE. However, the currents for dopamine at the ceftriaxone@GNP/GCE are higher than that of dopamine at the rinsed GNP/GCE, indicating that ceftriaxone catalyzes the redox of dopamine due to the formation of hydrogen bond between ceftriaxone and dopamine.

CVs of 10.0 mg L$^{-1}$ dopamine at GNP/GCE prepared by electrochemistry deposition and absorption method are shown in Fig. 9, compared with the oxidation currents (2.24 μA) at 0.185 V and reduction currents (2.12 μA) at

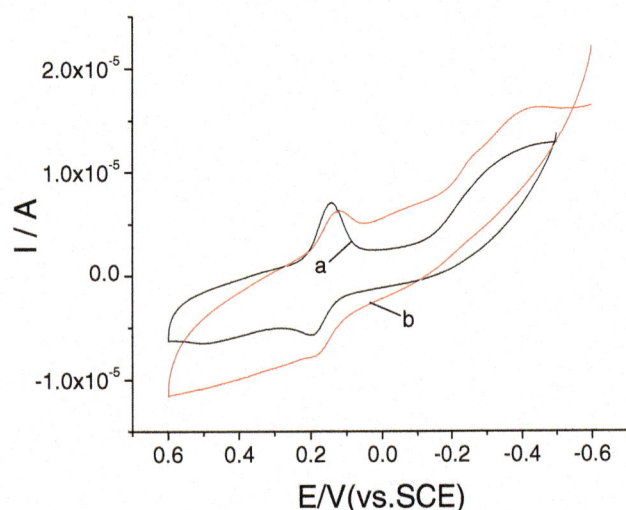

**Fig. 9** CVs of 10.0 mg L$^{-1}$ dopamine at GNP/GCE prepared by electrochemical deposition ($a$) and absorption method ($b$). Scan rate, 100.0 mV s$^{-1}$. Supporting electrolyte, 0.1 mol L$^{-1}$ PBS of pH 7.3

0.122 V for dopamine at GNP/GCE prepared by absorption method, both oxidation currents (3.60 μA) at 0.193 V and reduction currents (5.20 μA) at 0.142 V for dopamine at GNP/GCE prepared by electrochemical deposition increase, indicating that the GNPs on the surface of GCE prepared by electrochemical deposition catalyze dopamine well, the results may be ascribe to the well dispersion of GNPs on the surface of GCE prepared by electrochemical deposition.

On using the rinsed GNPs/GCE daily and storing under ambient conditions over a period of 2 months, and after stirring at 700 rpm/min with a magnetic stirring apparatus for 2 h, the electrode retained 96.5 % of its initial peak current response with relative standard deviation (RSD) of 2.3 % ($n=$ 25) for a dopamine concentration of 80.00 mg $L^{-1}$, which shows long-term stability of the film modifier on the surface of GCE. The results indicate that the rinsed GNPs/GCE has an excellent repeatability and reproducibility. However, the GNPs obtained from aqueous solution dropped on the surface of GCE as the adsorption method, followed by storage at 14 °C for 24 h, after stirring at 700 rpm/min with a magnetic stirring apparatus for 2 h the modified GCE retained 86.5 % of its initial peak current response with RSD of 4.3 % ($n=25$) for a dopamine concentration of 80.00 mg $L^{-1}$.

## Conclusions

The gold nanoparticles on the surface of GCE and in aqueous solution using ceftriaxone as a stabilizing reagent were prepared in this paper, and the catalysis of ceftriaxone@GNPs and GNPs for dopamine was demonstrated. The sensitivity of GNP/GCE for dopamine is lower than that GNP/DWCNT/GCE. However, the GNP/GCE is more stable than GNP/DWCNT/GCE, and the electrochemical synthesis of ceftriaxone@GNPs on the surface of glassy carbon is simple, cheap, and rapid. The relationships between the molecular structure of ceftriaxone and the dispersion of GNPs on the surface of GCE as well as the catalysis of GNPs for dopamine were discussed, and the rinsed GNPs/GCE has an excellent repeatability and reproducibility.

**Acknowledgments** The authors gratefully acknowledge the financial support of the National Science Foundation of China (grant nos. 51175245 and 51106061), the Science Foundation of Education of Jiangsu Province of China (grant no. JH10-48), the Open Science Foundation for Jiangsu Province Key Laboratory for Chemistry of Low-Dimensional Materials (grant no. JSKC11091), and the Science Foundation for Huaiyin Normal University (grant no. 11HSGJBZ13).

## References

1. Rashid MH, Bhattacharjee RR, Kotal A, Mandal TK (2006) Synthesis of spongy gold nanocrystals with pronounced catalytic activities. Langmuir 22:7141–7143
2. Mena ML, Yánz-Sedeño P, Pingarrõn JM (2006) A comparison of different strategies for the construction of amperometric enzyme biosensors using gold nanoparticle-modified electrodes. Anal Biochem 336:20–27
3. He P, Urban MW (2005) Phospholipid-stabilized Au− nanoparticles. Biomacromolecules 6:1224–1225
4. Qi ZM, Zhou HS, Matsuda NK, Honma I, Shimada K, Takatsu A, Kato K (2004) Characterization of gold nanoparticles synthesized using sucrose by seeding formation in the solid phase and seeding growth in aqueous solution. J Phys Chem B 108:7006–7011
5. Zubarev ER, Xu J, Sayyad A, Gibson JD (2006) Amphiphilic gold nanoparticles with V-shaped arms. J Am Chem Soc 128:4958–4959
6. Luty-Błocho M, Fitzner K, Hessel V, Löb P, Maskos M, Metzke D, Pacławski K, Wojnicki M (2011) Synthesis of gold nanoparticles in an interdigital micromixer using ascorbic acid and sodium borohydride as reducers. Chem Eng J 171:279–290
7. Koutsoulis NP, Giokas DL, Vlessidis AG, Tsogas GZ (2010) Alkaline earth metal effect on the size and color transition of citrate-capped gold nanoparticles and analytical implications in periodate-luminol chemiluminescence. Anal Chim Acta 669:45–52
8. Balasubramanian SK, Yang L, Yung L-Y L, Ong C-N, Ong W-Y, Yu LE (2010) Characterization, purification, and stability of gold nanoparticles. Biomaterials 31:9023–9030
9. Mikhlin Y, Karacharov A, Likhatski M, Podlipskaya T, Zubavichus Y, Veligzhanin A, Zaikovski V (2010) Submicrometer intermediates in the citrate synthesis of gold nanoparticles: new insights into the nucleation and crystal growth mechanisms. J Colloid Interf Sci 362:330–336
10. Seoudi AA, Fouda DA (2010) Synthesis, characterization and vibrational spectroscopic studies of different particle size of gold nanoparticle capped with polyvinylpyrrolidone. Physica B 405:906–911
11. He P, Zhu X (2007) Phospholipid-assisted synthesis of size-controlled goldnanoparticles. Mater Res Bull 42:1310–1315
12. Song YZ, Song Y, Zhong H (2011) L-cysteine-nano-gold modified glassy carbon electrode and its application for determination of dopamine hydrochloride. Gold Bull 44:107–111
13. Song Y, Song YZ, Zhu AF, Zhong H (2011) Indian J Chem A 50A:1006–1009
14. Arguedas A, Loaiza C, Perez A, Gutierrez A, Herrera ML, Rothermel CD (2003) A pilot study of single-dose azithromycin versus three-day azithromycin or single-dose ceftriaxone for uncomplicated acute otitis media in children. Curr Ther Res 64:16–29
15. Song YZ, Zhou JF, Song Y, Wei YG, Wang H (2005) Density-functional theory studies on standard electrode potentials of half reaction for L-adrenaline and adrenalinequinone. Bioor Med Chem Lett 15:4671–4680
16. Song YZ, Ruan M, Ye Y, Li YY, Xie W, Shen J, Shen AG (2008) Experimental and density functional theory and ab initio Hartree–Fock study on the vibrational spectra of 2-(4-fluorobenzylidenea-mino)-3-(4-hydroxyphenyl) propionic acid. Spectrochim Acta Part A 69:682–687
17. Song YZ (2007) Theoretical study on the electrochemical behavior of norepinephrine at Nafion multi-walled carbon nanotubes modified pyrolytic graphite electrode. Spectrochim Acta Part A 67:1169–1177
18. Song YZ, Zhang LL, Zhong H, Shi DQ, Xie JM, Zhao GQ (2008) Theoretical study on the geometry and vibration of 1-{6-(4-chlor-ophenyl)-1-[(6-chloropyridin-3-yl)methyl]-2-[(6-chloropyridin-3-

yl)methylsulfanyl]-4-methyl-1,6-dihydropyrimidin-5-yl}ethanone. Spectrochim Acta Part A 70:943–952

19. Shi DQ, Zhu XF, Song YZ (2008) Synthesis, crystal structure, insecticidal activity and DFT study on the geometry and vibration of O-(E)-1-{1-[(6-chloropyridin-3-yl) methyl]-5-methyl-1H-1,2,3-triazol-4-yl}ethyleneamino-O-ethyl-O-phenylphosphorothioate. Spectrochim Acta Part A 71:1011

20. Frisch MJ, Trucks GW, Schlegel HB, Scuseria GE, Robb MA, Cheeseman JR, Montgomery JA, Vreven JT, Kudin KN, Burant JC, Millam JM, Iyengar SS, Tomasi J, Barone V, Mennucci B, Cossi M, Scalmani G, Rega N, Petersson GA, Nakatsuji H, Hada M, Ehara M, Toyota K, Fukuda R, Hasegawa J, Ishida M, Nakajima T, Honda Y, Kitao O, Nakai H, Klene M, Li X, Knox JE, Hratchian HP, Cross JB, Bakken V, Adamo C, Jaramillo J, Omperts R, Stratmann RE, Yazyev O, Austin AJ, Cammi R, Pomelli C, Ochterski JW, Ayala PY, Morokuma K, Voth GA, Salvador P, Dannenberg JJ, Zakrzewski VG, Dapprich S, Daniels AD, Strain MC, Farkas O, Malick DK, Rabuck AD, Raghavachari K, Foresman JB, Ortiz JV, Cui Q, Baboul AG, Clifford S, Cioslowski J, Stefanov BB, Liu G, Liashenko A, Piskorz P, Komaromi I, Martin RL, Fox DJ, Keith T, Al-Laham MA, Peng CY, Nanayakkara A, Challacombe Gill MPM, Johnson B, Chen W, Wong MW, Gonzalez C, Pople JA (2004) Gaussian Inc, Wallingford

21. Joint Committee on Powder Diffraction Standards (1991) Diffraction Data File: JCPDS International Center for Diffraction Data. Swarthmore PA

22. Allpress JG, Sanders JV (1967) The structure and orientation of crystals in deposits of metals on mica. Surf Sci 7:1–25

23. Bradley JS, Tesche B, Busser W, Maase M, Reetz MT (2000) Surface spectroscopic study of the stabilization mechanism for shape-selectively synthesized nanostructured transition metal colloids. J Am Chem Soc 122:4631–4636

24. Puntes VF, Krishnan KM, Alivisatos AP (2001) Colloidal nanocrystal shape and size control: the case of cobalt. Science 291:2115–2117

25. Kirkland AI, Jefferson DA, Duff DG, Edwards PP, Gameson I, Johnson BFG, Smith D (1993) Structural studies of trigonal lamellar particles of gold and silver. Proc R Soc Lond Ser A 440:589–609

26. Ahmadi TS, Wang ZL, Green TC, Henglein A, El-Sayed MA (1996) Shape-controlled synthesis of colloidal platinum nanoparticles. Science 272:1924–1925

27. Cordente N, Respaud M, Senocq F, Casanove M-J, Amiens C, Chaudret B (2001) Synthesis and magnetic properties of nickel nanorods. Nano Lett 1:565–568

28. Kissinger PT, Heineman WR (1984) Laboratory techniques in electroanalytical chemistry Chapter 3. Marcel Dekker, New York

29. Zeng C, Armando I, Luo Y, Eisner GM, Felder RA, Jose PA (2008) Dysregulation of dopamine-dependent mechanisms as a determinant of hypertension: studies in dopamine receptor knockout mice. Am J Physiol Heart Circ Physiol 294:H551–H569

30. Hussain T, Lokhandwala MF (2003) Renal dopamine receptors and hypertension. Exp Biol Med (Maywood) 228:134–142

# 8

# Rate redox-controlled green photosynthesis of gold nanoparticles using $H_{3+x}PMo_{12-x}V_xO_{40}$

Ali Ayati · Ali Ahmadpour · Fatemeh F. Bamoharram · Majid M. Heravi · Mika Sillanpää

**Abstract** Stabilized and size-controlled gold nanoparticles were synthesized using molybdophosphoric acid ($H_3[PMo_{12}O_{40}]$, HPMo) and its vanadium-substituted mixed addenda ($H_{3+x}[PMo_{12-x}V_xO_{40}]$, ($x=0–3$); and HPMoV$_x$) under UV irradiation by a simple green method. In the process, HPMo and HPMoV$_x$ play the role of photocatalyst, reducing agent and efficient stabilizer. Control of gold nanoparticles size was achieved by variation of initial gold ions concentration and molar ratio of HPMo to gold ions. The synthesis rate of Au nanoparticles was found to be parallel to an increasing $x$ value in the order of: HPMoV$_3$>HPMoV$_2$ > HPMoV>HPMo, which leads to smaller and more uniform particles. Also, by substitution of vanadium instead of molybdenum in the HPMo formula, the morphology of nanoparticles was gradually changed from spherical-shaped nanoparticles in the presence of HPMo to nanorods in the case of HPMoV$_3$.

**Keywords** Molybdophosphoric acid · Vanadium · Gold · Nanoparticle · Nanorod · UV irradiation

A. Ayati · A. Ahmadpour (✉)
Department of Chemical Engineering,
Ferdowsi University of Mashhad,
Mashhad, Iran
e-mail: ahmadpour@um.ac.ir

A. Ayati · M. Sillanpää
Laboratory of Green Chemistry,
Lappeenranta University of Technology,
Jääkärinkatu 31,
50100 Mikkeli, Finland

F. F. Bamoharram
Department of Chemistry, Mashhad Branch,
Islamic Azad University,
Mashhad, Iran

M. M. Heravi
Department of Chemistry, School of Sciences, Alzahra University,
Tehran, Iran

## Introduction

One of the usual metal nanostructures is gold nanoparticles (Au NPs) which are used in optics, electrochemistry, catalysis, sensors, environmental engineering, and electronics because they are stable, relatively nontoxic, and biocompatible [1–5]. The size and shape of Au NPs strongly affect their physical and chemical properties and intense research has been devoted to the morphological control of these nanostructures in recent years. Finding the methods in which one could control the size and shape of the prepared nanoparticles is of great interest in order to maximize the particles efficiency [6].

A lot of techniques have been developed for the synthesis of these nanoparticles, such as electrochemical [7], chemical reduction [8], sonochemical [9], photochemical [10], and so on. To date, solution-based wet chemical synthesis is believed to be the best route to new nanostructures [11, 12]. In most of these procedures, the use of an organic environment and a relatively high temperature are common. Controlling size and shape of nanoparticles can be achieved through the control of nucleation and growth steps by varying synthesis parameters, including activity of reducing agents, type and concentration of precursors, and also nature and amount of protective agents [13–15]. However, the intervention of environmentally harmful and toxic chemicals in the Au NPs preparation procedures is inevitable.

Recently, the green synthesis or fabrication of Au nanostructures has been incomprehensively studied using harmless alternative polyoxometalates (POMs) [16–19]. Since the environmental care is one of the worldwide increasing worries, green chemistry has been defined as a set of principles which reduce or eliminate the use of hazardous substances or catalysts [20]. This fact encourages scientists to make efforts in finding processes working in this direction. For this reason, there is still a good scope for research towards finding green and ecofriendly materials, solvents, and catalysts in different reactions. Along this line, introducing clean processes and utilizing ecofriendly and green catalysts which can be simply recycled at the end of reactions have been under permanent attention and demands.

POMs as solid acid catalysts are green with respect to corrosiveness, safety, quantity of waste, recyclability, and separability. Therefore, using them in various processes is one of the innovative trends.

They are a unique class of molecularly defined inorganic metal–oxide clusters which have exceptional properties such as: strong Brönsted acidity, high hydrolytic stability (pH=0–12), high thermal stability, and operation in pure water without any additive [21, 22]. Attractively, POM's structures remain unchanged under stepwise and multielectron redox reactions and can be reduced by photochemical and electrochemical procedures using suitable reducing agents [23].

These compounds have been used as both reducing agents and stabilizers for the synthesis of metal nanoparticles such as Ag, Au, Pt, Se, and Pd upon illumination with UV/near-Vis light [6, 18, 24–28]. There are only limited reports regarding synthesis of gold nanoparticles (Au NPs) using these kinds of green materials. Troupis et al. used the photocatalytic process for Au NPs synthesis in the presence of $H_3[SiW_{12}O_{40}]$ [18]. Mandal et al. have synthesized more complicated nanostructures such as Au–Ag core–shell dimetallic compounds [29] and Au nanosheets [30]. Various crystalline gold nanostructures have also been synthesized using β-$[H_4PMo_{12}O_{40}]^{3-}$ [12] and transition metal monosubstituted POMs ($PW_{11}MO_{40}$, M=$Cu^{2+}$, $Ni^{2+}$, $Zn^{2+}$, $Fe^{3+}$) [31]. Moreover, in our previous work, we have synthesized Au NPs using Preyssler acid with a simple photoreduction technique [32].

Although, some Keggin, mixed valence, and Preyssler types of POMs have been used in the synthesis of Au NPs, to the best of our knowledge, the role of molybdophosphoric acid, HPMo, and its vanadium-substituted mixed addenda, $HPMoV_x$, has not been studied. HPMo, similar to the other types of POMs [18], can be reduced in the presence of oxidizable organic substrates like alcohols (e.g., propan-2-ol), under UV irradiation (Eq. (1)):

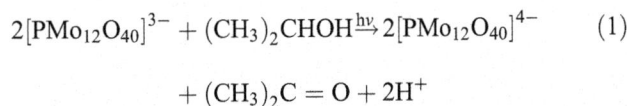

$$2[PMo_{12}O_{40}]^{3-} + (CH_3)_2CHOH \xrightarrow{h\nu} 2[PMo_{12}O_{40}]^{4-} \quad (1)$$

$$+ (CH_3)_2C = O + 2H^+$$

In contact with gold ions, $[PMo_{12}O_{40}]^{4-}$ is able to transfer electrons efficiently to gold ions and reduce them to $Au^0$. The color of the solution is then gradually turned from colorless to pink indicating the formation of $Au^0$. Equation (2) represents this reaction:

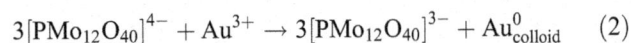

$$3[PMo_{12}O_{40}]^{4-} + Au^{3+} \rightarrow 3[PMo_{12}O_{40}]^{3-} + Au^0_{colloid} \quad (2)$$

According to Eqs. (1) and (2), HPMo ions can be utilized cyclically as oxidizing or reducing agent and propan-2-ol plays the role of sacrificial agent.

In the present work, we have investigated the synthesis of gold nanostructures generated by a green chemistry-type process, using HPMo and $HPMoV_x$ in the absence of any surfactant or seed. Also, the effect of gold ion concentration or molar ratio of HPMo to gold ion was studied on the size of Au NPs. Besides, vanadium-substituted mixed addenda of HPMo (i.e., $HPMoV_x$) was used for the synthesis of Au NPs and the effect of addition of vanadium atom ($x=0$–3) in the POMs structure was explored on the size and the shape of prepared NPs.

## Experimental

### Chemicals and apparatus

$H_3[PMo_{12}O_{40}]$ and other chemicals were purchased from Merck Company and used as received. $H_4[PMo_{11}VO_{40}]$, $H_5[PMo_{10}V_2O_{40}]$ and $H_6[PMo_9V_3O_{40}]$ were prepared according to the procedure reported in the literature [33]. UV-visible spectra were obtained using Avantes Avaspec-3648 single beam instrument. The synthesized Au NPs were characterized mainly by particle size distribution (PSD) using a ZetaSizer Nano ZS apparatus (Malvern Instruments Ltd.) as a laser particle sizer. The instrument allowed to measure particle size taking the advantage of optoelectronic systems. Also, nanoparticles were characterized using Transmission Electron Microscopy (Philips CM-120).

### Synthesis procedure of Au NPs

In a typical experiment, $5.5 \times 10^{-7}$ mol of $HPMoV_x$ was dissolved in 5 mL distilled water and then 10 mL $HAuCl_4$ ($5 \times 10^{-4}$ M) and 2 mL propan-2-ol were added. The solution was placed into a spectrophotometer cell and deaerated with $N_2$ gas. Then, the mixture was irradiated by UV light (125 W high pressure mercury vapor lamp) under continuous stirring. Reaction was performed in a constant room temperature, using water circulating around the cell. The color of the solution changed from colorless or pale yellow (at high HPMo concentration) to pink, indicating the formation of Au NPs. The nanoparticles were separated from the reaction mixture by a high-speed centrifuge (14,000 rpm), washed twice with water and redispersed in water before any analysis.

## Results and discussion

Polyoxometalates, regarding their redox abilities, can be divided into two groups of mono-oxo (type I) and cis-dioxo (type II). This classification is based on the number of terminal oxygen atoms attached to each addenda atom, e.g., molybdenum or tungsten, in the polyanion. Examples

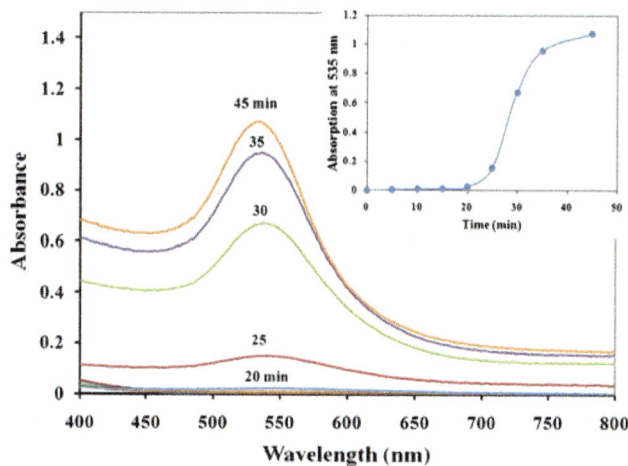

**Fig. 1** UV–vis spectra of HPMo (5.5×10⁻⁴ M)/propan-2-ol/Au³⁺ (5× 10⁻⁴ M) solution and the reaction progress

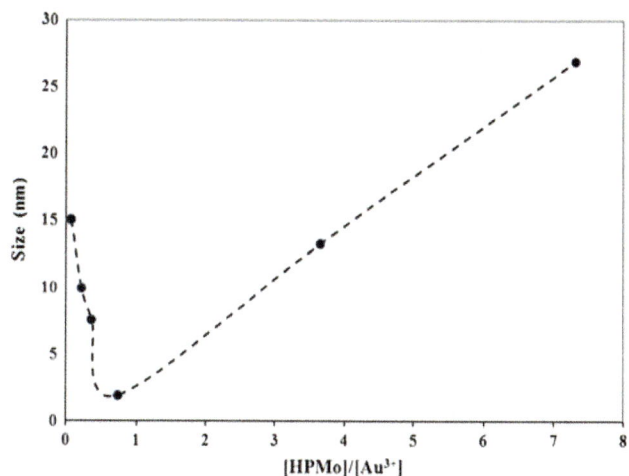

**Fig. 3** Effect of [HPMo]/[Au³⁺] ratio on the size of Au NPs ([Au³⁺]= 5×10⁻⁴ M, propan-2-ol=2 mL)

of type I polyanions are Keggins, Wells–Dawsons, and their derivatives that have one terminal oxygen atom M=O per each addenda atom. Type II polyanions can be represented by the Dexter–Silverton anion which has two terminal oxygens in *cis* positions on each addenda atom.

In type I octahedral $MO_6$, the lowest unoccupied molecular orbital (LUMO) is a nonbonding metal-centered orbital, whereas the LUMO for type II octahedral is antibonding with respect to the terminal M=O bonds. Consequently, type I polyoxometalates are reduced easily and often reversibly to form mixed-valence species, heteropoly blues, which can act as an oxidant. In contrast, type II polyoxometalates are reduced with more difficultly and irreversibly to complexes with yet unknown structures [34, 35]. For this reason, only type I heteropoly compounds, especially Keggins, are of interest for catalytic reactions. Therefore, $H_3[PMo_{12}O_{40}]$ with Keggin structure was selected since, to the best of our knowledge, the role of that has not been studied in the redox controlled synthesis of Au nanoparticles. Keeping in mind

that the introduction of vanadium (V) into the Keggin framework is beneficial for redox catalysis [36] and also it can shift its reactivity from acid-dominated to redox-dominated, we selected $H_{3+x}[PMo_{12-x}V_xO_{40}]$ (x=1–3).

The process was monitored by the visible absorption spectrometry. Figure 1 shows the UV/Vis spectra of the mixture at different treatment stages. It can be seen that primary solution does not have any distinct absorption band in the wavelength range of 400–800 nm. But, after 35 min, the absorption bands were observed in the SPR band of gold NPs at about 535 nm. These absorption bands caused by the excitation of surface–plasmon vibrations indicate formation of Au NPs. Furthermore, until 20 min irradiation, there is no absorption band at 535 nm, indicating that no nanoparticles were formed. In this time interval, the rate of Au NPs production reaction (Eq. 2) is negligible and Eq. 1 is in progress. From the figure, it can be observed that by increasing the time, the absorption band becomes sharper and the resonance intensity enhances due to the formation of Au NPs during the process.

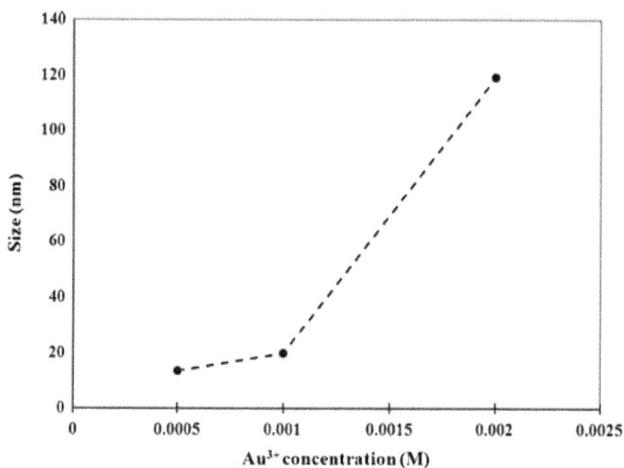

**Fig. 2** Effect of initial Au³⁺ concentration on the size of the synthesized Au NPs

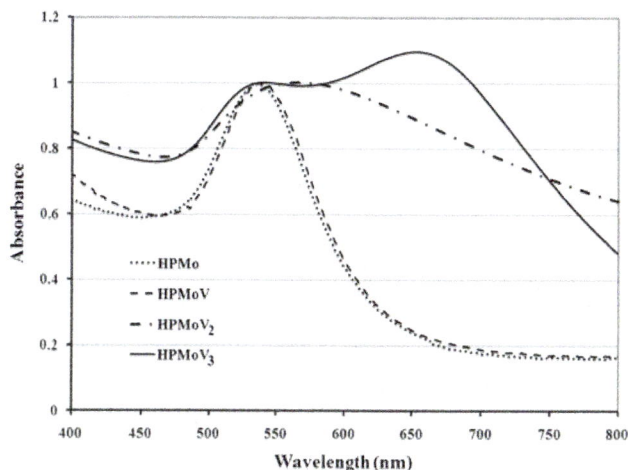

**Fig. 4** UV–vis spectra of Au³⁺/HPMoVx/propan-2-ol solutions. The spectra absorbances are normalized to unity

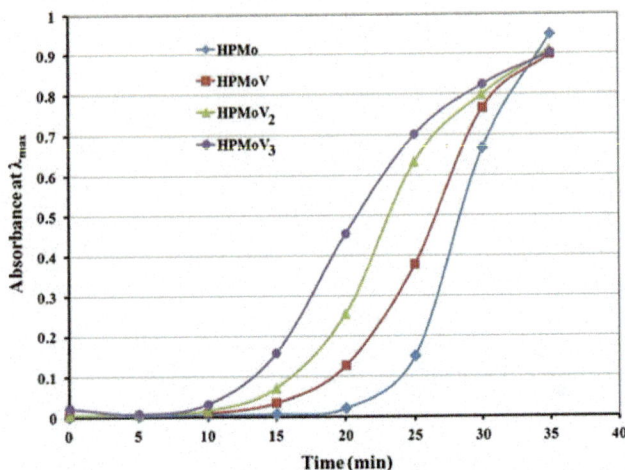

**Fig. 5** Synthesis rate of Au NPs in the presence of HPMoV$_x$ ($x$=0–3) ([Au3$^+$]=5×10$^{-4}$ M, HPMoV$_x$=5.5×10$^{-7}$ mol)

Besides the role as reducing agent, HPMo also plays the role of stabilizing agent in the above reactions. In our previous study, it was shown that in the absence of POMs, Au particles were precipitated in less than 2 days [32], but the resulting colloid in the presence of HPMo was stable without any precipitation for more than 3 months. It might be due to the adsorption of HPMo polyanions onto the surface of Au NPs which provide both steric stabilization and kinetic stabilization through coulombic repulsion between the negatively charged particles.

In the photolysis reaction, the propan-2-ol serves as a sacrificial agent for the photoformation of reduced HPMo, HPMo (e$^-$), which reacts with gold ions to produce Au NPs. A control experiment was performed in which 2 mL propan-2-ol was added to the deaerated aqueous solution of HAuCl$_4$ and irradiated for 6 h. There was no change in the color of solution and the characteristic gold absorption band was not observed. It indicates that the UV-irradiated propan-2-ol is not responsible for the reduction of Au$^{3+}$. On the other hand, our observations show that the amount of propan-2-ol affects the reaction rate which influences the size and uniformity of the synthesized Au NPs and by increasing the amount of the propan-2-ol, smaller and more uniform nanoparticles were obtained [15].

The rate of gold ions reduction (Eq. 2) affects strongly the initial nucleation and final size of nanoparticles. In fact, in this reaction, size control of Au NPs can be achieved via rate control of Eq. 2 by changing the experimental conditions. Faster reduction of gold ions leads to formation of smaller and more uniform nanoparticles. The initial Au$^{3+}$ ion concentration and also HPMo amount are two parameters which can influence the reaction rate in Eq. (2). Our findings show that increasing the initial Au$^{3+}$ concentration enhances reaction rate. As indicated in Fig. 2, an increase in concentration of gold ions results in the formation of larger nanoparticles. Also, our observations have shown that at higher gold ion concentrations, the stability of the prepared Au NPs decreases and they are precipitated after a short time. It might be due to (1) their bigger size or (2) increasing

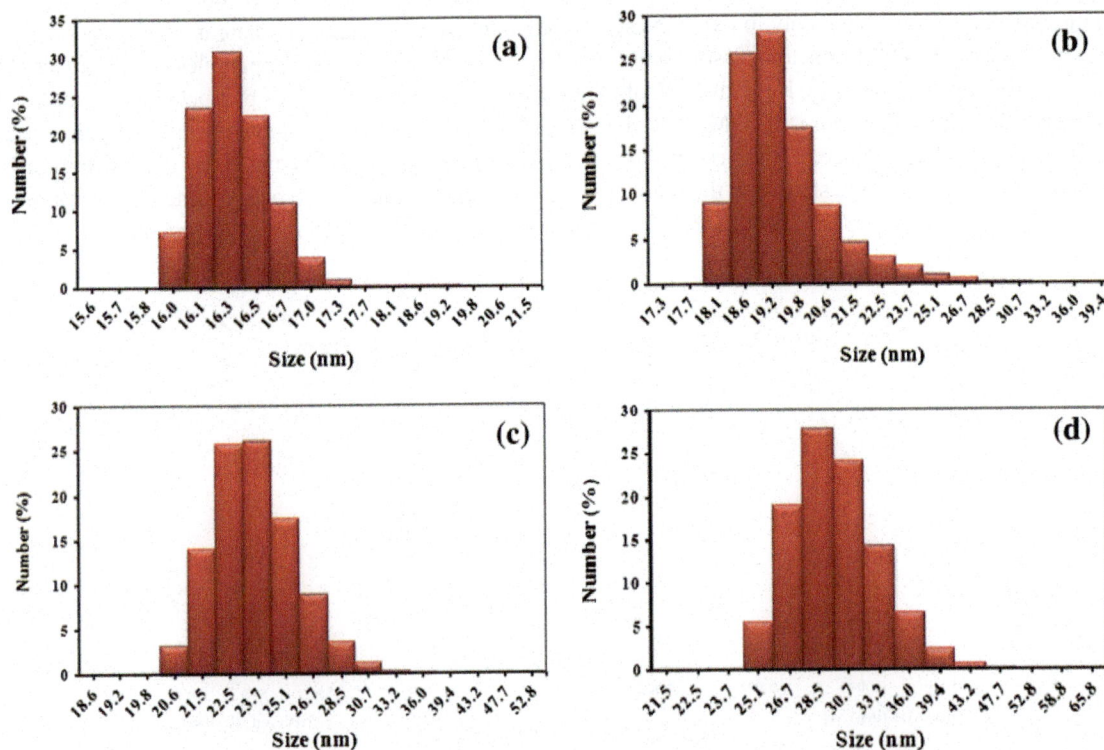

**Fig. 6** Effect of vanadium substitution on the size and uniformity of Au NPs using: **a** HPMo, **b** HPMoV, **c** HPMoV$_2$, **d** HPMoV$_3$ (HPMoV$_x$=5.5×10$^{-7}$ mol)

[Au$^{3+}$]/[HPMo] ratio in which the amount of HPMo might not be sufficient for Au NPs stabilization.

Also, we have found that the desired size of Au NPs can be achieved via changing the initial amount of HPMo. For this purpose, we have investigated the effect of [HPMo]/[Au$^{3+}$]=$\gamma$ on the size of synthesized gold NPs in which the initial concentration of Au$^{3+}$ was kept constant ($5 \times 10^{-4}$ M). The results are shown in Fig. 3. At low value of $\gamma$, less than 0.73, the mean diameter of Au NPs was decreased by increasing the $\gamma$ ratio. The fact that smaller Au NPs are formed with increasing the initial amount of HPMo implies that the nucleation process is enhanced more than the growth of nanoparticles.

Figure 3 also shows that by increasing the $\gamma$ above 0.73, the size of the synthesized NPs exhibited a contrary trend and larger NPs were formed through increasing the HPMo amount. The reason of the opposing trend of large Au NPs might be due to higher coverage of HPMo polyanions on the exterior surface of Au NPs at higher HPMo value that inhibits the reaction rate in Eq. 2.

For the HPMo value of $5.5 \times 10^{-6}$ M and 2 mL propan-2-ol, $\gamma$=0.73 acts as a critical amount in the synthesis of Au NPs in our experimental condition. This value depends on the type of metal ions, POM type, propan-2-ol amount and other operating conditions (temp., pH, ionic strength, etc.). This behavior is similar to that found in many chemical reduction approaches to nanosystems, because the nucleation and growth sequences are both affected by the relative concentrations of the reducing agent and the precursor [28].

Effect of vanadium substitution in HPMo

We have also investigated the effect of vanadium substitution in HPMo, i.e., HPMoV$_x$ ($x$=1–3), on the photosynthesis rate of Au NPs as well as the morphology of nanoparticles. Figure 4 shows the surface plasmon resonance spectra of the synthesized Au NPs solutions in the presence of HPMoV$_x$ ($x$=0–3) irradiated for 35 min. HPMoV$_x$ shift the peak of SPR bands gradually from 535 to 555 nm for $x$=2, which indicates Au NPs become larger using HPMoV$_2$.

When $x$=3, a new peak appears at ~690 nm, which is related to the synthesis of Au nanorods [7]. In fact, nanorods show two plasmon bands commonly ascribed to light

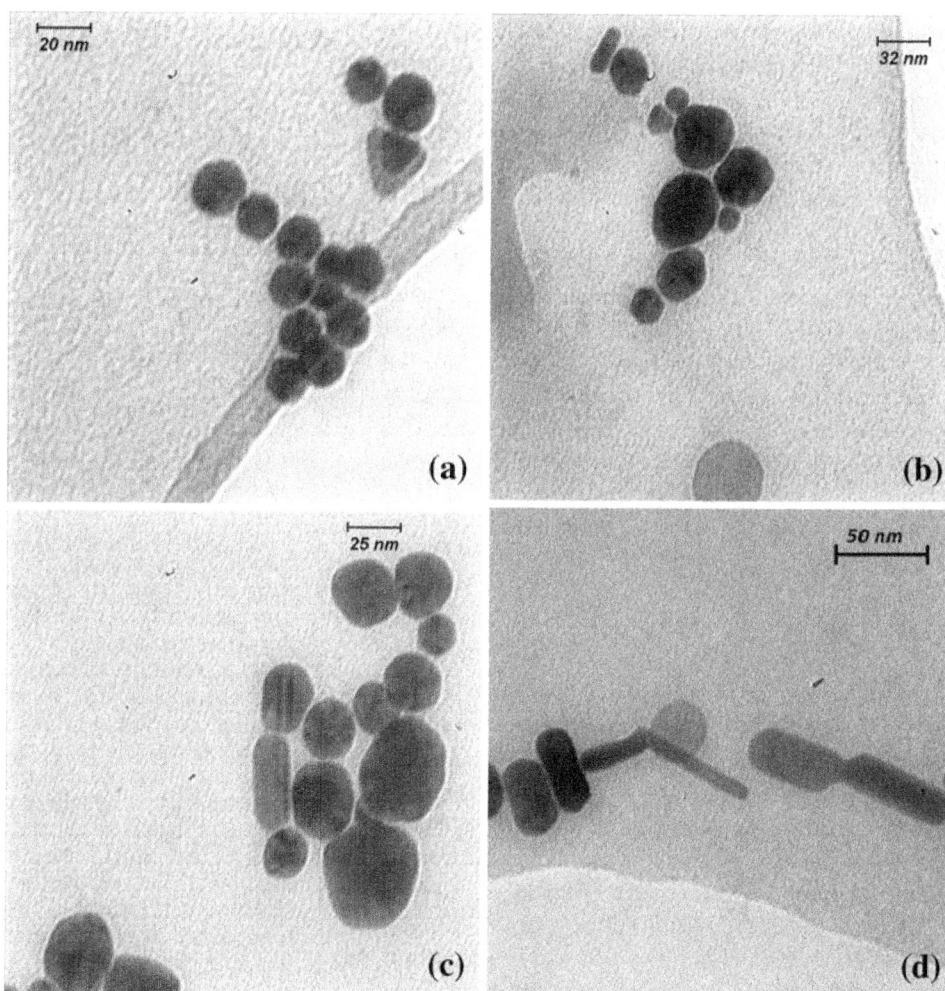

**Fig. 7** TEM images of synthesized Au NPs after 35 min irradiation using **a** HPMo, **b** HPMoV, **c** HPMoV$_2$, **d** HPMoV$_3$

absorption (and scattering) along both the long axis ("longitudinal plasmon band") and the short axis ("transverse plasmon band") of the colloid particles. As the aspect ratio increases, the position of the longitudinal plasmon band red shifts, and the transverse plasmon band position stay relatively invariable at ~520 nm [37]. The appearance of the longitudinal plasmon band at ~690 indicates the preparation of nanorods with aspect ratio of ~2 [7, 35].

Moreover, they affect the rate of synthesis reactions presented in Eqs. (1) and (2) (see Fig. 5). Faster kinetics can be observed by increasing the number of vanadium atoms. Figure 5 also demonstrates that increasing the number of vanadium atom substituted in HPMo ($x$ value) improve the redox potential of HPMo in the order of: $HPMoV_3$ $>HPMoV_2>HPMoV>HPMo$. Moreover, the synthesis reaction rate of Au NPs as well as the nucleation rate is enhanced in the same order, but larger NPs are produced. This fact is shown in Fig. 6a–d. The PSD analysis in these cases, demonstrate just the approximate size of synthesized nanoparticles. It is clear that at the same reaction conditions, the mean diameters of synthesized Au NPs become 16.3, 19.6, 23.8, and 30, by changing $x$ from 0 to 4, respectively.

Moreover, increasing $x$ has an interesting effect on the shape of prepared nanoparticles. Figure 7a shows TEM image of Au NPs synthesized using HPMo. The nanoparticles are seen to be hexagonal and spherical in shape. By substituting a vanadium atom in HPMo, a few anisotropic and irregularly shaped structures are observed (Fig. 7b). Also, by increasing the number of vanadium atom in HPMo, Au nanorods are formed, and for $x=3$ almost all hexagonal nanoparticles changed to nanorods. It has been also observed that all solutions including nanoparticles and nanorods are stable for few weeks. The TEM images confirm the PSD of the prepared Au nanoparticles using HPMo, but with increasing the $x$ value in the formula, the PSD histograms underestimate the size of true particles. It may be due to the formation of inhomogeneous nanoparticles in the case of $x=2$ and 3 and also nanorods in $x=3$. In fact, the PSD is unable to show a real distribution of Au nanorods.

## Conclusions

Molybdophosphoric acid and its vanadium-substituted products ($HPMoV_x$, $x=0$–3) were used as excellent photocatalysts, reducing agents, and stabilizers in the synthesis of gold nanoparticles. Uniform and size-controlled Au NPs were easily obtained by simple photolysis of $HPMoV_x$/ $Au^{3+}$/propan-2-ol solution at room temperature. Controlling the size of nanoparticles was achieved by changing the rate of $Au^{3+}$ reduction via variation of initial gold ions concentration and molar ratio of HPMo to gold ions. Faster reductions result in smaller and more uniform Au NPs as

exhibited by increasing the initial concentration of gold ions or the value of $x$. It was found that 0.73 is a critical ratio for $[HPMo]/[Au^{3+}]$, in which for its lower range, increasing the ratio leads to the formation of smaller nanoparticles and for its higher value the opposite trend is happened. Our findings have shown that the reduction of $Au^{3+}$ occurred in the order of: $HPMoV_3>HPMoV_2>HPMoV>HPMo$. It is suggested that, besides energy and composition of the LUMO, the presence of both Bronsted acidity and vanadium in the structure of mentioned heteropolyacids are responsible for catalytic activity. The greater protons number may lower the activation energy barrier, and the greater vanadium atoms may provide many sites for catalytic reaction. Also, by increasing the value of $x$ in the $HPMoV_x$ formula, spherical nanoparticles changed to nanorods.

## References

1. Ishida T, Haruta M (2007) Gold catalysts: towards sustainable chemistry. Angew Chem Int Ed 46(38):7154–7156
2. Castañeda MT, Merkoçi A, Pumera M, Alegret S (2007) Electrochemical genosensors for biomedical applications based on gold nanoparticles. Biosens Bioelectron 22(9–10):1961–1967
3. Rassaei L, Sillanpää M, French RW, Compton RG, Marken F (2008) Arsenite determination in the presence of phosphate at electro-aggregated gold nanoparticle deposits. Electroanalysis 20:1286–1292
4. Dubey S, Lahtinen M, Sillanpää M (2010) Tansy fruit mediated greener synthesis of silver and gold nanoparticles. Process Biochem 45:1065–1071
5. Dubey S, Lahtinen M, Särkkä H, Sillanpää M (2010) Bioprospective of *Sorbus acuparia* leaf extract in development of silver and gold nanocolloids. Colloids and Surfaces B: Biointerfaces 80:26–33
6. Troupis A, Triantis T, Hiskia A, Papaconstantinou E (2008) Rate-redox-controlled size-selective synthesis of silver nanoparticles using polyoxometalates. Eur J Inorg Chem 2008(36):5579–5586
7. Yu Y-Y, Chang S-S, Lee C-L, Wang CRC (1997) Gold nanorods: electrochemical synthesis and optical properties. J Phys Chem B 101(34):6661–6664
8. Jana NR, Gearheart L, Murphy CJ (2001) Evidence for seed-mediated nucleation in the chemical reduction of gold salts to gold nanoparticles. Chem Mater 13(7):2313–2322
9. Caruso RA, Ashokkumar M, Grieser F (2002) Sonochemical formation of gold sols. Langmuir 18(21):7831–7836
10. Kim F, Song JH, Yang P (2002) Photochemical synthesis of gold nanorods. J Am Chem Soc 124(48):14316–14317
11. Caixia K, Zhu X, Guanghou W (2006) Single-crystalline gold microplates: synthesis, characterization, and thermal stability. J Phys Chem B 110(10):4651–4656
12. Zhang G, Keita B, Biboum RN, Miserque F, Berthet P, Dolbecq A, Mialane P, Catala L, Nadjo L (2009) Synthesis of various crystalline gold nanostructures in water: the polyoxometalate β-

$[H_4PMo_{12}O_{40}]^{3-}$ as the reducing and stabilizing agent. J Mater Chem 19:8639–8644

13. Chen Y, Liew KY, Li J (2008) Size-controlled synthesis of Ru nanoparticles by ethylene glycol reduction. Mater Lett 62:1018–1021

14. Hostetler JM, Wingate EJ, Zhong JC, Harris EJ, Vachet RW, Clark RM (1998) Langmuir 14:17

15. Ayati A, Ahmadpour A, Bamoharram FF, Heravi MM, Rashidi H, Tanhaei B (2011) Application of molybdophosphoric acid in size-controlled synthesis of gold nanoparticles under UV irradiation. Int J Nanosci Nanotech 7(2):87–93

16. Bamoharram FF, Ahmadpour A, Heravi MM, Ayati A, Rashidi H, Tanhaei B (2011) Recent advances in application of polyoxometalates for the synthesis of nanoparticles. Synth React Inorg Met Org Chem.

17. Bamoharram FF (2011) Role of polyoxometalates as green compounds in recent developments of nanoscience. Synth React Inorg Met Org Chem 41(8):893–922

18. Troupis A, Hiskia A, Papaconstantinou E (2002) Synthesis of metal nanoparticles by using polyoxometalates as photocatalysts and stabilizers. Angew Chem Int Ed 41:1911–1913

19. Ayati A, Ahmadpour A, Bamoharram FF, Heravi MM, Rashidi H, Tanhaei B (2011) A new photocatalyst for preparation of silver nanoparticles and their photcatalysis of the decolorization of methyl orange. J Nanostruct Chem 2(1):15–22

20. Anastas PT, Warner JC (1998) Green chemistry: theory and practice. Oxford University Press, New York

21. Papaconstantinou E (1989) Photochemistry of polyoxometallates of molybdenum and tungsten and/or vanadium. Chem Soc Rev 18:1–31

22. Wang BE, Hu WC, Xu L (1998) Introduction to polyacid chemistry. Chemical Industry Press, Beijing, p 87

23. Weinstock AI (1998) Homogeneous-phase electron-transfer reactions of polyoxometalates. Chem Rev 98(1):113–170

24. Laurent R, Claire C-C, Sébastien S, Isabelle L (2008) Photocatalytic reduction of $Ag_2SO_4$ by Dawson-derived sandwich complex. Macromol Symp 270(1):117–122

25. Mandal S, Das A, Srivastava R, Sastry M (2005) Keggin ion mediated synthesis of hydrophobized Pd nanoparticles for multifunctional catalysis. Langmuir 21(6):2408–2413

26. Troupis A, Gkika E, Hiskia A, Papaconstantinou E (2006) Photocatalytic reduction of metals using polyoxometallates: recovery of metals or synthesis of metal nanoparticles. Comptes Rendus Chimie 9(5–6):851–857

27. Yang L, Shen Y, Xie A, Zhang B (2007) Facile size-controlled synthesis of silver nanoparticles in UV-irradiated tungstosilicate acid solution. J Phys Chem C 111(14):5300–5308

28. Triantis T, Troupis A, Gkika E, Alexakos G, Boukos N, Papaconstantinou E, Hiskia A (2009) Photocatalytic synthesis of Se nanoparticles using polyoxometalates. Catal Today 144(1–2):2–6

29. Mandal S, Selvakannan RP, Pasricha R, Sastry M (2003) Keggin ions as UV-switchable reducing agents in the synthesis of Au Core-Ag shell nanoparticles. J Am Chem Soc 125(28):8440–8441

30. Sanyal A, Mandal S, Sastry M (2005) Synthesis and assembly of gold nanoparticles in quasi-linear lysine-Keggin-ion colloidal particles. Adv Funct Mater 15(2):273–280

31. Niu C, Wu Y, Wang Z, Li Z, Li R (2009) Synthesis and shapes of gold nanoparticles by using transition metal monosubstituted heteropolyanions as photocatalysts and stabilizers. Frontiers of Chemistry in China 4(1):44–47

32. Ayati A, Ahmadpour A, Bamoharram FF, Heravi MM, Rashidi H (2011) Photocatalytic synthesis of gold nanoparticles using preyssler acid and their photocatalytic activity. Chin J Catal 32(6):978–982

33. Heravi MM, Benmorad T, Bakhtiari K, Bamoharram FF, Oskooie HH (2007) $H_{3+x}PMo_{12-x}V_xO_{40}$ (heteropolyacids)-catalyzed regioselective nitration of phenol to o-nitrophenol in heterogeneous system. J Mol Catal A: Chem 264(1–2):318–321

34. Pope MT (1983) Heteropoly isopoly oxometalates. Springer, Berlin

35. Pope MT, Müller A (1991) Polyoxometalate chemistry: an old field with new dimensions in several disciplines. Angew Chem Int Ed Eng 30(1):34–48

36. Mizuno N, Misono M (1994) Heteropolyanions in catalysis. J Mol Catal 86(1–3):319–342

37. Murphy CA, Thompson LB, Chernak DJ, Yang JA, Sivapalan ST, Boulos SP, Huang J, Alkilany AM, Sisco PN (2011) Gold nanorod crystal growth: from seed mediated synthesis to nanoscale sculpting. Current Opinion Colloid Interface Sci 16:128–134

# 9

# Gold nanoparticle-based fluorescence quenching via metal coordination for assaying protease activity

Se Yeon Park · So Min Lee · Gae Baik Kim · Young-Pil Kim

**Abstract** We report a gold nanoparticle (AuNP)-based fluorescence quenching system via metal coordination for the simple assay of protease activity. Carboxy AuNPs (5 nm in core diameter) functioned as both quenchers and metal chelators without requiring further modification with multidentate ligands; therefore, they were strongly associated with the hexahistidine regions of dye-tethered peptides in the presence of Ni(II) ions, leading to notable fluorescence quenching over the varying molar ratios of dye to AuNP. Upon the addition of matrix metalloproteinase-7 (MMP-7), the fluorescent intensity was efficiently recovered in one-pot mixture especially at 10:1–100:1 molar ratios of dye to AuNP. Consequently, the dequenching degree was dependent on the MMP-7 concentration in a hyperbolic manner, ranging from as low as 10 to 1,000 ngmL$^{-1}$. In this regard, we anticipate that the developed system will give us a general way to construct nanoparticle–dye conjugates and will find applications in the analyses of many other proteases mediating significant biological processes with low background and high sensitivity.

Se Yeon Park, So Min Lee and Gae Baik Kim contributed equally to this work.

S. Y. Park
Department of Chemical Engineering, Hanyang University, Seoul 133-791, South Korea

S. M. Lee
Department of Bio Engineering, Hanyang University, Seoul 133-791, South Korea

G. B. Kim · Y.-P. Kim (✉)
Department of Life Science, Hanyang University, Seoul 133-791, South Korea
e-mail: ypilkim@hanyang.ac.kr

G. B. Kim · Y.-P. Kim
Research Institute for Natural Sciences, Hanyang University, Seoul 133-791, South Korea

**Keywords** Gold nanoparticle · Quenching · Metal coordination · Protease · Matrix metalloproteinase

## Introduction

The interactions between gold nanoparticles (AuNP) and organic dyes have gained considerable interest in biochemical assay because they provide many advantages regarding quenching efficiency and photostability over the classical dye quencher system [1–6]. The ability of AuNPs to induce fluorescence quenching of proximal dyes is reported to be directed by a surface energy transfer process [7–9]; the rate of energy transfer from a dye to AuNP depends on the inverse of fourth power of the donor–acceptor separation, which triggers a much longer working distance (up to 22 nm) than that observed in a traditional fluorescence resonance energy transfer system (up to 10 nm, due to the inverse sixth power distance dependency). To this end, AuNP–organic dye couples have been implemented for the highly sensitive detection of oligonucleotides [10–12], proteins [13–17], and other small molecules [18–20].

Recently, the activities of enzymes such as proteases and nucleases have also been analyzed using activatable "switch-on" fluorescent nanoprobes in order to gain some insight into enzyme kinetics or biological activity [21–23]. In particular, proteases have been recognized as important targets due to their roles that are involved in multiple processes during malignant progression, including tumor angiogenesis, invasion, and metastasis [24, 25]. Protease-detecting methods, therefore, have been accomplished by incorporating relatively stable peptides between the AuNP and the dye via either biotin–avidin interaction [22, 26] or thiol-mediated coupling [27–29]. In addition to these conjugations, nickel–nitrilotriacetic acid (Ni(II)–NTA)-modified AuNPs were recently demonstrated by several groups [30, 31] since Ni(II)–NTA provides high binding affinity ($K_d = 10^{-13}$ M) for a hexahistidine tag at pH 8.0 [32], which has been widely used as one of the most useful affinity methods. However, these Ni affinity nanoconjugates have been only limited to capture or label proteins with polyhistidine tags, and there was little attempt for enzyme activity study. As an alternative, to enable an easy surface modification based on Ni affinity, Rao's group reported that NTA-free carboxy quantum dot (QD) could be conjugated to his-tagged luciferases in the presence of nickel ions, which developed a QD–bioluminescence resonance energy transfer system to assay protease activity [33]. On the basis of this observation, we envisioned that the use of his-tag-containing peptide affinity tag would allow a site-specific and multivalent conjugation to carboxy AuNPs and would generate a shorter distance between the AuNP and the organic dye, which is favorable for higher energy transfer efficiency regime than that of the biotin–avidin strategy.

Here, we demonstrate a simple fluorescence quenching system using carboxy AuNPs and dye-conjugated peptides and its application to protease assay. A facile conjugation of a dye-coupled peptide to the carboxy AuNP was made possible in the presence of Ni; the resultant AuNP–dye conjugate via metal affinity was used for the detection of matrix metalloproteinase (MMP) activity. We chose MMP as a model protease because MMPs play a crucial role in a wide variety of processes including tumor metastasis, inflammation, growth differentiation, and cell signaling [34–36]. To achieve the optimal fluorescence quenching and dequenching system by protease activity, the quenching efficiency and protease-induced recovery yield of AuNPs toward an organic dye (5(6)-carboxytetramethylrhodamine, TAMRA) was compared in terms of the dye-to-AuNP ratio. Details are reported herein.

## Experimental section

### Materials

Nickel(II) chloride hexahydrate (99.9 %, $NiCl_2 \cdot 6H_2O$), hydrogen tetrachloroaurate(III) trihydrate (99.9 % $HAuCl_4 \cdot 3H_2O$), sodium citrate dihydrate (trisodium salt, $C_6H_5Na_3O_7 \cdot 2H_2O$), and sodium borohydride (99 %, $NaBH_4$) were purchased from Sigma-Aldrich. Carboxy-$PEG_{12}$-thiol and methyl-$PEG_4$-thiol were purchased from Thermo Scientific. Active matrix metalloproteinase-7 (MMP-7) enzyme was purchased from Merck4Biosciences. The TAMRA-labeled peptide (TAMRA-GPLGMRGLHHHHHH) was synthesized from Peptron, Inc. (Korea). All chemicals were of analytical grade and were used as received.

### Synthesis of AuNPs

AuNPs were synthesized by reduction and stabilization with citrate. Briefly, 100 µL of a stock solution containing 300 mM of $HAuCl_4 \cdot 3H_2O$ was added to 100 mL of distilled water to give a final concentration of 300 µM followed by vigorous stirring. To this solution, 2 mL of 30 mM sodium citrate dihydrate was added at a final concentration of 600 µM (the molar ratio of tetrachloroaurate to sodium citrate is 1:2) and stirred. For the fast reduction and formation of gold colloids, 100 µL of a stock solution containing 300 mM of $NaBH_4$ was quickly added to the reaction solution followed by stirring. The clustering of AuNPs was checked by UV–Visible spectroscopy (Cary 60 UV-Vis, Agilent Technologies), and the average size of AuNPs was estimated to be $5.1 \pm 1.4$ nm ($n = 100$) using a field emission transmission electron microscope (FE-TEM; FEI Tecnai G2F30S-TWIN, the Netherlands). Surface modification of the synthesized AuNPs was performed with the 1:1 mixture of methyl-$PEG_4$-thiol and carboxy-$PEG_{12}$-thiol (total, 100 µM), which was added to the citrate-stabilized AuNP solution (final, 50 nM). Since the used 5-nm AuNP is estimated to have 3,858 Au atoms in total and 984 Au atoms at its surface based on the reported calculation method [37, 38], the 2,000:1 ligand-to-AuNP molar ratio was used to ensure the complete surface modification of AuNPs. After 2 h of incubation under convection, the carboxy-modified AuNPs were purified using an Amicon® Ultra Centrifugal Filter Unit (50 kDa, MWCO) and centrifugation (8,000×$g$ for 10 min). The final concentration of the AuNPs in solution was calculated using the molar extinction coefficient ($1.2 \times 10^7 M^{-1} cm^{-1}$) at 520 nm.

### Analysis of fluorescence quenching

For quenching experiments, the TAMRA peptide (2.5 µL at 10 µM) was mixed with varying amounts of carboxy AuNPs (2.5–25 µL at 1 µM) at a 100:1–1:1 ratio of the TAMRA peptide and AuNP in the absence or presence of $NiCl_2$ (10 µL at 1 mM). All reactions were performed at a final volume of 100 µL in 20 mM Tris buffer (pH 7.5) at RT. After 30 min of incubation, the fluorescence spectra were measured at an excitation wavelength of 550 nm using a spectrofluorometer (FS-2, Sinco, South Korea). We initially tested the self-quenching and detection range of the TAMRA-conjugated peptide. Self-quenching was significant at more than 10 µM TAMRA; therefore, the final concentration of the TAMRA-conjugated peptide in this study was determined to be 250 nM.

### Protease assay

In a one-pot method, AuNPs (2.5 or 5 µL at 1 µM), TAMRA peptide (2.5 µL at 10 µM), $NiCl_2$ (10 µL at

1 mM), and MMP-7 protease (10 μL at different stock concentrations) were mixed at a time in 20 mM Tris buffer (pH 7.5) to give a final volume of 100 μL and incubated at 37 °C for 2 h. It was followed by monitoring the emission spectra of the solution using a spectrofluorometer. In a two-step method, TAMRA peptide (2.5 μL at 10 μM), MMP-7 (10 μL at different stock concentrations), and 20 mM Tris buffer (77.5 μL) were initially mixed and incubated at 37 °C for 2 h, followed by the addition of AuNPs (2.5 or 5 μL at 1 μM) and NiCl$_2$ (10 μL at 1 mM). After additional incubation at room temperature for 30 min, the AuNP mixture was subjected to fluorescence scanning. Fluorescence intensity was normalized to the background intensity from the control solution without protease.

## Results and discussion

To construct an efficient quenching system, peptide substrates comprising red dyes (TAMRA) at their N-termini and hexahistidines at their C-termini were mixed with carboxy AuNPs in the presence of Ni(II) ions (Scheme 1a). As a consequence, a dye-to-AuNP quenching was induced by a strong association between polyhistidine residues of the TAMRA peptide and the carboxy groups of the AuNPs via the coordination of Ni(II) metal ions (electron pair acceptors; Scheme 1b). Although common metal-chelating agents

including NTA, iminodiacetic acid (IDA), carboxymethylated aspartic acid, and tris(carboxymethyl)ethylenediamine are available and being widely used for binding polyhistidine tag [39], the chelator-free metal affinity here can be achieved by only the surface-exposed carboxyl groups on the AuNPs. Since highly compacted carboxyl groups in the nanostructured surface can function like multidentate chelators, the binding affinity of Ni(II)-his-tagged carboxy AuNP is likely to be comparable to that of chelator-mediated conjugation (e.g., Ni(II)-his-tagged NTA; our unpublished data), which allowed for the higher quenching efficiency of the dye to AuNP due to their close proximity. In addition, site-specific conjugation and the simplicity of Ni(II)-his-tagged carboxy AuNP were further advantageous for protease assay. Based on this conjugation principle, fluorescence quenching and dequenching were strongly induced in the absence and the presence of protease, respectively (Scheme 1c, d).

The carboxy AuNPs were synthesized from citrate-stabilized AuNPs by conjugating carboxy-PEG-thiol and were characterized using a UV–Vis spectrophotometer and FE-TEM, which represented a strong surface plasmon resonance band near 520 nm and around 5 nm in diameter (Fig. 1). The extinction and fluorescent emission spectra of the TAMRA dye were also displayed in Fig. 1a. To check the quenching efficiency of the AuNP, different concentrations of the carboxy AuNPs, while maintaining the concentration of the

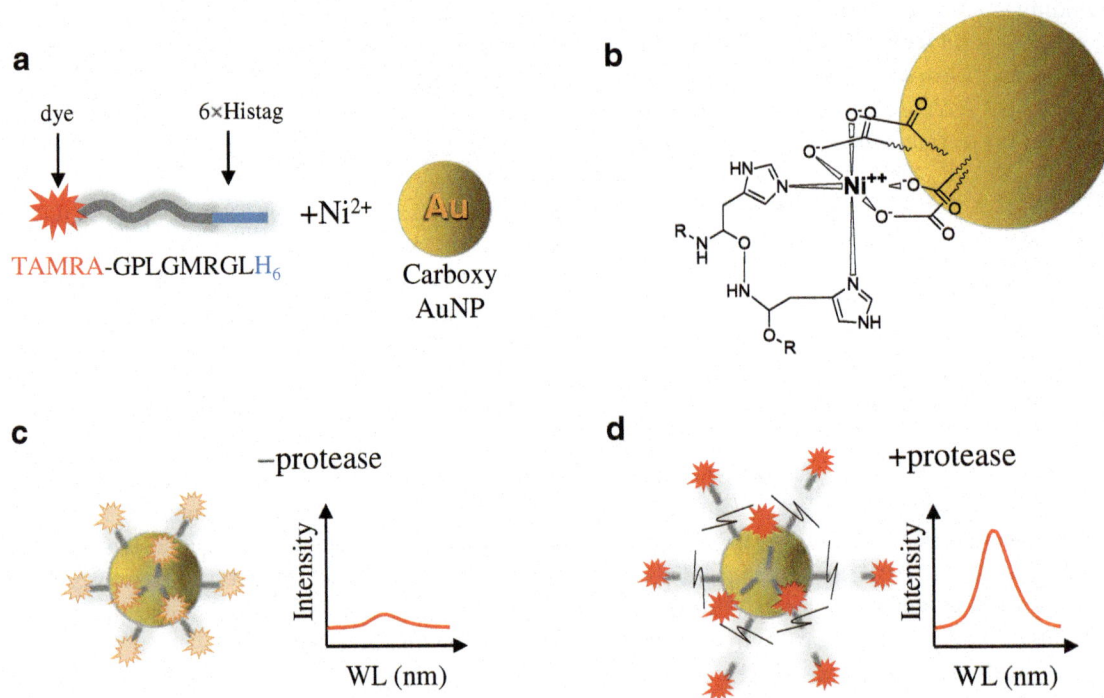

Scheme 1 **a** Schematic representation of the component of the AuNP–dye conjugate for protease assay. **b** Schematic of the coordination of the Ni(II) ion with the histidines of peptides and the carboxyl groups on the AuNP. **c** Resultant fluorescence quenching. **d** Dequenching of the dye-conjugated AuNPs illustrated in the absence and the presence of protease

**Fig. 1** **a** Normalized extinction spectra of AuNPs (*black solid*) and the TAMRA dye (*black dotted*) and emission spectrum for the TAMRA dye (*red dashed*) showing considerable overlap of AuNP extinction and TAMRA emission. **b** High-resolution TEM image of carboxy AuNPs with a diameter of $5.1 \pm 1.4$ nm ($n=100$)

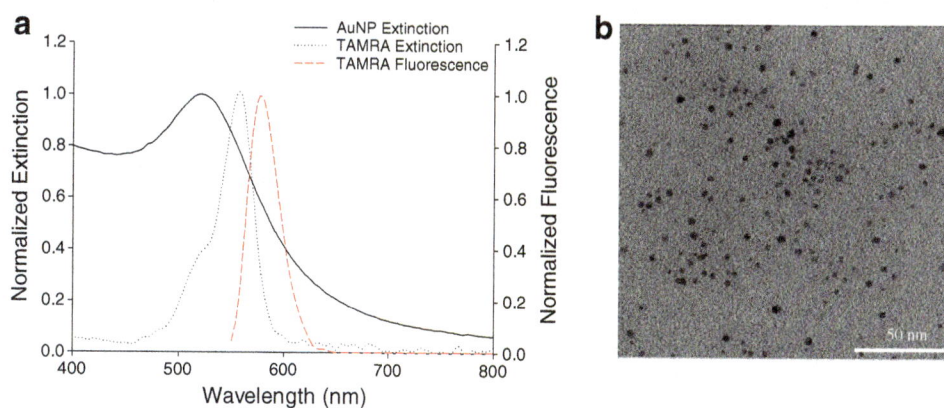

TAMRA peptide constant, were added to the TAMRA peptide ($\text{TAMRA-GPLGMRGLH}_6$) in the absence or the presence of Ni. As shown in Fig. 2a, the fluorescence intensity declined as the TAMRA-to-AuNP molar ratio decreased from 100:1 to 1:1 even in the absence of Ni. Since the amount of TAMRA was fixed at varied concentrations of AuNPs, the changes in fluorescence intensity were attributed to a quenching effect by the AuNPs. However, this result is probably due to a dynamic collisional quenching effect rather than an affinity-induced one because Ni-free TAMRA peptide can also be adsorbed on the AuNP by a diffusion-driven electrostatic interaction. In contrast, the addition of Ni facilitated a strong quenching effect by the proximate conjugation between the AuNP and dye, leading to a relatively large decrease in fluorescence intensity over all molar ratios (Fig. 2b). It was shown in Fig. 2c that the Ni(II) ion induced very effective quenching between the TAMRA peptide ($\text{His}_6$) and the carboxy AuNP (Fig. 2c) compared to the TAMRA dye without AuNPs.

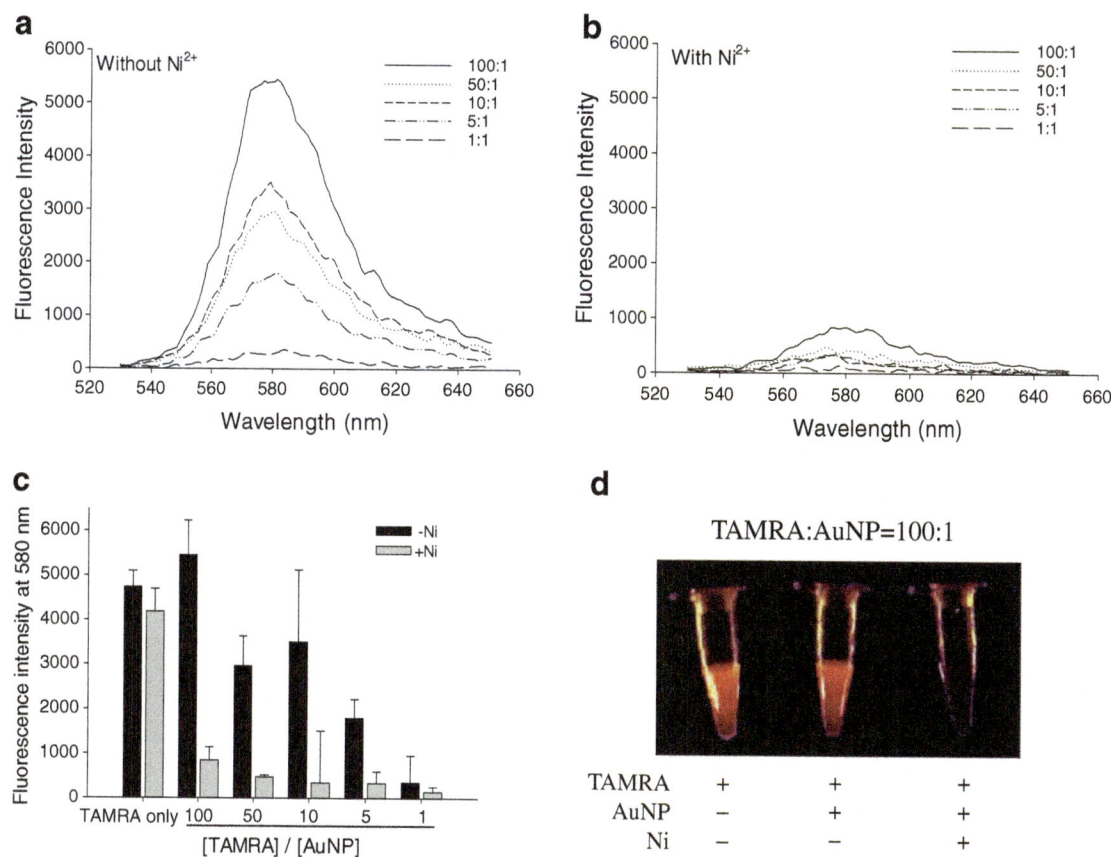

**Fig. 2** Fluorescence spectra of the TAMRA peptide at different concentrations of AuNPs in the absence (**a**) or the presence (**b**) of Ni. The molar ratios of TAMRA to AuNP were varied from 100:1 to 1:1 (from *top* to *bottom*). **c** Maximal fluorescence intensities at 580 nm from (**b**) and (**c**) were compared as *bar graphs* in the absence or the presence of Ni. **d** Fluorescent images of the TAMRA peptide with and without AuNP or Ni

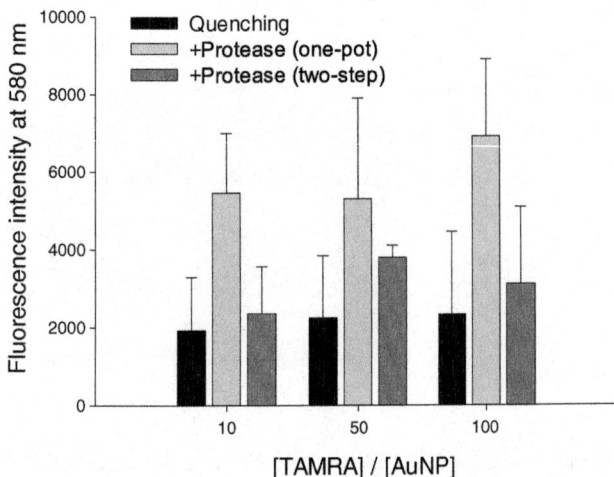

**Fig. 3** Changes in the fluorescence intensities of the AuNP–quenched TAMRA conjugate before and after enzyme reaction at different molar ratios of TAMRA to AuNP. Protease reaction was performed in two methods: one-pot (*light gray*) and two-step reaction (*dark gray*). The enzyme (MMP-7) concentration was 1 μgmL$^{-1}$

Fluorescent images also represented the effect of Ni addition to the AuNP-based fluorescence quenching (Fig. 2d). This result strongly indicates that the his-tagged-dye and the carboxy AuNP were tightly associated via nickel coordination, giving rise to a higher fluorescence quenching. The quenching efficiency was calculated using the following equation: $100 \times \left(1 - F_{\text{Ni/AuNP addition}}/F_{\text{Ni/AuNP-free}}\right)$, where $F_{\text{Ni/AuNP addition}}$ is the fluorescence intensity of the TAMRA peptide in the presence of AuNP and Ni and $F_{\text{Ni/AuNP-free}}$ is the fluorescence intensity of the TAMRA peptide in the absence of AuNP and Ni. Particularly, the most significant difference in fluorescence intensity before and after the addition of Ni(II) ion was observed at a 100:1 ratio of TAMRA to AuNP, where the quenching

efficiency is 82.2 %; in other cases, the corresponding quenching efficiencies were 90.0 % for 50:1, 92.6 % for 10:1, 92.8 % for 5:1, and 96.9 % for 1:1. It is worth noting that this quenching efficiency at a 100:1 ratio was much higher than that observed in the direct adsorption of rhodamine dye to the citrate-capped AuNPs [40], supporting the metal affinity interactions of carboxyl AuNPs in the present study.

To gain some insight into the dequenching effect by protease activity, enzyme reaction was attempted with varying ratios of the dye to AuNP in two different ways: a one-pot reaction (all components were mixed at a time) and a two-step reaction (the TAMRA peptide initially reacted with the protease, followed by the addition of other components). MMP-7 was employed as a model protease. As shown in Fig. 3, a strong recovery of fluorescence intensity was observed for 10:1–100:1 quenched solutions by the one-pot enzyme reaction, where the signal intensity increased by 2.8-fold (10:1), 2.4-fold (50:1), and 3.0 fold (100:1) to the quenching state (black and light gray bars in Fig. 3). Compared to that in Fig. 2c, the background intensity in Fig. 3 slightly increased after the quenched solution was subjected to the enzyme reaction condition (2 h at 37 °C). Importantly, the one-pot reaction was found to be much more efficient than the two-step reaction over the differing TAMAR-to-AuNP ratios, indicating that freely moving TAMRA peptide (His$_6$) in the initial step of the two-step method is expected to be either not much cleaved by the protease or induce high nonspecific binding to the AuNP after cleavage. It is postulated that the one-pot method enabled histidines to be captured initially by the carboxy AuNP in the presence of Ni$^{2+}$, providing the optimal orientation and structural stability of the peptide–AuNP complex for protease reaction. Unlike the one-pot reaction, when the

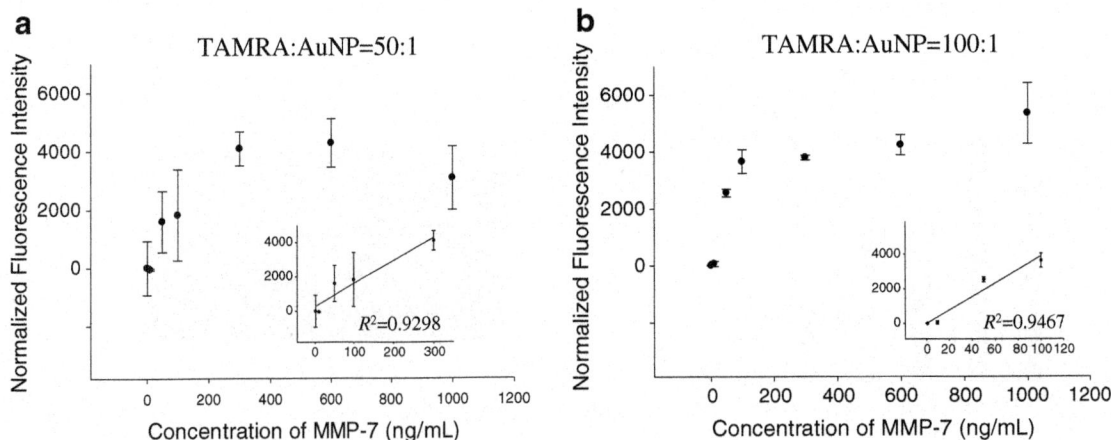

**Fig. 4** Plot of the fluorescence intensity of the TAMRA peptide (His$_6$)/ Ni(II)/AuNP as a function of MMP-7 concentration (1–1,000 ngmL$^{-1}$) at different ratios of TAMRA/AuNP: 50:1 (**a**) and 100:1 (**b**). Peak intensities at 580 nm were normalized to the control set without MMP-

7. *Error bars* represent the standard deviation from two repeated experiments. The *inset* indicates the linearity between the fluorescence and MMP-7 concentration over the different dynamic ranges

pre-quenched probe was subjected to the same enzyme reaction, no significant signal recovery was observed (data not shown). Despite the similarity in fluorescence recovery, in the case of 100:1, a high binding number and the close packing density of TAMRA peptides on the AuNP surface appear to allow the recovery yield to be slightly improved. These results suggest that the mixing type and timing between the AuNP reactant and enzyme would be very critical for the enzyme reaction, and the all-in-one reaction would be very suited to analyze the protease activity in terms of saving detection time.

To check for enzyme-dependent signal intensity in this system, the protease activity was monitored as a function of the MMP-7 concentration (Fig. 4). When the dequenching intensity was normalized to the control set in the absence of MMP-7, a hyperbolic curve was similarly observed both at a 50:1 (Fig. 4a) and at a 100:1 (Fig. 4b) ratio of the dye to AuNP, ranging from as low as 10 to 1,000 $ngmL^{-1}$ in terms of enzyme concentration. Although there was a slight decrease at a high concentration of MMP-7 in the case of a 50:1 ratio of dye to AuNP, the signal recovery showed a plateau after 300 $ngmL^{-1}$, which corresponds to approximately 70 % of the maximum intensity of the TAMRA peptide displayed in Fig. 2a. This reveals that all of the peptides were not likely to be fully cleaved by MMP-7 at a high concentration. Additionally, there was a considerable linearity in the 50:1 ($R^2=0.9298$ for 10–300 $ngmL^{-1}$) and 100:1 ratios ($R^2=0.9467$ for 10–100 $ngmL^{-1}$), where the 100:1 ratio condition showed a relatively improved sensitivity and reproducibility over the tested range based on the standard deviation. The detection sensitivity was comparable to those of other assay systems reported previously [22, 41]. Although the dynamic range seems to cover only one order range of the MMP-7 concentration, this result indicates that our developed system is well suited to detect the low concentration range of MMPs. While the AuNP-based colorimetric assay has been well developed [42], such AuNP-based fluorescence detection can offer greater sensitivity in terms of targeting DNA and proteins.

The AuNP-quenched strategy presented in this study has several advantages over the conventional dye-to-quencher system. In addition to the superior quenching effect of AuNPs, the simple and easy fabrication of fluorophore-tethered peptides to the carboxy AuNPs via metal affinity can be achieved without requiring further complicated modifications of the AuNP by multidentate ligands, such as NTA and IDA. This strategy, therefore, enables fluorescent proteins fused to peptides and polyhistidine to be simply conjugated to the AuNP surface. Moreover, since the AuNPs can be generally employed as common quenchers, several fluorophores with different colors could be applied to the AuNPs for a multiplex assay with extremely low background signal.

## Conclusion

In conclusion, we demonstrated the simple assay of protease activity using the AuNP-based fluorescence quenching system via metal affinity. Simple and rapid association between the carboxy groups of AuNPs and the hexahistidine regions of the dye-tethered peptides was observed in the presence of Ni(II) ions, leading to notable fluorescence quenching over varying molar ratios (100:1–1:1) of the dye to AuNP. When MMP-7 was added to the AuNP–dye solution, significant fluorescence dequenching was found, especially at 10:1–100:1 dye-to-AuNP molar ratios, where the detection limit was as low as 10 $ngmL^{-1}$. By combining fluorophores with different colors, this developed system will have great potential to study the protease activity with low background and high sensitivity.

**Acknowledgment** This work was supported by the Basic Science Research Program (2012-0008222), the Bio-Signal Analysis Technology Innovation Program (2012-0006053), and the Nano Material Technology Development Program (2012035286) through the National Research Foundation of Korea (NRF) funded by the Ministry of Education, Science and Technology.

## References

1. Dubertret B, Calame M, Libchaber AJ (2001) Single-mismatch detection using gold-quenched fluorescent oligonucleotides. Nat Biotechnol 19(7):680–681
2. Dulkeith E, Morteani AC, Niedereichholz T, Klar TA, Feldmann J, Levi SA, van Veggel FCJM, Reinhoudt DN, Moller M, Gittins DI (2002) Fluorescence quenching of dye molecules near gold nanoparticles: radiative and nonradiative effects. Phys Rev Lett 89(20):203002
3. Maxwell DJ, Taylor JR, Nie SM (2002) Self-assembled nanoparticle probes for recognition and detection of biomolecules. J Am Chem Soc 124(32):9606–9612
4. Rosi NL, Mirkin CA (2005) Nanostructures in biodiagnostics. Chem Rev 105(4):1547–1562
5. Bunz UHF, Rotello VM (2010) Gold nanoparticle–fluorophore complexes: sensitive and discerning "noses" for biosystems sensing. Angew Chem Int Edit 49(19):3268–3279
6. Acuna GP, Bucher M, Stein IH, Steinhauer C, Kuzyk A, Holzmeister P, Schreiber R, Moroz A, Stefani FD, Liedl T, Simmel FC, Tinnefeld P (2012) Distance dependence of single-fluorophore quenching by gold nanoparticles studied on DNA origami. ACS Nano 6(4):3189–3195
7. Sen T, Sadhu S, Patra A (2007) Surface energy transfer from rhodamine 6G to gold nanoparticles: a spectroscopic ruler. Appl Phys Lett 91(4):2762283
8. Yun CS, Javier A, Jennings T, Fisher M, Hira S, Peterson S, Hopkins B, Reich NO, Strouse GF (2005) Nanometal surface

energy transfer in optical rulers, breaking the FRET barrier. J Am Chem Soc 127(9):3115–3119

9. Jennings TL, Singh MP, Strouse GF (2006) Fluorescent lifetime quenching near *d*=1.5 nm gold nanoparticles: probing NSET validity. J Am Chem Soc 128(16):5462–5467

10. Kim JH, Estabrook RA, Braun G, Lee BR, Reich NO (2007) Specific and sensitive detection of nucleic acids and RNases using gold nanoparticle–RNA–fluorescent dye conjugates. Chem Commun 42:4342–4344

11. Obliosca JM, Wang PC, Tseng FG (2012) Probing quenched dye fluorescence of Cy3–DNA–Au-nanoparticle hybrid conjugates using solution and array platforms. J Colloid Interface Sci 371:34–41

12. Wang WJ, Chen CL, Qian MX, Zhao XS (2008) Aptamer biosensor for protein detection using gold nanoparticles. Anal Biochem 373 (2):213–219

13. Kim YP, Oh YH, Kim HS (2008) Protein kinase assay on peptide-conjugated gold nanoparticles. Biosens Bioelectron 23(7):980–986

14. Mayilo S, Kloster MA, Wunderlich M, Lutich A, Klar TA, Nichtl A, Kurzinger K, Stefani FD, Feldmann J (2009) Long-range fluorescence quenching by gold nanoparticles in a sandwich immunoassay for cardiac troponin T. Nano Lett 9(12):4558–4563

15. Guirgis BSS, Cunha CSE, Gomes I, Cavadas M, Silva I, Doria G, Blatch GL, Baptista PV, Pereira E, Azzazy HME, Mota MM, Prudencio M, Franco R (2012) Gold nanoparticle-based fluorescence immunoassay for malaria antigen detection. Anal Bioanal Chem 402(3):1019–1027

16. Hu PP, Chen LQ, Liu C, Zhen SJ, Xiao SJ, Peng L, Li YF, Huang CZ (2010) Ultra-sensitive detection of prion protein with a long range resonance energy transfer strategy. Chem Commun 46 (43):8285–8287

17. Kim GB, Kim YP (2012) Analysis of protease activity using quantum dots and resonance energy transfer. Theranostics 2 (2):127–138

18. Lee H, Lee K, Kim IK, Park TG (2008) Synthesis, characterization, and in vivo diagnostic applications of hyaluronic acid immobilized gold nanoprobes. Biomaterials 29(35):4709–4718

19. Chen WY, Lan GY, Chang HT (2011) Use of fluorescent DNA-templated gold/silver nanoclusters for the detection of sulfide ions. Anal Chem 83(24):9450–9455

20. Jin LH, Shang L, Guo SJ, Fang YX, Wen D, Wang L, Yin JY, Dong SJ (2011) Biomolecule-stabilized Au nanoclusters as a fluorescence probe for sensitive detection of glucose. Biosens Bioelectron 26(5):1965–1969

21. Huang Y, Zhao SL, Liang H, Chen ZF, Liu YM (2011) Multiplex detection of endonucleases by using a multicolor gold nanobeacon. Chem-Eur J 17(26):7313–7319

22. Kim YP, Oh YH, Oh E, Ko S, Han MK, Kim HS (2008) Energy transfer-based multiplexed assay of proteases by using gold nanoparticle and quantum dot conjugates on a surface. Anal Chem 80 (12):4634–4641

23. Swierczewska M, Lee S, Chen XY (2011) The design and application of fluorophore-gold nanoparticle activatable probes. Phys Chem Chem Phys 13(21):9929–9941

24. Welser K, Adsley R, Moore BM, Chan WC, Aylott JW (2011) Protease sensing with nanoparticle based platforms. Analyst 136 (1):29–41

25. Turk B (2006) Targeting proteases: successes, failures and future prospects. Nat Rev Drug Discov 5(9):785–799

26. Lowe SB, Dick JAG, Cohen BE, Stevens MM (2012) Multiplex sensing of protease and kinase enzyme activity via orthogonal coupling of quantum dot peptide conjugates. ACS Nano 6 (1):851–857

27. Mu CJ, LaVan DA, Langer RS, Zetter BR (2010) Self-assembled gold nanoparticle molecular probes for detecting proteolytic activity in vivo. ACS Nano 4(3):1511–1520

28. Free P, Shaw CP, Levy R (2009) PEGylation modulates the interfacial kinetics of proteases on peptide-capped gold nanoparticles. Chem Commun 33:5009–5011

29. Lee S, Cha EJ, Park K, Lee SY, Hong JK, Sun IC, Kim SY, Choi K, Kwon IC, Kim K, Ahn CH (2008) A near-infrared-fluorescence-quenched gold-nanoparticle imaging probe for in vivo drug screening and protease activity determination. Angew Chem Int Edit 47(15):2804–2807

30. Hainfeld JF, Liu WQ, Halsey CMR, Freimuth P, Powell RD (1999) Ni–NTA–gold clusters target his-tagged proteins. J Struct Biol 127 (2):185–198

31. Swartz JD, Gulka CP, Haselton FR, Wright DW (2011) Development of a histidine-targeted spectrophotometric sensor using Ni(II)NTA-functionalized Au and Ag nanoparticles. Langmuir 27(24):15330–15339

32. Hochuli E, Bannwarth W, Dobeli H, Gentz R, Stuber D (1988) Genetic approach to facilitate purification of recombinant proteins with a novel metal chelate adsorbent. Nat Biotechnol 6(11):1321–1325

33. Yao HQ, Zhang Y, Xiao F, Xia ZY, Rao JH (2007) Quantum dot/bioluminescence resonance energy transfer based highly sensitive detection of proteases. Angew Chem Int Edit 46(23):4346–4349

34. Sternlicht MD, Werb Z (2001) How matrix metalloproteinases regulate cell behavior. Annu Rev Cell Dev Biol 17:463–516

35. Egeblad M, Werb Z (2002) New functions for the matrix metalloproteinases in cancer progression. Nat Rev Cancer 2(3):161–174

36. Page-McCaw A, Ewald AJ, Werb Z (2007) Matrix metalloproteinases and the regulation of tissue remodelling. Nat Rev Mol Cell Biol 8(3):221–233

37. Cumberland SL, Strouse GF (2002) Analysis of the nature of oxyanion adsorption on gold nanomaterial surfaces. Langmuir 18 (1):269–276

38. Lewis DJ, Day TM, MacPherson JV, Pikramenou Z (2006) Luminescent nanobeads: attachment of surface reactive Eu(III) complexes to gold nanoparticles. Chem Commun 13:1433–1435

39. Block H, Maertens B, Spriestersbach A, Brinker N, Kubicek J, Fabis R, Labahn J, Schafer F (2009) Immobilized-metal affinity chromatography (IMAC): a review. Method Enzymol 463:439–473

40. Stobiecka M, Hepel M (2011) Multimodal coupling of optical transitions and plasmonic oscillations in rhodamine B modified gold nanoparticles. Phys Chem Chem Phys 13(3):1131–1139

41. Kim YP, Oh YH, Oh E, Kim HS (2007) Chip-based protease assay using fluorescence resonance energy transfer between quantum dots and fluorophores. Biochip J 1(4):228–233

42. Wang LH, Zhang J, Wang X, Huang Q, Pan D, Song SP, Fan CH (2008) Gold nanoparticle-based optical probes for target-responsive DNA structures. Gold Bull 41(1):37–41

# 10

# Modulation of anisotropic middle layer on the plasmon couplings in sandwiched gold nanoshells

DaJian Wu · ShuMin Jiang · Ying Cheng · XiaoJun Liu

**Abstract** The influence of the spherical anisotropy (SA) of a middle layer on the plasmon resonance couplings in the sandwiched gold nanoshell (Au/SA/Au) has been investigated by means of a modified Mie theory. It is found that the plasmon couplings in the Au/SA/Au nanoshells are more sensitive to the permittivity along the radial direction of SA layer than the permittivity along the tangential direction. With increasing the anisotropic value of the middle layer, the dipole peaks of antisymmetric $\omega_-^-$ mode and symmetric $\omega_-^+$ mode both show blue-shifts, while the shift of the antisymmetric $\omega_-^-$ mode is larger than that for the symmetric $\omega_-^+$ mode. The larger anisotropic value of the SA layer induces the stronger near-field outside the nanoparticles for the antisymmetric $\omega_-^-$ mode, while the smaller anisotropic value makes the larger near-field for the symmetric $\omega_-^+$ mode. We further have found that the middle SA layer with smaller anisotropic value is helpful to obtain larger electric fields inside the nanoshells, which may be useful for their potential applications in nonlinear optics.

**Keywords** Spherical anisotropy · Sandwiched gold nanoshells · Plasmon coupling · Near-field enhancement

## Introduction

Metallodielectric layered nanoparticles and nanostructures have attracted increasing scientific and technological interest in recent years due to their importance in the fundamental physics and the potential applications in nano-electronics, biomedical imaging, nano-optical device, and optical sensing [1–5]. A special interesting structure is metal–dielectric–metal three-layered nanoparticle, whose optical properties highly depend on the couplings between the plasmons of inner core and outer shell [6–8]. Changing the internal geometry of this nanoparticle not only shifts its resonance frequencies, but also strongly modifies the relative magnitudes of the absorption and scattering cross sections [5]. Mukherjee et al. [9] have found that an asymmetric core in Au/SiO$_2$/Au can lead to an additional high multipolar Fano resonance. A significant superscattering phenomenon also has been found in the plasmonic–dielectric–plasmonic layered nanorod [3] and nanosphere [4].

Among the metallodielectric layered nanoparticles, the spherical anisotropic (SA) material–metal nano-composite shows a growing interest because of the applications in optical devices, surface-enhanced Raman spectroscopy (SERS) and optical nonlinearity enhancement [10–12]. Spherical anisotropy indicates that the tensor for dielectric function is radially anisotropic, i.e., the dielectric function is uniaxial in spherical coordinate with $\varepsilon_r$ along the radial direction and $\varepsilon_t$ along the tangential direction. Gao et al. [10] have reported that the adjustment of the dielectric anisotropy in the core or shell could result in large enhancements of the second harmonic generation and induced third harmonic generation susceptibilities at surface plasmon resonant frequencies. Yin et al. [11] have found that the introduction of spherical anisotropy into the core or the shell provides a novel approach to tailor the surface plasmon resonant frequencies and enhanced SERS peaks. It is further found that the spherically anisotropy of inner core can strongly modulate the Fano resonance in the Ag nanoshell with a spherically anisotropic core [12]. However, the modulation of spherical anisotropy on the plasmon couplings in metallodielectric layered nanoparticles is seldom reported.

D. Wu · Y. Cheng · X. Liu (✉)
School of Physics, Nanjing University,
Nanjing 210093, China
e-mail: liuxiaojun@nju.edu.cn

D. Wu · S. Jiang
Faculty of Science, Jiangsu University,
Zhenjiang 212013, China

In this paper, we have investigated the optical properties of the sandwiched gold nanoshell with an SA middle layer (Au/SA/Au) by using a modified Mie theory. Spherical anisotropy was indeed found in phospholipid vesicle systems [13] and in cell membranes containing mobile charges [14]. Lucas et al. [15] have easily established the spherically anisotropic materials by using the graphitic multishells. We focus on the influence of the spherical anisotropy of the middle layer on the plasmon couplings between the inner core and outer shell. In addition, the dependence of the local electric field enhancement on the spherical anisotropy of the middle layer has been further discussed in detail.

## Electromagnetic scattering model

Figure 1 shows the specific geometry of a sandwiched gold nanoshell. The nanoparticle consists of a gold core with a radius of $r_1$, an SA middle layer with a radius of $r_2$, and a gold shell with a radius of $r_3$. The dielectric constant of embedding medium is $\varepsilon_4$. The SA middle layer is characterized by constitutive tensors of permittivity

$$\overleftrightarrow{\varepsilon_2} = \begin{pmatrix} \varepsilon_{2r} & 0 & 0 \\ 0 & \varepsilon_{2t} & 0 \\ 0 & 0 & \varepsilon_{2t} \end{pmatrix}. \tag{1}$$

$\varepsilon_{2r}$ is along the radial direction and $\varepsilon_{2t}$ is along the tangential direction. The dielectric functions of inner metal core $\varepsilon_1$ and outer metal shell $\varepsilon_3$ have real and imaginary frequency-dependent components, which are affected by the scattering of the conduction electrons in the particle surfaces. Thus, $\varepsilon_1$ and $\varepsilon_3$ are usually accounted by replacing the ideal Drude part in the dielectric function with a size-dependent one [8]. Light scattering by a spherical particle can be expressed through Debye potentials [16]. For spherical anisotropic material, the electric $\psi_{TM}$ and magnetic $\psi_{TE}$ Debye potentials are presented as [17]

$$\frac{\varepsilon_r}{\varepsilon_t}\frac{\partial^2 \psi_{TM}}{\partial r^2} + \frac{1}{r^2 \sin\theta}\frac{\partial}{\partial\theta}\left(\sin\theta\frac{\partial \psi_{TM}}{\partial\theta}\right) + \frac{1}{r^2\sin^2\theta}\frac{\partial^2 \psi_{TM}}{\partial\varphi^2}$$
$$+ k_0^2 \varepsilon_r \mu_t \psi_{TM} = 0, \tag{2}$$

$$\frac{\mu_r}{\mu_t}\frac{\partial^2 \psi_{TE}}{\partial r^2} + \frac{1}{r^2 \sin\theta}\frac{\partial}{\partial\theta}\left(\sin\theta\frac{\partial \psi_{TE}}{\partial\theta}\right) + \frac{1}{r^2\sin^2\theta}\frac{\partial^2 \psi_{TE}}{\partial\varphi^2}$$
$$+ k_0^2 \varepsilon_t \mu_r \psi_{TE} = 0. \tag{3}$$

Here, $k_0 = 2\pi/\lambda$ is the wave vector in vacuum and $\mu_r = \mu_t = 1$. The electromagnetic waves are expanded to spherical partial waves using vector spherical harmonics, and then, Maxwell's boundary conditions are applied to resolve the unknown expansion coefficients of the scattered and interior waves. According to the Mie scattering theory, the obtained extinction efficiency $Q_{ext}$, scattering efficiency $Q_{sca}$, and absorption efficiency $Q_{abs}$ can be expressed as [16]

$$Q_{ext} = \frac{2}{(k_4 r_3)^2} \sum_{n=1}^{\infty} (2n+1)\mathrm{Re}(a_n + b_n), \tag{4}$$

$$Q_{sca} = \frac{2}{(k_4 r_3)^2} \sum_{n=1}^{\infty} (2n+1)\left(|a_n|^2 + |b_n|^2\right), \tag{5}$$

$$Q_{abs} = Q_{ext} - Q_{sca}. \tag{6}$$

Here, $k_4 = k_0\sqrt{\varepsilon_4}$, $a_n$ and $b_n$ are the scattering coefficients. The information about spherical anisotropic middle layer is presented by the order of spherical Bessel functions [17]. The order can be expressed as

$$v = \left[n(n+1)\frac{\varepsilon_{2t}}{\varepsilon_{2r}} + \frac{1}{4}\right]^{1/2} - \frac{1}{2}. \tag{7}$$

## Results and discussion

Figure 2a shows the extinction spectra of the Au/SA/Au nanoshells for the conditions of $\rho<1$. We assume $\rho$ as the anisotropic value of the middle layer $\varepsilon_{2t}/\varepsilon_{2r}$. Here, $r_1$, $r_2$, and $r_3$ are fixed at 35, 50, and 65 nm, respectively. To discuss the influence of $\varepsilon_{2r}$ on the plasmon resonances, $\varepsilon_{2t}$ value is fixed at 2.04. The embedding medium is assumed to

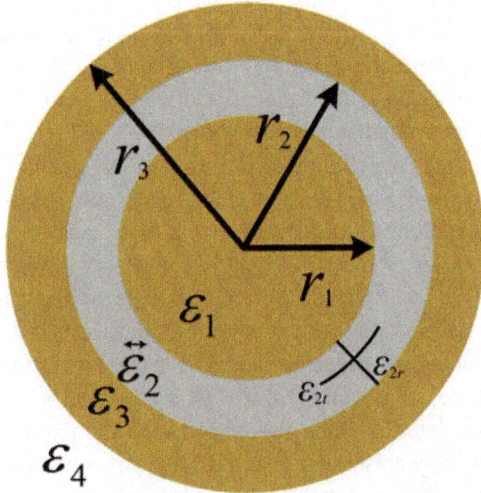

**Fig. 1** Geometry diagram of a sandwiched gold nanoshell

be water ($\varepsilon_4 = 1.7689$). The plasmon resonances in the Au/SA/Au nanoshell can be considered as an interaction between the plasmons of a gold nanosphere and a gold nanoshell [5, 8, 18]. As shown in Fig. 2a, for $\varepsilon_{2t}/\varepsilon_{2r} = 1$, one dipole peak appears at 838 nm, which corresponds to the antisymmetric coupling ($\omega_-^-$ mode) between the plasmons of inner Au core ($\omega_s$) and the symmetric mode of outer Au nanoshell ($\omega_-$). Another dipole peak appears at 577 nm, which corresponds to the symmetric coupling ($\omega_-^+$ mode) between $\omega_s$ and $\omega_-$ modes. The decreased $\rho$ value means the increase of $\varepsilon_{2r}$ value. With increasing $\varepsilon_{2r}$ value, the dipole peak of $\omega_-^+$ mode shows a red-shift from 577 nm at $\rho = 1$ to 625 nm at $\rho = 1/5$, while the dipole peak of $\omega_-^-$ mode shows a very large red-shift from 838 nm to 1,511 nm. The increased $\varepsilon_{2r}$ value should decrease the induced charges in the inner and outer surfaces of the middle layer [19]. In this case, the plasmon resonance energies of $\omega_s$ and $\omega_-$ modes decrease, and hence, the red-shifts of the $\omega_-^-$ and $\omega_-^+$ modes. According to plasmon hybridization theory [19–21], the $\omega_-^-$ mode is sensitive to the $\omega_-$ mode, while the $\omega_-^+$ mode depends on the $\omega_s$ mode. The large red-shift of the $\omega_-^-$ mode indicates that the variation of $\varepsilon_{2r}$ has more effect on the $\omega_-$ mode than the $\omega_s$ mode. Figure 2b shows the extinction spectra of the Au/SA/Au nanoshells for the conditions of $\rho > 1$. Here, the $\varepsilon_{2r}$ value is fixed at 2.04. The increased $\rho$ value means the increase of $\varepsilon_{2t}$ value. With increasing $\varepsilon_{2t}$ value, the dipole peak of $\omega_-^+$ mode shows a red-shift from 577 nm at $\rho = 1$ to 611 nm at $\rho = 5$, while the

dipole peak of $\omega_-^-$ mode shows a red-shift from 838 to 859 nm. The small red-shifts of the $\omega_-^-$ and $\omega_-^+$ modes suggest that the increased $\varepsilon_{2t}$ value has little effect on the decrease of the induced charges in both surfaces of the middle layer. The red-shift of the mode is larger than that of the $\omega_-^-$ mode, which indicates the suppressed coupling between the $\omega_s$ and $\omega_-$ modes due to the increased $\varepsilon_{2t}$ value.

Figure 3a, d shows the distributions of the electric field enhancement in the sandwiched nanoshell ($\omega_{2t}/\varepsilon_{2r} = 1$) at wavelengths of 577 nm ($\omega_-^+$) and 838 nm ($\omega_-^-$), respectively. The distributions of the electric field enhancement show the typical dipole resonance properties [22]. The large electrical field on the shell occurs along the incident polarization and only locates within a few nanometer of the shell surface, while the inner core also exhibits a similar dipole pattern. In Fig. 3d, a large electric field is observed near the core mainly due to the antisymmetric coupling between the $\omega_s$ and $\omega_-$ modes, which leads to different kinds of charges induced in inner and outer surfaces of the middle layer [18]. Figure 3b, e shows the contour plots of the electric field enhancement in the Au/SA/Au with $\rho = 1/1.1$ (divided by the results of the Au/SA/Au with $\rho = 1$). The calculation wavelengths are fixed at 577 and 838 nm for subpanels b and e of Fig. 3, respectively. In Fig. 3b, the decrease of the induced charge in outer surface of the middle layer is larger than that in inner surface of the middle layer. For the $\omega_-^-$ mode, the increased $\varepsilon_{2r}$ value induces the large electric field inside the nanoparticle. Figure 3c, f shows the contour plots of the electric field enhancement in the Au/SA/Au with $\rho = 1.1$ at wavelengths of 577 and 838 nm (divided by the results of the Au/SA/Au with $\rho = 1$), respectively. The increased $\varepsilon_{2t}$ value should lead to the decrease of the induced charges on the inner and outer surfaces of the SA layer. It is obvious that the effects of the variation of $\varepsilon_{2t}$ on the induced charges and the coupling between $\omega_s$ and $\omega_-$ modes are weaker than those due to the variation of $\varepsilon_{2r}$.

To further clarify the role of anisotropy, we keep the geometric average of dielectric components $\varepsilon_i = \varepsilon_{2r}/3 + 2\varepsilon_{3t}/3$ unchanged for the SA middle layer [12], while $\varepsilon_{2t}/\varepsilon_{2r}$ value is varied. Figure 4 shows the extinction spectra of the Au/SA/Au nanoshells with various $\rho$ values. Here, $r_1$, $r_2$, and $r_3$ are fixed at 35, 50, and 65 nm, respectively. The $\varepsilon_i$ value is fixed at 2.04. With an increasing $\rho$ value, the dipole peak of the $\omega_-^-$ mode shows a distinct blue-shift from 1,589 nm at $\rho = 1/5$ to 722 nm at $\rho = 5$, while the strength of the peak increases. At the same time, the dipole peak of the $\omega_-^+$ mode shows a blue-shift from 606 nm at $\rho = 1/5$ to 503 nm at $\rho = 5$, but the strength of the peak decreases. The influence of the variation of anisotropic value on the plasmon resonances in the Au/SA/Au nanoshell with $\rho < 1$ is stronger than that for $\rho > 1$. Figure 5a shows the dependences of the E-field enhancement in the Au/SA/Au nanoshell as a function of $\rho$ value, which are calculated at the dipole resonance wavelengths of $\omega_-^+$ mode. In Fig. 5a, the dashed

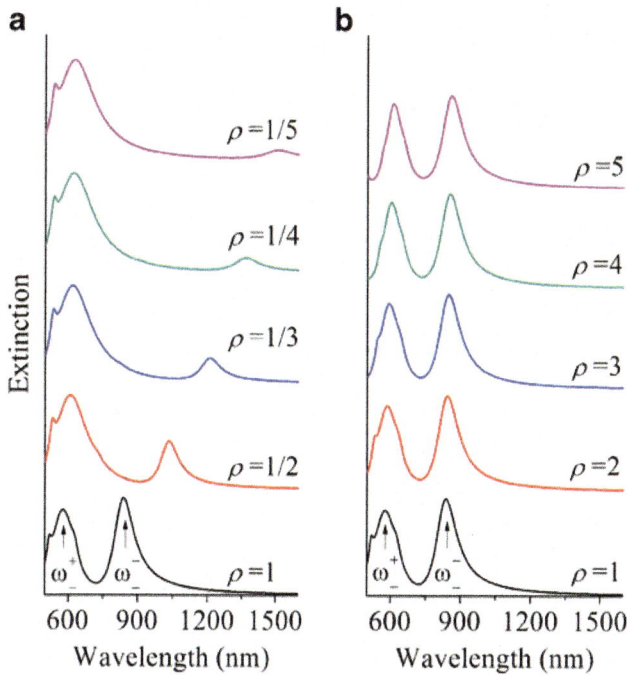

**Fig. 2** Extinction spectra of the Au/SA/Au nanoshells with various anisotropic values of the middle layer for **a** $\varepsilon_{2t}/\varepsilon_{2r} < 1$ and $\varepsilon_{2t} = 2.04$ and **b** $\varepsilon_{2t}/\varepsilon_{2r} > 1$ and $\varepsilon_{2r} = 2.04$. Here, $r_1$, $r_2$, and $r_3$ are fixed at 35, 50, and 65 nm, respectively

**Fig. 3** Contour plots of the
electric field enhancements in
Au/SA/Au nanoshells with **a**
$\rho=1$ ($\varepsilon_{2r}=\varepsilon_{2t}=2.04$) at 577 nm,
**b** $\rho=1/1.1$ ($\varepsilon_{2t}=2.04$) at
577 nm, **c** $\rho=1.1$ ($\varepsilon_{2r}=2.04$) at
577 nm, **d** $\rho=1$ ($\varepsilon_{2r}=\varepsilon_{2t}=2.04$)
at 838 nm, **e** $\rho=1/1.1$ ($\varepsilon_{2t}=$
2.04) at 838 nm, and **f** $\rho=1.1$
($\varepsilon_{2r}=2.04$) at 838 nm. Here, $r_1$,
$r_2$, and $r_3$ are fixed at 35, 50,
and 65 nm, respectively

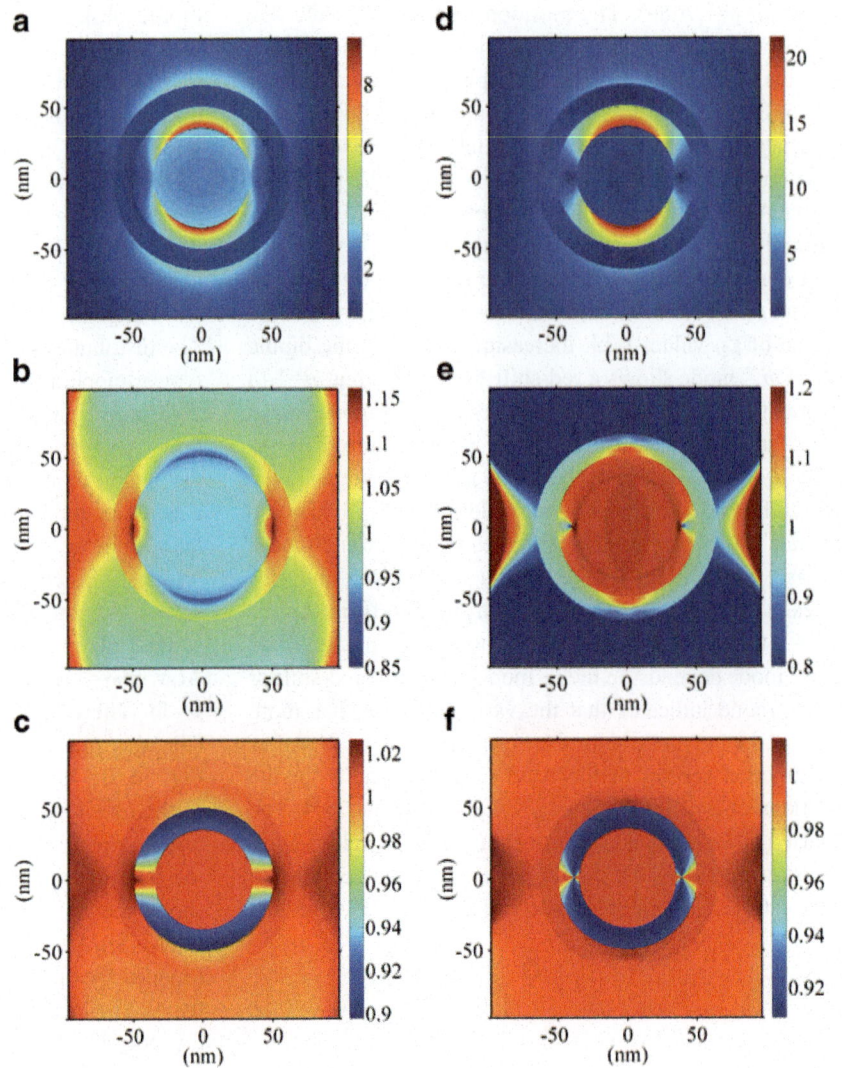

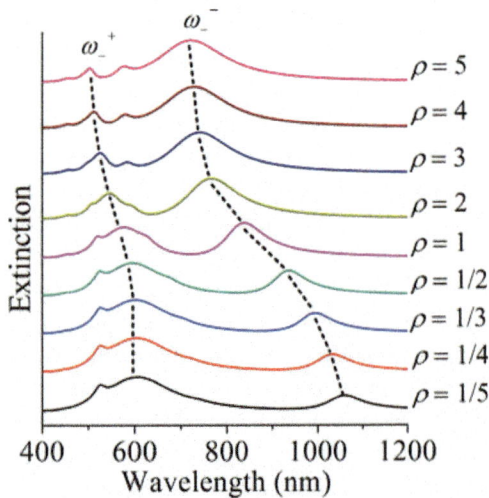

**Fig. 4** Extinction spectra of the Au/SA/Au nanoshells with various $\rho$
values ($\varepsilon_i=2.04$). Here, $r_1$, $r_2$, and $r_3$ are fixed at 35, 50, and 65 nm,
respectively

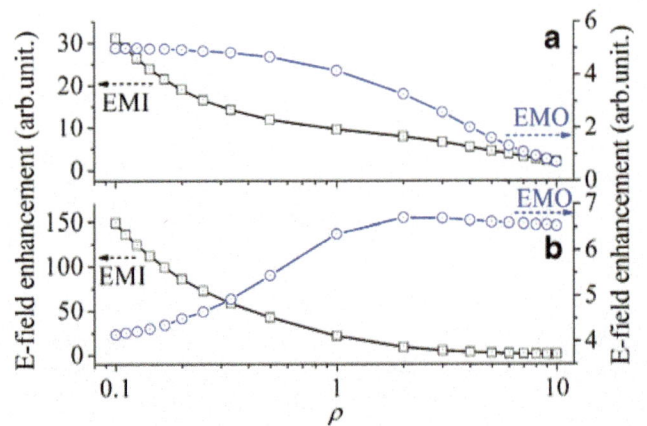

**Fig. 5** Dependences of the E-field enhancement maximum for **a** $\omega_-^+$
mode and **b** $\omega_-^-$ mode in Au/SA/Au nanoshells as a function of $\rho$
value. Here, $\varepsilon_i=2.04$, $r_1=35$ nm, $r_2=50$ nm, and $r_3=65$ nm. The
dashed circle line for right-hand scale and dashed square line for left-
hand scale represent the variations of EMO and EMI, respectively

circle line for right-hand scale represents the variation of E-field enhancement maximum outside the nanoshell (EMO), which often locates on the outer surface of the particle and at the poles along the incident polarization. The dashed square line for left-hand scale shows the variation of the E-field enhancement maximum inside the nanoshell (EMI), which can be obtained on the surface of inner gold core and at the poles along the incident polarization. In Fig. 5b, the dashed circle line for right-hand scale and dashed square line for left-hand scale represent the variations of EMO and EMI, respectively, which are calculated at the dipole resonance wavelengths of $\omega_-^-$ mode. It is found with the increase in $\rho$ value that the EMO for the $\omega_-^+$ mode decreases from 4.97 at $\rho = 1/10$ to 0.68 at $\rho = 10$, while the EMI decreases from 31.15 at $\rho = 1/10$ to 1.92 at $\rho = 10$. At the same time, the EMO for the $\omega_-^-$ mode increases first from 4.13 at $\rho = 1/10$ to 6.68 at $\rho = 2$ and then decreases to 6.51 at $\rho = 0$, while the EMI decreases from 149.00 at $\rho = 1/10$ to 0.85 at $\rho = 10$. It is found that the smaller $\rho$ value is helpful to obtain larger electric field inside the nanoparticle. The larger $\rho$ value of the middle SA layer can induce the larger near-field outside the nanoparticle for the $\omega_-^-$ mode and the smaller $\rho$ value makes larger near-field for the $\omega_-^+$ mode.

## Conclusion

We have investigated the plasmon resonance properties of the Au/SA/Au nanoshells. The extinction spectra and the electric field enhancement of the Au/SA/Au nanoshells have been calculated based on a modified Mie scattering theory. We focus on the influence of the spherical anisotropy of the middle layer on the plasmon resonance couplings in the Au/SA/Au nanoshells. It is found that the permittivity along the radial direction plays a dominant role on the plasmon couplings in Au/SA/Au nanoshells, and the permittivity along the tangential direction leads to significant modulations. With the increase of $\rho$, both the dipole peaks of the $\omega_-^-$ and $\omega_-^+$ modes show blue-shifts. The variation of the $\omega_-^-$ mode is stronger than the $\omega_-^+$ mode. The large $\rho$ value of the middle SA layer can induce the large near-field outside the nanoparticle for the $\omega_-^-$ mode and the small $\rho$ value can induce the large near-field for the $\omega_-^+$ mode. Such enhanced near-field can be used to the enhanced Raman excitation and emission. Furthermore, the small $\rho$ value is helpful to obtain larger electric field inside the nanoparticle, which may be helpful for their potential applications in nonlinear optics.

**Acknowledgments** This work was supported by the National Basic Research Program of China under grant no. 2012CB921504, the National Natural Science Foundation of China (11174113, 10904052, 11274171, 11104319, and 11204129), and project funded by the Priority Academic Program Development of Jiangsu Higher Education Institutions.

## References

1. Hao F, Sonnefraud Y, Dorpe PV, Maier SA, Halas NJ, Nordlander P (2008) Symmetry breaking in plasmonic nanocavities: subradiant LSPR sensing and a tunable Fano resonance. Nano Lett 8:3983–3988
2. Kodali AK, Schulmerich MV, Palekar R, Llora X, Bhargava R (2010) Optimized nanospherical layered alternating metal-dielectric probes for optical sensing. Opt Express 18:23302–23313
3. Ruan ZC, Fan SH (2010) Superscattering of light from subwave-length nanostructures. Phys Rev Lett 105:013901
4. Ruan ZC, Fan SH (2011) Design of subwavelength superscattering nanospheres. Appl Phys Lett 98:043101
5. Bardhan R, Mukherjee S, Mirin NA, Levit SD, Nordlander P, Halas NJ (2010) Nanosphere-in-a-nanoshell: a simple nanomatryushka. J Phys Chem C 114:7378–7383
6. Xia XH, Liu Y, Backman V, Ameer GA (2006) Engineering sub-100 nm multi-layer nanoshells. Nanotechnology 17:5435–5440
7. Hu Y, Fleming RC, Drezek RA (2008) Optical properties of gold-silica-gold multilayer nanoshells. Opt Express 16:19579–19591
8. Wu DJ, Jiang SM, Liu XJ (2011) Tunable Fano resonances in three-layered bimetallic Au and Ag nanoshell. J Phys Chem C 115:23797–23801
9. Mukherjee S, Sobhani H, Lassiter JB, Bardhan R, Nordlander P, Halas NJ (2010) Fanoshells: nanoparticles with built-in Fano resonances. Nano Lett 10:2694–2701
10. Gao L, Yu XP (2007) Second- and third-harmonic generations for a nondilute suspension of coated particles with radial dielectric anisotropy. Eur Phys J B 55:403–409
11. Yin YD, Gao L, Qiu CW (2011) Electromagnetic theory of tunable SERS manipulated with spherical anisotropy in coated nanoparticles. J Phys Chem C 115:8893–8899
12. Wu DJ, Jiang SM, Liu XJ (2012) A tunable Fano resonance in silver nanoshell with a spherically anisotropic core. J Chem Phys 136:034502
13. Lange B, Aragon SR (1990) Mie scattering from thin anisotropic spherical shells. J Chem Phys 92:4643–4650
14. Ambjörnsson T, Mukhopadhyay G, Apell SP, Käll M (2006) Resonant coupling between localized plasmons and anisotropic molecular coatings in ellipsoidal metal nanoparticles. Phys Rev B 73:085412
15. Lucas AA, Henrard L, Lambin P (1994) Computation of the ultraviolet absorption and electron inelastic scattering cross section of multishell fullerenes. Phys Rev B 49:2888–2896
16. Bohren CF, Huffman DR (1983) Absorption and scattering of light by small particles. Wiley, New York
17. Luk'yanchuk BS, Qiu CW (2008) Enhanced scattering efficiencies in spherical particles with weakly dissipating anisotropic materials. Appl Phys A 92:773–776
18. Radloff C, Halas NJ (2004) Plasmonic properties of concentric nanoshells. Nano Lett 4:1323–1327
19. Prodan E, Lee A, Nordlander P (2002) The effect of a dielectric core and embedding medium on the polarizability of metallic nanoshells. Chem Phys Lett 360:325–332
20. Prodan E, Radloof C, Halas NJ, Nordlander P (2003) A hybridization model for the plasmon response of complex nanostructures. Science 302:419–422
21. Prodan E, Nordlander P (2004) Plasmon hybridization in spherical nanoparticles. J Chem Phys 120:5444–5454
22. Kelly KL, Coronado E, Zhao LL, Schatz GC (2003) The optical properties of metal nanoparticles: the influence of size, shape, and dielectric environment. J Phys Chem B 107:668–677

# 11

# Formation of nanometer-sized Au particles on USY zeolites under hydrogen atmosphere

**Kazu Okumura · Chika Murakami · Tetsuya Oyama ·
Takashi Sanada · Ayano Isoda · Naonobu Katada**

**Abstract** Gold was deposited on ultrastable Y (USY) zeolite using a newly developed method: just mixing an aqueous solution of $HAuCl_4$ and zeolite at 353 K in which the $NH_4^+$ cation reacted with the $Cl^-$ in $HAuCl_4$. Treatment of the Au-loaded USY zeolite in the atmosphere of hydrogen resulted in the formation of $Au^0$ nanoclusters with 1.8 nm diameter at 773 K. The size of Au particle was dependent on the type of zeolite support, composition of gas atmosphere, and temperature of calcination, which was correlated with catalytic performance. This study demonstrated the potential use of zeolites with strong Brønsted acid character as gold supports.

**Keywords** Gold cluster · Nanoparticle · USY zeolite · Brønsted acid · Homocoupling

## Introduction

Since the pioneering work by Haruta and co-researchers, efforts have increasingly been directed toward the development of supported Au catalysts [1–4]. Interest in these catalysts is fuelled by their unique catalytic performance in various reactions [5, 6]. The catalytic performance of Au particles is sensitive to the Au particle size, and consequently, the regulation of the size of Au particles is of fundamental importance for the application to catalytic reactions [7]. In particular, high catalytic performance has been reported for Au particles smaller than several nanometers in size [1]. Preceding reports primarily focused on the use of reducible metal oxides such as $TiO_2$ and $Fe_2O_3$ as supports for Au. Other researchers attempted to use zeolites as the supports for Au. For instance, Au loaded on Y-type zeolites has been applied to CO oxidation [8]. Reduction of NO by $H_2$ has been performed over $Au^0$ or Au(I) loaded on $Na^+$–Y zeolite [9], and the formation of electron-deficient gold particles inside H–Y cavities has been observed [10]. Propene epoxidation by $H_2$ and $O_2$ was carried out over Au/TS-1 [11]. We have also utilized H–Y zeolites as a support for Au [12, 13]. Zeolites are promising supports for metals because the presence of strongly acidic sites in these materials promotes high dispersion of metal particles. In fact, metal clusters of Pd and Pt with high dispersion were obtained on zeolites having Brønsted acid sites [14, 15]. Ultrastable Y (USY) zeolites contain strongly acid sites [16] (ca. 150 kJ mol$^{-1}$) and also exhibit high thermal stability. These acidic sites may be exploited for interaction with Au, and their thermal stability makes USY zeolites attractive for use as Au catalyst supports in high-temperature applications.

To date, several methods including precipitation deposition [17], chemical vapor deposition [18], and cation adsorption [19] have been applied for the preparation of Au/$TiO_2$ catalysts. Cation exchange using $[Au(en)_2]^{3+}$ and incipient wetness impregnation have been employed for the loading of zeolites [20, 21].

The purpose of this study is to establish a new method for the preparation of Au nanoclusters on zeolite supports, demonstrating that nanometer-sized $Au^0$ particles having a narrow size distribution were obtained after calcination at

K. Okumura (✉) · C. Murakami · T. Oyama · T. Sanada ·
N. Katada
Department of Chemistry and Biotechnology,
Graduate School of Engineering, Tottori University,
4-101 Koyama-cho Minami,
Tottori 680-8552, Japan
e-mail: okmr@chem.tottori-u.ac.jp

T. Sanada · A. Isoda
Research Department, NISSAN ARC, LTD.,
Yokosuka 237-0061, Japan

temperature higher than 573 K. Moreover, the influences of the concentration of hydrogen in the gas phase and the Brønsted acid strength on the size of Au particles were investigated. Preliminary data were recently reported elsewhere as a communication [22]. In the present study, USY zeolite is used as a support for Au. First, we developed a new method for loading Au on the zeolite support in which Au was loaded on $NH_4^+$-type USY zeolites using an aqueous solution of $HAuCl_4$. Then, the as-prepared Au/USY was thermally treated in an atmosphere of $H_2$ from 353 to 773 K. Changes in the local structure and size of Au were analyzed by X-ray absorption fine structure (XAFS), X-ray diffraction (XRD), and transmission electron microscopy (TEM). Homocoupling of phenylboronic acid was performed as the catalytic reaction. The reason of the choice of this reaction was that the acid sites present in zeolites were supposed not to be influential on the homocoupling reaction.

## Experimental

### Sample preparation

Gold was loaded on an $NH_4$-type USY zeolite (Tosoh, HSZ-341NHA, Si/Al$_2$=7.7) using $HAuCl_4 \cdot 4H_2O$ (Wako Chemical Co.) as the precursor. The USY (1 g) was immersed in an aqueous solution of $HAuCl_4$ (250 mL, $6.1 \times 10^{-4}$ mol L$^{-1}$), and the slurry was kept at 343 K for 1 h with constant stirring. The Au-loaded USY zeolite was filtered, washed with deionized water, and dried in an oven at 323 K. The loading of Au was most commonly 3 wt% as measured by inductively coupled plasma. The obtained Au/USY was typically treated with a stream of 6 % $H_2$ diluted with Ar at a given temperature for 30 min. Temperature ramping rate for the treatment was 5 K min$^{-1}$. Au/USY samples are henceforth denoted as Au/USY-x in which x represented the calcination temperature. Loading of Au on CaNH$_4^+$-type Y and NH$_4^+$-type Y, NH$_4$NO$_3$-treated USY was carried out in a similar way as described for the preparation of Au/USY. Concentration of Ca in CaH–Y was 1.0 mol kg$^{-1}$. Precipitation–deposition method was employed for the loading of 3 wt% Au on H-ZSM-5 (JRC-Z5-90H(1),Si/Al$_2$=90), H-mordenite (JRC-Z-HM15, Si/Al$_2$=15), and TiO$_2$ (JRC-TIO-11), which were supplied by the Catalysis Society of Japan.

### TEM, XAFS, and XRD data collection and analysis

TEM images were acquired using a HITACHI H-9000UHR microscope with an acceleration voltage of 300 kV. Synchrotron radiation experiments (XAFS) were performed at the BL01B1 station with the approval of the Japan Synchrotron Radiation Research Institute (Proposal No. 2011A1106,

2011B1095). A Si(111) single crystal was used to obtain a monochromatic X-ray beam. Measurements were recorded in the quick mode at room temperature. For the collection of Au $L_3$-edge data, ion chambers filled with $N_2$ and a mixture of $N_2$ (50 %)/Ar (50 %) were used for $I_0$ and $I$, respectively. The energy was calibrated using an Au foil. The data were analyzed using the REX2000 Ver. 2.5.9 program (Rigaku Co.). Fourier transform of $k^3\chi(k)$ data was performed in a $k$ range 30–160 nm$^{-1}$ for the analysis of the Au $L_3$-edge extended X-ray absorption fine structure (EXAFS) spectra. The inversely Fourier filtered data were analyzed using a common curve-fitting method. The empirical phase shift and amplitude functions for Au–O and Au–Au were extracted from the data of FEFF code (ver. 8) and Au foil, respectively. The crystalline structure was analyzed by XRD under ambient conditions using a Rigaku Ultima IV X-ray diffractometer with Cu Kα radiation.

### Catalytic reactions

The samples were pretreated in 6 % $H_2$ at a given temperature before being used in homocoupling reaction of phenylboronic acid. The treated samples were stored in the dark prior to the reaction. Homocoupling reactions were carried out over the Au catalysts. Phenylboronic acid (0.25 mmol; Tokyo Kasei Chemicals Ltd., Japan), $K_2CO_3$ (0.75 mmol; Wako Chemicals Ltd., Osaka, Japan), deionized water (solvent, 5 mL), and Au-loaded catalyst (16.4 mg) were used for the reaction. Molar amount of gold was 1 mol% with respect to phenylboronic acid. A sample bottle (20 mL) was placed on a magnetic stirrer and subjected to vigorous stirring. The reaction was performed at 300 K under atmospheric conditions for 0.5 h. The reaction was quenched by the addition of ethyl acetate. The product was extracted with ethyl acetate. Following evaporation of the ethyl acetate, the residue was analyzed by gas chromatography [Shimadzu 2010 Gas Chromatograph equipped with an InertCap 1 (30 m) capillary column (Shimadzu Corp., Kyoto, Japan)]. Tridecane was used as the internal standard.

## Results and discussion

### Influence of the calcination temperature on Au dispersion

In the preparation of Au/USY, Au was loaded on NH$_4$–USY using $HAuCl_4$ as the Au precursor. The solution of $HAuCl_4$ had been yellow and it was decolorized, while the zeolites became a pale orange color within 1 h after temperature of the solution reached 343 K. These changes in color indicate the loading of Au on USY. Formation of NH$_4$Cl in the filtered solution was confirmed by means of IR, XRD, and TG analysis. The molar amount of NH$_4$Cl dissolved in the

**a**

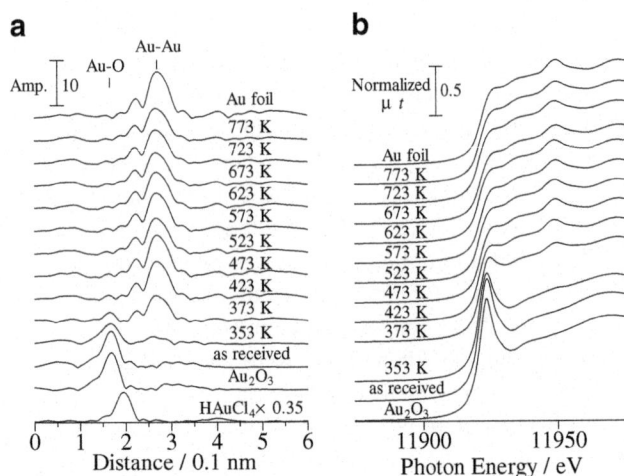

**b**

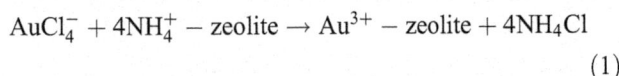

**Fig. 1** Au-$L_3$ edge EXAFS $k^3\chi(k)$ Fourier transforms (**a**) and XANES (**b**) of 3 wt% Au/USY treated at different temperatures under 6 % $H_2$ atmosphere and reference samples. Data collection was carried out at room temperature

solution was four times larger than that of Au, suggesting the following stoichiometry (Eq. 1).

$$AuCl_4^- + 4NH_4^+ - zeolite \rightarrow Au^{3+} - zeolite + 4NH_4Cl \tag{1}$$

The $Au^{3+}$ zeolite was hydrated by water to give $Au_2O_3$ simultaneously according to Eq. 2:

$$2Au^{3+} - zeolite + 3H_2O \rightarrow Au_2O_3 + 6H^+ - zeolite \tag{2}$$

The ion exchange sites of USY zeolite should be partially exchanged with $H^+$. Formation of $Au_2O_3$ in the as-prepared sample was confirmed by EXAFS, as mentioned later. Deposition of Au also proceeded smoothly on $NH_4$–Y as confirmed by ICP analysis. In addition, Au was readily supported on $NH_4^+$-type USY even at higher loadings of 5 wt%. In marked contrast to the $NH_4^+$-type USY, Au was not deposited on either $H^+$-type USY or $Na^+$-type Y in a same method; this indicated that the presence of $NH_4^+$ cation is indispensable for the loading of Au. Loading of Au on $NH_4^+$-type ZSM5 and mordenite by a same procedure was also attempted. However, deposition of Au was not successful on other kinds of zeolites, except for Y-type materials. Therefore, deposition of Au specifically took place on the $NH_4^+$-type Y and USY zeolites.

Figure 1a shows Au-$L_3$ edge EXAFS of 3 wt% Au/USY. The Fourier transform of the as-received sample was identical to those of $Au_2O_3$. The Au–O bond appears at 0.17 nm (phase shift uncorrected). The features of spectrum in the range 0.25–0.40 nm were also similar to those of $Au_2O_3$. The peak corresponding to the Au–Cl was not observed in the Fourier transform. The absence of Au–Cl can be confirmed by comparison with $HAuCl_4$ in which the Au–Cl bond appears at ca. 0.20 nm (phase shift uncorrected),

indicating that Au was successfully loaded on USY as aggregated $Au_2O_3$. The Au–O peak was decreased by $H_2$ treatment with temperature elevation, and the peak disappeared at 473 K. Alternatively, a new peak appeared at 0.27 nm, which was assignable to the nearest neighbor Au–Au bond in $Au^0$ metal on the basis of comparison with the spectrum of Au foil. The complete disappearance of the Au–O peak at 473 K indicates the formation of $Au^0$ as a result of $H_2$ reduction at this temperature. Formation of $Au^0$ at 473 K was also confirmed from the X-ray absorption near edge structure (XANES) regions; the shapes of the XANES profiles were markedly similar to that of Au foil, as shown in Fig. 1b. At temperatures above 473 K, the intensity of EXAFS peak corresponding to the Au–Au bond decreased slightly (Fig. 1a). This slight change suggests a decrease in the size of Au particles. The coordination numbers (CNs) of the nearest neighboring Au–Au bond in Au/USY samples treated at 473 and 773 K were calculated to be $11.8 \pm 1.6$ and $9.6 \pm 1.4$, respectively. The change in CNs of Au–Au atoms suggested the decrease in the size of the Au particles at higher treatment temperatures. The particle size of the Au/USY-773 K was estimated to be 2 nm on the basis of the CN=9.6, assuming a cuboctahedron shape [23].

Because EXAFS analysis is rather insensitive to CNs when the particle size of a metal is larger than several nanometers, Au/USY was analyzed by XRD, which is sensitive to changes in size of Au particles larger than several nanometers. Figure 2 shows the XRD patterns of the 3 and 5 wt% Au/USY samples treated at various temperatures. The peaks at $2\theta=38.2°$ and $44.4°$ were due to the diffractions from the (111) and (200) planes of $Au^0$, respectively. Weak diffractions due to the USY zeolites appeared at $38.1°$ and $44.2°$, which overlapped with the diffractions from $Au^0$. The intensity of the reflections from the $Au^0$(111) and (200) planes increased with increasing the treatment temperature from room temperature to 473 K on the 3 wt% $H_2$-treated

**a**

**b**

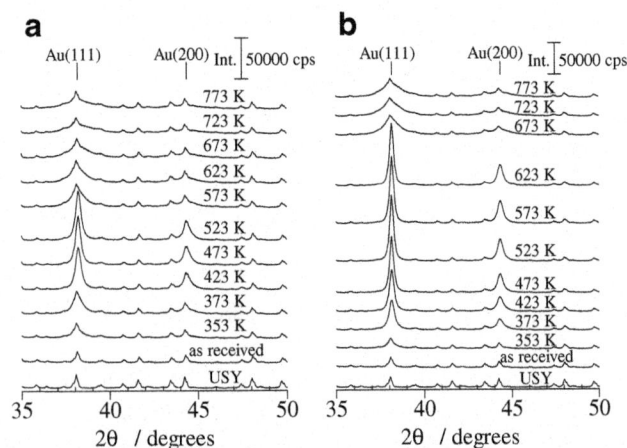

**Fig. 2** XRD patterns of Au/USY treated at different temperatures under a 6 % $H_2$ atmosphere: **a** 3 wt% and **b** 5 wt%

**Fig. 3** TEM images of Au/USY calcined at **a** 473 K and **b** 773 K under an atmosphere of 6 % $H_2$. Loading of Au was 3 wt%

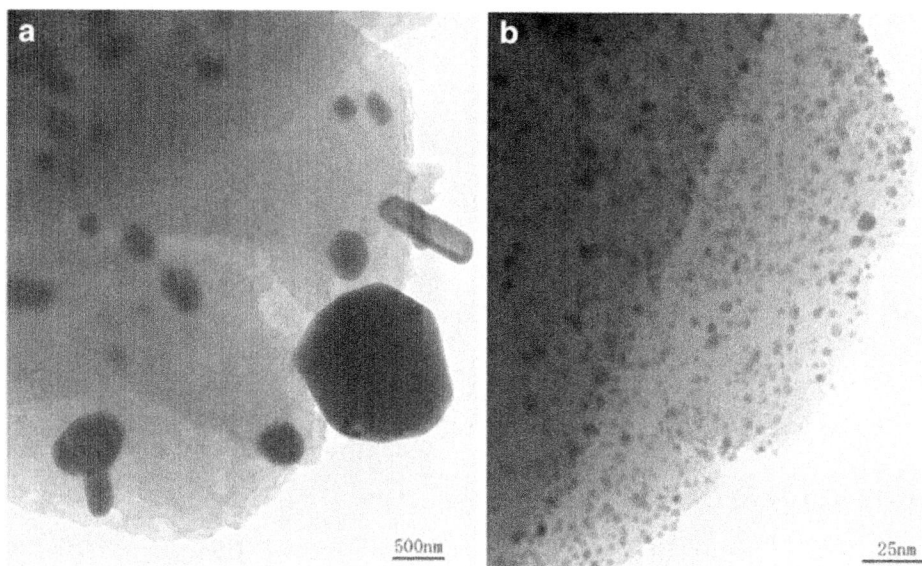

samples (Fig. 2a). This change can be attributed to the formation of $Au^0$ as confirmed by Au-$L_3$ edge XAFS (EXAFS and XANES). On further increase of the temperature, the intensity of these diffractions decreases, accompanied by peak broadening. A marked reduction of the peaks can be observed at 573 K indicating that highly dispersed Au particles were obtained when the treatment temperature was increased from 473 to 773 K. The XRD patterns of 5 wt% Au-loaded USY (Fig. 2b) exhibit a similar reduction of the Au(111) and Au(200) peaks indicative of dispersion of the $Au^0$ particles. However, the reduction in peak intensity was observed in the 5 wt% sample at 673 K, which was 100 K higher in the temperature than that for 3 wt% Au/USY, suggesting that the extent of Au loading impacts the behavior of Au.

TEM analysis was employed to directly observe the $Au^0$ particles on Au/USY in order to confirm the dispersion of Au. The TEM images of Au/USY treated at 473 and 773 K under a 6 % $H_2$ atmosphere are shown in Fig. 3a, b, respectively. Large $Au^0$ particles with a variation of size were observed in the image of Au/USY treated at 473 K (Fig. 3a). The shape of the Au particles was not uniform. In marked contrast to this, and in agreement with the EXAFS and XRD, the particles of $Au^0$ in Au/USY-773 K (Fig. 3b) were much smaller and seem to have a homogeneous size distribution compared to the sample treated at 473 K. Figure 4 shows the particle size distribution of $Au^0$ determined from the TEM images. The $Au^0$ particle sizes ranged from 10 to 150 nm with an average size of 25 nm for Au/USY-473 K. The average size of $Au^0$ in Au/USY-773 K was 1.8 nm, 1/14 of that of Au/USY-473 K. The estimated size is consistent with the Au $L_3$-edge EXAFS analysis. The mean diameter of Au (1.8 nm) is slightly larger than the diameter of super cages of in Y-type zeolite (1.3 nm), suggesting that most of the loaded Au was located on the

external surface of zeolite. The distribution of $Au^0$ particle size in Au/USY-773 K ranges from 0.4 and 5.9 nm, significantly narrower as compared to that in Au/USY-473 K.

The formation of highly dispersed $Au^0$ at 773 K thus demonstrated by TEM coupled with EXAFS and XRD is unusual, because supported metal particles generally undergo sintering to form aggregates at high temperature in reductive atmosphere. It has been known that oxidative atmosphere can keep small particles of metal oxides; dispersion of Pd and Pt as cations has been observed under an atmosphere of oxygen [24]. We have also reported the spontaneous dispersion of molecular-like PdO on H-ZSM-5 [25]. In contrast to the above-mentioned examples, the dispersed $Au^0$ was obtained on USY zeolite support under an atmosphere of $H_2$ at high temperature in the present study. The origin of dispersion of $Au^0$ on USY zeolites has not been fully elucidated on this stage; however, one hypothesis is that the interaction of Au and $H^+$ generated as a result of thermal decomposition of $NH_4^+$ causes the dispersion of Au on the external surface of USY zeolites. As a matter of fact, Sachtler et al. proposed the possibility of the generation of $[Pd_n H]^+$ adducts encaged in zeolite pores [26,

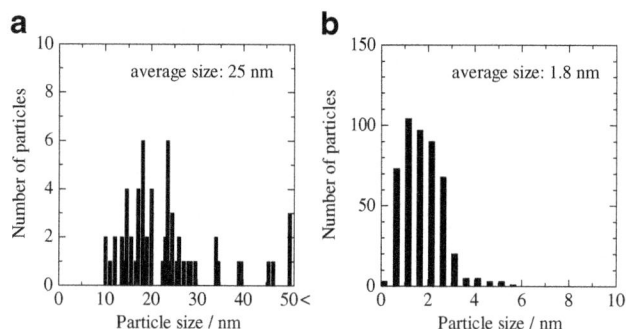

**Fig. 4** Distribution of Au/USY calcined at **a** 473 K and **b** 773 K under an atmosphere of 6 % $H_2$. Loading of Au was 3 wt%

**a**

**b**

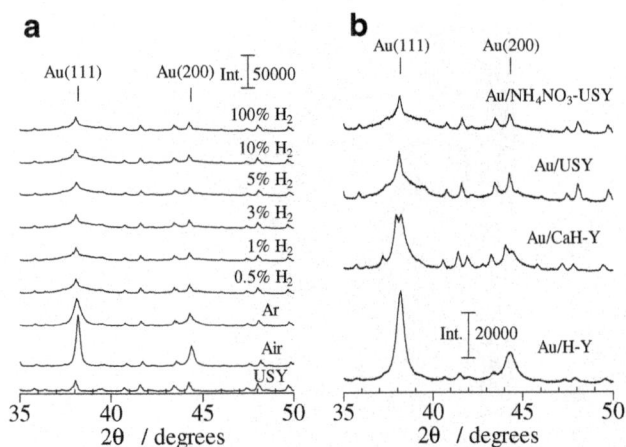

**Fig. 5** XRD patterns of **a** Au/USY treated at 773 K in different atmosphere for 0.5 h, **b** Au loaded on Y-type zeolites calcined at 773 K in 6 % $H_2$

27]. Similarly, Fraissard et al. reported the chemical anchoring of Au clusters by the Brønsted sites of the support in Au/Y-type zeolites [20].

## Influence of the concentration of $H_2$ and time on the dispersion of Au

Figure 5a shows XRD patterns of 3 wt% Au/USY treated under different atmospheres at 773 K. Steep diffraction peaks appeared in the samples calcined in air or Ar, namely 0 % $H_2$. Addition of 0.5 % $H_2$ to the atmosphere resulted in the formation of dispersed Au as confirmed from the broadening of diffraction peaks due to metal Au. No further change was observed with increasing the hydrogen concentration up to 100 %, leading us to conclude that the presence of low levels of hydrogen is sufficient to give rise to the dispersion to give nanometer-sized Au clusters. Figure 6 shows TEM images and particle size distribution of 3 wt% Au/USY treated under an atmosphere containing Ar and 0.5 % $H_2$, corresponding to the samples in Fig. 5a, respectively. The sizes of Au particles in the sample treated in the atmosphere of Ar are distributed over 4 to 59 nm with an average diameter of 14.6 nm. In marked contrast to Au/USY treated in Ar, the distribution (1–8 nm) of the size of Au treated under an atmosphere of 0.5 % $H_2$ is narrower and the average size (3.7 nm) is appreciably smaller. The treatment times of Au/USY were varied from 10 min to 10 h under an atmosphere comprising 6 % $H_2$ at 773 K. No differences were observed in the diffraction patterns obtained for these samples, implying that dispersion of Au occurs immediately when the temperature reaches 773 K. Although the role of hydrogen is not clearly understood at this stage, the TEM images of Au/USY treated in different atmospheres indicate that the presence of hydrogen has a profound influence on the dispersion of Au.

## Influence of the acid strength of supports on the dispersion of Au

The acid strength of Brønsted acid sites present in Y-type zeolites can be finely tuned by (a) the introduction of divalent cations such as $Ca^{2+}$ and (b) by dealumination in combination with posttreatment employing an aqueous solution of ammonium salts [16]. The influence of the acid strength on the dispersion of Au was examined by employing four types of Y-type zeolite supports: H–Y, CaH–Y, USY, and $NH_4NO_3$-treated USY. The highest acid strengths (heat of ammonia desorption) of the Brønsted acid sites present in H–Y, CaH–Y, USY, and $NH_4NO_3$-treated USY are 109, 122, 140, and 157 kJ $mol^{-1}$, respectively. The acid strength increased in the following order: H–Y<CaH–Y<USY<$NH_4NO_3$-treated USY. In the preparation of CaH–Y support, 30 % of $NH_4^+$ present in $NH_4^+$-type Y was exchanged with $Ca^{2+}$ using $Ca(NO_3)_2$. Figure 5b shows XRD patterns of 3 wt% Au loaded on these supports. The samples were calcined at 773 K under a 6 % $H_2$ atmosphere. It can be seen that the intensity of the diffractions assignable to the Au(111) and (200) decrease with an increase in the intensity of acid strength of Y-type supports, suggesting that the greater the acid strength, the smaller the Au particles become.

Figure 7 displays TEM images of Au loaded on different Y-type zeolites, corresponding to the XRD patterns of

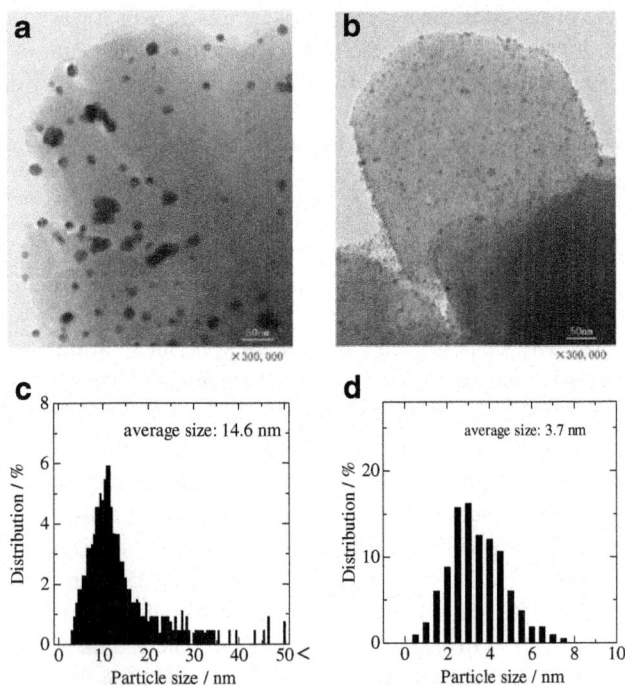

**a**

**b**

**c**

**d**

**Fig. 6** TEM images of Au/USY calcined in **a** Ar and **b** 0.5 % $H_2$ at 773 K. Distribution of Au particles in 3 wt% Au loaded on USY calcined in (**c**) Ar and (**d**) 0.5 % $H_2$

**Fig. 7** TEM images of 3 wt%
Au loaded on **a** H–Y, **b** CaH–Y,
**c** USY, and **d** NH$_4$NO$_3$–USY

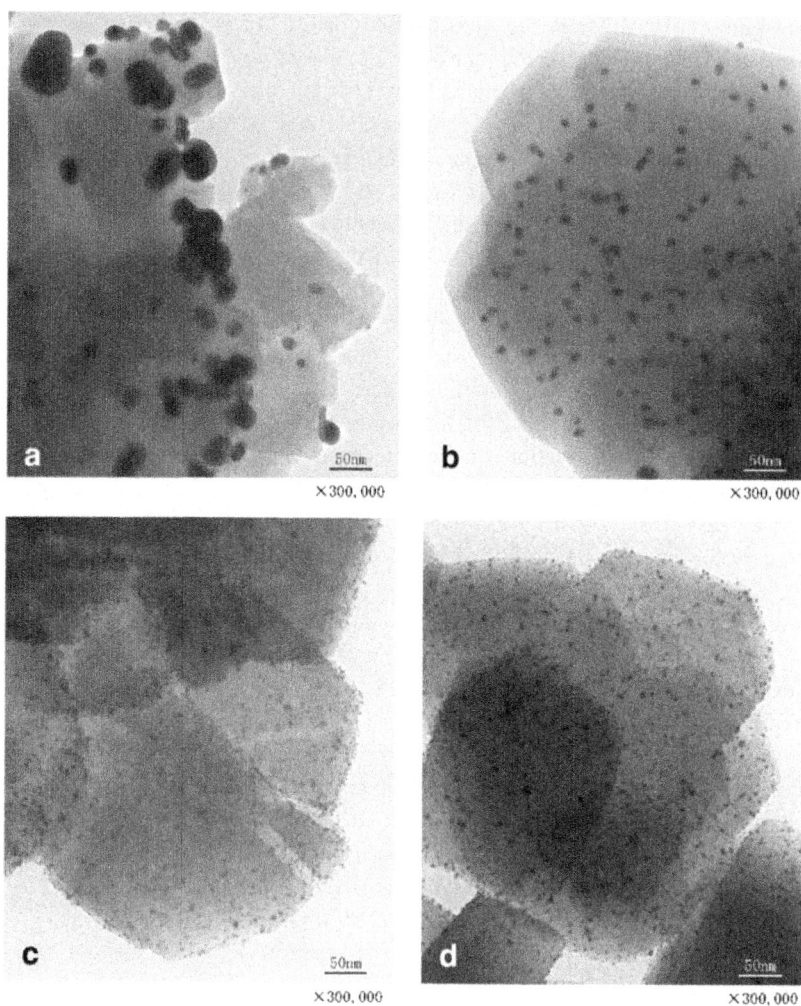

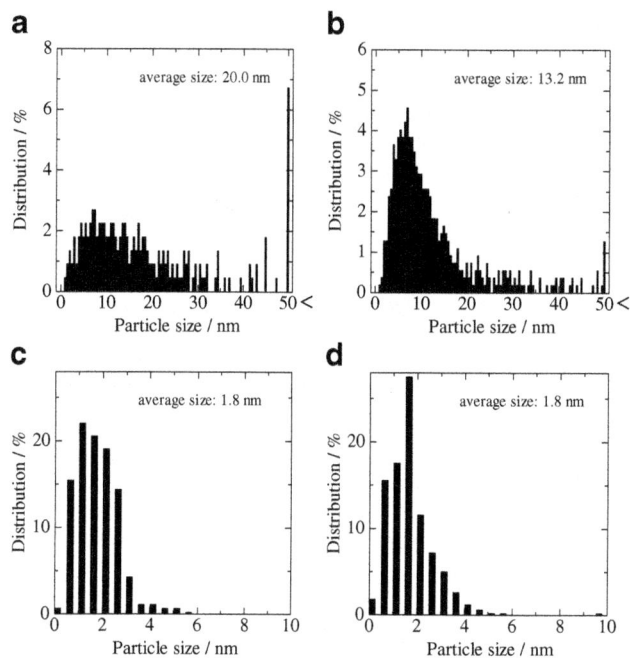

**Fig. 8** Distribution of Au particles in 3 wt% Au loaded on **a** H–Y, **b** CaH–Y, **c** USY, and **d** NH$_4$NO$_3$–USY

Fig. 5b. The distribution of the size of Au particles in these samples is summarized in Fig. 8. An inhomogeneous distribution of large Au particles can be seen in the image of Au/H–Y (Fig. 7a). The size distribution plot confirms this observation (Fig. 8a). In the case of Au/CaH–Y, uniform Au particles ca. 10 nm in diameter can

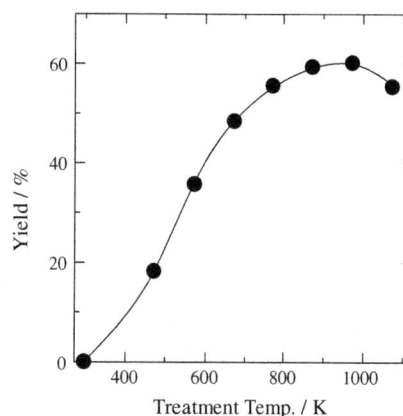

**Fig. 9** Dependence of the yield of product (biphenyl) on the calcination temperature of 3 wt% Au/USY treated in a stream of 6 % H$_2$

be seen (Fig. 7b). Although the size distribution of Au particles in Au/CaH–Y is much narrower compared to that of Au/H–Y, some large particles can also be observed in the TEM images. The average particle size of Au loaded on USY and NH₄NO₃-treated USY is 1.8 nm, a significantly smaller value than those of Au particles loaded on H–Y and CaH–Y. Despite the identical average size of Au in the USY and NH₄NO₃–USY (1.8 nm), the Au/NH₄NO₃–USY shows a narrower distribution compared to Au/USY. The TEM images in combination with XRD patterns prove that the size of the Au particles is dependent on the acid strength of Y-type zeolites, indicating that the presence of the strong Brønsted acid sites induces the formation of nanometer-sized Au particles with a narrow size distribution.

Catalytic performance in the homocoupling reaction

The homocoupling reaction of phenylboronic acid was carried out over the Au-loaded catalysts. Figure 9 shows the yield of biphenyl product plotted as a function of calcination temperature of 3 wt% Au/USY. The yield of biphenyl increased as the calcination temperature increased. The highest yield (60 %) was obtained when the catalyst was calcined at 773–1,073 K. The catalytic activity of the Au/USY was equal to the performance of Au nanoclusters protected by PVP polymers [28]. The change in the particle size of Au can be seen to influence the catalytic activity; the yield of biphenyl increases as the particle size of Au particles decreases. Formation of biphenyl was negligible over USY zeolites without the loading of Au.

Catalytic reaction was carried out over 1.8-nm-sized Au-loaded USY in which the acid sites were changed to NH₄⁺ through the exposure of NH₃. The yield of biphenyl over the Au loaded on NH₄⁺-type USY (60 %) was close to that obtained in Au loaded on H⁺-type USY, meaning the acid sites were not influential on the catalytic performance of Au.

Au was loaded on different types of support using the deposition precipitation method. The yields of biphenyl obtained with Au/TiO₂, ZSM-5, mordenite, and Na–Y were 35, 35, 36, and 4 %, respectively. The yields obtained with Au/H–Y and CaH–Y were and 16 and 17 %, respectively, lower than that obtained with Au/USY-773-1,073 K; this indicated that Au loaded on the USY zeolite afforded the highest activity. Calcination atmosphere affected the catalytic activity significantly as expected from the difference in the dispersion. The yields of biphenyl obtained with Au/USY calcined in Ar and air were 18 and 6 % respectively, which was much lower than that obtained after calcination in the atmosphere of H₂ (60 %).

## Conclusions

Reduction of Au₂O₃ on USY zeolites exhibiting strong Brønsted acid character, in the presence of hydrogen, afforded nanometer-sized Au clusters with mean diameter of 1.8 nm. Various factors including temperature (>573 K), composition of gas phase (hydrogen), and acid strength (USY zeolites exhibiting strong Brønsted acid) of the support were shown to greatly influence the dispersion of Au. For instance, the average size of Au particles in Au/USY (1.8 nm) was 11 times smaller than that of Au/H–Y (20 nm). The catalytic activity of Au/USY in the homocoupling reaction of phenylboronic acid was in good correlation with the dispersion of Au. That is to say, the yield of biphenyl increased from 16 % (Au/H–Y) to 60 % (Au/USY) by employing the USY supports having strong acid character ca. 150 kJ mol⁻¹. The simplicity of technique used for loading of Au should be emphasized; stirring a solution of HAuCl₄ and NH₄⁺-type USY zeolite at 343 K afforded Au₂O₃. This study shows the possibility of regulating the size of metal particles by making use of the strong interactions between Au and zeolite supports.

## References

1. Haruta M (1997) Catal Today 36:153
2. Daniel MC, Astruc D (2004) Chem Rev 104:293
3. Haruta M (2004) Gold Bulletin 37:27
4. Bond GC, Thompson DT (1999) Catal Rev 41:319
5. Haruta M, Date M (2001) Appl Catal A 222:427
6. Corma A, Garcia H (2096) Chem Soc Rev 2008:37
7. Subramanian V, Wolf EE, Kamat PV (2004) J Am Chem Soc 126:4943
8. Fierro-Gonzalez JC, Gates BC (2004) J Phys Chem B 108:16999
9. Salama TM, Ohnishi R, Shido T, Ichikawa M (1996) J Catal 162:169
10. Guillemot D, Borovkov VY, Kazansky VB, PolissetThfoin M, Fraissard J (1997) J Chem Soc Faraday Trans 93:3587
11. Huang JH, Lima E, Akita T, Guzman A, Qi CX, Takei T, Haruta M (2011) J Catal 278:8
12. Okumura K, Yoshino K, Kato K, Niwa M (2005) J Phys Chem B 109:12380
13. Shimizu K, Yamamoto T, Tai Y, Okumura K, Satsuma A (2011) Appl Catal A 400:171
14. Okumura K, Yoshimoto R, Uruga T, Tanida H, Kato K, Yokota S, Niwa M (2004) J Phys Chem B 108:6250
15. Treesukol P, Srisuk K, Limtrakul J, Truong TN (2005) J Phys Chem B 109:11940

16. Okumura K, Tomiyama T, Morishita N, Sanada T, Kamiguchi K, Katada N, Niwa M (2011) Appl Catal A 405:8
17. Hayashi T, Tanaka K, Haruta M (1998) J Catal 178:566
18. Okumura M, Nakamura S, Tsubota S, Nakamura T, Azuma M, Haruta M (1998) Catal Lett 51:53
19. Zanella R, Giorgio S, Henry CR, Louis C (2002) J Phys Chem B 106:7634
20. Riahi G, Guillemot D, Polisset-Thfoin M, Khodadadi AA, Fraissard J (2002) Catal Today 72:115
21. Delannoy L, El Hassan N, Musi A, Le To NN, Krafft JM, Louis C (2006) J Phys Chem B 110:22471
22. Sanada T, Okumura K, Murakami C, Oyama T, Isoda A, Katada N (2012) Chem Lett 41:337
23. Jentys A (1999) PhysChemChemPhys 1:4059
24. Bera P, Patil KC, Jayaram V, Subbanna GN, Hegde MS (2000) J Catal 196:293
25. Okumura K, Amano J, Yasunobu N, Niwa M (2000) J Phys Chem B 104:1050
26. Homeyer ST, Karpinski Z, Sachtler WMH (1990) J Catal 123:60
27. Bai XL, Sachtler WMH (1991) J Catal 129:121
28. Tsunoyama H, Sakurai H, Ichikuni N, Negishi Y, Tsukuda T (2004) Langmuir 20:11293

# 12

# Effects of Au nanoparticles on thermoresponsive genipin-crosslinked gelatin hydrogels

Ana L. Daniel-da-Silva · Ana M. Salgueiro · Tito Trindade

**Abstract** Gold gelatin hydrogel nanocomposites cross-linked with genipin have been prepared, and the effect of citrate capped Au nanoparticles (NPs) as nanofillers in the crosslinking and swelling of gelatin and release of a model drug (methylene blue) from gelatin nanocomposites have been investigated. The citrate-capped Au NPs prevented the crosslinking reaction between the gelatin and genipin and resulted in less crosslinked hydrogels. Although less cross-linked, the Au gelatin nanocomposites swelled less than the unfilled crosslinked gelatin. The gelatin composites were optically active and thermo-sensitive in a temperature range acceptable for living cells. In vitro release studies demonstrated that the irradiation of the composite gels with monochromatic green light ($\lambda$=532 nm, 100 mW) increases the release of the encapsulated methylene blue, most likely due to the photothermal effect of Au nanoparticles. This opens the possibility to explore the application of these nanocomposites as carriers in remotely controlled light-triggered drug release.

**Keywords** Gold nanoparticles · Gelatin · Genipin · Biomaterials

## Introduction

Recent years have witnessed growing importance of hydrogel nanocomposites for bio-applications [1,2]. These materials bring together the intrinsic functionalities of inorganic nanoparticles (NPs) and the properties of tridimensional networks offered by hydrogels. These hydrogels can be obtained from biopolymers that in controlled experimental conditions provide hydrophilicity, soft consistency, and ability to accommodate biomolecules. All these properties are attractive to promote the efficient transport of pharmaceuticals in living systems, and as such, biocompatible hydrogels have been widely used in drug delivery procedures. A recent trend in the development of hydrogels for drug delivery has been the implementation of multifunctionality, thus leading to smart drug carriers responsive to external stimuli, such as applied magnetic gradients [3,4], monochromatic light irradiation [5] or temperature effects [6]. These endeavors have relied in large extent on the use of inorganic NPs such as gold NPs that are used as functional nanofillers in the hydrogel matrix.

Gelatin is a thermoresponsive hydrogel derived from collagen that has been widely used in bio-applications, either on its own or combined with inorganic phases [7,8]. For example, gelatin composites filled with calcium phosphate have found application in tissue engineering [7] and gelatin fibers containing Ag NPs have been used in antibacterial wound-dressing materials [8]. Gelatin gels have a triple-helical structure that can be crosslinked to improve the thermal and mechanical stability of the network. Gelatin crosslinking has been achieved mainly by using chemical agents, most commonly glutaraldehyde whose major drawback is toxicity [9]. Alternatively, genipin (Fig. 1) is a chemical crosslinker, with very low toxicity, obtained from the natural occurring product geniposide that is extracted from *Gardenia jasminoides* fruits. Genipin has attracted great attention due to its ability to crosslink polymer chains containing primary amines, namely proteins such as gelatin [10–12] and polysaccharides such as chitosan [10,13]. The cytotoxicity of genipin is ca. 10,000 times lower than that of glutaraldehyde [14] and therefore genipin is extremely valuable in the development of biocompatible materials.

A. L. Daniel-da-Silva (✉) · A. M. Salgueiro · T. Trindade
Department of Chemistry, CICECO, Aveiro Institute
of Nanotechnology, University of Aveiro,
3810-193 Aveiro, Portugal
e-mail: ana.luisa@ua.pt

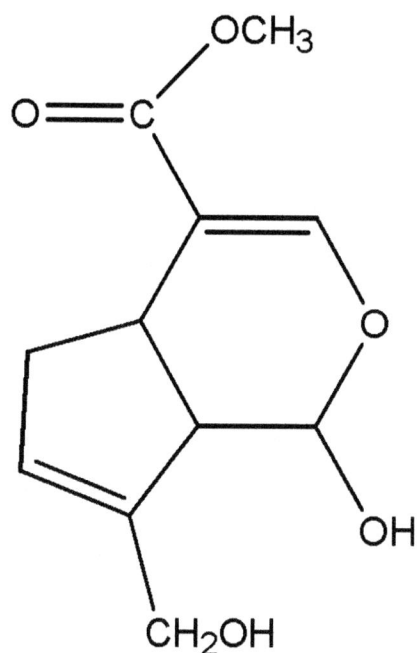

**Fig. 1** Chemical structure of genipin

The recent interest in Au hydrogel nanocomposites for biomedical applications relies mostly on their optical properties due to the plasmonic behavior of Au NPs dispersed in the matrix. For Au NPs, the surface plasmon resonance (SPR) band is located in the visible and is sensitive to several parameters, including particle size and shape, dielectric constant of the dispersing medium [15]. In addition, Au NPs act as heat dissipators by absorbing light at a frequency matching that one of the SPR band. This effect has been exploited to induce localized temperature gradients that can be used in photothermal therapy [16] and to trigger the release of encapsulated drugs from thermosensitive hydrogels [5,17]. Examples of Au hydrogel nanocomposites are those of chitosan [17,18] and alginate [19]. Although there are a few reports on the preparation of Au gelatin nanocomposites [20,21], the effect of light-irradiated Au NPs on the release properties of the hydrogel has not been investigated. Moreover, there are no reports on the use of genipin for preparing crosslinked Au gelatin nanocomposites, despite its relevance in the context of new formulations for biocompatible hydrogels.

In the sequence of our current work on the development of hydrogel nanocomposites for controlled release [22,23], we report here the preparation of genipin-crosslinked Au gelatin nanocomposites. This work aims to investigate the effect of colloidal gold nanoparticles on the crosslinking, thermal, swelling, and release properties of genipin-crosslinked gelatin hydrogels. Selected hydrogel nanocomposites were then tested for in vitro release of methylene blue, a commonly used model, with and without light irradiation of the gels.

## Materials and methods

### Materials

Porcine gelatin (type A, 300 bloom, Sigma-Aldrich), genipin ($C_{11}H_{14}O_5$) (98 %, Chengdu King-tiger Pharm-chem. Tech.Co., Ltd), tetrachloroauric acid ($HAuCl_4 \cdot 3H_2O$) (99.9 %, Sigma-aldrich), trisodium citrate dihydrate ($HOC(COONa)(CH_2COONa)_2 \cdot 2H_2O$) (99 %, Sigma-Aldrich), methylene blue (($C_{16}H_{18}ClN_3S$) (Riedel-de-Häen), and phosphate-buffered saline (PBS) solution (pH 7.4, Sigma-Aldrich) were used as received without any further purification.

### Synthesis of Au nanoparticles

Gold NPs were prepared by reduction of a gold(III) complex using sodium citrate as reducing agent. Typically, 3.76 ml of an aqueous solution of sodium citrate (96.8 mM) was added to a mixture of 95 ml of ultrapure water and 3.7 ml of $HAuCl_4 \cdot 3H_2O$ aqueous solution (12.95 mM) at 80 °C, under vigorous stirring and reflux, and allowed to react over 1 h. A deep-red hydrosol was obtained, which indicated the formation of the gold colloid.

### Preparation of Au/gelatin nanocomposites

The nanocomposites were prepared by blending the Au NPs with the gelatin as follows. The colloidal Au (1.25, 2.5, and 3.75 ml) was added to an aqueous solution of gelatin previously prepared, at 45 °C. The volume of water was adjusted to perform a total of 15 mL and the gelatin concentration was 15 $gL^{-1}$. After achieving a homogenous dispersion of the nanoparticles, 1.25 ml of the selected model drug, methylene blue 0.3 $gL^{-1}$, was added. The mixture was cool down until 40 °C and 2.25 ml of genipin solution 4 $gL^{-1}$ was added and stirred for 30 min. Afterwards, 2.5 ml of the composite mixture was transferred to a cylindrical glass vial (Ø 17 mm) which was sealed and left for incubation for 48 h at 25 °C. The content of genipin in the composites was 0.4 wt.%, related to the gelatin. The Au NPs load in the composites was calculated assuming complete conversion of the gold complex into metal Au NPs and was found to be 52, 104, and 156 ppm. After incubation, the nanocomposite discs presented a dark blue appearance which is related to the level of crosslinking of gelatin. The gel samples were frozen at −5 °C for 24 h and lyophilized. The final freeze dried discs had approximately 15 mm diameter and 8 mm height.

### Swelling studies

The swelling measurements were carried out by immersion of lyophilized hydrogel discs in 50 ml PBS 0.01 M pH 7.4 at 37 °C. At the required intervals of time, the samples were removed from the solution and wiped with filter paper to remove the excess of water before being weighted. The swelling ratio ($Q$) was calculated from Eq. 1:

$$Q = \frac{W_s - W_d}{W_d} \qquad (1)$$

where $W_d$ and $W_s$ are the weight of the lyophilized and swollen gel, respectively. The equilibrium swelling ratio ($Q_{equil}$) was determined at the point the hydrated gels achieved a constant weight value. The swelling experiments were performed in triplicate.

Materials characterization

*FTIR analysis* Fourier transform infrared (FTIR) spectra of the lyophilized gelatin and gelatin nanocomposites were collected using a spectrometer Mattson 7000 coupled to a horizontal attenuated total reflectance (ATR) cell, accumulating 256 scans and using a resolution of 4 cm$^{-1}$. FTIR spectrum of the Au nanoparticles was collected after drying the colloidal suspension in a KBr pellet. The spectrum was collected with 256 scans and 4 cm$^{-1}$ resolution.

*Transmission electron microscopy* Transmission electron microscopy (TEM) analysis of Au nanoparticles was performed using a transmission electron microscope JEOL 200CX operating at an accelerating voltage of 30 kV. Samples for TEM analysis were prepared by evaporating dilute suspensions of the nanoparticles on a copper grid coated with an amorphous carbon film.

*Differential scanning calorimetry* The gel-sol transitions of hydrogels were determined by differential scanning calorimetry (DSC) using a Shimadzu DSC-50 calorimeter. With sample masses of ca. 25 mg, 30-μL aluminum pans were used. Samples were heated from 25 to 80 °C at 2 °Cmin$^{-1}$. An empty pan was used as reference.

*Zeta potential measurements* The surface charge of the Au NPs was assessed by zeta potential measurements, using a Zetasizer Nanoseries instrument from Malvern Instruments (UK).

*UV–vis spectrophotometry* The optical properties of Au NPs and gelatin Au nanocomposites were investigated by UV–vis analysis of aliquots of the samples. A Jasco V 560 UV/Vis spectrophotometer (Jasco Inc., USA) was used for recording the UV/vis absorption spectra of the aliquots.

In vitro MB release studies

Methylene blue (MB) was used as a model drug and was loaded during the stage of the preparation of the nanocomposites as described above. MB has been used as a model

drug namely because it is a water-soluble dye that allows an immediate visual inspection of the test. The release experiments were performed with and without laser irradiation of the samples.

The release experiments carried out without laser irradiation were performed in a thermostatic orbital shaker KS 4000I Control from IKA at the physiological temperature 37 °C and 120 rpm. A lyophilized disc was introduced in a glass beaker containing 50 ml PBS 0.01 M pH7.4 and 0.05 % ($w/v$) sodium azide as preserving agent. After predetermined intervals, 1.0 ml of the release medium was drawn and analyzed by UV–vis spectroscopy ($\lambda$=663 nm) to determine the amount of MB released at each time point and replaced by 1 ml of fresh PBS to maintain the original volume. The cumulative released fraction at time $t$ ($m_t/m_0$) was calculated using Eq. 2:

$$\frac{m_t}{m_0} = \frac{50 \times C_n + \sum_{i=0}^{n-1} C_i}{m_0} \qquad (2)$$

where $m_t$ is the cumulative mass of MB released at time $t$, $m_0$ is the original mass of MB loaded, $C_i$ is the mass concentration of MB (per milliliter) of the aliquot, $C_n$ is the mass concentration of MB (per milliliter) of the aliquot at time $t$, and $n$ is the total number of aliquots extracted until time $t$.

The release experiments carried out with laser irradiation were performed at room temperature (ca. 11 °C). In a typical experiment, the crosslinked gelatines containing methylene blue were prepared in a glass tube ($\emptyset_{int}$=0.6 cm), following the procedure described above. The final volume of the gelatin samples was 0.25 mL per tube. A volume of 0.7 mL of PBS 0.01 M pH7.4 was added to each tube. The gelatines were irradiated with the laser spot positioned immediately below the interface gel/PBS using a CW diode pumped solid state laser ($\lambda$=532 nm, 100 mW) located at a distance of 10 cm from the sample. After irradiation, the PBS solution was analyzed by UV–vis spectroscopy ($\lambda$ = 663 nm) to determine the amount of MB released. For comparison, release tests were also performed in identical conditions, in the absence of laser irradiation.

**Results and discussion**

Characterization of Au colloid

Gold NPs were prepared by reduction of a gold(III) complex in aqueous medium using sodium citrate as reducing and stabilizing agent. The resulting aqueous Au colloid

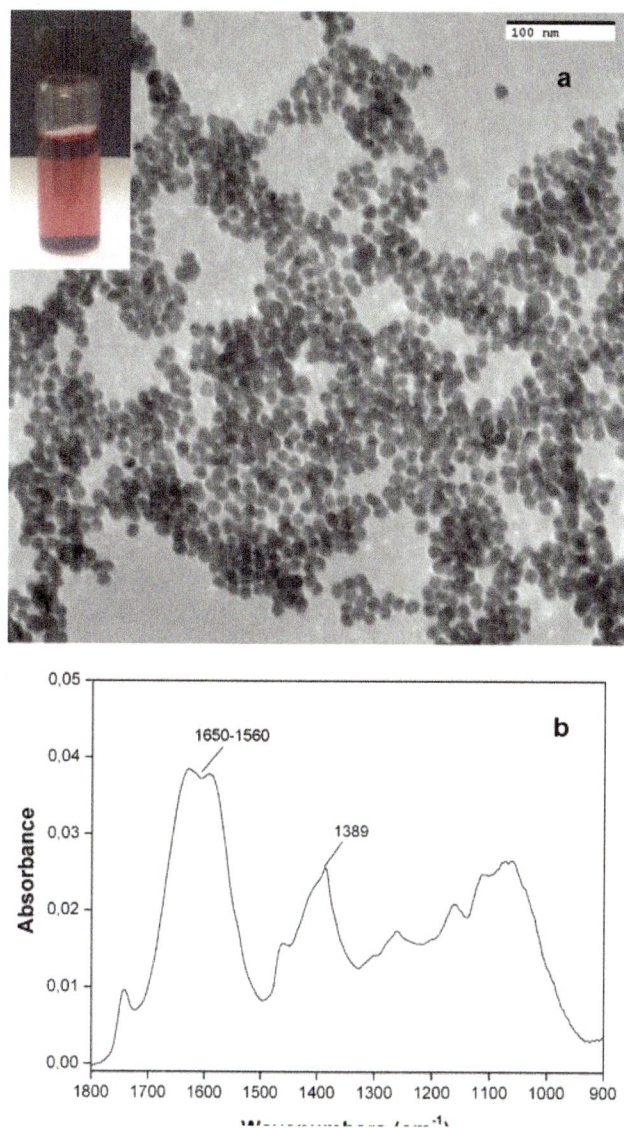

Fig. 2 **a** TEM micrograph and **b** FTIR spectrum of colloidal Au NPs used as dispersed phase

exhibited a red color (Fig. 2a, inset) due to the SPR band peaked about 520 nm. The average particle size for Au NPs was found to be $10 \pm 2$ nm as determined by transmission electron microscopy (Fig. 2a).

Gold colloids as prepared above are stable because surface chemisorbed citrate ions impart a net negative charge to the particles causing electrostatic repulsion. This was confirmed by zeta potential measurements ($-40.1 \pm 0.8$ mV, pH 6.8) performed on the aqueous Au colloids. FTIR spectrum of Au colloid (Fig. 2b) shows two broad bands, one in the region 1,560–1,650 $cm^{-1}$ and the other centered at 1,389 $cm^{-1}$ corresponding, respectively, to the asymmetric and symmetric stretching vibrations of carboxylate groups [24,25], thus confirming that the Au NPs are capped with citrate anions.

## Chemical and thermal characterization of Au and gelatin nanocomposites

The ATR-FTIR spectrum of the non-crosslinked gelatin (Fig. 3) shows the vibrations amide I and amide II characteristic from polypeptides. The band in the amide I region is centered at 1,629 $cm^{-1}$ and corresponds to the C=O stretching vibration in the amide group coupled to the in-phase bending of the N–H bond and the C–N stretching vibration [26]. The band in the amide II region that corresponds to the N–H bending vibration coupled to stretching C–N vibrations was identified at 1,524 $cm^{-1}$. The spectrum of genipin-crosslinked gelatin is very similar to that of non-crosslinked sample. However, the amide II band is slightly less intense than in the non-crosslinked gelatin. This indicates the reduction of the number of primary amine groups due to the reaction with genipin molecules [27].

The ATR-FTIR spectra of the genipin-crosslinked Au gelatin nanocomposites are shown in Fig. 3. These spectra are dominated by the vibrational bands ascribed to gelatin due to the minor amount of Au NPs present in the composite. Nevertheless, a small shift of the amide II band from 1,524 $cm^{-1}$, in the unfilled crosslinked gelatin, to 1,536 $cm^{-1}$ in the composites is observed in Fig. 3. This band shift suggests that Au NPs interact with amine groups from gelatin, most likely via electrostatic interactions between carboxylate and protonated amine groups. At the pH of the composite mixture (pH 5.3), the amine groups from gelatin are protonated and can interact electrostatically with citrate groups capping the Au NPs. These results are in agreement with previous observations that report changes in the IR spectra for the amide II region of different proteins, after the addition of citrate-stabilized Au colloidal nanoparticles [28]. Other workers have proven the electrostatic

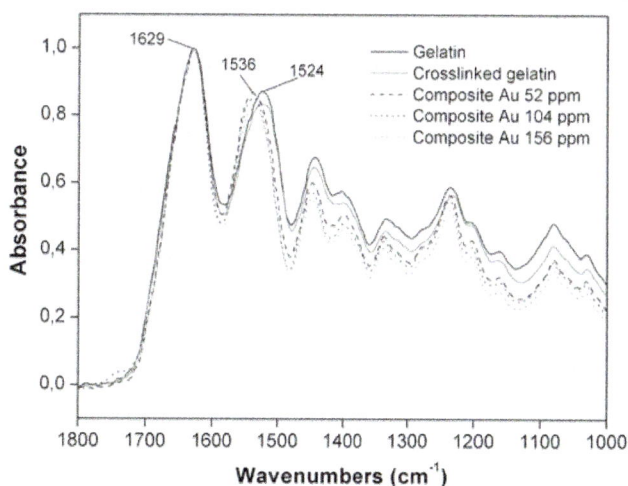

Fig. 3 FTIR-ATR spectra of uncrosslinked gelatin, genipin-crosslinked gelatin, and Au nanocomposites. The intensity of spectra was normalized in relation to the amide I vibration band

binding of bovine serum albumin to citrate-capped gold NPs due to interactions between carboxylate and protonated amine groups using non-spectroscopic techniques [29].

The addition of Au NPs at a concentration of 52 ppm resulted also in the increase of the intensity of the Amide II band in the nanocomposite when compared to the unfilled crosslinked gelatin. This suggests that, due to the interaction of citrate/amine groups, the citrate-capped Au NPs prevent the reaction between the gelatin and genipin, hence restraining the reticulation of gelatin. This is in agreement with the observed decrease of the amide II band for increasing amounts of Au NPs in the nanocomposites (Fig. 3). In fact, similar effects of citrate-capped Au NPs on the chemical reactivity of the amino groups of proteins have been reported for the glycation by fructose [24].

The DSC thermograms of genipin-crosslinked gelatin and derived Au nanocomposites (Fig. 4) show a broad endothermic peak that corresponds to the denaturation (helix-to-coil transition) of the gelatin [11]. The values for the denaturation temperature ($T_D$), defined here as the temperature of the main peak, are depicted in Table 1. For the neat hydrogel, $T_D$ is 78.9 °C and then decreases by addition of Au NPs to the hydrogel, reaching the value 65.1 °C for the nanocomposite with 156 ppm Au content. Adding Au NPs also resulted in a slight decrease of the area of the endothermic peak, which is directly related to the denaturation enthalpy ($\Delta H_D$) (Table 1).

The decrease of $T_D$ might be ascribed to reduction of covalent crosslinking of the hydrogels, since an increased extent of crosslinking of gelatin implies higher $T_D$ [11]. Thus, the DSC results indicate that the incorporation of Au NPs restrain the formation of covalent crosslinks and are in agreement with the FTIR spectroscopic data presented above. Noteworthy, these observations demonstrate that

**Table 1** Denaturation temperature ($T_D$) and enthalpy ($\Delta H_D$) of genipin crosslinked gelatin hydrogels and Au nanocomposites calculated from DSC experiments

| Au NPs (ppm) | $T_D$ (°C) | $\Delta H_D$ (kJ/g) |
|---|---|---|
| 0 | 78.9 | −2.0 |
| 52 | 74.5 | −1.7 |
| 104 | 69.2 | −1.8 |
| 156 | 65.1 | −1.6 |

the nanocomposites are thermo-sensitive in the 37–45 °C temperature range, i.e., the temperature range acceptable for living cells. Therefore, these composites have potential for the release of encapsulated molecules triggered by thermal stimuli.

Optical properties of the nanocomposites

The visible spectrum of the aqueous Au colloid showed a well-defined SPR band centered at 523 nm (Fig. 5a), characteristic of nearly monodispersed Au NPs, whose

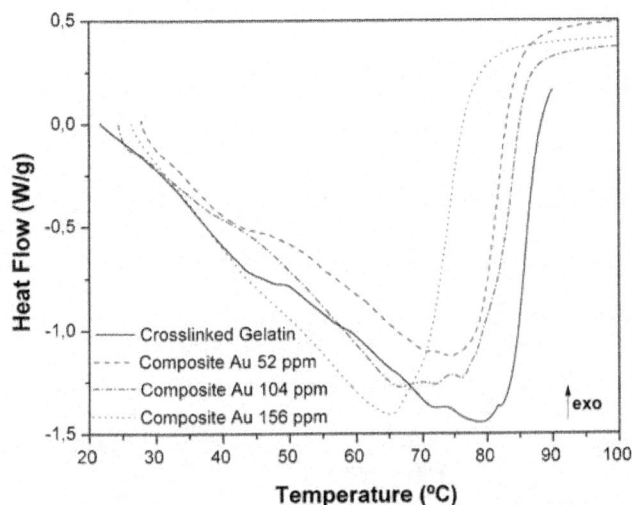

**Fig. 4** DSC thermograms of genipin-crosslinked gelatin and derived Au nanocomposites

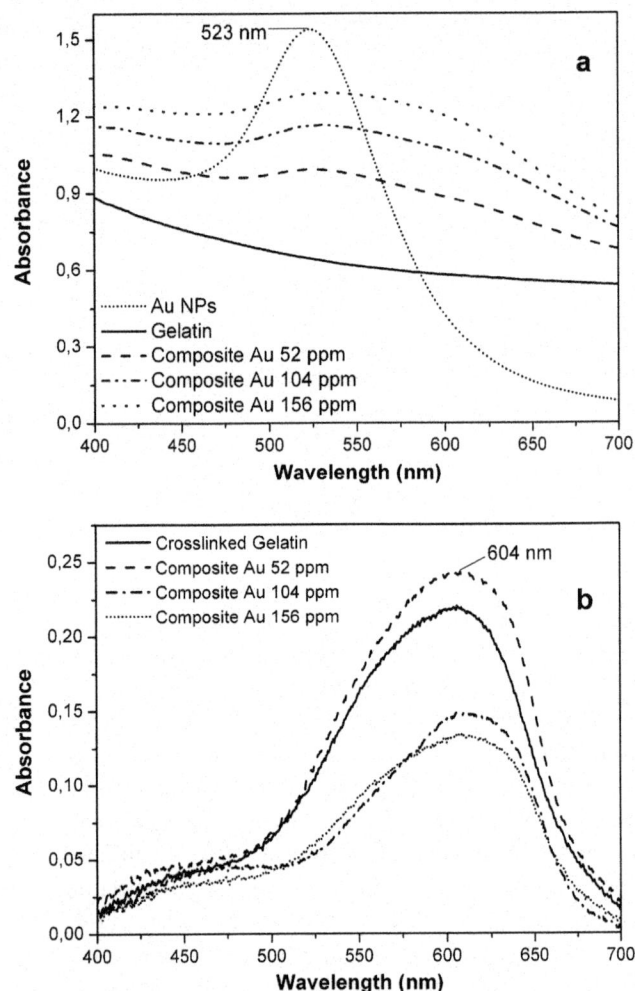

**Fig. 5** Absorption spectra of the gelatin nanocomposites with variable Au NPs content **a** before reticulation with genipin and **b** reticulated with genipin

morphological characteristics are shown in Fig. 2. On the other hand, the visible spectra of gelatin Au composites prior reticulation with genipin (Fig. 5a) display a broader band and a shoulder at ca. 600 nm, the latter becoming more pronounced for increasing Au content. The observation of a shoulder extending to higher wavelength in relation to the SPR band of the original colloid is an indication of Au NPs aggregation, probably mediated by the biopolymer chains, due to interparticle plasmon coupling. This behavior apparently contrasts with the stabilizing effect previously reported for gelatin [20]. However, it should be noted that this stabilizing effect was reported for in situ prepared Au NPs, using gelatin as a reducing and stabilizing agent, thus in the absence of citrate anions [20]. Thus, it might be inferred that the Au NPs aggregation observed in this case arises from the interaction between the gelatin and the citrate capping. These results are in line with previous observations indicating that the adsorption of proteins onto the surface of citrate-capped Au NPs induce their aggregation [29]. At the temperature used for preparing the composites (45 °C), the gelatin polypeptide chains exist predominantly in the form of flexible, unfold coils in solution, and the adsorption of these chains onto the Au NPs surface might also be expectable, thus promoting the NPs aggregation.

For the case of crosslinked gelatin, the optical spectra also evidence the effect of genipin. In fact, these hydrogels exhibit a characteristic blue color that arises from products of the reaction of genipin with amino acids of gelatin [27,30]. The blue pigments are most likely formed through the oxygen radical-induced polymerization and dehydrogenation of intermediate compounds [31]. As a result, the optical spectra of the crosslinked gelatin shows a broad band centered at 600 nm that for the nanocomposites spectra probably have some contribution from the broadening of the SPR band of Au NPs. Figure 5b shows that the overall trend in the composites is the decay of the absorbance peaks at ca. 600 nm with increasing amount of Au NPs. This indicates that less genipin has reacted in the presence of citrate-capped Au NPs and is well in line with FTIR and DSC observations. Crosslinked composites containing 52 ppm Au NPs show however a slight increase of the absorbance peak at 600 nm in relation to the unfilled crosslinked gelatin that contrasts with the overall trend. This can be ascribed to the contribution of the aggregation of the Au NPs that leads to an increase of the absorbance at these wavelengths, as seen in Fig. 5a.

Swelling properties

Figure 6 displays the swelling ratio in function of time, for genipin-crosslinked gelatin hydrogels in PBS, and for variable Au NPs load. The unfilled hydrogels swelled smoothly over a period of 60 h, until reaching the equilibrium

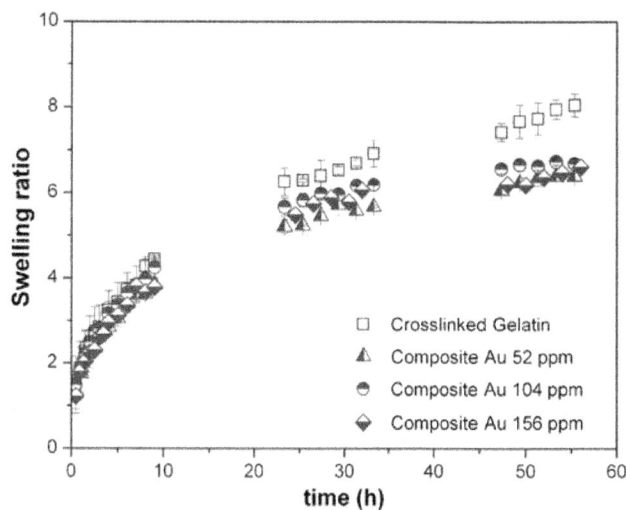

**Fig. 6** Swelling ratio ($Q$) genipin of crosslinked gelatin and derived Au nanocomposites in PBS at 37 °C as a function of time

swelling ratio ($Q_{equil}$) of $8.1 \pm 0.3$. The Au composite hydrogels swelled slightly slower than the unfilled gelatin and the incorporation of Au NPs resulted in a decrease of $Q_{equil}$ from 8.1 to ca. 6.5, regardless the Au NPs content.

Gelatin swelling usually decreases with the extension of crosslinking [12,32,33]. The FTIR and DSC results presented above have shown a decrease on crosslinking when Au NPs have been added to the hydrogel. Therefore, more swelling would be expected for these nanocomposites as compared to unfilled gelatin. Indeed, the opposite effect was observed, which indicates that water diffusion into the gel was limited by the Au NPs dispersed in the network. A similar effect has been reported for montmorillonite (MMT) gelatin nanocomposites [34]. Although MMT hinders the formation of chemical crosslinkages, MMT gelatin nanocomposites crosslinked with dextran dialdehyde has been described as swelling less than the unfilled crosslinked gelatin [34]. The reduction on swelling with MMT content was ascribed to a barrier effect of the MMT NPs to the diffusion of solvent molecules into the gelatin [34].

In vitro MB release

In order to assess the potential of the nanocomposites for controlled release procedures, hydrogels with the highest Au NPs content (156 ppm) have been selected for in vitro MB release studies. The amount of MB released was monitored by measuring the absorbance of the release medium at 663 nm, i.e., at the wavelength of maximum absorbance for MB. The release experiments were carried out with and without laser irradiation ($\lambda = 532$ nm) of the samples.

Figure 7a displays the MB release profiles in PBS from the genipin-crosslinked gelatin and the Au nanocomposite, at 37 °C, in absence of laser irradiation. Both hydrogels

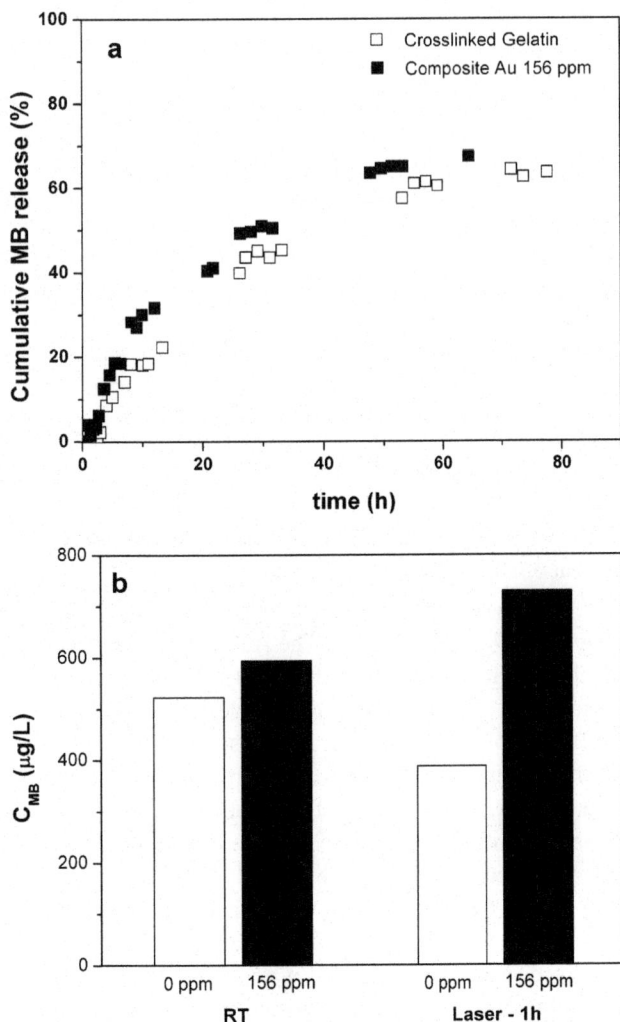

**Fig. 7** In vitro MB release in PBS from unfilled crosslinked gelatin (0 ppm) and derived Au nanocomposites (156 ppm): **a** at 37 °C without irradiation, as function of time; **b** at RT, with and without exposure to green light ($\lambda$=532 nm) after 1 h

surface, as discussed above. In this case, the gelatin polypeptide chains might enfold the Au NPs and thus prevent the sorption of MB molecules onto the NPs surfaces.

Figure 7b shows the amount of MB released to the surrounding medium after exposure of the gels to the laser light ($\lambda$=532 nm) for a period of 1 h. Under exposure to the green laser, the MB released from the Au composite (156 ppm) is approximately twice than in the unfilled gelatin. For comparison, the MB released was also quantified in identical experimental conditions, without exposing the gels to the green light. In absence of laser exposure, both gels (blank and Au composite) release equivalent amounts of MB. This is in agreement with the results above presented (Fig. 7a) for the release experiments performed at physiological temperature without irradiation. The results show an increment of the MB released from the Au composites in 25 % after exposed to the green light, an effect that might be ascribed to the photothermal conversion of Au nanoparticles. Indeed, the wavelength of the incident light ($\lambda$=532 nm) matches the maximum absorption of the SPR band of the Au nanoparticles, and therefore, it is expected that the absorption of this light by the nanoparticles will increase locally the temperature. Since genipin-crosslinked gelatin is sensitive to temperature, as confirmed by DSC results (Fig. 4), it is expected that this heating will induce local transformations in the gel network, thus promoting the release of the encapsulated MB molecules, as observed. It should be noted that, in opposition to the behavior observed in the Au composites, the irradiation of the blank hydrogel seems to decrease the amount of MB released (Fig. 7b), most probably due to photobleaching of some of the MB molecules trapped in the gelatin matrix, as previously reported [38].

## Conclusions

The effect of colloidal Au NPs as nanofillers in the crosslinking and swelling of genipin-crosslinked gelatin hydrogels and methylene blue (MB) release from Au gelatin nanocomposites have been investigated. The citrate-capped Au NPs prevent the reaction between the gelatin and genipin and hinder the formation of the gelatin crosslinking network, most probably due to the interaction of carboxylate groups with protonated amine groups from gelatin residues. Although less crosslinked, the Au gelatin nanocomposites swell less than the unfilled crosslinked gelatin, which suggest that Au NPs may act as barrier for the diffusion of solvent molecules into the gel network. Nevertheless, the incorporation of Au NPs did not affect the MB release from gelatin in isothermal conditions. The gelatin composites were thermo-sensitive at the physiological temperatures (37–45 °C) and revealed optical features from Au NPs. The irradiation of the composite gels with monochromatic

exhibited a sustainable MB release for a period of 70 h. Figure 7a shows a profile in which there is a fast release at an initial stage, followed by a slower rate until a steady concentration is achieved. The release profiles are identical in the first 10 h, and minor differences are observed afterwards. Apparently, the decrease of the swelling ratio in the composites does not affect the MB release profile. Also it appears that the Au NPs do not limit the release of MB from the gelatin matrix, conversely to the effect reported for the MB release from Au nanocomposites prepared with silicone elastomers [35] and κ-carrageenan [36]. The decrease of MB release in these composites was ascribe to the interaction between Au NPs and MB, most probably by chemisorption of MB onto Au NPs surface via sulfur atoms [37], that limits the diffusion of MB from the polymer matrix. The absence of this effect on the gelatin composites supports the assumption that the gelatin chains are strongly adsorbed onto the Au NPs

green light ($\lambda$=532 nm) in the range of the Au NPs SPR band promoted the release of the encapsulated methylene blue, most probably due to the photothermal effect of Au nanoparticles. Thus, applications of these nanocomposites as carriers in light-triggered drug release can be envisaged.

**Acknowledgments** The authors acknowledge FCT—Fundação para a Ciência e Tecnologia (ERA-Eula/0003/2009, Pest-C/CTM/LA0011/ 2011), FSE, and POPH for funding. We thank the RNME (National Electronic Microscopy Network) for TEM images. The authors are very grateful to M.Sc. M.C. Azevedo and Dr. A.V. Girão (University of Aveiro, Chemistry Department) for technical support and to M.Sc. Marco Peres and Prof. Jorge Soares (University of Aveiro, Physics Department) for their assistance in laser experiments.

# References

1. Trindade T, Daniel-Silva AL (2011) Nanocomposite particles for bio-applications: materials and bio-interfaces. Pan Stanford Publishing Pte. Ltd, Singapore
2. Schexnailder P, Schmidt G (2009) Nanocomposite polymer hydrogels. Colloid Polym Sci 287:1–11
3. Satarkar NS, Hilt JZ (2008) Magnetic hydrogel nanocomposites for remote controlled pulsatile drug release. J Control Release 130:246–251
4. Brazel CS (2009) Magnetothermally-responsive nanomaterials: combining magnetic nanostructures and thermally-sensitive polymers for triggered drug release. Pharm Res 26:644–656
5. Kang H, Trondoli AC, Zhu G, Chen Y, Chang Y-J, Liu H, Huang Y-F, Zhang X, Tan W (2011) Near-infrared light-responsive core-shell nanogels for targeted drug delivery. ACS Nano 5:5094–5099
6. Daniel-da-Silva AL, Ferreira L, Gil AM, Trindade T (2011) Synthesis and swelling behavior of temperature responsive $\kappa$-carrageenan nanogels. J Colloid Interf Sci 355:512–517
7. Chen K-Y, Yao C-H (2011) Repair of bone defects with gelatin-based composites: a review. Biomedicine 1:29–32
8. Rujitanaroj P-O, Pimpha N, Supaphol P (2008) Wound-dressing materials with antibacterial activity from electrospun gelatin fiber mats containing silver nanoparticles. Polymer 49:4723–4732
9. Sisson K, Zhang C, Farach-Carson MC, Chase DB, Rabolt JF (2009) Evaluation of cross-linking methods for electrospun gelatin on cell growth and viability. Biomacromolecules 10:1675–1680
10. Chiono V, Pulieri E, Vozzi G, Ciardelli G, Ahluwalia A, Giusti P (2008) Genipin-crosslinked chitosan/gelatin blends for biomedical applications. J Mater Sci Mater Med 19:889–898
11. Bigi A, Cojazzi G, Panzavolta S, Roveri N, Rubini K (2002) Stabilization of gelatin films by crosslinking with genipin. Biomaterials 23:4827–4832
12. Yao C-H, Liu B-S, Chang C-J, Hsu H-S, Chen Y-S (2004) Preparation of networks of gelatin and genipin as degradable biomaterials. Mater Chem Phys 83:204–208
13. Muzzarelli RAA (2009) Genipin-crosslinked chitosan hydrogels as biomedical and pharmaceutical aids. Carbohyd Polym 77:1–9
14. Mi FL, Tan YC, Liang HC, Huang RN, Sung HW (2001) In vitro evaluation of a chitosan membrane crosslinked with genipin. J Biomater Sci, Polym Ed 12:835–850
15. Daniel M-C, Astruc D (2004) Gold nanoparticles: assembly, supramolecular chemistry, quantum-size-related properties, and applications toward biology, catalysis, and nanotechnology. Chem Rev 104:293–346
16. Pissuwan D, Cortie CH, Valenzuela SM, Cortie MB (2007) Gold nanosphere-antibody conjugates for hyperthermal therapeutic applications. Gold Bull 40:121–129
17. Choi WI, Kim JY, Kang C, Byeon CC, Kim YH, Tae G (2011) Tumor regression in vivo by photothermal therapy based on gold-nanorod-loaded, functional nanocarriers. ACS Nano 5:1995–2003
18. Matteini P, Ratto F, Rossi F, Centi S, Dei L, Pini R (2010) Chitosan films doped with gold nanorods as laser-activatable hybrid bioadhesives. Adv Mater 22:4313–4316
19. Lim SY, Lee JS, Park CB (2010) *In Situ* growth of gold nanoparticles by enzymatic glucose oxidation within alginate gel matrix. Biotechnol Bioeng 105:210–214
20. Zhang J-J, Gu M-M, Zheng T-T, Zhu J-J (2009) Synthesis of gelatin-stabilized gold nanoparticles and assembly of carboxylic single-walled carbon nanotubes/Au composites for cytosensing and drug uptake. Anal Chem 81:6641–6648
21. Neupane MP, Park IS, Bae TS, Yin HK, Uo M, Watari F (2011) Titania nanotubes supported gelatin stabilized gold nanoparticles for medical implants. J Mater Chem 21:12078–12082
22. Daniel-da-Silva AL, Moreira J, Neto R, Estrada AC, Gil AM, Trindade T (2012) Impact of magnetic nanofillers in the swelling and release properties of $\kappa$-carrageenan hydrogel nanocomposites. Carbohyd Polym 87:328–335
23. Daniel-da-Silva AL, Fateixa S, Guiomar AJ, Costa BFO, Silva NJO, Trindade T, Goodfellow BJ, Gil AM (2009) Biofunctionalized magnetic hydrogel nanospheres of magnetite and $\kappa$-carrageenan. Nanotechnology 20:355602
24. Singha S, Dasgupta A, Kr DH (2011) Gold nanoparticle induces masking of amines and some therapeutic implications. J Nanosci Nanotechnol 11:7744–7752
25. Coleman MM, Lee JY, Painter PC (1990) Acid salts and the structure of ionomers. Macromolecules 23:2339–2345
26. Hashim DM, Man YBC, Norakasha R, Shuhaimi M, Salmah Y, Syahariza ZA (2010) Potential use of Fourier transform infrared spectroscopy for differentiation of bovine and porcine gelatins. Food Chem 118:856–860
27. Butler MF, Y-F NG, Pudney PDA (2003) Mechanism and kinetics of the crosslinking reaction between biopolymers containing primary amine groups and genipin. J Polym Sci, Part A: Polym Chem 41:3941–3953
28. Brewer SH, Glomm WR, Johnson MC, Knag MK, Franzen S (2005) Probing BSA binding to citrate-coated gold nanoparticles and surfaces. Langmuir 21:9303–9307
29. Lacerda SHDP, Park JJ, Meuse C, Pristinski D, Becker ML, Karim A, Douglas JF (2010) Interaction of gold nanoparticles with common human blood proteins. ACS Nano 4:365–379
30. Lee S-W, Lim J-M, Bhoo S-H, Paik Y-S, Hahn T-R (2003) Colorimetric determination of amino acids using genipin from *Gardenia jasminoides*. Anal Chim Acta 480:267–274
31. Park J-E, Lee J-Y, Kim H-G, Hahn T-R, Paik Y-S (2002) Isolation and characterization of water-soluble intermediates of blue pigments transformed from geniposide of *Gardenia jasminoides*. J Agric Food Chem 50:6511–6514
32. Bigi A, Cojazzi G, Panzavolta S, Rubini K, Roveri N (2001) Mechanical and thermal properties of gelatin at different degrees of glutaraldehyde crosslinking. Biomaterials 22:763–768
33. Goutam T, Analava M, Rousseau M, Basak A, Sarkar S, Pal K (2011) Crosslinking of gelatin-based drug carriers by genipin induces changes in drug kinetic profiles in vitro. J Mater Sci: Mater Med 22:115–123
34. Li P, Zheng JP, Ma YL, Yao KD (2003) Gelatin/montmorillonite hybrid nanocomposite. II. Swelling behavior. J Appl Polym Sci 88:322–326

35. Perni S, Piccirillo C, Pratten J, Prokopovich P, Chrzanowski W, Parkin IP, Wilson M (2009) The antimicrobial properties of light-activated polymers containing methylene blue and gold nanoparticles. Biomaterials 30:89–93

36. Salgueiro AM, Daniel-da-Silva AL, Fateixa S, Trindade T (2013) κ-Carrageenan hydrogel nanocomposites with release behavior mediated by morphological distinct Au nanofillers. Carbohyd Polym 91:100–109

37. Narband N, Uppal M, Dunnill CW, Hyett G, Wilson M, Parkin IP (2009) The interaction between gold nanoparticles and cationic and anionic dyes: enhanced UV–visible absorption. Phys Chem Chem Phys 11:10513–10518

38. Yunus WMM, Sheng CK, Yunus WMZW (2003) Study on photobleaching of methylene blue doped in PMMA, PVA and gelatin using photoacoustic technique. J Nonlinear Opt Phys 12:91–100

# 13

# Effect of gold alloying on stability of silver nanoparticles and control of silver ion release from vapor-deposited Ag–Au/polytetrafluoroethylene nanocomposites

N. Alissawi · V. Zaporojtchenko · T. Strunskus ·
I. Kocabas · V. S. K. Chakravadhanula · L. Kienle ·
D. Garbe-Schönberg · F. Faupel

**Abstract** In the present study, nanocomposites containing Ag–Au alloy nanoparticle ensembles with various compositions on polytetrafluoroethylene were prepared by physical vapor deposition. After a certain time of immersion of the samples in water, oxidation and dissolution of the Ag nanoparticles (AgNPs) occurred, and changes in the morphology, optical properties and composition of the nanocomposites were examined using transmission electron microscopy, ultraviolet-visible spectroscopy and X-ray photoelectron spectroscopy (XPS), respectively. The composition-dependence and the time-dependence of the silver ion release were studied, and the concentration of the silver ions in water was detected using inductively coupled plasma mass spectrometry. The results indicate that with increasing gold fraction in the Au–Ag alloy nanoparticles, a strong improvement of the oxidation resistance of the AgNPs occurs. The dissolution of Ag is rapid at the first contact of the sample with water until a saturation state is reached. XPS synchrotron measurements with different excitation energies show that the depletion of Ag from the nanoparticles does not lead to the formation of a Au-rich shell and that the atomic mobility is high enough in the small nanoparticles of about 5 nm average size to equilibrate any concentration gradient.

**Keywords** Au–Ag alloy nanoparticles · Silver ion release · Surface plasmon resonance · Sputtered PTFE · PVD · ICP-MS

N. Alissawi · V. Zaporojtchenko · T. Strunskus · I. Kocabas ·
F. Faupel (✉)
Institute for Materials Science, Multicomponent Materials, Faculty of Engineering, Christian-Albrechts-University (CAU) Kiel, Kiel, Germany
e-mail: ff@tf.uni-kiel.de

V. S. K. Chakravadhanula · L. Kienle
Institute for Materials Science, Synthesis and Real Structure, Faculty of Engineering,
Christian-Albrechts-University (CAU) Kiel,
Kaiser Str. 2,
24143 Kiel, Germany

D. Garbe-Schönberg
Institute of Geosciences/ICP-MS Lab,
Christian-Albrechts-University (CAU) Kiel,
Ludewig-Meyn-Str. 10,
24118 Kiel, Germany

V. S. K. Chakravadhanula
Helmholtz Institute Ulm (HIU) Electrochemical energy storage,
Karlsruhe Institute of Technology (KIT),
Albert-Einstein-Allee 11,
89081, Ulm, Germany

## Introduction

Bimetallic structures such as alloyed silver and gold nanoparticles have attractive catalytic, electronic, and optical properties that can vary with respect to size and composition of the different metals within the nanostructure [1–3]. Formation of Ag–Au alloy structures takes place by mixing them on atomic level without distinct phase boundaries [4]. Due to their plasmonic properties, structural changes upon particle medium interactions in Ag and Au nanoparticles can be easily screened. Since their surface plasmon resonance peaks (SPR) are highly sensitive to the particle's size, shape, and surrounding, they can give a lot of information about the status of the nanoparticles [5] and tunable physical properties that can be obtained by changing composition ratio of the alloy nanoparticles [6].

Moreover, the usage of noble metals with silver has shown different silver dissolution behavior. An early study done by Forty examined the oxidation reaction of a bulk silver/gold alloy system in aqueous environment under the phenomena of selective dissolution. Silver is less noble than gold and releases easily from the gold alloy system, by leaving a high content of gold residue behind and creating surface vacancies and disordered areas near the surface. When all the surface of alloy is dominated by gold atoms, alloy becomes passivated, and volume and surface diffusions start to play an important role [7].

Recently, several bimetallic/polymer nanocomposites were developed showing promising properties in the silver ion release studies as it has been demonstrated that galvanic coupling of silver with platinum [8–10] or gold [11] strongly increases the silver ion release and its antimicrobial activity.

Barcikowski et al. have examined copper/silver and gold/silver nanocomposites containing nanoparticles dispersed in a polymer matrix without direct contact generated by laser ablation in liquids. In the Cu/Ag system, they found that electrochemical oxidation reaction between the two different metals enhanced the ion release of the less noble Cu by the ion-mediated electrochemical reaction, and silver ion release was retarded because, when two noble metals with different standard electrochemical potential are in contact with each other in an electrolyte, one of the metals oxidizes or corrodes. This corrosion process is called galvanic corrosion (galvanic coupling), and it enhances the dissolution of the less noble metal atom in alloys, thin films, or sputtered targets. The more noble metal acts as a cathode, and the less noble metal acts as an anode in the electrolyte, thus it provides ion transfer between anode and cathode. Whereas in the case of Au–Ag nanocomposite films, no effect on silver ion release was observed due to the absence of mobile gold ions [12, 13].

On the other hand, Besner et al. have observed a strong increase in the oxidation resistance with the increase in the Au fraction inside the Ag–Au alloy NPs which could probably result from significant modifications of the electrochemical properties of the NPs. Ag and Au significantly differ in their electronic properties. In this respect, colloidal AgNPs are known to act as an electron storage (donor) medium. Hence, they would easily dissociate through an anodic reaction to release an electron and an Ag ion. In contrast, Au is known to possess some electron acceptor properties. The addition of Au atoms inside the AgNPs would then contribute to the formation of internal electron traps, inhibiting the dissolution of the AgNPs and the release of Ag ions. Hence, the formation of nanoalloys which exhibit more stability in comparison to Ag and higher plasmonic response in comparison to Au could also be beneficial for surface-enhanced Raman scattering applications [14].

We have investigated the silver ion release properties of well-defined model systems consisting of two-dimensional Ag nanoparticle ensembles on top of a polytetrafluoroethylene (PTFE) films which are either directly accessible or covered by polymer layers of well-defined thickness and composition. We observed a correlation between changes in surface plasmon resonance and kinetics of Ag ion release and that the strong dependence of the silver ion release on the particle size leads to a significant redistribution of the composite morphology and suppression of the Ag ion release rate with time. It was also shown that a polymer barrier stabilizes the morphology of the composites and can be applied to control the Ag ion release rate [15]. Besides, in situ electrochemical impedance spectroscopy–ultraviolet-visible spectroscopy (UV-vis) and AFM studies have been also preformed on the same nanocomposite system by our collaborative group indicating the stability of the AgNPs after covering by a PTFE barrier and the slowing of the silver ion release rate [16].

In this work, we report our investigations of silver ion release kinetics from silver–gold alloy nanoparticle ensembles on an insulating hydrophobic polymer substrate and compare these model systems to composites with only pure AgNPs. We examined the changes that occurred after immersing the samples in water for certain time. Two-dimensional ensembles of 5.5 nm nominal thickness consisting of Ag–Au alloy NPs of 5 nm average size were prepared on a highly cross-linked sputtered PTFE substrate using physical vapor deposition (PVD) techniques. The fractions of Ag and Au were varied, and the optical properties, composition, and morphology of the nanocomposites were examined using UV-vis, X-ray photoelectron spectroscopy (XPS), and transmission electron microscopy (TEM), respectively. The influence of varying the fraction of the two metals in the alloy NPs on the nanocomposite properties and on the silver ion release rate was also investigated. The time-dependent silver ion release was measured by inductively coupled plasma mass spectrometry (ICP-MS). The results show that gold alloying strongly improves the oxidation resistance of the AgNPs. The dissolution of Ag quickly slows down after first exposure to water and reaches a saturation state. No concentration gradient is established upon Ag depletion due to the high atomic mobility in the small NPs.

## Experimental

### Preparation of the model system

The 20-nm thin films of PTFE were deposited on quartz ($1 \times 1$ cm$^2$), silicone ($0.5 \times 0.5$ cm$^2$), and carbon-coated copper TEM grids by sputtering from a 50 mm diameter polymer target by RF magnetron source to prevent charging of the target. Preparation of nanocomposites based on sputtering of polymers was reported by the group of Biederman before [17]. The sputtering process of PTFE results in the deposition of a highly crosslinked flouropolymer film where surface structure and contact angle determination were shown in our group's previous work of Schürmann et al. [18] and Hassel et al. [19]. The detailed procedure of polymer vapor deposition is described in our previous work [20]. The experiment was done in a homemade stainless steel vacuum chamber, which was initially evacuated to a pressure below $10^{-6}$ mbar. A RF power of 20 W and a deposition rate of about 4 nm/min were used resulting in the deposition of a highly crosslinked flouropolymer film. Later on, samples were mounted in a homemade thermal evaporation chamber and beads of silver (99.99 %, Sigma Aldrich) and gold (99.99 %, Goodfellow) were loaded into two separated alumina crucibles, and the chamber was evacuated to a pressure below $10^{-6}$ mbar by a vacuum pump system. The 5.5 nm nominal thickness of Ag–Au alloy NPs with various compositions were deposited on top of the polymer film by simultaneous thermal evaporation from the two separated crucibles. The silver evaporator current was increased slowly in order to avoid thermal shocks by raising the voltage 5 mV/s up to 6.4 V and a current of 2.2 A with a deposition rate of 0.5 nm/min. The deposition rate of gold was varied for the different compositions. For 10 % Au samples, gold evaporator was heated up to 15.8 V and 3.5 A with a deposition rate of 0.05 nm/min. For 30 % Au samples, the evaporator was heated up to 18 V and 3.7 A with a deposition rate of 0.25 nm/min, while for 50 % Au samples the evaporator was heated up to 18.5 V, 3.8 A with a deposition rate of 0.5 nm/min. A STM−100/ MF quartz crystal microbalance system was used for deposition rate and film thickness monitoring. A DEK-TAK 8000 profilometer was used in the calibration of the film thickness with a radius of diamond stylus as 12.5 μ and a stylus tracking force factory, set to 50 mg).

Characterization

*Transmission electron microscopy (TEM)*

The morphology of the nanocomposites was examined using a Tecnai F30G$^2$ ST (FEI) transmission electron microscope operating at 300 kV. Samples were prepared on carbon-coated copper grids. The size distribution and average size of the NPs were evaluated by Gatan Digital Micrograph software (by measuring the diameter of each particle in a representative image). Energy-filtered TEM and scanning TEM-HAADF were used to carry out the TEM studies.

*UV-visible spectroscopy (UV-vis)*

The UV-visible absorption spectra of the prepared nanocomposites on quartz glass were recorded using a PerkinElmer Lambda 900 UV-vis/NIR spectrophotometer from 300 to 800 nm with 2 nm resolution. A 20 nm PTFE on quartz sample was used as a reference sample, and all the UV-vis plots are presented in this paper after linear background correction. Due to the small size of the NPs in our model system, the scattering was neglected, and only absorption was considered.

*X-ray photoelectron spectroscopy (XPS)*

Films deposited on silicon wafers were used for XPS measurements which were performed by XPS full lab setup from Omicron Nanotechnology GmbH, in its ultra high vacuum analytical chamber where the pressure is usually in $10^{-9}$ mbar range. The analytical chamber consists of sample holder, X-ray source, a hemispherical electron analyzer (VSW EA 125), a detection system which counts the number of photoelectrons, and a data acquisition and processing system. An aluminum anode of the X-ray source (VG Microtech XR3E2) was used here. XPS was used to determine elemental composition of alloy systems taking into account that the information depth is approximately $2\lambda$ (2–3 nm) where $\lambda$ is the elastic mean free path. More details about using XPS for the study of metal/polymer covering were discussed by Zaporojtchenko et al. [21].

In addition, selected samples were analyzed by XPS with different excitation energies at the PM-4 beamline of the synchrotron facility HZB-BESSY II using the SurICat endstation. Measurements were done with an exit slit of 100 μm and with photon energies ranging from 500 to 1,500 eV. Spectra were taken in normal electron emission using a hemispherical electron analyzer (Scienta SES 100) operated at a pass energy of 50 eV. The spot size on the sample is in the range of 0.5 mm×0.2 mm. All binding energies and intensities were referenced to a pure gold foil permanently mounted in the vacuum system below the sample holder.

*Inductively coupled plasma spectrometry mass (ICP-MS)*

ICP-MS is a sensitive method to detect a wide range of elements with a concentration down to nanograms per liter (ppt; parts per trillion) and below. In this work, an Agilent 7500cs instrument was used with liquid sample introduction by a PFA micro-nebulizer. Calibration was against aqueous multi-element solutions using indium and rhenium for internal standardization. The procedure detection limit was 10 ng/l (ppt) Ag. All results are blank-subtracted averages of three replicate measurements. Analytical quality control was monitored by multiple analyses of procedural blanks, unknown samples, and certified reference materials NIST 1643e, IAEA W4. The analytical error for the Ag measurements was <1 % RSD (1 sigma). The actual values for replicate measurements of two samples were 0.4 and 0.8 %RSD, respectively.

Silver ion release measurements

The Ag ion release rate was studied by observing the changes that occur as samples were immersed in air saturated 10 mL deionized water (pH=7, $\sigma$=0.06 $\mu$S/cm at T=13 °C) in small bottles made of high-density polyethylene at room temperature. More details about the pH control can be found in our previous publication [15]. UV-vis and XPS measurements were carried out on the samples during an immersion time of seven days starting from 40 min to check the variations, and the water was checked by ICP-MS to detect the Ag ions concentrations. Additionally, TEM measurements were done after immersion of the coated TEM grids in water for 3 days.

## Results and discussion

### Au–Ag alloy NP systems

According to the nucleation and growth model of metal atoms on polymeric surfaces shown in our previous work [22], metal atoms initially adsorb on the polymeric surface until they meet with other atoms on their diffusing paths, and then random or preferred nucleation at defects starts. Initially, the growth of metal nanoparticles on polymeric surface starts by forming small nuclei at certain sites which grow by the continued deposition of the metal via direct impingement and surface diffusion. For alloy NPs, silver and gold were evaporated simultaneously, and a complete alloy formation is expected as gold and silver are in the same group of periodic table and have similar atomic radius and same structures that make them fully miscible in each other in the solid solutions [23]. Figure 1a, b shows a bright field (BF) TEM image for 5.5 nm of 70 % Ag–30 % Au alloy NPs on PTFE and its corresponding particle size distribution. A uniform distribution could be observed with an average diameter of about 5 nm, except for some larger-diameter particles having irregular shape contribute to coalescence growth of the particles.

The 5.5-nm nominal thickness Ag and Au alloy nanoparticles have almost a similar size distribution and an average diameter of about 5 nm as the pure AgNPs prepared in the same way and same amount as was shown in the previous work [15]. The position of the plasmon absorption peak of these alloy NPs, however, depends linearly on the composition of the alloy particles when expressed in terms of the gold fraction as reported previously [2, 14, 24, 25]. The absorption spectra obtained by UV-vis for 5.5 nm pure Ag and Ag–Au alloy NPs on PTFE with various compositions are shown in Fig. 2. The alloying of Ag–Au NPs on PTFE can be concluded from the resulted optical absorption spectra which show a single plasmon resonance peak, whereas two bands would be expected for the case of phase

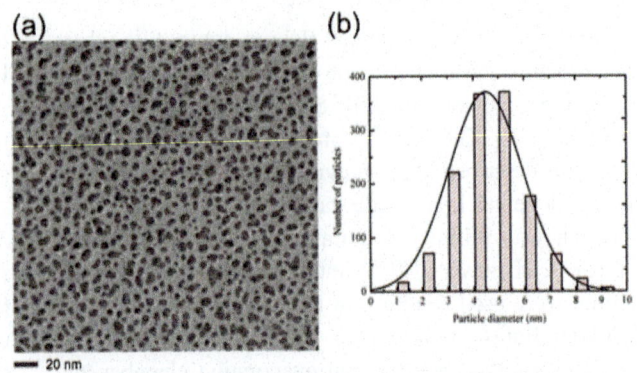

**Fig. 1** BF-TEM image (**a**) and the corresponding particle size distribution (**b**) of 5.5 nm 70 % Ag–30 % Au alloy NPs on PTFE

separated NPs [26, 27]. Alloy formation of Ag–Au NPs caused a plasmon maximum in the UV-vis spectrum between the absorption maximum of Ag and Au that varies with respect to composition distribution. The plasmon peak position was red-shifted due to the continuous change of the d-band energy level that contributes to the interband transition term in the dielectric function as a result of an increase in Au composition, and a damping of the absorbance maxima is seen attributed to the higher electron scattering by foreign atoms upon alloying and the gold d-$_{sp}$ interband transition [24, 26, 28].

### Silver ion release studies

The Ag ion release kinetics was studied by observing the changes as samples were immersed in water for different time intervals starting from 40 min up to 7 days. The concentration of the released silver ions from each sample was measured by ICP-MS where results showed a negligible Au ion concentration. Additionally, UV-vis and XPS measurements were

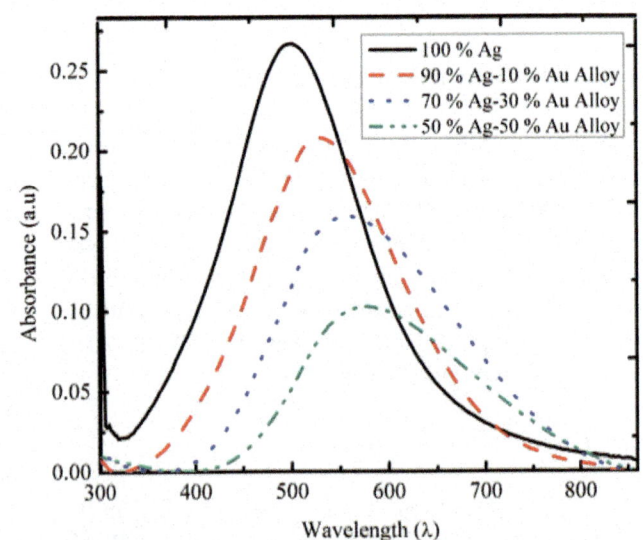

**Fig. 2** UV-vis spectra for 5.5 nm nominal thickness of various composition of Ag–Au alloy NPs

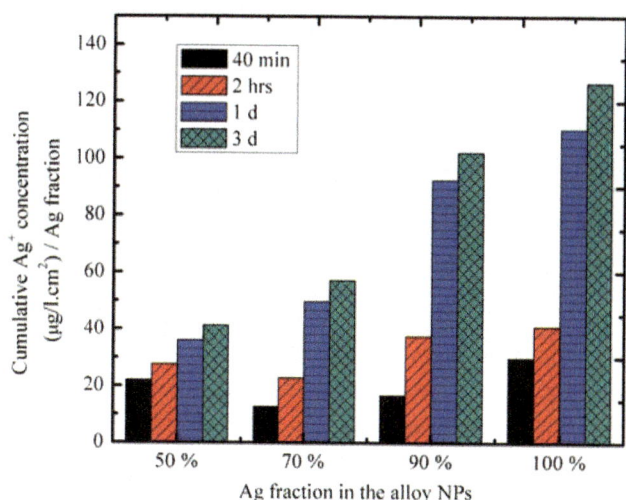

**Fig. 3** Cumulative concentration of silver ions released from Ag–Au alloy systems at different immersion times in water and different Ag fraction

carried out for all samples after immersion into aqueous medium. The cumulative concentrations of the released Ag ions (micrograms per liter per square centimeter) normalized to the silver fraction in the alloy nanocomposites were measured at several time periods as shown in Fig. 3. All silver ion concentrations are the mean values of three ICP-MS measurements.

One notices that the release of the silver ions is not just proportional to the amount of silver present in the nanoparticles. In that case, all four diagrams should look the same. But one notes significant differences for short and long release times. In all systems, there was a rapid release of Ag ions at the beginning (determined after 40 min) due to the first interaction of the nanoparticles with water and the Ag dissolution from the surface. The fastest release of the silver is here observed for the pure silver system and the 50 % alloy, while the relative release is smallest for the 70 % alloy system. This can be explained with the entropy of mixing which is most favorable for a 50 % alloy and decreases from there for higher silver fractions [29]. Thus, the initial silver release of the 50 % alloy is not slowed down so much, and the initial release of the 70 % and 90 % alloy is even below the 50 % alloy. But, after 1 day, the relative silver release scales with the amount of silver present, and the pure silver shows the largest silver release, whereas the release is significantly reduced for the alloys. Here, the slowing down of the silver release is directly correlated with the gold fraction in the alloy. The more gold is present in the alloy the more the release is slowed down. For the 90 % alloy system, one observes still a big difference between the release after 2 h and 1 day, whereas the change for the 50 % alloy system has become already very small. To explain this behavior, one has to consider the change in composition caused by the silver release. For the nanoparticles with

initially 90 % silver fraction, the composition will move towards 50 % with a still-higher silver than gold fraction. Here, entropic contributions favor further silver release. The situation is different for the 50 % alloy. Here, the entropic contribution due to change in composition is not favorable for further silver release, and thus, the silver release will be slowed down.

Additionally, increasing the Au fraction inside the NPs could lead to a significant change of the electron transfer properties of the alloy NPs with the composition. This interaction between Ag–Au causes depletion and reduces the silver ion release. Thus, a rapid dissolution of Ag occurred only at the first contact of the sample with water, and when the composition of the alloy NPs becomes Au-rich, a saturation state is approached. This could be an advantage for potential applications where the Ag ion release is required to be rapid and effective at the beginning then a slower dissolution rate, leading to a continuous release of Ag ions for long-term application.

XPS analysis was also performed to examine changes which occurred for the 70 % Ag–30 % Au alloy NPs on PTFE film deposited on Si substrate after immersion in water. Intensity of each element and area under the XPS peaks give quantitative information about the elements within the composite. The sensitivity factors of Ag and Au are approximately the same [30], so the alloy composition can be determined from the intensity of XPS signal which is the area under the Ag and Au peaks.

Variations in Ag and Au peaks intensities were investigated before and after immersion of samples for several days in water. Ag and Au intensities were normalized with respect to the carbon intensity in order to eliminate time-dependent changes in the intensity of the XPS spectra, as

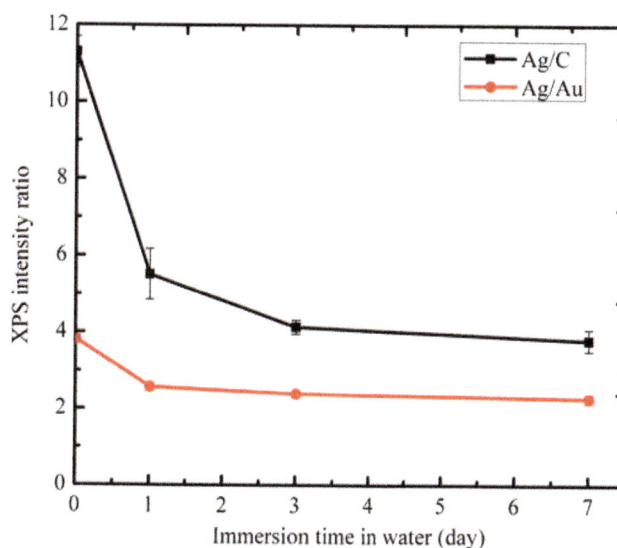

**Fig. 4** Change in the XPS Ag and Au intensities for 5.5 nm 70 % Ag–30 % Au alloy NPs on PTFE after immersion in water

**Table 1** Summary of energy-dependent XPS analysis

| Photon energy (eV) | Kinetic energy at Ag 3d lines (eV) | Inelastic mean free path (probing depth) (nm)[a] | Decrease of Ag 3d signal after 3 days immersion in water (%) |
|---|---|---|---|
| 500 | 130 | 0.43 (0.86) | 55.3±3 |
| 650 | 280 | 0.61 (1.21) | 58.3±3 |
| 900 | 530 | 0.89 (1.78) | 53.5±3 |
| 1,200 | 830 | 1.21 (2.41) | 52.8±3 |
| 1,500 | 1,130 | 1.50 (3.00) | 53.2±3 |

[a] calculated from the data provided by Tanuma et al. assuming a 70 % Ag and 30 % Au composition [31]

shown in Fig. 4. Before immersion in water, the ratio of Ag to Au was a little higher than the expected value. This could be caused by a slightly higher deposition rate of the silver during the co-evaporation of the alloy NPs. Note that, prior to co-evaporation, the deposition rates were calibrated independently, and no cross-influence of the evaporators was taken into account. In addition, the deviation from the expected value could be due to systematic errors in the determination of the peak areas, in particular, for the weaker gold lines.

Due to dissolution of silver after immersion of samples in water, the Ag/C and Ag/Au intensity ratios decrease rapidly at the beginning, and after the first day, the decrease in the ratios becomes slower with time indicating the approach of a saturation state. One could speculate that this saturation states occurs because silver ions are released only from the outer layers of the nanoparticles and that a gold shell is formed subsequently by the gold that is left behind.

In order to check this possibility, energy-dependent XPS spectra were measured on one set of the 5.5 nm layer of the 70 % Ag–30 % Au alloy nanoparticles on sputtered PTFE. One sample was measured as prepared and the other one after 3 days of immersion in water. By varying the photon energy, the probing depth is varied (see Table 1). The spectra were analyzed by normalizing the Au 4f lines (at

84.0 and 87.6 eV) to the same intensity for all spectra, and then the decrease of the signal of the Ag 3d lines (at 368.3 and 374.2 eV) due to immersion in water was determined for the different photon energies, i.e., for different probing depths (see Table 1).

The data show no significant variation of the silver depletion with XPS probing depth, indicating that the silver distribution within the nanoparticles remains essentially unchanged, i.e., it remains alloy-like. Therefore, the formation of a gold shell can be ruled out, and the observed slowing down of the silver release of the gold containing alloys must be due to a shift of the chemical potential with increasing gold fraction as has been suggested above and in different words also by Besner et al. [14].

At first glance, this result could appear somewhat surprising since for bulk Ag–Au alloy systems the dissolution of silver leads to formation of porous gold structures and for larger nanoparticles by a galvanic exchange reaction indeed hollow nanoparticles with a gold shell can be produced [32]. But as suggested by Shibata et al. [33], the situation is different for smaller nanoparticles, in particular, if they have defects. Here, the vacancies created in the alloy nanoparticles by the released silver atoms should allow a fast redistribution of the silver in a range of several nanometers, thus leading to an always almost homogeneous spatial distribution of the silver in the small nanoparticles.

TEM measurements were also done after 3 days of immersion of the sample deposited on the TEM grid in water. Figure 5a, b shows changes in the morphology and in the surface amount of the 5.5 nm alloy NPs of 70 % Ag–30 % Au before and after immersion in water. The nanoparticles density looks smaller after 3 days in water, and the interparticle distance increases as a result of the reduction of the Ag

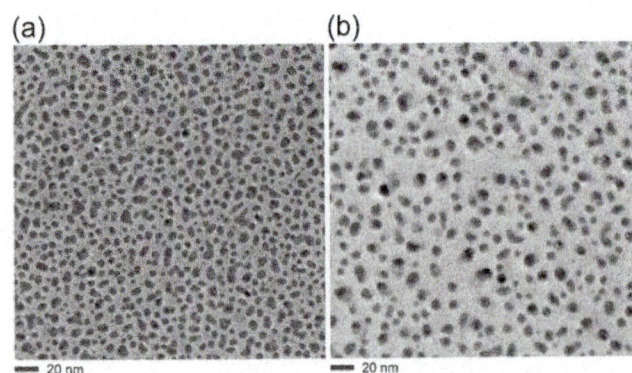

(a)                                    (b)

**Fig. 5** 5.5 nm 70 % Ag–30 % Au alloy NPs on PTFE BF-TEM image before immersion in water (**a**) and after 3 days in water (**b**)

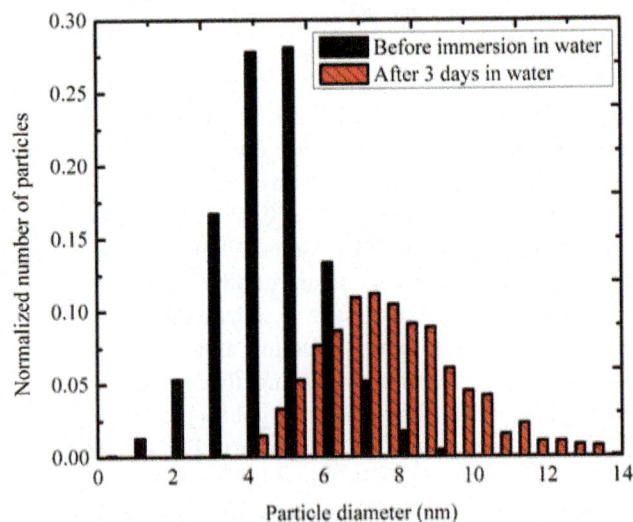

**Fig. 6** Particle size distribution of the 5.5 nm 70 % Ag–30 % Au alloy NPs on PTFE before and after immersion the TEM grid in water for 3 days

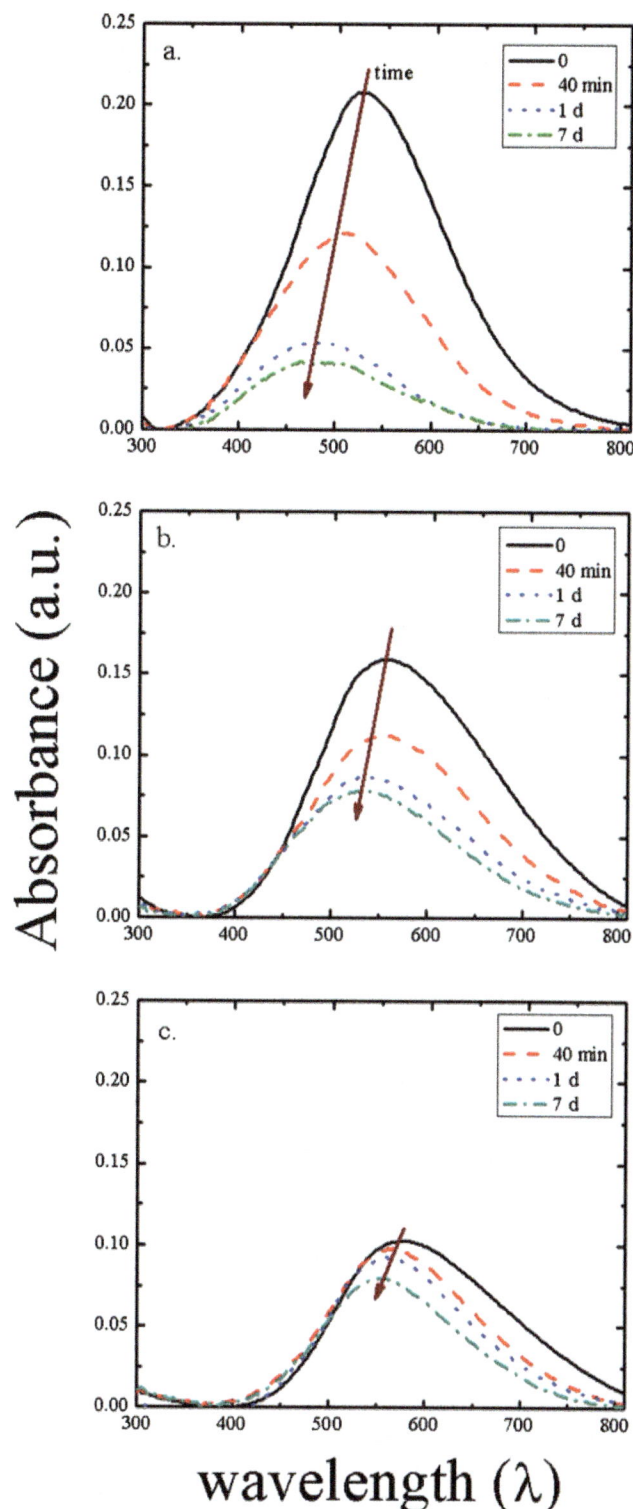

**Fig. 7** UV-vis spectra for 5.5 nm of **a** 90 % Ag–10 % Au; **b** 70 % Ag–30 % Au; and **c** 50 % Ag–50 % Au alloy NPs on PTFE at different immersion times in water

dependence of its standard electrode potential, in a way similar to what we observed before for pure AgNPs [15, 16]. Moreover, no Au shell formation was observed here either. The change in the particle size distribution is shown in Fig. 6, where the number of particles was normalized to the total number of particles for each case.

Further investigations were performed using the UV-vis technique by observing the changes in the absorbance spectra of 5.5 nm alloy nanocomposite with various compositions as a function of the immersion time in water; see Fig. 7. After immersion in water, SPR peak positions and absorbance maxima values were changed due to the oxidation and the release of silver ions into aqueous media. The shape of the absorbance peaks for each time period differs from each other too. One expects a red shift towards the typical gold plasmon absorption band due to the increase of the Au fraction in the nanoalloy [2, 25]. In contrast, a dominant shift to smaller wavelength occurred due to changes in the morphology, size distribution, and the interparticle distance. These changes are similar to the changes that occurred in the pure AgNPs system shown in our previous work [15]. However, the rate of the change in the SPR peak absorbance maxima values and its position after immersion of the samples in water is slower than that for the pure silver due to the increase of the Au fraction on the expense of Ag in the alloy, in accordance with the results from the ICP-MS and XPS measurements.

**Conclusions**

Model systems consisting of two-dimensional ensembles of 5.5 nm nominal thickness of Ag–Au alloy NPs on a 20 nm highly cross-linked sputtered PTFE substrate were prepared using PVD techniques to investigate the effect of gold alloying on the silver ion release upon immersion in water. Alloying of Au increases the oxidation resistance of the nanoparticles and results in a reduced absolute release rate. We noticed that the release of the silver ions is proportional to the amount of silver present in the alloy nanoparticles with significant differences for short and long release times. In all systems, a rapid dissolution of Ag occurred only at the first contact of the sample with water, and later on, a saturation state is approached which is also affected by concentration-dependent entropic contributions. XPS and energy-dependent XPS spectra analysis were also performed to examine changes which occurred for the 70 % Ag–30 % Au alloy NPs system before and after immersion of samples in water, and results showed that Ag depletion does not lead to a concentration gradient but rather to a homogeneous drop in Ag concentration in the small NPs of about 5 nm average size investigated. TEM measurements for the same

fraction due to Ag ion release which is more favorable from the smaller particles. Additionally, other NPs may undergo Ostwald ripening process, i.e., growth of large particles by dissolution of small particles driven by the particle size

alloy system showed changes in the morphology and in the surface amount of the alloy NPs, and no Au shell formation was observed here either. The results suggest that Au alloying can be made instrumental to tailor silver ion release in silver-based nanocomposites.

**Acknowledgment** We acknowledge the financial support for the initial part of the work by the World Gold Council under project GROW RP 07–07. The work was continued by financial support of the German Research Foundation (Deutsche Forschungsgemeinschaft; DFG) under grant number (FA 234/20-1). The authors are grateful for Dipl. Ing. S. Rehders for constructing the deposition chambers and for his expertise in solving technical problems. Many thanks to Dipl. Ing. U. Westernströer for her help in performing the ICP-MS measurements. Moreover, we would like to thank M.Sc. B. Erkartal and Dr. U. Schürmann for TEM measurements. Thanks go also to M. Bauer and Dr. R. Ovsyannikov for support of the XPS synchrotron experiments.

# References

1. Fleger Y, Rosenbluh M (2009) Surface plasmons and surface enhanced Raman spectra of aggregated and alloyed gold-silver nanoparticles. Res lett Opt 2009:1–5
2. Link S, Wang ZL, El-Sayed MA (1999) Alloy formation of gold-silver nanoparticles and the dependence of the plasmon absorption on their composition. J Phys Chem B 103:3529–3533
3. Hubenthal F, Ziegler T, Hendrich C, Alschinger M, Träger F (2005) Tuning the surface plasmon resonance by preparation of gold-core/silver-shell and alloy nanoparticles. Eur Phys J D 34:165–168
4. Zhang Q, Xie J, Lee JY, Zhang J, Boothroyd C (2008) Synthesis of Ag@AgAu metal core/alloy shell bimetallic nanoparticles with tunable shell compositions by a galvanic replacement reaction. Small 4(8):1067–1071
5. Ashby MF, Ferreira Paulo JSG, Schodek-Daniel L (2009) Nanomaterials, nanotechnologies and design. Butterworth-Heinemann, Amsterdam; Boston
6. He ST, Xie SS, Yao JN, Gao HJ, Pang SJ (2002) Self-assembled two-dimensional superlattice of Au-Ag alloy nanocrystals. Appl Phys Lett 81:150–152
7. Forty AJ (1981) Micromorphological studies of the corrosion of gold alloys. Gold Bull 14(1):25–35
8. Dowling DP, Betts AJ, Pope C, McConnell ML, Eloy R, Arnaud MN (2003) Anti-bacterial silver coatings exhibiting enhanced activity through the addition of platinum. Surf Coat Technol 163–164:637–640
9. Betts AJ, Dowling DP, McConnell ML, Pope C (2005) The influence of platinum on the performance of silver-platinum anti-bacterial coatings. Mater Des 26:217–222
10. Kumar R, Howdle S, Münstedt H (2005) Polyamide/silver antimicrobials: effect of filler types on the silver ion release. J Biomed Mater Res 75B(2):311–319
11. Zaporojtchenko V, Podschun R, Schürmann U, Kulkarni A, Faupel F (2006) Physico-chemical and antimicrobial properties of co-

12. sputtered Ag-Au/PTFE nanocomposite coatings. Nanotechnology 17:4904–4908
13. Hahn A, Brandes G, Wagener P, Barcikowski S (2011) Metal ion release kinetics from nanoparticle silicone composites. J Controlled Release 154:164–170
14. Hahn A, Günther S, Wagener P, Barcikowski S (2011) Electrochemistry-controlled metal ion release from silicone elastomer nanocomposites through combination of different metal nanoparticles. J Mater Chem 21:10287–10289
15. Besner S, Meunier M (2010) Femtosecond laser synthesis of AuAg nanoalloys: photoinduced oxidation and ions release. J Phys Chem C 114:10403–10409
16. Alissawi N, Zaporojtchenko V, Strunskus T, Hrkac T, Kocabas I, Erkartal B, Chakravadhanula VSK, Kienle L, Grundmeier G, Garbe-Schönberg D, Faupel F (2012) Tuning of the ion release properties of silver nanoparticles buried under a hydrophobic polymer barrier. J Nanopart Res 14:928–939
17. Yliniemia K, Ozkaya B, Alissawi N, Zaporojtchenko V, Strunskus T, Wilson BP, Faupel F, Grundmeier G (2012) Combined in situ electrochemical impedance spectroscopy-UV/Vis and AFM studies of Ag nanoparticle stability in perfluorinated films. Mater Chem Phys 134:302–308
18. Biederman H (2000) RF sputtering of polymers and its potential application. Vacuum 59:594–599
19. Schürmann U, Hartung W, Takele H, Zaporojtchenko V, Faupel F (2005) Controlled syntheses of Ag-PTFE nanocomposite thin films by co-sputtering from two magnetron sources. Nanotechnology 16:1078–1082
20. Hassel AW, Milenkovic S, Schürmann U, Greve H, Zaporojtchenko V, Adelung R, Faupel F (2007) Model systems with extreme aspect ratio, tunable geometry, and surface functionality for a quantitative investigation of the lotus effect. Langmuir 23:2091–2094
21. Faupel F, Zaporojtchenko V, Greve H, Schürmann U, Chakravadhanula VSK, Hanisch C, Kulkarni A, Gerber A, Quandt E, Podschun R (2007) Deposition of nanocomposites by plasmas. Contrib Plasma Phys 47(7):537–544
22. Zaporojtchenko V, Behnke K, Strunskus T, Faupel F (2000) Condensation coefficients of noble metals on polymers: a novel method of determination by X-ray photoelectron spectroscopy. Surf Interface Anal 30:439–443
23. Zaporojtchenko V, Strunskus T, Behnke K, Von Bechtolsheim C, Kiene M, Faupel F (2000) Metal/polymer interfaces with designed morphologies. J Adhesion Sci Technol 14:467–490
24. Palanna OG (2009) Engineering chemistry. Tata McGraw-Hill Education Pvt. Ltd., New Delhi
25. Chen DH, Chen CJ (2002) Formation and characterization of Au-Ag bimetallic nanoparticles in water-in-oil microemulsions. J Mater Chem 12:1557–1562
26. Beyene HT, Chakravadhanula VSK, Hanisch C, Elbahri M, Strunskus T, Zaporojtchenko V, Kienle L, Faupel F (2010) Preparation and plasmonic properties of polymer-based composites containing Ag-Au alloy nanoparticles produced by vapor phase co-deposition. J Mater Sci 45:5865–5871
27. Kreibig U, Vollmer M (1995) Optical properties of metal clusters. Springer series in Material Sciences, vol 25. Springer, Berlin
28. Chen Y, Wu H, Li Z, Wang P, Yang L, Fang Y (2012) The study of surface plasmon in Au/Ag core/shell compound nanoparticles. Plasmonics.
29. Feng L, Gao G, Huang P, Wang K, Wang X, Luo T, Zhang C (2010) Optical properties and catalytic activity of bimetallic gold-silver nanoparticles. Nano Biomed Eng 2(4):258–267
30. Haasen P (1996) Physical metallurgy. Translated by Janet Mordike; Cambridge: Cambridge University Press, 3rd ed

30. Moulder F, Stickle WF, Sobol PE, Bomben K (1992) Handbook of X-ray photoelectron spectroscopy. Prairie, MN: Perkin-Elmer Corporation, 2nd ed

31. Tanuma S, Shiratori T, Kimura T, Goto K, Ichimura S, Powell CJ (2005) Experimental determination of electron inelastic mean free paths in 13 elemental solids in the 50 to 5000 eV energy range by elastic-peak electron spectroscopy. Surf Interface Anal 37:833–845

32. Petri MV, Ando RA, Camargo PHC (2012) Tailoring the structure, composition, optical properties and catalytic activity of Ag-Au nanoparticles by the galvanic replacement reaction. Chem Phys Lett 531:188–192

33. Shibata T, Bunker BA, Zhang Z, Meisel D, Vardeman CF II, Gezelter JD (2002) Size-dependent spontaneous alloying of Au-Ag nanoparticles. J Am Chem Soc 124(40):11989–11996

# 14

# Gold highlights at the Third International NanoMedicine Conference, in Coogee Beach, Sydney, Australia, 2–4 July 2012

Sónia Carabineiro

This meeting was attended by 240 participants from diverse fields of nanomedicine. Most presentations were from Australia; however, around 40 delegates were from other countries (including New Zealand, Thailand, Taiwan, China, Hong Kong, India, UK, Hungary, Latvia, Czech Republic, Germany, Italy, Netherlands, France, Portugal and USA). Most attendants were academics but some participants from the industry were also present. Gold-related subjects had an importance presence, showing that gold is very promising for nanomedicine, which aims to treat and prevent disease and traumatic injury, relieving pain and improving human health, using molecular tools and molecular knowledge of the human body (European Science Foundation definition). The topics at the conference ranged from theranostics (a proposed process of personalised medicine consisting on diagnostic therapy for individual patients), drug delivery (administration of a pharmaceutical compound to achieve a therapeutic effect), nanomedicine analytics (techniques used in nanomedicine to detect, analyse, quantify and characterise), macromolecular design (synthesis of large molecules for several nanomedicine applications) and regeneration (process of renewal, restoration and growth to fight disturbance or damage), among others. A brief description is indicated below.

## Delivery

In an oral presentation entitled "Nanoparticles for cancer treatment", Nicole Bryce (University of Sydney, Australia)

S. Carabineiro (✉)
Laboratory of Catalysis and Materials,
Department of Chemical Engineering, Faculty of Engineering,
University of Porto,
Rua Dr. Roberto Frias, s/n,
4200-465 Porto, Portugal
e-mail: scarabin@fe.up.pt

showed that sterically stabilised nanoparticles could penetrate spheroid solid tumour models prepared from DLD-1 (colorectal adenocarcinoma) cancer cells. Moreover, the sterically stabilised particles could facilitate the penetration of the chemotherapeutic drug doxorubicin into the spheroid. The speaker demonstrated that the type of steric stabiliser used was the most important, whereas the nature of the solid core particles was relatively unimportant with cores of iron oxide, silica and gold all behaving in a similar manner when similarly stabilised.

Cyrille Boyer (University of New South Wales, Australia) gave an invited oral presentation on "Engineering polymeric nanoparticles for nanomedicine". The speaker showed that soft core–shell polymeric nanoparticles have potential advantages in the sustained and targeted delivery of therapeutic payloads and could offer significant improvements in the temporal and spatial control of drug delivery. The synthesis of new hybrid organic/inorganic nanomaterials, based on iron oxide, gold and gadolinium, was reported for use as contrast agents. These compounds can be used to improve the visibility of internal body structures, after oral or intravenous administration.

Mariana Beija (University of New South Wales, Australia) presented a poster dealing with "Gold nanoparticles as dual contrast agents for magnetic resonance imaging/X-ray computed tomography (MRI/CT)". It is important to understand that MRI is a medical imaging technique that uses nuclear magnetic resonance to visualise internal structures of the body in detail and is especially adequate to soft tissues. CT uses computer-processed X-rays to produce tomographic images or 'slices' of specific areas of the body, especially bones and lungs; the cross-sectional images are used for diagnostic and therapeutic purposes in various medical disciplines. If both techniques are combined, a more complete diagnosis can be achieved and the administration of contrast agents into patients with the aim of enhancing the signal-to-background ratio is standard practice, and that is what these authors intended. In that scope, they developed macromolecular carriers that improved the relaxivity properties of Gd(III) complexes [1] and prepared gold nanoparticles that were modified with these macromolecular carriers through a 'grafting to' strategy (Fig. 1). These nanohybrids can act as dual contrast agents for MRI and CT and the addition of gold can enhance contrast six times more than when 2,2',2''-(1,4,7,10-tetraazacyclododecane-1,4,7-triyl) triacetate monoamide–Gd(III) complex is used.

The poster of Jane Phui Mun Ng (University of Technology, Sydney, Australia) dealt with "Cellular interactions and cytotoxicity of gold nanoparticles with endothelial cells". It was explained that gold nanoparticles not only provide high contrast in X-ray imaging systems, but are also inert to oxidation, can be modified with fluorescent tags for optical microscopy and can be stabilised against aggregation and modified for increased biocompatibility. Moreover, as gold is non-toxic, the toxicity of gold nanoparticles is dependent on the surface modifying material and size. Once introduced into the human body, they will have contact with the main barriers and lining surfaces, such as the epithelial and endothelial cells, which are commonly employed for the study of cytotoxicity. Therefore, this work focused on antibody CD31 (or platelet endothelial cell adhesion molecule-1, a glycoprotein membrane mediating endothelial cell–cell adhesion involved in angiogenesis for the formation of new vessels). Surface-modified 20 nm Au nanoparticles (NPs) were prepared through self-assembly of polyethylene glycol (PEG) with thiol(-SH) groups for gold. The resulting PEG-modified nanoparticulate systems exhibited superior colloidal stability and limited toxicity, and binding to non-specific proteins and cells was minimised, enabling their delivery to targeted sites.

## Nanomedicine analytics

Mariam Darestani (University of Sydney, Australia) gave an oral presentation on "Label-free biosensing of protein/antibody interaction". She explained that biosensors are devices designed to detect or quantify biochemicals and that electrical biosensors, including voltammetric, amperometric/coulometric and impedance sensors, are very promising and affordable diagnostic detectors, as they are cheap, low power and easy to miniaturise. Moreover, electrochemical impedance spectroscopy (EIS) technique is a label-free process that does not need any reagent (and impedance is measured as a ratio of current and voltage).

In Darestani's work, protein A (a surface protein found in the cell wall of *Staphylococcus aureus*) was covalently immobilised on the surface of a self-assembled monolayer on gold and directly on a gold surface. The binding processes were captured by EIS that was able to detect and characterise all the steps of the chemistry and protein/antibody interactions. The direct binding of protein A on gold surface resulted in a capacitance decline at low frequencies, as shown in Fig. 2. This figure also shows that almost 90 % of the binding process occurred in the first 2 h and the binding process was completed in less than 9 h.

Alice Kar Lai Yang (Chinese University of Hong Kong) gave an oral presentation on "Detection of Methicillin-resistant *Staphylococcus aureus* bacteria (MRSA) deoxyribonucleic acid (DNA) by isothermal amplification (LAMP) with gold nanoparticles by resistive pulse sensing". It was explained that loop-mediated isothermal amplification (LAMP) is a powerful technique to amplify target DNA from pathogens [2] and also a high sensitive and rapid detection method as it can amplify genes within 30 min at ~60 °C that can measure complementary DNA or RNA quantitatively.

In Yang's work, Au NPs, acting as carriers or probes for biosensing, were coated with DNA and used for detecting the leukocidin toxic gene (*pvl*) of MRSA. After LAMP, the *pvl* amplicons were mixed with the DNA probes on the surface of Au NPs. As a result of the DNA-DNA hybridization, Au NPs formed aggregations. Analysis by resistive pulse sensing provided information on the size and dynamics of the agglometated Au NPs. This sensitive technique could detect up to 100 copies of MRSA templates within a short period of time. The LAMP-based AuNP-RSP showed

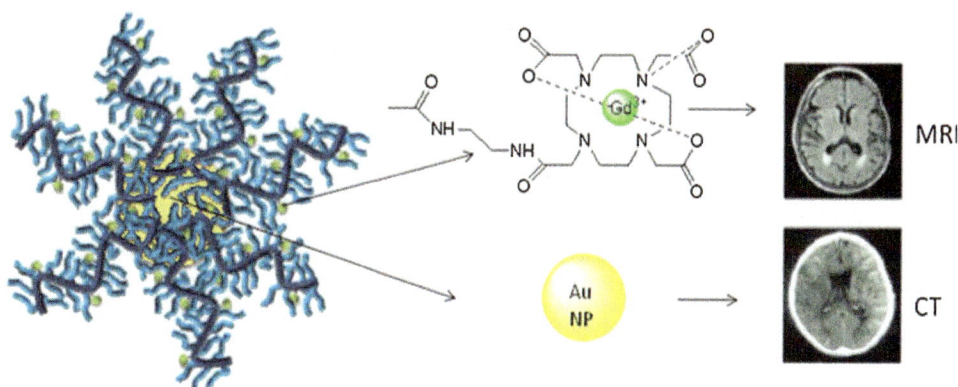

**Fig. 1** Structure of the dual contrast agent for MRI and CT (adapted from the abstract), consisting in gold nanoparticles modified with 2,2',2''-(1,4,7,10-tetraazacyclododecane-1,4,7-triyl) triacetate monoamide–Gd(III) complexes

**Fig. 2** Capacitance profile of *S. aureus* protein—a binding process in a gold surface (adapted from the abstract)

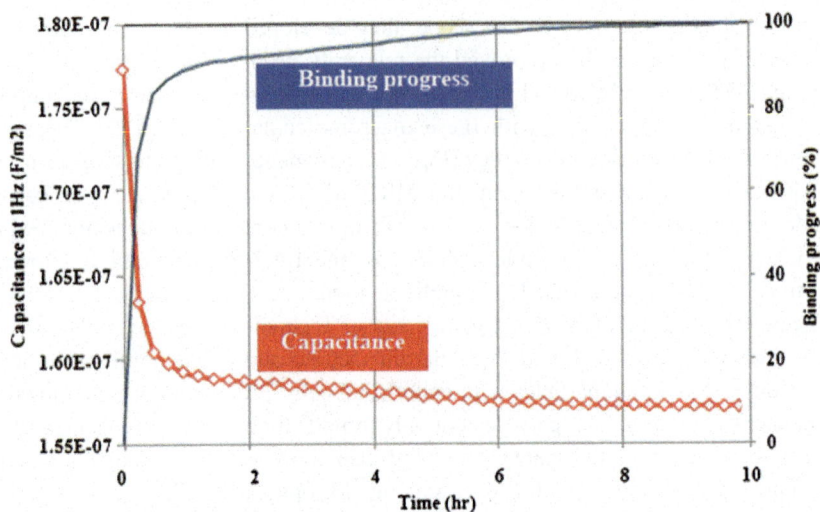

to be a simple and sensitive way to detect different kinds of antigens.

Maitreyee Roy (National Measurement Institute, Australia) gave an oral presentation on "Asymmetric flow–field flow fractionation technique for separation and characterisation of gold nanoparticles in aqueous media". Roy explained that Au NPs are a preferred material for biomedical research into challenging applications, such as advanced diagnostics, drug delivery and therapeutics and that particle size and size distribution are important physical–chemical key characteristics required for describing the nanoparticle systems. Moreover, the asymmetric flow–field flow fractionation (AF-FFF) is a high-resolution elution technique that separates suspension constituents based on hydrodynamic size and a powerful size-based separation method of the components of complex particle suspensions, facilitating determination of chemical composition [3].

In Roy's work, a tri-modal suspension of citrate-stabilised Au NPs, created by mixing three different sizes (20, 40 and 60 nm), at ratios chosen to generate equal volume fractions, was measured with the AF-FFF system. Distinct populations of gold nanoparticles were separated chromatographically from the tri-modal mixture as shown in Fig. 3.

William Olds (Queensland University of Technology, Australia) presented a poster dealing with "Deep Raman spectroscopy for sub-surface probing within biological tissue". Until recently, Raman spectroscopy was considered only a surface analysis technique, but new methods have extended the working depth of Raman spectroscopy to several millimetres, and up to ~1 to 2 cm, beneath diffuse opaque samples. This explains why Raman is a potential tool for performing non-invasive biomedical spectroscopy and diagnosis.

In Olds's work, the authors constructed a optical probe for conducting Raman spectroscopy through layers of biological media using gold nanoparticles to provide surface-

enhanced Raman scattering (SERS), thereby enhancing the signals of buried analytes. In the pilot tests, small samples of chicken skin (both with and without fat layer intact) were used as a model tissue and methylene blue was used as the detection analyte. A strong enhancement of the Raman signals of the methylene blue was observed in the presence of gold nanoparticles. Without the nanoparticles and resultant SERS enhancement, the methylene blue signals could not be detected behind the skin due to the low concentration. Although further work is required to increase the sensitivity of the constructed Raman probe and to optimise the SERS enhancements provided by the gold nanoparticle substrate, these results indicate the possibility of using Raman as a non-invasive biomedical analysis technique, with potential applications in disease diagnosis, blood analyte and drug delivery monitoring.

Chi-Chang Lin (Tunghai University, Taiwan) presented a poster dealing with "Rapid discrimination of bacteria

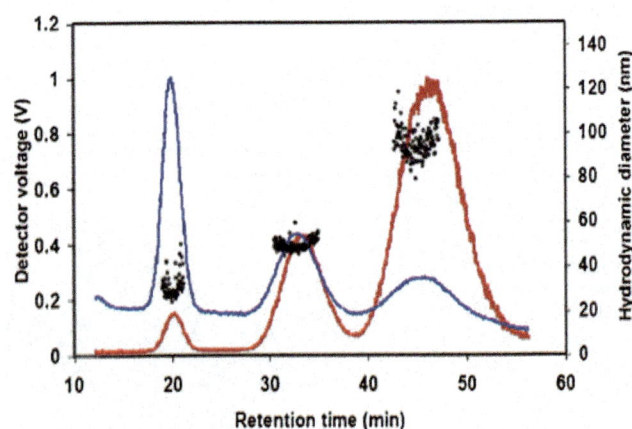

**Fig. 3** Fractograms for a tri-modal sample of citrate-stabilised gold nanoparticles. *Red line* multi-angle static light scattering at 90°; *blue line* IV–Vis absorption at 520 nm; *circles* hydrodynamic diameter measured as a function of time by the dynamic light scattering detector (adapted from the abstract)

fingerprint by using SERS". Lin explained that since infectious diseases are one of the major causes of morbidity and mortality in humans throughout the world, rapid and more accurate methods of screening, identification and susceptibility testing of bacteria are needed.

In Lin's work, a filter-like substrate with high SERS functions made by metal nanoparticles-embedded mesoporous silica was applied for discrimination of bacteria fingerprint under SERS analysis. Gelatine with amide groups ($-CO-NH_2$) was introduced as organic template of mesoporous silica and may also act as the protecting agent for stabilising Au NPs and to prevent their aggregation. The method was simple, low cost, and a strong adsorption affinity to *S. aureus* and *Escherichia coli* was found. The intensity of *S. aureus* SERS spectrum was enhanced more than 700 times compared with normal Raman spectra upon a 13 wt% Au loading.

The poster of Wen-Fu Lai (Taipei Medical University, Taiwan) dealt with "Early detection of prostate cancers and their bone metastasis using near-infrared fluorescent imaging". It was explained that prostate cancer is one of the leading causes of cancers to death in men worldwide, and the metastasis of prostate cancer to bone is the most significant cause of mortality. Therefore, early detection of prostate cancer metastasis to bone is needed for better diagnosis of the disease and available imaging strategies for assessing bony lesions are still lacking.

In Lai's work, a near-infrared fluorescent (NIRF) bisphosphonate derivative probe was designed with ideal optical properties and hydroxyapatite (HA) binding efficacy (HA was selected as a target for a molecular imaging probe because HA deposits at tumour sites of bone metastasis of prostate cancer). Pamidronate (Fig. 4) was selected as the backbone for the molecular probe because of its high affinity towards HA as well as for its small molecular weight. This molecule was conjugated with a nanogold cluster bearing a NIFR fluorescent ligand that can be detected using a in vivo imaging system with adequate filtering. Molecular imaging in animal model was done with NIFR, a technique

that allows a high-resolution image to be taken of the probe due to its ability to penetrate the skin layer and absorb the fluorescence emitted by the probe. The results showed that this probe could specifically and efficiently detect the new bone formation by metastatic prostate cancer in animal models.

Grainne Moran (University of New South Wales, Australia) presented a poster dealing with "Analytical and particle characterisation techniques applied to gold nanoparticles in aqueous media". In a collaboration between the National Measurement Institute Nanometrology Group and the University of New South Wales, techniques for characterising modified Au nanoparticles in aqueous media were investigated. Particle sizes, size distributions and aggregation behaviour were monitored by nanoparticle tracking analysis and differential centrifugal sedimentation, as well as transmission electron microscopy and optical spectroscopy. Cysteine-modified Au NPs with primary particle diameters ranging from 25 to 50 nm were used as model systems. The information provided by each technique was distinct and complementary. The authors concluded that methods need to be validated using appropriate standards, where available or, at least, cross-referenced using multiple techniques.

## General

Sónia Carabineiro (Universidade do Porto, Portugal) presented a poster dealing with "Uses and applications of gold in nanomedicine". A historical perspective on the use of colloidal gold in medicine throughout the ages was shown, starting from ancient times, where alchemists worked on the development of a gold elixir, which supposedly had the ability to restore health and youth. More recently, gold nanoparticles have been used in the treatment of rheumatoid arthritis due to their anti-inflammatory properties and seem promising for the treatment of Alzheimer's disease. Colloidal methods are often employed for their synthesis (Fig. 5).

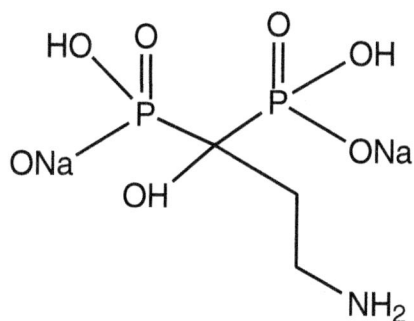

**Fig. 4** Structure of pamidronate, a nitrogen-containing bisphosphonate that was conjugated with a nanogold cluster bearing a near-infrared fluorescent molecular ligand, to be used as a molecular probe

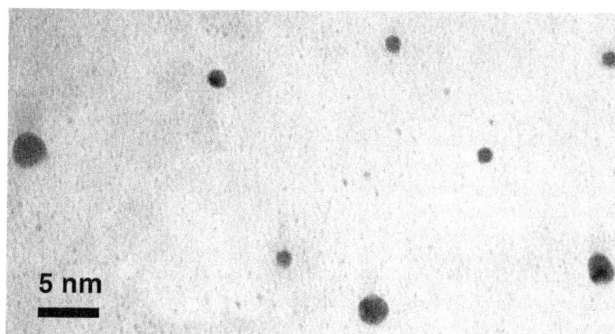

**Fig. 5** Au NPs synthesised by a colloidal method using tetrakis (hydroxymethyl) phosphonium chloride to reduce a $HAuCl_4$ solution

Examples from a recent review [4] were given on the use of gold as a contrast agent in biological electron microscopy or computed tomography; in cancer research, to target tumours and provide detection and also as a drug carrier. Au NPs showed to be effective for the inhibition of pathogenic bacteria cell growth, including Gram-positive bacteria, Gram-negative bacteria and antibiotic-resistant bacteria and have demonstrated effective photothermal destruction of cancer cells and tissue. They can act as antennas, providing enhanced radiation targeting with lower radiation doses, consequently avoiding damage to healthy tissues. This shows that gold has already an important role in nanomedicine. Its potential is enormous and certainly future results will lead to more practical and commercial applications.

**Acknowledgments**   I am grateful to Nicole Bryce and to Carla Gerbo for all the information and assistance provided. Fundação para a Ciência e Tecnologia (FCT) is acknowledged for funding (CIENCIA 2007 programme and project PTDC/QUI-QUI/100682/2008, financed by FCT and FEDER in the context of Programme COMPETE).

## References

1. Li Y, Beija M, Laurent S, van der Elst L, Muller RN, Duong HTT, Lowe AB, Davis TP, Boyer C (2012) Macromolecular ligands for gadolinium MRI contrast agents. Macromolecules 45:4196–4204
2. Tomita N, Mori Y, Kanda H, Notomi T (2008) Loop-mediated isothermal amplification (LAMP) of gene sequences and simple visual detection of products. Nat Protoc 3:877–882
3. Cho TJ, Hackley VA (2003) Fractionation and characterization of gold nanoparticles in aqueous solution: asymmetric-flow field flow fractionation with MALS, DLS, and UV–Vis detection. Anal Bioanal Chem 398:2003–2018
4. SAC Carabineiro (2012) Synthesis of colloidal and supported gold nanoparticles. In: Jarnagin A, Halshauser L (eds) Gold nanoparticles: synthesis, optical properties and applications for cancer treatment, Nova Science, New York (in press)

# 15

# Gold nanoparticles on wool in a comparative study with molecular gold catalysts

Thomas Borrmann · Teck Hock Lim · Hannah Cope ·
Kerstin Lucas · Michael Lorden

**Abstract** The catalytic activity of gold chloride nanoparticles is compared to the activity of two molecular gold(I) chloride phosphine complexes for the addition of methanol to 3-hexyne. The phosphines are triphenylphosphine and the bispidinone related bulky 6,8-bis-(4-dimethylamino-phenyl)-3-methyl-9-oxo-7-phenyl-3-aza-7-phospha-bicyclo[3.3.1]nonan-1,5-dicarboxylic acid dimethyl ester. Use of the bulky ligand made the addition reaction selective towards the enol product, meaning that no addition of methanol or water to alkenes, which were produced during the reaction, occurred. In contrast, use of triphenylphosphine gold(I) chloride resulted in the synthesis of a variety of products. The phosphines decomposed during reaction leading to the formation of gold nanoparticles, which were found to be catalytically inactive. Artificially produced gold nanoparticles also proved to be inactive. In contrast, gold chloride nanoparticles deposited on wool were active comparable to the gold phosphine-containing catalysts tested previously. Overall activities observed were low compared to results from the literature suggesting that the operating conditions chosen could be optimised.

**Keywords** Gold-catalysed addition · Gold nanoparticles · Gold chloride nanoparticles · Gold phosphine · Homogeneous catalysis · Heterogeneous catalysis

T. Borrmann · T. H. Lim · H. Cope · K. Lucas · M. Lorden
School of Chemical and Physical Sciences, Victoria University
of Wellington, PO Box 600, 6140 Wellington, New Zealand

*Present Address:*
T. Borrmann (✉)
Othbergstr. 10,
37632 Eschershausen, Germany
e-mail: aoc@gmx.li

## Results and discussion

Gold has been shown to be catalytically active [1–4]. Several review papers provide an excellent overview of the field, types of catalysts used and ligand effects [1–5].

In a preliminary study, we investigated the addition of methanol to 3-hexyne using triphenylphosphine gold(I) chloride, $[(H_6C_5)_3PAu]Cl$. Triphenylphosphine was synthesised by reacting dimethylsulfide gold(I) chloride with triphenylphosphine (Scheme 1).

While the gold compound was catalytically active, it decomposed during the reaction (Scheme 2) resulting in the formation of gold nanoparticles. Such decomposition was reported in the literature [1, 2]. One solution reported is the addition of electron withdrawing ligands. As this route has already been explored, it was not investigated.

Furthermore, the reaction was not selective resulting in a mixture of various products, the main two being 3-methoxy 3-hexyne and 3,4-dimethoxy hexane. Consequently, other catalysts were studied.

In attempt to increase selectivity of the catalytic reaction and to increase the stability of the catalyst, a more stable and more bulky phosphine was sought. Due to some positive experiences in prior research [6] the ligand 6,8-bis-(4-dimethylamino-phenyl)-3-methyl-9-oxo-7-phenyl-3-aza-7-phospha-bicyclo[3.3.1]nonan-1,5-dicarboxylic acid dimethyl ester, $C_{34}H_{40}N_3O_5P$ (Scheme 3) was chosen.

The reaction of the ligand $C_{34}H_{40}N_3O_5P$ with dimethylsulfide gold(I) chloride resulted in the formation of clear yellow crystals (Scheme 4). The crystal structure and details of the synthesis will be reported shortly in a separate article.

$[C_{34}H_{40}N_3O_5PAu]Cl$ was tested in the catalytic addition of methanol to 3-hexyne. Results were compared the ones achieved using $[(H_6C_5)_3PAu]Cl$ as catalyst. The activity of $[C_{34}H_{40}N_3O_5PAu]Cl$ was comparable to that of $[(H_6C_5)_3PAu]Cl$ indicating that the bulk of the ligand did not impede the reaction and hence that limiting step in the proposed addition reaction mechanism is the

$$AuCl_3 + HCl \longrightarrow H[AuCl_4]$$

$$H[AuCl_4] + S(CH_3)_2 \longrightarrow [(H_3C)_2SAuCl_3] + HCl$$

$$[(H_3C)_2SAuCl_3] + S(CH_3)_2 + H_2O \longrightarrow [(H_3C)_2SAu]Cl + (H_3C)_2SO$$

$$[(H_3C)_2SAu]Cl + (H_5C_6)_3P \longrightarrow [(H_5C_6)_3PAu]Cl + S(CH_3)_2$$

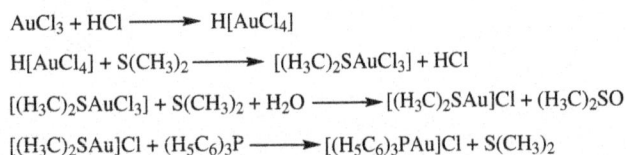

**Scheme 1** Synthesis of triphenylphosphine gold(I) chloride

dissociation of the product from the catalyst (step IV in Scheme 5).

However, in terms of selectivity the use of $[C_{34}H_{40}N_3O_5$-PAu]Cl resulted in only 3-methoxy 3-hexyne being produced. No other products were detected after completion of the reaction. This indicates that the bulky ligand had a directing effect and is in agreement with the assumption that the dissociation of the product from the catalyst is the determining step in terms of the reaction kinetics

It was hoped that introducing a bulky ligand would improve the stability of the catalyst. This was found not to be the case. As observed for $[(H_6C_5)_3PAu]Cl$, the catalyst decomposed during the reaction and the formation of gold nanoparticles was observed. In return, the phosphines were oxidised as indicated by $^{31}P$-NMR. The catalyst could be used for three conversion reactions before it became completely inactive. The reason for the instability of even the bulky gold compound $[C_{34}H_{40}N_3O_5PAu]Cl$ was considered to be due to its structure (Scheme 6, Fig. 1) adopting a linear structure for ligand–gold–chloride, with the PAuCl angle being 180° and no bonding to nitrogen being observed, comparable to the structure found in other phoshines such as $[(H_6C_5)_3PAu]Cl$ [7]. The linear structure in regards to the environment around the gold centre of $[C_{34}H_{40}N_3O_5PAu]Cl$ was confirmed by single crystal X-ray analysis

A colorimetric and electron microscopic investigation showed that the gold nanoparticles formed increased in size during the reaction and over several uses of the catalyst. This raised the possibility that the catalytic activity observed was due to the presence of the nanoparticles or some intermediate form. However, gold nanoparticles of various sizes produced in-house [8] showed no measurable catalytic activity. Gold nanoparticles were used suspended in solution and supported on wool (Fig. 2). As reported in the literature [1–4], gold(I) or gold(III) needs to be present for any catalytic activity to occur.

In another project [9], it was found that gold chloride nanoparticles could be stabilised by depositing them on wool. The gold oxidation state in these nanoparticles is +1.

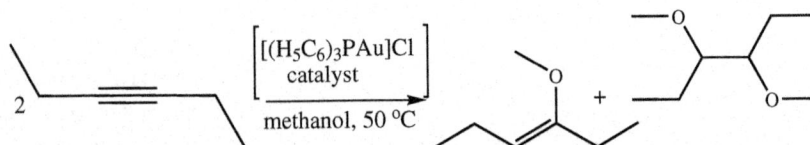

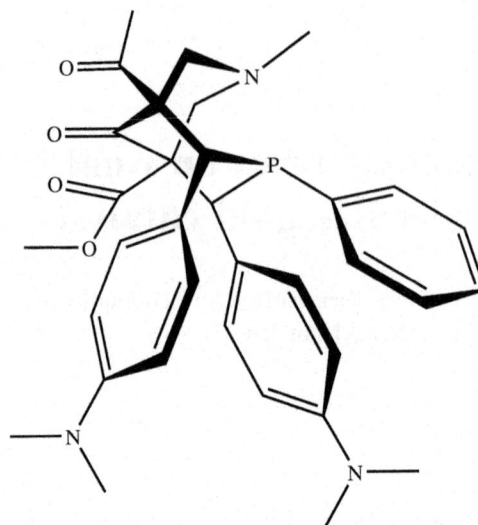

**Scheme 3** The ligand 6,8-bis-(4-dimethylamino-phenyl)-3-methyl-9-oxo-7-phenyl-3-aza-7-phospha-bicyclo[3.3.1]nonan-1,5-dicarboxylic acid dimethyl ester, $C_{34}H_{40}N_3O_5P$

A sample of 50-nm gold chloride nanoparticles on wool is shown in Fig. 2.

The sample of gold chloride nanoparticles was also tested in the addition of methanol to 3-hexyne. The nanoparticles were found to be a lot less active per gold ion present (Table 1) than the molecular catalysts tested before. They were as selective as $[C_{34}H_{40}N_3O_5PAu]Cl$. In comparison to $[C_{34}H_{40}N_3O_5PAu]Cl$ and $[(H_6C_5)_3PAu]Cl$, the gold chloride nanoparticles did not show any signs of decomposing or changing during reaction. Even after eight cycles, no change in activity was observed. A comparison of the catalytic activities of the three catalysts tested is shown in Fig. 3 (see also Table 1). First-order kinetics are observed, which supported the proposed mechanism shown in Scheme 5. As promising as the performance of the gold chloride nanoparticles was, the overall activity of this catalyst was poor compared to the phosphines investigated earlier.

One fact has to be considered though: Only a few gold centres in the gold nanoparticles were catalytically active. To truly compare the performance of the gold nanoparticles with the molecular gold catalysts, a few assumptions and contemplations have to be made.

The amount of gold added in the form of molecules in the other experiments involving $[C_{34}H_{40}N_3O_5PAu]Cl$ and $[(H_6C_5)_3PAu]Cl$ was calculated. Then it was attempted to match this amount of active gold by calculating the surface

**Scheme 2** Catalytic addition of methanol to 3-hexyne at 50 °C in excess methanol using the catalyst $[(H_6C_5)_3PAu]$ Cl

$$[(H_3C)_2SAu]Cl + C_{34}H_{40}O_5N_3P \longrightarrow [C_{34}H_{40}O_5N_3PAu]Cl + S(CH_3)_2$$

**Scheme 4** Synthesis of 6,8-bis-(4-dimethylamino-phenyl)-3-methyl-9-oxo-7-phenyl–3-aza-7-phospha-bicyclo[3.3.1]nonan-1,5-dicarboxylic acid dimethyl ester gold(I) chloride [$C_{34}H_{40}N_3O_5PAu$]Cl

area of nanoparticles. 5.0 µmol of gold were used in the cases of [$C_{34}H_{40}N_3O_5PAu$]Cl and [$(H_6C_5)_3PAu$]Cl. The particle size of the nanoparticles was 10 nm in diameter. Particles were assumed to be spherical, which is only a rough approximation as different shapes were observed by electron microscopy. A 10 nm particle was calculated to contain about 12,000 gold ions, of which about 1,500 were located on its surfaces. A packing factor of 0.68 was considered based on the body-centred tetragonal crystal structure of AuCl. It was assumed that about half of the surface gold will not be accessible due to facing the wool surface or other obstructions. This meant that for each active gold, there were assumed to be 15 inactive ones. One gram of wool contained 0.016 g of gold equivalent to 81 µmol of which 5.0 µmol were assumed to be active. This meant that 1 g of wool covered in gold nanoparticles should have been required to match the activity of the molecular gold catalysts used. Experiments showed that only 0.1 g of gold(I) covered wool were required. This means that the turn over frequency per catalytic centre of the nanoparticles is about 290 mol of product per mole active centre per hour, which is actually quite comparable to the results achieved for the phosphines.

Findings by Haruta [10] suggest that the size of the nanoparticles investigated in this study was not optimal

**Scheme 6** Structure of [$C_{34}H_{40}N_3O_5PAu$]Cl

and could have limited the activity observed significantly. Smaller gold chloride nanoparticles (about 5 nm in diameter) on wool should be able to achieve higher turnover frequencies. Furthermore, Haruta found that electron-rich supports in the form of oxide particles are required to achieve good results in catalytic reactions. It is likely that the wool acted as electron rich substrate in this case, a fact that should be investigated in future projects.

Overall, all catalytic conversions studied were not as efficient as some of the ones reported in the literature [3, 4]. It is likely that the reaction conditions chosen were to

**Scheme 5** Proposed 4-step catalytic cycle for the addition of methanol to 3-hexyne employing a gold catalyst based on transition states and intermediates suggested by Asao et al. [2, 3]

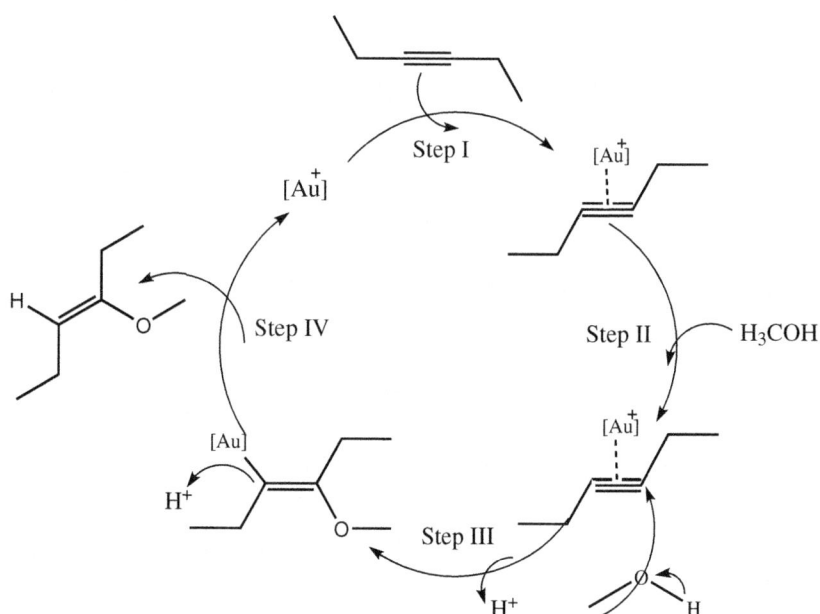

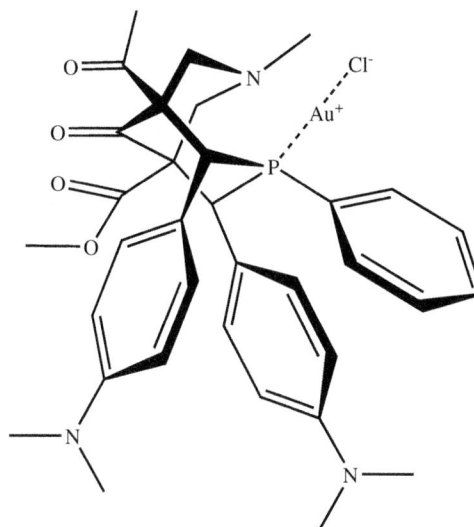

**Fig. 1** Ortep diagram showing
the structure of
[$C_{34}H_{40}N_3O_5PAu$]Cl

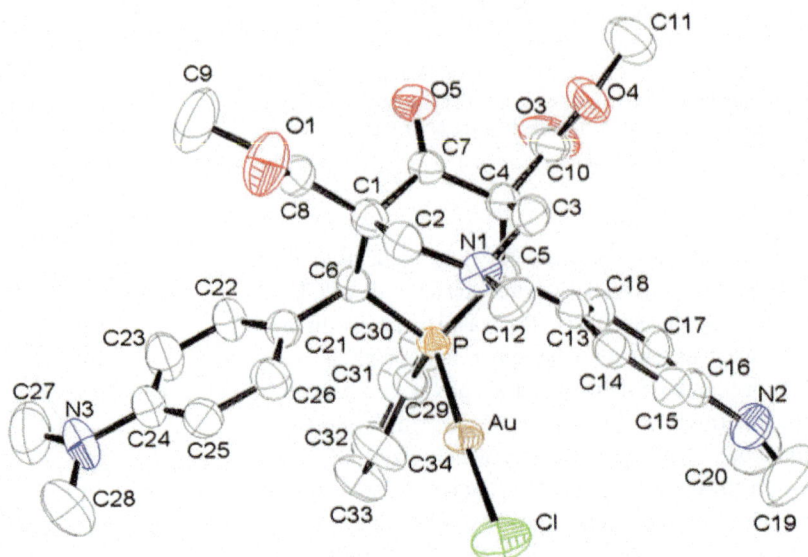

some degree the limiting factor in the activity of the catalysts. Further studies are required to optimise reaction conditions.

Experimental

All reagents unless specifically noted were purchased from Sigma Aldrich. All experiments unless otherwise noted were carried out in purified solvents open to air at ambient temperature. All spectra were acquired at room temperature. Instruments: NMR, Varian UNITY Inova Plus spectrometer operating at 300Hz; MS, PE Biosystem Mariner 5158 TOF in positive ion mode; IR, Bio-RAD FTS-7 (resolution of 4 cm$^{-1}$). GC: HP 540, Column DB-WAX. The Analytical Lab of Otago University carried out elemental analyses.

Synthesis of [Au(S(CH$_3$)]Cl [11]

HCl (150 μL, 1.75 mmol) was added to solid AuCl$_3$ (0.49 g, 1.65 mmol) concentration followed by addition of water (2.3 ml). Ethanol (11.5 ml) was added to the resulting

solution. Two equal portions of S(CH$_3$)$_2$ (2×142 μL, 3.85 mmol) were added. A bright yellow precipitate was observed immediately after the first portion of S(CH$_3$)$_2$ was added. The second portion was added 3 min after the first portion and the resulting mixture was allowed to stir for 15 min. Snowy white crystals precipitated, which were recovered by vacuum filtration and washed with ethanol (1.1 ml) twice. Recrystallization from absolute ethanol gave a white solid 0.25 g (50 %). $^1$H NMR (acetone–D$_6$): δ (2.86, s). MS: $m/z$=553 (dimer ([Au(S(CH$_3$)]Cl)$_2$, $-^{37}$Cl), $m/z$=555 (dimer ([Au(S(CH$_3$)]Cl)$_2$, $-^{35}$Cl).

Synthesis of triphenylphosphine gold chloride

Triphenylphosphine (6.6 mg, 25 mmol) was dissolved in a degassed solution of [Au(S(CH$_3$)]Cl (7.4 mg, 25 mmol) in methanol (10 ml). The resulting mixture was heated and stirred under nitrogen and reflux for 1 h. Then it was allowed to cool to room temperature. Yellow–white needle-shaped crystals precipitated, which were recovered by vacuum filtration and washed with cold methanol. $^{31}$P

**Fig. 2** Gold nanoparticles on wool of various sizes; the particle size decreases from the left to the right for the first six samples. The seventh sample, on the far right is comprised of gold chloride nanoparticles deposited on wool

**Table 1** Overview of catalytic tests; conversion (percent) after 1 h

| Catalyst | Cycle | | | | | | | | TOF |
|---|---|---|---|---|---|---|---|---|---|
| | 1 | 2 | 3 | 4 | 5 | 6 | 7 | 8 | |
| $[(H_6C_5)_3PAu]Cl$ | 16.2[a] | 1.51[a] | 0.00 | | | | | | 286[b] |
| $[C_{34}H_{40}N_3O_5PAu]Cl$ | 16.3[a] | 15.1[a] | 4.3 [a] | 0.00 | | | | | 287[b] |
| Gold chloride nanoparticles on wool | 16.9 | 16.1 | 17.0 | 16.2 | 17.1 | 16.5 | 16.7 | 16.7 | 18[b] |

[a] Formation of gold nanoparticles was noticed.

[b] TOF (based only on the first cycle) in mol product per mole of gold per hour. For the nanoparticles, this calculation is only partially valid (see below).

NMR ($CDCl_3$): $\delta$, $-5.05$ (s)ppm. Elemental analysis in percent—calculated (measured): Au, 39.81 (39.80); C, 43.70 (43.70); H, 3.06 (3.06); P, 6.26 (6.27); Cl, 7.15 (7.17).

Preparation of $[C_{34}H_{40}N_3O_5PAu]Cl$

The ligand $C_{34}H_{40}N_3O_5P$ was prepared as reported by Vagana [6]. A solution of $C_{34}H_{40}N_3O_5P$ (18.3 mg, 25 mmol) in chloroform (6 ml) was added dropwise to a solution of $[Au(S(CH_3)]Cl$ (7.4 mg, 25 mmol) in chloroform (5 ml). The resulting clear yellow solution was allowed to stir for 2 h before the solvent was removed to give a semitransparent yellowish glass-like material, which was crystallised from a minimum of chloroform. $^1H$ NMR ($CDCl_3$): $\delta$, 2.8 [$s$, 12H, C14]; $\delta$, 3.0 [$d$, $J_{H-H}$ 13 Hz, 2 H, C2/C4]; $\delta$, 3.6 [$s$, 6 H, C15]; $\delta$, 3.8 [$d$, $J_{H-H}$ 13 Hz, 2 H, C2/C4]; $\delta$, 4.4 [$d$, $J_{P-H}$ 16 Hz, 2 H, C6/C8], $\delta$6.5 ["$d$", $J_{P-H}$ 7 Hz, 4 H, C12]; $\delta$ (7.1–7.4) [$m$, C11/C18/C19]; $\delta$, 7.6 [$m$, 2 H, C12]ppm. IR, $\nu$=1,742 $cm^{-1}$ C=O. $^{31}P$ NMR ($CDCl_3$): $\delta$, 42.3 (s)ppm.

Preparation of gold(I) chloride nanoparticles on wool [9]

Gold(III) chloride (2.46 mg, 81 μmol) were dissolved in 40 mg 1.0 $molL^{-1}$ HCl. Ten millilitre distilled water were added to this solution. Suspended in the solution was 0.1 g wool. The suspension was shaken for 72 h during which the nanoparticles developed. Then the pH value of the reaction mixture was adjusted to pH 11.2 using a 0.1 $molL^{-1}$ solution of potassium hydroxide to create an environment, where the nanoparticles remained unaffected while the cysteine and cysteine moieties of the wool were oxidised. This step was necessary to prevent reduction of the gold chloride nanoparticles by the wool. The mixture was heated and shaken at 50 °C for 7 days to ensure conversion of all active groups present in the wool. Then the wool was recovered by filtration, washed extensively with water and air-dried. The sample was analysed by electron microscopy and by analysis of the residual solution (determination of the gold uptake by atomic absorption).

Catalytic investigations

In a typical experiment, a catalyst was dissolved or suspended in methanol (10.0 mL, 0.791 $gcm^{-3}$, 0.247 mol) in a two-neck round-bottom flask fitted with a condenser. $n$-Decane (0.1 mL, 0.730 $gcm^{-3}$, 0.51 mmol) was added as internal standard. The resulting solution or mixture was heated to 50 °C. Then 3-hexyne (1.00 mL, 0.723 $gcm^{-3}$, 8.80 mmol) was added. Immediately, a sample (1 μL) was taken and analysed by gas chromatography. In 10-min intervals, further samples were taken and also analysed. After 2 h, the reaction was stopped. The catalyst was recovered by distilling off the liquids under nitrogen or with a pair of tweezers in the case of wool samples.

$[(H_6C_5)_3PAu]Cl$

Tripheylphosphine gold chloride (24.7 mg, 50.0 μmol) was dissolved in methanol (0.500 mL, 0.791 $gcm^{-3}$, 12.3 mmol). Of this solution, 0.050 mL was used as catalyst.

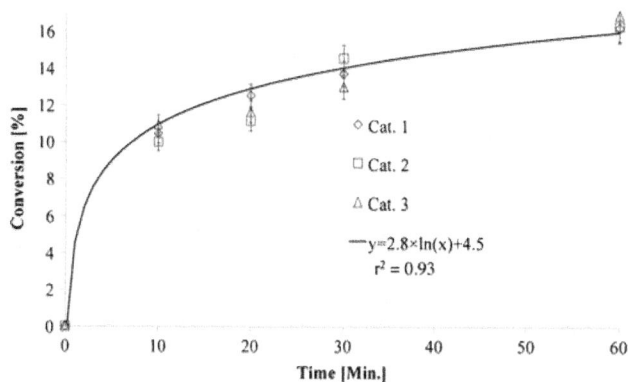

**Fig. 3** Catalytic activity of $[C_{34}H_{40}N_3O_5PAu]Cl$ (Cat. 1), $[(H_6C_5)_3PAu]Cl$ (Cat. 2) and gold chloride nanoparticles on wool (Cat. 3)—a plot of conversion vs time

$[C_{34}H_{40}N_3O_5PAu]Cl$

$[C_{34}H_{40}N_3O_5PAu]Cl$ (20.8 mg, 25.0 μmol) was dissolved in chloroform (0.200 mL, 1.48 g cm$^{-3}$, 2.48 mmol). Of this solution, 0.020 mL was used as catalyst.

Gold chloride nanoparticles on wool

A solid sample (0.100 g) of gold chloride nanoparticles deposited on wool was suspended in methanol.

# References

1. Gorin DJ, Sherry BD, Toste FD (2008) Ligand effects in homogeneous Au catalysis. Chem Rev 108:3351–3378

2. Gamelin FX, Baquet G, Berthoin S, Thevenet D, Nourry C, Nottin S, Bosquet L (2009) Effect of high intensity intermittent training on heart rate variability in prepubescent children. Eur J Appl Physiol 105:731–738.

3. Li Z, Brouwer C, He C (2008) Gold-catalyzed organic transformations. Chem Rev 108:3239–3265

4. Hashmi ASK, Hutchings GJ (2006) Gold catalysis. Angew Chem Int Ed 45:7896–7936

5. Hashmi ASK ASK (2007) Homogeneous gold catalysis: the role of protons. Catal Today 122:211–214

6. Zayya AI, Vagana R, Nelson MRM, Spencer JL (2012) Synthesis and characterisation of 3-aza-7-phosphabicyclo[3.3.1]nonan-9-ones. Tetrahedron Lett 53(8):923–926.

7. Liau R-Y (2003) Contributions to the chemistry of gold(I) cyanide, isocyanide and acetylide complexes. Dissertation, Technical University, Munich

8. Johnston JH, Lucas KA (2012) Nanogold synthesis in wool fibres: novel colourants. Gold Bull 44(2):85–89.

9. Johnston JH, Burridge KA, Kelly FM, Small AC (2010) NZ patent application 589498.

10. Haruta M (2003) When gold is not noble: catalysis by nanoparticles. The Chemical Record 3(2):75–87.

11. Kaesz HD (ed) (1989) Inorganic syntheses vol. 26. Wiley, New York

# 16

# The combination effects of trivalent gold ions and gold nanoparticles with different antibiotics against resistant *Pseudomonas aeruginosa*

**Zeinab Esmail Nazari · Maryam Banoee ·
Abbas Akhavan Sepahi · Fatemeh Rafii ·
Ahmad Reza Shahverdi**

**Abstract** Despite much success in drug design and development, *Pseudomonas aeruginosa* is still considered as one of the most problematic bacteria due to its ability to develop mutational resistance against a variety of antibiotics. In search for new strategies to enhance antibacterial activity of antibiotics, in this work, the combination effect of gold materials including trivalent gold ions ($Au^{3+}$) and gold nanoparticles (Au NPs) with 14 different antibiotics was investigated against the clinical isolates of *P. aeruginosa*, *Staphylococcus aureus* and *Escherichia coli*. Disk diffusion assay was carried out, and test strains were treated with the sub-inhibitory contents of gold nanomaterial. Results showed that Au NPs did not increase the antibacterial effect of antibiotics at tested concentration (40 µg/disc). However, the susceptibility of resistant *P. aeruginosa* increased in the presence of $Au^{3+}$ and methicillin, erythromycin, vancomycin, penicillin G, clindamycin and nalidixic acid, up to 147 %. As an individual experiment, the same group of antibiotics was tested for their activity against clinical isolates of *S. aureus*, *E. coli* and a different resistant strain of *P. aeruginosa* in the presence of sub-inhibitory contents of $Au^{3+}$, where $Au^{3+}$ increased the susceptibility of test strains to methicillin, erythromycin, vancomycin, penicillin G, clindamycin and nalidixic acid. Our finding suggested that using the combination of sub-inhibitory concentrations of $Au^{3+}$ and methicillin, erythromycin, nalidixic acid or vancomycin may be a promising new strategy for the treatment of highly resistant *P. aeruginosa* infections.

**Keywords** $Au^{3+}$ · Au NPs · Antibiotic resistance · Combination effect · *Pseudomonas aeruginosa*

Z. E. Nazari
Institute of Physics and Nanotechnology, Aalborg University,
Aalborg, Denmark

M. Banoee · A. A. Sepahi
Azad University of North Tehran Branch,
Tehran, Iran

F. Rafii
Division of Microbiology,
National Center for Toxicological Research,
Jefferson, AR, USA

A. R. Shahverdi (✉)
Department of Pharmaceutical Biotechnology and Biotechnology
Research Center, Faculty of Pharmacy,
Tehran University of Medical Sciences,
Tehran, Iran
e-mail: shahverd@sina.tums.ac.ir

## Introduction

The emerging use of antibiotics has lead to microbial resistance which is still considered as a major problem in chemotherapy of many infectious diseases [1]. Many mechanisms are involved in the process of antibiotic resistances which mainly include enzymatic degradation and modification of the antibiotic agent [2], modification of the target site of the drug [3], and active reflux and reduced uptake of the drug [4]. *Pseudomonas aeruginosa* is an opportunistic human pathogen Gram negative bacterium which is responsible for infections such as blood stream nosocomial infections, especially in immunocompromised patients and the elderly population of industrial societies [5]. The organism has reputation for having minimal nutritional requirements, the ability to tolerate a wide variety of physical conditions and the ability to resist against new antibiotics such that it has been addressed as "the worst nightmare" of microbiologists and clinical pharmacists for its highly resistant nature [6]. The antibiotic resistance of this microorganism is due to a number of mechanisms which mainly include co-operation of multidrug efflux pumps such as MexAB-OprM, a pump system that removes β-lactams, chloramphenicol, fluoroquinolones, macrolides, novobiocin, sulfonamides, tetracycline and trimethoprim, as well as various dyes and detergents [7]; low impermeability of the membrane to drugs and the fact that it readily develops mutational resistance to most antibacterial agents [8]. The problem of multidrug resistance of *P. aeruginosa* has urged many scientists and pharmaceutical companies to search for new potential therapies for this Gram negative bacterium [9–11]. Therefore, studies on new strategies to combat *P. aeruginosa* would be of great value.

An alternative strategy to overcome the problem of resistance is the use of commonly used antibiotics in combination with different natural or chemical agents [12, 13]. To date, many organic and inorganic compounds have been reported to enhance the antibacterial activity of different antibiotics against many bacteria and fungi resistant test strains [14]. Metallic ions and metallic nanoparticles including zinc and silver have particularly shown promise when used in combination with a number of antibiotics such as ciprofloxacin [15], penicillin G, amoxicillin, erythromycin, clindamycin, vancomycin [16], ampicillin, kanamycin, erythromycin and chloramphenicol [17] in different Gram positive or Gram negative test strains. In a recent approach, Rai et al. reported the antibiotic mediated synthesis of Au NPs with potent antimicrobial activity for application in antimicrobial coatings [18].

During our previous attempts to explore new agents that modulate antibiotic resistant, we screened different chemical substances and natural products and reported a number of organic and inorganic compounds which reduced the resistance of various Gram positive and negative bacteria including *Staphylococcus aureus*, *Clostridium difficile* and *Aspergillus* sp. [14, 19]. In particular, our team reported that some monoterpenes (one of the major components of essential oils) enhanced antibacterial activity of nitrofurantoin against resistant strains of *Enterobacteria* [19]. We also showed that the antibacterial activity of fluconazol against different species of *Aspergillus* increased in the presence of different concentrations of *Sarcococca saligna* ethanol extract [14]. In our very recent work, we demonstrated the enhanced activity of ciprofloxacin in the presence of ZnO nanoparticles [15]. We also reported that silver nanoparticles enhance antibacterial activity of different antibiotics against *S. aureus* and *Escherichia coli* [16]. However, our search for finding new compounds that enhance the activity of antibiotics against *P. aeruginosa* was not successful during past years.

To date, gold-based drugs have shown great promise in treatment of various diseases such as auranofin for the treatment of arthritis and triphenylphosphinegold (I) complexes for the treatment of cancer tumors, psoriasis and HIV

infections [20]. Moreover, gold complexes have shown considerable cytotoxic and antimicrobial activity [21–23]. In this study, we report the combination effect of $Au^{3+}$ on antibacterial activity of different antibiotics against clinically resistant strains of *P. aeruginosa*, *S. aureus* and *E. coli*.

## Material and methods

### Gold materials

Chloroauric acid was purchased from Merck, Darmstadt, Germany. Au NPs used during this investigation were prepared by previously described tannin-free ethanol extract of black tea (*Camellia sinensis*) method [24]. Briefly, an aqueous chloroauric acid solution ($10^{-3}$ M) was added separately to the reaction vessel containing the tannin-free ethanol extract of black tea (10 %v/v), and the resulting mixture was allowed to stand for 15 min at room temperature. The reduction of the $Au^{3+}$ ions by tannin-free ethanol extract of black tea in the solutions was monitored by sampling the aqueous component (2 ml) and measuring the UV–visible spectrum of the solutions. This sample was diluted three times with distilled water, and the UV–visible spectrum of this sample was measured on a Labomed Model UVD-2950 UV–Vis Double Beam PC Scanning spectrophotometer, operated at a resolution of 2 nm. Furthermore, Au NPs were characterized by transmission electron microscopy (model EM 208 Philips). The gold colloid solution was centrifuged ($12000 \times g$) for 60 min. Subsequently the setteled Au NPs were washed three times with deionised water. A stock colloid solution (100 mg/ml) was prepared and reserved in 4°C for further experiments.

### Antimicrobial assay

The antibacterial activity of $Au^{3+}$ and Au NPs was evaluated at different contents (31.25, 62.5, 125, 250, 500, 1000, 2000 and 4000 µg/disc) on Müeller–Hinton agar (MHA) (Difco, Germany) using conventional disk diffusion method against *P. aeruginosa*, *S. aureus* and *E. coli*. Minimum inhibitory content was defined as the lowest content of $Au^{3+}$ or Au NPs creating clear zone of inhibition after 24 h at 35°C. This disk diffusion susceptibility test was also carried out on MHA plates in order to examine the antibacterial activity of candidate antibiotics against resistant test strains. Standard antibiotics disks, listed in Table 1, were purchased from Mast Co., UK. In order to explore a possible combination effect of gold materials and antibiotics, each standard paper disk was impregnated with the sub-inhibitory content of 40 µg/disk of $Au^{3+}$ and Au NPs. A resistant strain of *P. aeruginosa* was obtained from Imam University Hospital (Tehran, Iran) and identified by conventional microbiological

**Table 1** The antibacterial activity of $Au^{3+}$ ions and gold nanoparticles against *Pseudomonas aeruginosa*, *Staphylococcus aureus* and *Escherichia coli*

| Compounds (μg/disk) | Zone of inhibition diameter (mm) | | |
|---|---|---|---|
| | *E. coli* | *P. aeruginosa* (I) | *S. aureus* |
| $Au^{3+}$ | | | |
| 4000 | 32 | 30 | 34 |
| 2000 | 28 | 24 | 30 |
| 1000 | 21 | 20 | 22 |
| 500 | 14 | 17 | 16 |
| 250 | 10 | 14 | 13 |
| 125 | 9 (MIC) | 12 | 11 |
| 62.5 | – | 9 (MIC) | 9 (MIC) |
| 31.25 | – | – | – |
| AU NPs | | | |
| 4000 | 14 | 14 | 13 |
| 2000 | 10 | 11 | 10 (MIC) |
| 1000 | 9 (MIC) | 9 (MIC) | – |
| 500 | – | – | – |
| 250 | – | – | – |
| 125 | – | – | – |
| 62.5 | – | – | – |
| 31.25 | – | – | – |

Minimum inhibitory content (MIC) was defined as the lowest content of $Au^{3+}$ or Au NPs creating clear zone of inhibition after 24 h at 35°C

and biochemical methods. A single colony of test strains was grown overnight on Mueller–Hinton broth (MHB) medium on a rotator shaker (200 rpm) at 35°C. The inocula were prepared by diluting the cultures with 0.9 % NaCl to a 0.5 McFarland standard and were applied to the plates along with the standard and test disks containing 40 μg/disk of $Au^{3+}$ and Au NPs. After incubation at 35°C for 24 h, the zones of inhibition were measured. The mean surface area of each inhibition zone (square millimeter) was calculated from the mean diameter of each tested antibiotic. The percent of increase in the inhibition zone areas for different antibiotics against *P. aeruginosa* was calculated as $(b^2 - a^2)/a^2 \times 100$ where *a* is the inhibition zone in the presence of antibiotic only, and before addition of $Au^{3+}$ and Au NPs, while *b* represents the inhibition zone in the presence of antibiotic plus $Au^{3+}$ or Au NPs. The same procedure was used for combination of $Au^{3+}$ and Au NPs with antibiotics against additional test strains. All experiments were performed in triplicate.

Additionally, a different clinical isolate of *P. aeruginosa* and two test strains of *S. aureus* and *E. coli* were obtained from Ghods Polyclinic Laboratory (Tehran, Iran), and the same procedure was repeated. To compare the antibacterial activity results of test and control samples and in order to avoid possible errors, a parallel test was run for pure antibiotics with conditions similar to those for antibiotic–gold material

combination. The enhancing effect of $Au^{3+}$ with different antibiotics was further determined against mentioned test strains using the method described above (Table 1).

## Results and discussion

In this study, the Au NPs were prepared using a tannin-free ethanol extract of black tea. The inset to Fig. 1 shows the tubes containing this tannin-free extract before (tubes A) and after the reaction with $Au^{3+}$ for 15 min (tubes B). The gold-containing solutions (tubes A and B) that were a transparent yellow at first turned into purple on completion of the reaction by the tannin-free ethanol extract of black tea (tube B). These reaction mixtures were further characterized by UV–visible spectroscopy. As illustrated in Fig. 1, a strong surface plasmon resonance maximum was observed at ca. 527 nm. This peak is assigned to a surface plasmon phenomenon that is well documented for various metal nanoparticles with sizes ranging from 2 to 100 nm [25–27]. Figure 2 shows representative TEM images recorded from the drop-coated film of the as-prepared Au NPs, synthesized by treating the chloroauric acid solution with a tannin-free ethanol extract of black tea (left picture) after 15 min. The particle size histogram of these spherical gold particles, produced with this tannin-free ethanol extract (right illustration) in Fig. 2), shows that the particles range in size from 1.25 to 17.5 nm. It should be mentioned that almost 60 % of prepared Au NPs were in the range 2–6 nm.

The antibacterial effect of $Au^{3+}$ and Au NPs against *P. aeruginosa*, *S. aureus* and *E. coli* has been determened by

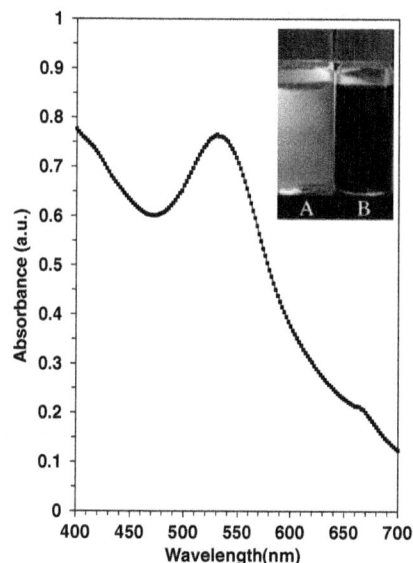

**Fig. 1** UV–visible spectrum of gold colloid. Spectrum recorded after adding the tannin-free ethanol extract of black tea (10 ml) to 90 ml of a chloroauric acid solution (1 mM). The curve is recorded after a period of 15 min. The *inset* shows the solution of chloroauric acid (1 mM) before (**a**) and after exposure to the tannin-free ethanol extract of black tea (**b**)

**Fig. 2** Transmission electron micrograph recorded from a small region of a drop-coated film of chloroauric acid solution treated with the tannin-free ethanol extract of black tea (*left-side picture*) for 15 min (scale bars correspond to 50 nm). The related particle size distribution histogram (*right-side picture*) obtained after measuring the size of 350 individual particles

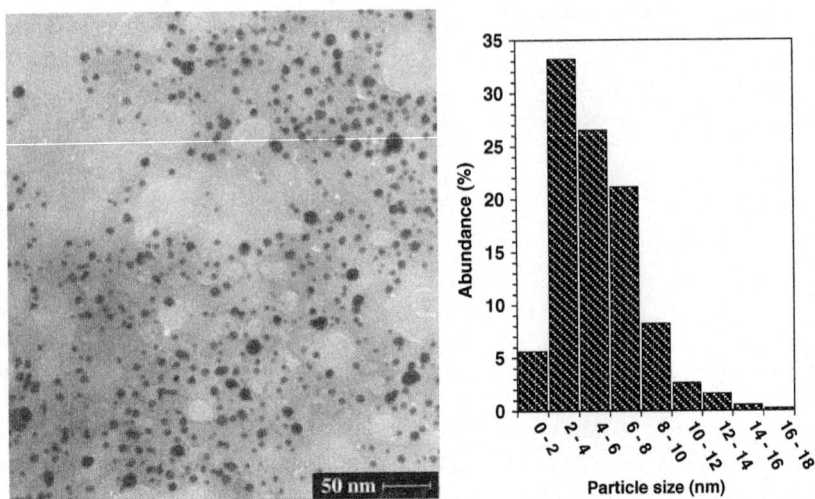

disk diffusion method and reported in Table 1. Higher concentrations of both $Au^{3+}$ and Au NPs (1000–4000 µg/disc) showed antibacterial activity aganist above test strains. As antibacterial agent, the $Au^{3+}$ ions were considarably more potent than inert Au NPs against all test strians. Lowest MICs were obtained for $Au^{3+}$ against *P. aeruginosa*, *S. aureus* and *E. coli* (Table 1). In the next step, the antibacterial activity of sub-inhibitory contents of $Au^{3+}$ and Au NPs was investigated in combination with a number of commonly used antibiotics against resistant strains of *P. aeruginosa*, *S. aureus* and *E. coli*. The diameters of inhibition zones (millimeter) in antibiotic disks both in the presence and absence of sub-inhibitory contents of $Au^{3+}$ and Au NPs were calculated. It should be noted that no antibacterial activity was observed for both $Au^{3+}$ and Au NPs at concentartions lower than 62.5 µg/disk. In this investigation the sub-inhibitory content of 40 µg/disk was chosen which is much lower than the MIC value required to produce antibacterial effect. Therefore, any increase in the antibacterial effect of antibiotics could be attributed not to the cytotoxic effect of $Au^{3+}$, but to the combination effect of $Au^{3+}$ with antibiotics.

Table 2 shows the inhibition zones (square millimeter) of candidate antibiotics against two different strains of *P. aeruginosa* both in presence and absence of sub-inhibitory content of 40 µg/disk $Au^{3+}$. As shown in the Table 2, different antibiotics showed different activities in the presence of $Au^{3+}$. In both resistant *P. aeruginosa* isolates, the antibacterial activity of penicillin G, methicillin, erythromycin, vancomycin, clindamycin and nalidixic acid increased, while no enhancing effect was observed for the remaining antibiotics. In detail, the surface area of inhibition zones (percent) in resistant *P. aeruginosa* strain 1 plates containing either of methicillin, erythromycin, vancomycin, penicillin G, clindamycin and nalidixic acid increased by 147, 147, 147, 104, 125 and 147 %, respectively.

Furthermore, the effect of $Au^{3+}$ was evaluated using the same set of antibiotics against a different clinical isolate of

*P. aeruginosa* and the clinical isolates of *S. aureus* and *E. coli*. In *P. aeruginosa* strain 2 group, the sensitivity of the

**Table 2** Increase in inhibition zone area (%) of candidate antibiotics against two resistant test strains of *Pseudomonas aeruginosa*, the clinical isolates of *Staphylococcus aureus* and *Escherichia coli* in the presence of $Au^{+3}$ at sub-inhibitory content of 40 µg/disk

| Antibiotics (µg/disk) | Increase in inhibition zone area (%) | | | |
|---|---|---|---|---|
| | *P. aeruginosa* (1) | *P. aeruginosa* (2) | *S. aureus* | *E. coli* |
| Penicillin G 10 | 104 | 65 | 19 | 0 |
| Amoxicillin 10 | 0 | 0 | 0 | 0 |
| Methicillin 5 | 147 | 146 | 0 | 0 |
| Cephalexin 30 | 0 | 0 | 0 | 30 |
| Cefixime 5 | 0 | 0 | 0 | 0 |
| Erythromycin 5 | 147 | 104 | 0 | 0 |
| Gentamicin 10 | 0 | 0 | 7 | 0 |
| Amikacin 30 | 0 | 0 | 7 | 9 |
| Tetracycline 30 | 0 | 0 | 10 | 7 |
| Ciprofloxacin 5 | 0 | 0 | 7 | 0 |
| Clindamycin 2 | 125 | 65 | 13 | 65 |
| Nitrofurantoin 300 | 0 | 0 | 8 | 21 |
| Nalidixic acid 30 | 147 | 146 | 39 | 0 |
| Vancomycin 30 | 147 | 104 | 0 | 147 |

Mean surface area of the inhibition zone ($mm^2$) was calculated from the mean diameter of each tested antibiotic. The percent of increase in the inhibition zone areas in presence of sub-inhibitory contents of $Au^{+3}$ for different antibiotics against *Pseudomonas aeruginosa* was calculated as $(b^2 - a^2)/a^2 \times 100$ where $a$ is the inhibition zone in the presence of antibiotic only, and before inoculation of $Au^{3+}$, and $b$ represents the inhibition zone in the presence of antibiotic plus $Au^{3+}$. No significant inhibition was observed when the combination of antibiotic–Au NPs was used under similar condition

new isolate of *P. aeruginosa* to methicillin, erythromycin, vancomycin, penicillin G, clindamycin and nalidixic acid increased by 146, 104, 104, 65, 65 and 146 %, respectively. However, the inhibition zone was not the same in different clinical isolates of *P. aeruginosa* (104 % and 65 %). In *S. aureus* group, $Au^{3+}$ slightly enhanced the antibacterial activity of penicillin G, gentamicin, amikacin, tetracycline, ciprofloxacin, clindamycin, nitrofurantoin and nalidixic acid. The most enhancing effects in this group were observed for nalidixic acid (39 % increase) and clindamycin (13 % increase). In *E. coli* group, $Au^{3+}$ had enhancing effect on antibacterial activity of cephalexin, amikacin, tetracycline, clindamycin, nitrofurantoin and vancomycin, and the most enhancing effect was observed for cephalexin (30 % increase), clindamycin (65 % increase), nitrofurantoin (21 % increase) and vancomycin (147 % increase).

The test was also repeated with the same set of antibiotics, using the same sub-inhibitory content of 40 µg/disk Au NPs against resistant *P. aeruginosa* which showed that Au NPs did not have a significant effect on antibacterial activity of antibiotics at a content level of 40 µg (results not shown).

In 2007, Grace and Pandium reported that Au NPs did not have antibacterial effects against a number of microorganisms including *P. aeruginosa*, *S. aureus* and *E. coli*, while coating of Au NPs with antibiotics increased their antibacterial activity [28]. Furthermore, recent studies by Burygin et al. revealed that Au NPs have no enhancing effect on the antibacterial activity of gentamycin [26]. In this study, no significant difference was observed between antibacterial activity of antibiotics alone and their mixture with Au NPs. However, as suggested by Burygin et al., it seems that Au NPs enhance antibacterial activity of antibiotics only when antibiotic is chemically attached on the surface of Au NPs and formed stable conjugates with particles rather than when used in combination with antibiotics as a mixture [29]. Therefore, comparing Au NPs and $Au^{3+}$, it is plausible that $Au^{3+}$ form more potent mixtures with antibiotics concerning the fact that they are more actively involved in reactions due to their ionic nature. This probably best explains our finding that when used together with antibiotics, Au NPs had no significant effect on antibacterial activity of antibiotics, while $Au^{3+}$ (which form more potent complexes) enhanced the inhibition zone of bacterial growth.

Comparing four groups, it seems that the enhancing effect of $Au^{3+}$ was more considerable in Gram negative *P. aeruginosa* and *E. coli* rather than Gram positive *S. aureus*. It is notable that studies of Marques et al. on MIC (minimum inhibitory concentration) value of $Au^{3+}$ metal compounds in complex with sulphamethoxazole in *P. aeruginosa*, *E. coli* and *S. aureus* suggested no difference between the response of Gram positive and Gram negative bacteria treated with Au compounds in combination with sulphamethoxazole [30], while other studies suggest that the enhancing effect

of $Au^{3+}$ is better in Gram negative bacteria [31]. However, more investigations should be carried out at the molecular level to clarify whether $Au^{3+}$ is more effective on Gram negative rather than Gram positive bacteria.

Fourteen candidate antibiotics were carefully chosen since they represent major classes of antibiotics (penicillins, cephallosporins, macrolides, aminoglicosides, tetracyclines, fluoroquinolones, lincomycin derivatives, nitrofurans and glycopeptides). However, different antibiotics showed different activities in the presence of $Au^{3+}$. In *P. aeruginosa*, a considerable enhancing effect was observed for methicillin, erythromycin, vancomycin, penicillin G, clindamycin and nalidixic acid, while in *S. aureus* $Au^{3+}$ slightly enhanced the antibacterial activity of penicillin G, gentamicin, amikacin, tetracycline, ciprofloxacin, clindamycin, nitrofurantoin and nalidixic acid. In *E. coli* group, $Au^{3+}$ had an enhancing effect on antibacterial activity of cephalexin, amikacin, tetracycline, clindamycin, nitrofurantoin and vancomycin.

It is already known that $Au^{3+}$ may form a coordination complex with available donor groups such as nitrogen, sulfur and phosphor. Therefore, in the presence of antibiotics such as penicillin, nalidixic acid and clindamycin, $Au^{3+}$ may form coordination complexes with ring nitrogen on these compounds. This interaction may have induced changes in morphology of these compounds, thereby increasing their efficiency. The same mechanism has been proposed for the interaction of gold (III) with zeatin [32]. In beta-lactam and cephalosporins such as penicillin and methicillin, this coordination may have happened between $Au^{3+}$ and free electrons on sulphur and nitrogen donor groups. However, this was not the case for all beta-lactam and cephalosporins tested (amoxicillin and cefixime).

It is also deducible that the enhancing effect of $Au^{3+}$ was more significant in Gram negative *P. aeruginosa* and *E. coli* rather than the Gram positive *S. aureus*. This is potentially interesting since studies by Chudasama et al. on core–shell silver nanostructures [33] and Nomiya et al. on gold (I) complexes [21] also revealed that gold complexes are considerably more effective on Gram negative rather than Gram positive bacteria. However, further investigations are suggested to be carried out on other Gram positive and Gram negative strains to see whether the structure of cell wall in bacteria affects the enhancing effect of gold materials.

So far, a number of elements including copper, lead and zinc have been studied for their interaction with *P. aeruginosa* [28, 34]. As mentioned earlier, the antibiotic resistance of *P. aeruginosa* is primarily due to the co-operation of multidrug efflux pumps and impermeability of bacterial membrane [6]. Therefore, it is possible that $Au^{3+}$ may sensitize *P. aeruginosa* cells by either interfering in the function of these efflux pumps or increasing the permeability of the bacterial membrane. Moreover, *P. aeruginosa* is well known for its resistance to heavy metals through reduction of metal ions. For instance, it actively resists against $Au^{3+}$ through reducing it to its metallic form [35, 36]. However, when treated with the combination of antibiotic–$Au^{3+}$, *P. aeruginosa* was more sensitive to the antibiotics. This could be due to the possible new interactions that may form between antibiotics and $Au^{3+}$ which literally hinder *P. aeruginosa* from reducing it to metallic Au and the fact that the organism may not be able to reduce heavy metal ions in the presence of antibiotics. However, more experiments are required to be done in order to determine the underlying mechanism of $Au^{3+}$ in the presence of antibiotics. The mechanism underlying the enhancing effect of $Au^{3+}$ is probably multi-factorial, and a molecular approach is necessary to identify the determinants of this effect.

## Conclusion

The result of this work demonstrates that using $Au^{3+}$, it is possible to enhance the efficacy of a number of commonly used antibiotics against *P. aeruginosa* up to 146 %. This finding is of particular value since resistant strains of *P. aeruginosa* are considered as a major problem in chemotherapy of many infectious diseases. This enhancing effect also occurred in other resistant microorganisms including *S. aureus* and *E. coli*. This suggests that the combination therapy of gold materials such as $Au^{3+}$ could be considered as a new approach and the common antibiotics may have an even broader range of medical applications in the future.

**Acknowledgments**   This work was supported by the Pharmaceutical Sciences Research Center, Faculty of Pharmacy, Tehran University of Medical Sciences, Tehran, Iran. The views presented in this article do not necessarily reflect those of the U.S. Food and Drug Administration.

## References

1. Levy SB, Marshall B (2004) Antibacterial resistance worldwide: causes, challenges and responses. Nat Med 10:122–129
2. Wright GD (2005) Bacterial resistance to antibiotics: enzymatic degradation and modification. Adv Drug Deliver 57:1451–1470
3. Lambert PA (2005) Bacterial resistance to antibiotics: modified target sites. Adv Drug Deliver 57:1471–1485
4. Kumar A, Schweizer HP (2005) Bacterial resistance to antibiotics: active efflux and reduced uptake. Adv Drug Deliver 57:1486–1513
5. Wisplinghoff H et al (2003) Current trends in the epidemiology of nosocomial bloodstream infections in patients with hematological malignancies and solid neoplasms in hospitals in the United States. Clin Infect Dis 36:1103–1110
6. Livermore DM (2002) Multiple mechanisms of antimicrobial resistance in *Pseudomonas aeruginosa*: our worst nightmare? Antimicrobial Resistance 34:634–640
7. Poole K (2001) Multidrug efflux pumps and antimicrobial resistance in *Pseudomonas aeruginosa* and related organisms. J Mol Microbiol Biotechnol 3:255–264
8. Li XZ, Zhang L, Poole K (2000) Interplay between the MexA-MexB-OprM multidrug efflux system and the outer membrane barrier in the multiple antibiotic resistance of *Pseudomonas aeruginosa*. J Antimicrob Chemother 45:433–436
9. Cornelis P (2008) *Pseudomonas*: genomics and molecular biology. Caister Academic, Brussels
10. Poole K (2004) Efflux-mediated multiresistance in Gram-negative bacteria. Clin Microbiol Infec 10:12–26
11. Obritsch MD et al (2005) Nosocomial infections due to multidrug-resistant *Pseudomonas aeruginosa*: epidemiology and treatment options. Pharmacotherapy 25:1353–1364
12. Tan YT, Tillett DJ, Mackay LA (2000) Molecular strategies for overcoming antibiotic resistance in bacteria. Mol Med Today 6:309–314
13. Wright GD (2000) Resisting resistance: new chemical strategies for battling superbugs. Chem Biol 59:7–16
14. Mollazadeh Moghaddam K et al (2010) The Antifungal activity of *Sarcococca saligna* ethanol extract and its combination effect with fluconazole against different resistant *Aspergillus* species. Appl Biochem Biotech 162(1):127–133
15. Banoee M et al (2010) ZnO nanoparticles enhanced antibacterial activity of ciprofloxacin against *Staphylococcus aureus* and *Escherichia coli*. J Biomed Matter Res B 93:557–561
16. Shahverdi AR et al (2007) Synthesis and effect of silver nanoparticles on the antibacterial activity of different antibiotics against *Staphylococcus aureus* and *Escherichia coli*. Nanomedicine 3:168–171
17. Fayaz AM et al (2010) Biogenic synthesis of silver nanoparticles and their synergistic effect with antibiotics: a study against Gram-positive and Gram-negative bacteria. Nanomedicine 6:103–109
18. Rai A, Prabhune A, Carole C (2010) Antibiotic mediated synthesis of gold nanoparticles with potent antimicrobial activity and their application in antimicrobial coatings. J Mater Chem 20:6789–6798
19. Rafii F, Shahverdi AR (2007) Comparison of essential oils from three plants for enhancement of antimicrobial activity of nitrofurantoin against Enterobacteria. Chemotherapy 53:21–25
20. Bowman MC et al (2008) Inhibition of HIV fusion with multivalent gold nanoparticles. J Am Chem Soc 130:6896–6897
21. Nomiya K, Noguchi R, Oda M (2000) Synthesis and crystal structure of coinage metal (I) complexes with tetrazole (Htetz) and triphenylphosphine ligands, and their antimicrobial activities. A helical polymer of silver (I) complex [Ag (tetz)(PPh3)2]n and a monomeric gold (I) complex [Au(tetz)(PPh3)]. Inorg Chim Acta 298:24–32
22. Nomiya K et al (2000) Synthesis, crystal structure and antimicrobial activities of two isomeric gold(I) complexes with nitrogen-containing heterocycle and triphenylphosphine ligands, [Au(L)(PPh3)] (HLspyrazole and imidazole). J Inorg Biochem 78:363–370
23. Noguchi R, Hara A, Sugie NK (2006) Synthesis of novel gold(I) complexes derived by AgCl-elimination between [AuCl(PPh3)] and silver(I) heterocyclic carboxylates, and their antimicrobial activities. Molecular structure of [Au(R, S-Hpyrrld)(PPh3)] (H2pyrrld2-pyrrolidone-5-carboxylic acid). Inorg Chem Commun 9:355–359
24. Banoee M et al (2010) The green synthesis of gold nanoparticles using the ethanol extract of black tea and its tannin free fraction. Iran J Mater Sci Eng 7:48–53
25. Henglein A (1993) Physicochemical properties of small metal particles in solution: "microelectrode" reactions, chemisorption, composite metal particles, and the atom-to-metal transition. J Phys Chem B 97:5457–5471

26. Sastry M et al (1997) pH dependent changes in the optical properties of carboxylic acid derivatized silver colloidal particles. Colloid Surf A 27:221–228

27. Sastry M et al (1998) Electrostatically controlled diffusion of carboxylic acid derivatized silver colloidal particles in thermally evaporated fatty amine films. J Phys Chem B 102:1404–1410

28. Nirmala Grace A, Pandian k (2007) Antibacterial efficacy of aminoglycosidic antibiotics protected gold nanoparticles: a brief study. Colloid Surface A 297: 63–70

29. Burygin GL et al (2009) On the enhanced antibacterial activity of antibiotics mixed with gold nanoparticles. Nanoscale Res Lett 4:794–801

30. Marques LL et al (2007) New gold(I) and silver(I) complexes of sulfamethoxazole: synthesis, X-ray structural characterization and microbiological activities of triphenylphosphine(sulfamethoxazolato-N2)gold(I) and (sulfamethoxazolato)silver(I). Inorg Chem Commun 10:1083–1087

31. Teitzel GM, Parsek MR (2003) Heavy metal resistance of biofilm and planktonic Pseudomonas aeruginosa. Appl Environ Microb 69:2313–2320

32. Fowles CC, Smoak EM, Banerjee IA (2010) Interactions of zeatin with gold ions and biomimetic formation of gold complexes and nanoparticles. Colloid Surf B 78:250–258

33. Chudasama B et al (2009) Enhanced antibacterial activity of bifunctional Fe3O4–Ag core–shell nanostructures. Nano Res 2:955–965

34. Aendekerk S, Ghysels B, Cornelis P, Baysse C (2002) Characterization of a new efflux pump, MexGHI-OpmD, from *Pseudomonas aeruginosa* that confers resistance to vanadium. Microbiology 148:2371–2381

35. Campbell SC et al (2001) Biogenic production of cyanide and its application to gold recovery. J Ind Microbiol Biotechnol 26 (3):134–139

36. Durán N (2007) Antibacterial effect of silver nanoparticles produced by fungal process on textile fabrics and their effluent treatment. J Biomed Nanotechnol 3:203–208

# 17

# Subcellular localization of gold nanoparticles in the estuarine bivalve *Scrobicularia plana* after exposure through the water

Yolaine Joubert · Jin-Fen Pan · Pierre-Emmanuel Buffet ·
Paul Pilet · Douglas Gilliland · Eugenia Valsami-Jones ·
Catherine Mouneyrac · Claude Amiard-Triquet

**Abstract** Nanoparticles are extensively used particularly in biomedical and industrial applications. Because of their colloidal stability, gold nanoparticles (AuNPs) are suspected being persistent in aquatic ecosystem. Thus, the potential toxicity of gold nanoparticles is addressed by using a bivalve model *Scrobicularia plana*. Using AuNPs in a range of sizes (5, 15, and 40 nm), we examined their subcellular localization in gills and digestive gland. Clams were exposed to AuNPs stabilized with citrate buffer and then diluted in seawater at the concentration of 100 $\mu gL^{-1}$. After 16 days water-borne exposure, using transmission electron microscopy, few particles were observed in gills, distributed as free in the cytoplasm, or associated with vesicles. In the digestive gland, the most striking feature was the presence of individual or small aggregates 40 nm sized within the nuclei colocalized with DNA. Depending on the size, individual or small aggregates (40 nm AuNPs) or more aggregated NPs (5 and 15 nm) were observed, with at least one of the dimensions (40–50 nm) allowing the passage through nuclear pores. Disorganization of chromatin was marked with an increase in filamentous structures. In some parts no chromatin was visible. Moreover, the perinuclear space from nuclei was enlarged in contaminated clams when compared to controls.

**Keywords** Gold nanoparticles · *Scrobicularia plana* · Transmission electron microscopy · Chromatin

Y. Joubert · J.-F. Pan · P.-E. Buffet (✉) · C. Mouneyrac ·
C. Amiard-Triquet
Université de Nantes, MMS EA 2160,
LUNAM Université, 9, Rue Bias,
44035 Nantes, France
e-mail: pe.buffet@yahoo.fr

J.-F. Pan
College of Environmental Science and Engineering,
Ocean University of China, Qingdao 266100 China

P.-E. Buffet · C. Mouneyrac
Université Catholique de l'Ouest, MMS EA 2160,
LUNAM Université, 3, Place André Leroy,
49000 Angers, France

P. Pilet
School of dental Surgery, INSERM U791,
1 Place Alexis Ricordeau,
44042 Nantes Cedex 1, France

D. Gilliland
Institute For Health and Consumer Protection European
Commission—DG JRC, Via E. Fermi,
19 I-21027 Ispra, VA, Italy

E. Valsami-Jones
School of Geography, Earth and Environmental Sciences,
University of Birmingham, Edgbaston, Birmingham B15 2TT, UK

E. Valsami-Jones
Department of Earth Sciences, Natural History Museum London,
Cromwell Road,
London SW7 5BD, UK

## Introduction

Nanotechnology is an emerging field exploiting different materials at the nanometer scale. A wide range of nanomaterials such as iron, silver, carbon, titan, diamond, and gold have been engineered. Among these nanomaterials, nanoparticles (NPs) have been broadly defined as having one size range of 1–100 nm diameter. Due to their size, they have provoked an enormous interest for both industrial and biomedical applications. Gold nanoparticles (AuNPs) show a great potential for cell imaging, targeted drug delivery, cancer diagnostics, and therapeutics. Recently, several groups have demonstrated that AuNPs possess an enormous potential to improve the efficiency of clinical diagnosis [1] and of cancer treatment [2–5]. As the field continues to develop, the impact of AuNPs on human and environmental health remains unclear. Understanding and controlling the interactions between NPs and living cells will be important for assessing their designated functions since NPs may cause undesirable interactions with biological systems. Moreover, the engineering of large quantities of nanoparticles may lead to unintended contamination of terrestrial and aquatic ecosystems [6]. Thus, they could also represent a potential source of emerging contaminants in the environment. Only a few studies deal with their behavior or impact on the environment [7–12].

At the nanometric scale, NPs acquire novel physico-chemical properties that may influence bioavailability. Size, shape, surface chemistry, stability, concentration, and time of exposure are reported to induce different effects (see reviews [13, 14]. Despite showing little or no cytotoxicity via several standard assays, AuNPs may be internalized in the cells and cause cellular damage (see reviews [13, 15, 16]. Most investigators studied specific nanoparticle interactions with single cellular system in which parameters can be controlled, even though this type of model is artificial. Up to date, no consensus exist in regard with the subcellular location of AuNPs (reviewed in Khlebtsov and Dykman [17]): freely dispersed in cytoplasm [18–20] clustered in vesicles [7, 11, 12, 18, 21, 22]. Some studies showed a high fraction of radioactive AuNPs linked to DNA [23], an aggregation of small AuNPs (2 nm) within the nuclei which were damaged [24], a nuclear fragmentation [25, 26].

The main molecular mechanism of nanotoxicity is the induction of oxidative stress by free radical formation [27]. Recent literature contains conflicting data regarding oxidative stress [7, 13, 28] and cytotoxicity of AuNPs [15, 24, 29]. Tissues have potential defense mechanisms, including intracellular antioxidants and antioxidant enzymes [30] such as glutathione S-transferase (GST), superoxide dismutase (SOD), catalase (CAT), and metallothionein proteins (MTs). Our previous investigations [31] showed that activities/concentration of these biomarkers increased following exposure to AuNPs of different sizes on the marine bivalve *Scrobicularia plana* which is an intertidal deposit-feeder organism widely used in ecotoxicological studies [32, 33]. With regard to these results, in the present study we explored the cellular impact of these gold nanoparticles on *S. plana*. Clams were exposed for 16 days to AuNPs of size 5, 15, and 40 nm initially stabilized in citrate buffer (2.5 mM, pH 6.3; 2.5 mM, pH 6.1; and 0.5 mM, pH 6.9 for 5, 15, and 40 nm AuNPs, respectively) as described by Turkevich et al. [34], then diluted in seawater at a concentration of 100 µg AuL$^{-1}$, concentration used in our previous work [31]. The goal of this study was to determine the subcellular localization of AuNPs in *S. plana* by using transmission electron microscopy (TEM). AuNPs are electronically dense due to their elevated extinction coefficient that allows their detection by TEM. Targeted organs were gills since in bivalves they are the first organs in contact with particles, and digestive gland as a key organ for metal metabolism.

## Methods

### Animal collection and acclimation

*S. plana* with shell length of 2.5 cm were collected from the top 20 cm depth intertidal mudflat in March 2010 from the bay of Bourgneuf, located on the French Atlantic coast (1°59′ 04.80″ W, 47°01′50.35″ N). This area is comparatively low in contaminant bioavailabilities according to the results of the French national biomonitoring network RNO [35]. Then clams were transported to the laboratory in cool boxes covered with seaweeds. They were immediately transferred to aerated seawater and allowed to acclimate to the laboratory conditions for 48 h at the same temperature as in the field (10 °C).

### Nanoparticle preparation and characterization

More details on the characterization methods are described in Pan et al. [31]. Briefly, AuNPs of three different sizes were prepared at Joint Research Center, Ispra, Italy as a suspension of 98.5 mgL$^{-1}$ in citrate buffer. Gold nanoparticle suspension was characterized [31] by UV–vis spectroscopy and dynamic light scattering. Following addition of the gold to the seawater, the samples were mixed and agitated for a period of 24 h. Electrostatic charge of nanoparticles were defined in citrate buffer and seawater using a ZetaSizer Nano Zs (Malvern Instruments). Samples were transferred to a zeta cell (Malvern Instruments) and measured at 25 °C using an applied voltage of 150 V. Data are expressed as means ± standard error (SE) performed in five replicates.

Particle size and morphology were characterized using a Jeol JEM 1010 (80 kV) equipped with a camera system (Orius 200w Gatan Inc. USA). For sample preparation carbon-coated copper 200 meshes TEM grid (Agar Scientific, UK) were placed onto a drop of 50 µL of citrate-AuNPs for 1 min, and dried at room temperature. Electron micrographs were digitized and analyzed using a Digital Micrograph (Gatan Inc.). For each sample, the size of 200 particles was measured to obtain histograms of particle size distribution.

Elemental analysis was performed on the grids using an X-ray energy dispersive system (ISIS, Oxford Instruments, England) coupled to the TEM.

### Nanoparticle exposure

The nanoparticle semi-static exposures were carried out using pre-filtered natural seawater (0.45 µm), with one control and three AuNP treatments each containing one size of AuNPs (5, 15, and 40 nm, respectively) as described earlier in Pan et al. [31]. For each of the three sizes, the exposure concentration was $100 \, \mu g \, Au \, L^{-1}$. For each condition, clams ($n=36$) were distributed into three polypropylene tanks, each containing 2.0 L exposure medium (12 individuals per tank). Exposure tests were carried out for 16 days at 10 °C in a dark conditioned cabin to avoid light disturbance of endobenthic bivalves. The experimental media (water and NPs) were renewed every other day to ensure oxygen saturation and readjust NP concentration in the water column. Bivalves remained unfed during the whole experiment (16 days) to eliminate the potential food interference and working with lower toxicant conditions than if conducting a shorter test.

Sample preparation for TEM

Following exposure, for each condition (control and NPs of each different size), three clams were collected from three replicated experimental tanks. The isolated tissues (gills and digestive gland) were cut into small parts to obtain fine pieces. They were rinsed in cold phosphate buffer and placed in a fixing solution of glutaraldehyde (2.5 %) and cacodylate buffer (0.1 M) for 2 h at 4 °C and post-fixed in 1 % osmium tetroxide and cacodylate buffer for 1 h at 4 °C. After fixation, samples were rinsed with cacodylate buffer

and dried with increasing concentrations of ethanol and propylene oxide. Samples were embedded in EMBed-812 resin (Agar Scientific, UK) and polymerized. Ultrathin sections were performed with an ultramicrotome (Ultracut E, Leica Microsystems, Germany) for TEM were prepared with a diamond knife (Diatome, Switzerland), collected on copper grids and contrasted with uranyl acetate and lead citrate. Samples were observed using a transmission electron microscope (Jeol JEM 1010, Japan).

## Results

### Elemental analysis

As shown by Zeta potential analysis, AuNPs were negatively charged when suspended in citrate buffer (mean value for 5, 15, and 40 nm, $-70\pm2$ mV) and in seawater $-18\pm5$ mV).

The nanosize determined by TEM is reported in Fig. 1a–c. Particle sizes are almost homogeneous with respect to size as indicated by the scale bar of 50 nm. Size distribution reported as histograms were respectively $5.3\pm1.3$, $14.1\pm1.4$, and $31\pm8$ nm. The observations of thin sections in biological tissues described below (Figs. 2, 3, 4, 5, 6, and 7) reveal the presence of NPs showing the same sizes and shape than those described in suspensions (Fig. 1) used for experimental contaminations of bivalves.

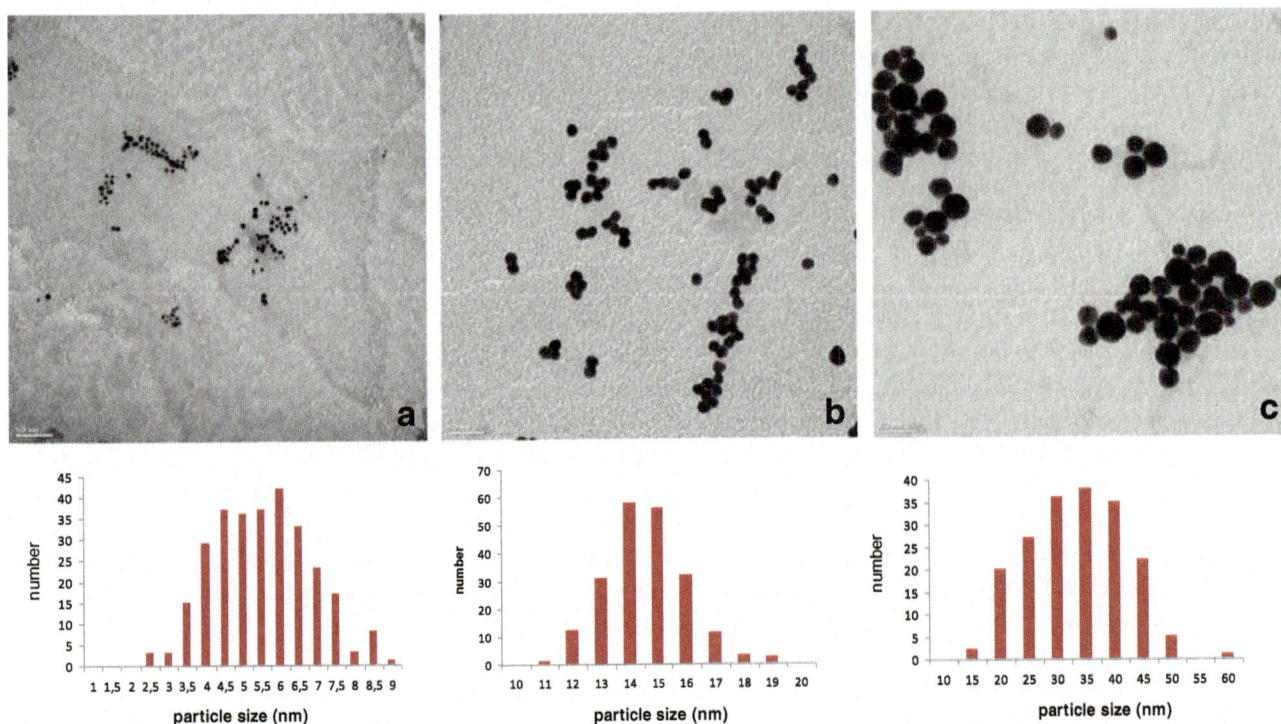

**Fig. 1** TEM images of 5 nm (**a**), 15 nm (**b**), and 40 nm (**c**) AuNPs on carbon-coated grids. In each panel, *scale bars* denoting 50 nm and histogram of AuNP diameters determined by analysis of approximately 200 AuNPs located at different regions of the grid

**Fig. 2** Electron micrograph of a transverse section through the surface of the border wall of gill in *S. plana*. The presence of cilia (*asterisks*) which penetrate the cytoplasm indicates the apical side of the tissue. **a** localization of 40 nm AuNPs, **b** higher magnification in which three AuNPs are visible. Note the presence of numerous vesicles of different sizes (*currency signs*), **c** localization of a 40 nm AuNP free in the cytoplasm (*arrow*) just near the cell membrane; visualization of rough reticulum endoplasmic (*RER*) and of free ribosomes surrounded by circles in the cytoplasm (*circles*)

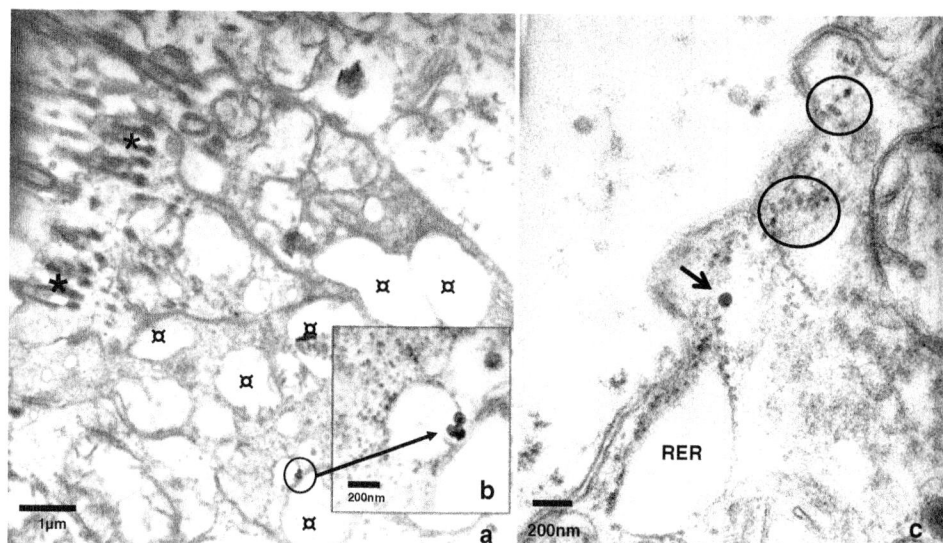

## Gills

AuNPs were detected close to the basal side of microvilli, therefore demonstrating the ability of AuNPs to penetrate this epithelium (Fig. 2a). Inside the tissue, few particles were found free in the cytoplasm (Fig. 2a, b). Vesicles were more numerous in exposed specimens than observed in controls and their size was highly variable (Fig. 2a). The border wall of microvilli showed no AuNPs retained outside the cell membrane (Fig. 2c). A 40 nm AuNP was visible just near the cell membrane (Fig. 2c). TEM examinations did not reveal any structural disturbance of the plasma membrane. We can observe the rough endoplasmic reticulum (RER) and free ribosomes in the cytoplasm. Since no endocytosis

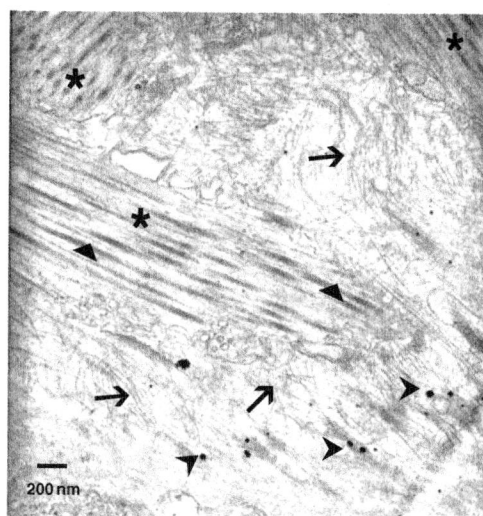

**Fig. 3** Electron micrograph through the microvillus border of the digestive gland. Straight microvilli (*asterisks*) are longitudinally oriented and anchored in the apical cytoplasm. The central microvillus core composed of a dense fibrillar meshwork (*triangles*) is surrounded by a microtubule zone (*arrows*) near the apical web. AuNPs (15 nm) are associated with these microtubules (*arrowheads*)

was observed in contaminated gills, TEM examinations cannot give any information about the nature of these vesicles. Although many ultrathin sections from several experimental samples were analyzed, the contamination and the subsequent bioaccumulation of AuNPs in the gills remained still very weak.

## Digestive gland

Electron micrograph of a section through the microvillous border and the apical cytoplasm of the absorptive tissue showed the footlet of a microvillus surrounded by a dense fibrillar meshwork (Fig. 3). These cytoplasmic microtubules shaped a filamentous area out of several microtubules orientated as longitudinal sections. Not a single AuNPs could be observed in contact with the microvilli border outside the plasma membrane. No endocytosis figures (i.e., vesicles formed by invagination of the plasma membrane) were found in the apical plasma membrane. AuNPs were detected close to the basal side near microvilli inside epithelial digestive gland. Unlike AuNPs found in gills, those found in digestive gland were never located inside vesicles.

## Exposure to 40 nm AuNPs

When comparing the morphological features between controls and contaminated samples, nuclei contrasted markedly. In controls, chromatin condensed as heterochromatin was distributed all over the nucleus (Fig. 4a). In experimental *S. plana*, TEM examination revealed the localization of 40 nm AuNPs within the nuclei in the digestive gland tissue. Single particles or small clusters of three to five AuNPs were distributed all around the nucleus (Fig. 4b). In clams exposed to AuNPs, visually the amount of heterochromatin was generally less abundant and few amount of chromatin was visible in the central part of the nucleus 1 (Fig. 4b).

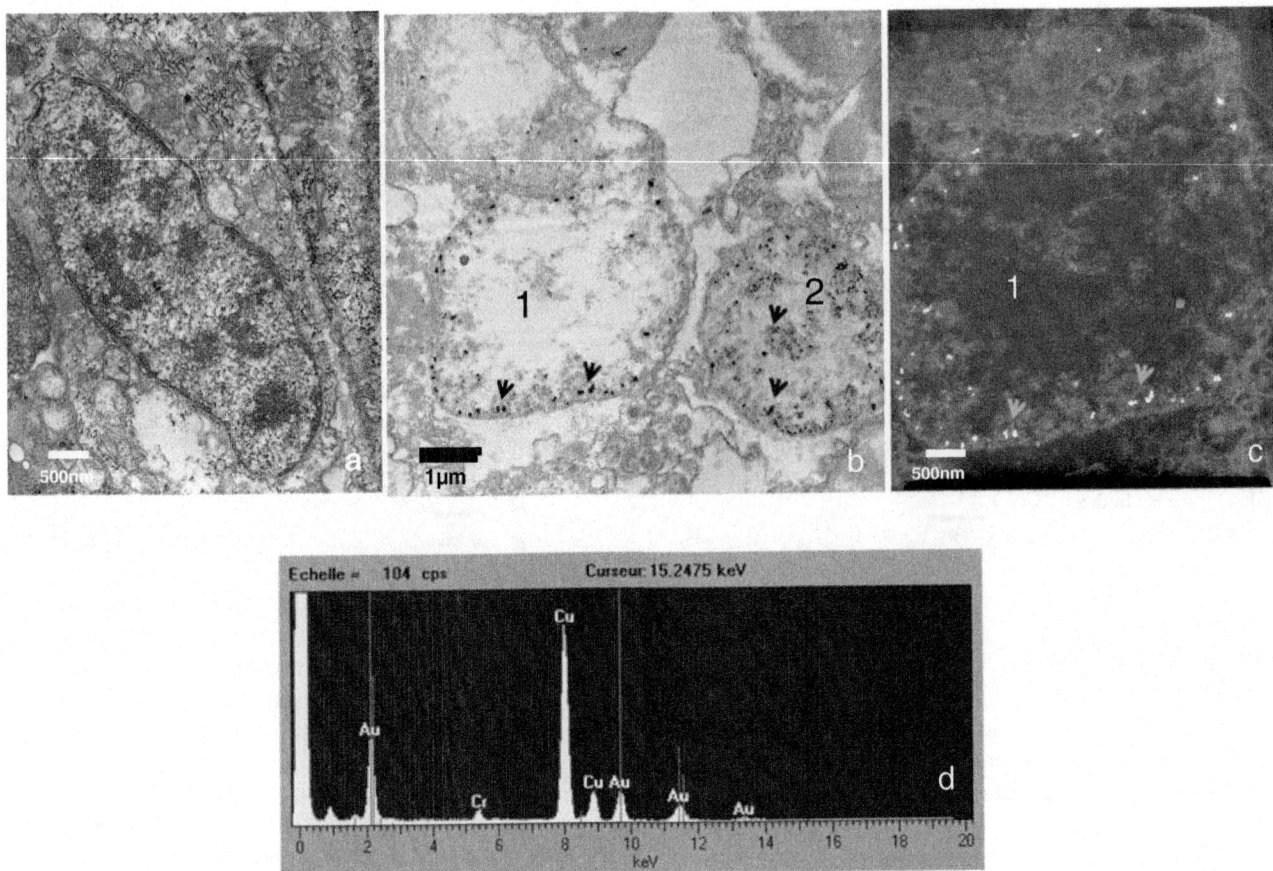

Fig. 4 **a** In controls, chromatin appears as heterochromatin distributed all over the nucleus, **b** in contaminated *S. plana* 40 nm AuNPs are visualized within two nuclei (*1, 2*), AuNPs are always associated with chromatin (*arrowheads*), the degradation of which seems to be higher in nucleus 1 than in nucleus 2, **c** negative film of the nucleus 1, AuNPs associated with chromatin are more visible and appear as *white dots* (*arrowheads*). **d** Elemental composition of NPs collected through an analysis X by EDS that shows the presence of Au as indicated by the three peaks corresponding to the gold M shell (2.2 keV) and L shells (9.7 and 11.5 keV)

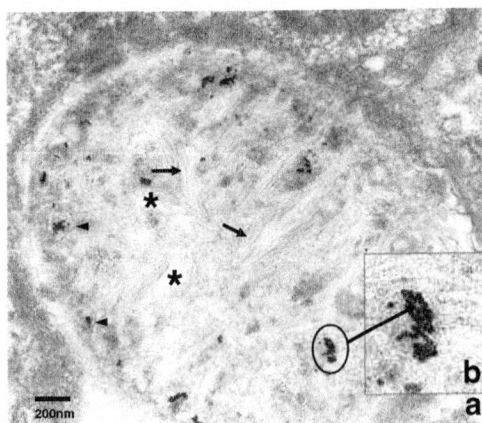

Everytime, AuNPs were localized in the vicinity of chromatin. However, AuNP uptake within cell nuclei was not homogeneous in the whole tissue. TEM digestive gland cells examination revealed two nuclei invaded by AuNPs (Fig. 4b). Figure c is the negative film from the figure b, in which AuNPs appear as white dots. These nuclei were always visualized within the apical cytoplasm adjacent to the plasma membrane. No endocytosis process was visible at the apical surface. Cytoplasmic membrane, mitochondria, and cytoplasmic reticulum seemed morphologically intact (not shown).

Elemental analysis using an X-ray energy dispersive system on the ultrathin sections (Fig. 4d) proved the presence of Au as indicated by the three peaks corresponding to the gold M shell (2.2 keV) and L shells (9.7 and 11.5 keV).

## Exposure to 15 nm AuNPs

As for 40 nm AuNPs, 15 nm AuNPs were mainly located within digestive gland cell nuclei (Fig. 5a) but particles often aggregated up to a number of about 15–20 (Fig. 5b).

Fig. 5 **a** In contaminated *S. plana*, AuNPs (15 nm) are aggregated to each other. The number of NPs inside these formations seems to be variable (*arrowheads*). They are localized within the chromatin which appears strongly altered. At some places, no more chromatin can be seen (*asterisks*) and fibrillar material was observed instead of condensed chromatin (*arrows*). Nucleus volume is swollen, **b** higher magnification of an AuNP aggregate with an approximate width of 40–50 nm

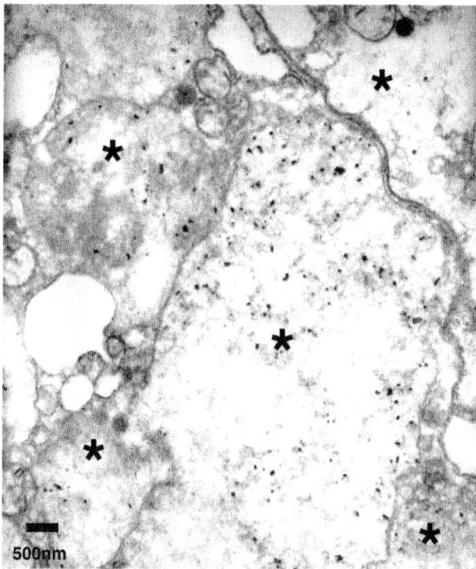

**Fig. 6** In contaminated *S. plana*, 15 nm AuNPs are localized within chromatin of several nuclei (*asterisks*) belonging to different cells. The number of contaminated nuclei is higher than that observed with 40 nm AuNPs

At a high magnification, width of aggregates was measured (Fig. 5b). When referred to the scale bar, the width was about 40–50 nm. AuNPs accumulated only in the vicinity of chromatin, the ultrastructure of which was strongly altered compared to controls and 40 nm AuNP exposed tissues. There was less condensed chromatin which seemed to be more dispersed. At some places, fibrillar material was observed instead of condensed chromatin. Moreover, a number of nuclei seemed to be swollen. As in 40 nm AuNPs exposed tissues, nuclei were visualized within the apical cytoplasm (Fig. 6). In the case of 15 nm AuNP exposed tissues, the number of nuclei in which AuNPs were visible, was

higher than observed with 40 nm AuNPs (Fig. 6). Aggregated nanoparticles were attached to the microtubules described above (Fig. 3). No endocytosis vacuoles were seen near this terminal web. Although we noticed an increase in intracellular vesicles, there was not AuNPs invasion within these organelles.

Exposure to 5 nm AuNPs

TEM examinations of ultra-thin sections of 5 nm AuNPs exposed epithelial cells of the digestive gland did not reveal any structural disturbance of the plasma membrane, mitochondria, and endoplasmic reticulum (not shown). However, as mentioned previously for 15- and 40-nm AuNP experiments, our results showed the ability of AuNPs to penetrate digestive gland epithelium. Aggregates of AuNPs detected in this tissue remained localized in nuclei (Fig. 7a). Chromatin appeared to be more disorganized than in digestive glands exposed to 15 and 40 nm AuNPs. Chromatin amount was strongly decreased and DNA disorganization extended to the peripheral area (Fig. 7a). We noticed an increase in filamentous structure formation. Therefore, some parts seemed to be devoid of chromatin. AuNPs were condensed into large aggregates which decorated the chromatin. The width of these aggregates was about 40–50 nm (Fig. 7b). The nuclear membrane appeared ruffling (Fig. 7a) and the perinuclear space seemed to be enlarged (Fig. 7a) compared to control (Fig. 7c).

**Discussion**

As a whole organism is much more complex than a single cell, in vivo toxicological studies are required to assess the safety of nanoparticles. Uptake of AuNPs was shown in the

**Fig. 7 a** In contaminated *S. plana*, 5 nm AuNP aggregates with an approximate width of 40–50 nm are localized within chromatin which appears strongly disorganized. That is attested by an increase in filamentous structure. Some parts of the nucleus seem to be devoid of chromatin (*asterisks*), **b** higher magnification of an aggregate. The perinuclear space is enlarged (*up down arrows*) in contaminated *S. plana* (**a**) when compared to the control (**c**). Note the morphological change in chromatin between control (**a**) and contaminated *S. plana* (**b**)

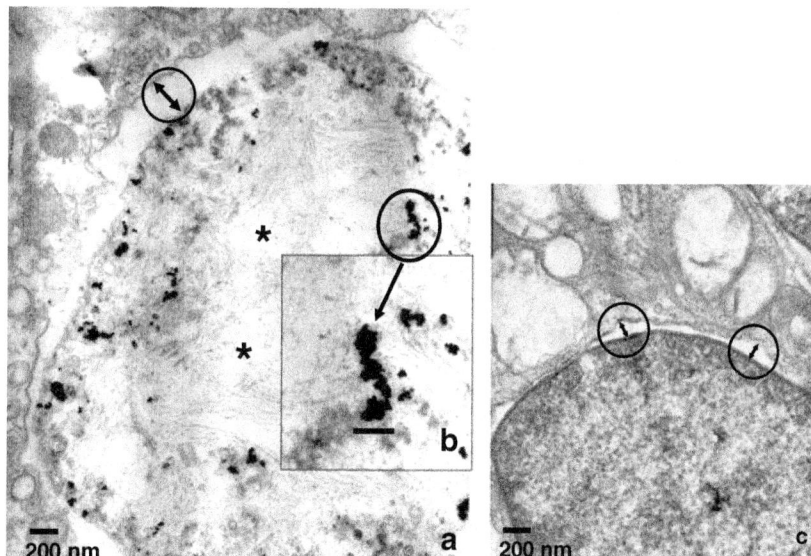

whole soft tissues of the bivalve *S. plana* exposed in vivo to AuNPs of different sizes [31].

The AuNPs examined in the present study have been characterized by Pan et al. [31]. It has been shown that aggregation occurred in seawater for all the three different sizes of AuNPs, increasing from nanosize 5–40 nm to size >700 nm. These findings are in agreement with the loss of charge measured with ZetaSizer for AuNPs (5, 15, and 40 nm) suspended in seawater in the present work, contributing to the aggregation.

The present study reveals that all of these bioaccumulated AuNPs were localized almost exclusively in the digestive gland confirming the results obtained in two other bivalves, namely *Mytilus edulis* [11] and *Corbicula fluminea* [7, 36]. Particulate matter and AuNP aggregates deposited on the bottom of the experimental tank are ingested through the inhalant siphon of the clams, subsequently transported to the mouth, then to the digestive tract and the digestive gland for intracellular digestion [37]. Such a location of AuNPs in digestive gland is not surprising as this organ is known to be a key site of metal detoxification [38].

AuNPs were detected inside digestive epithelium but also inside gill epithelium, demonstrating their ability to cross these barriers. AuNPs had different cellular localization when comparing gills and digestive gland. AuNPs which had penetrated gill cells seemed to be free in the cytoplasm. AuNPs were never observed outside this border and no damage or invagination of the plasma membrane suggesting endocytosis were visible. So, the mechanism allowing AuNPs to enter the cells could not be established from by our observations. However, from several in vitro [16, 19, 39] and in vivo studies [7, 11], it was reported that particles entering cells were trapped in vesicles.

In the digestive gland, once particles have crossed the microvillous border, our TEM observations indicated the presence of AuNPs associated with filaments supporting the apical web near the outer surface, suggesting that AuNPs could be passively transported all along these contractile structures toward the nuclear membrane.

The 40 nm AuNPs entered the digestive gland cells and were exclusively localized within cell nuclei (Fig. 4b and c). In these nuclei, single or two to three aggregates were distributed along the chromatin and seemed to be specially linked to the chromatin. The loss of condensed chromatin is evident, indicating that the condensed DNA of the nucleus had been damaged. No more dense chromatin was observed in some part of the nucleus. Despite this ultrastructural abnormality, mitochondria or plasmic membranes seemed to be still intact. The 15 and 5 nm AuNPs were also localized within the nuclei. However, for these sizes of AuNPs, some striking differences have been noticed. There are only a few single particles linked to chromatin, most of them were aggregated into patches of different sizes. The

morphological modifications of chromatin previously mentioned were more pronounced and DNA appeared as fibrillar. AuNPs could be counted and sized within these aggregates. Whatever number of AuNPs in aggregates the shape of them seemed to be defined. When measuring the size of these aggregates, we noticed that the width was always between 40 and 50 nm. As the AuNPs of the three sizes (5, 15, and 40 nm) are able to cross the nuclear membrane, it appears that AuNPs may have pass through the nuclear pores which have a central channel of a patent diameter of 40 nm [40]. Based on this assumption, single 40 nm and aggregates of 5 and 15 nm AuNPs have to be flexible for crossing the nuclear pores. Chitrani et al. [21] showed that the maximum cellular uptake occurred at a nanoparticle size of 50 nm. However, these authors claimed that particles were trapped in vesicles and did not enter the nucleus.

Our previous data [31] showed that in seawater an aggregation occurred for AuNPs of the three sizes. The diameter size of these aggregates was identical and peaked at 600 nm for the three types of NPs. That implies that following uptake, aggregates will be likely broken down by the action of the cilia present all along the gills and on the microvillous border of the digestive gland. Moreover, aggregates could be dissociated chemically in the digestive tract under acid pH 4.5 [41].

In bivalves, bioaccumulation and cytotoxicity of AuNPs was reported by Renault et al. [7] and Tedesco et al. [11, 12] without any clear demonstration of nuclear localization. AuNPs have been found inhibiting cell proliferation by down-regulating cell cycle genes [19]. Panessa-Warren et al. [24] claimed that only small clusters of 2 nm NPs were seen at the nuclear membrane and within the nucleus of lung epithelial cells, whereas, the 10 nm AuNPs were not seen within nuclei. They suggested that the larger core size may not allow their crossing through the nuclear channel measuring 9 nm. But it seems to result from a misinterpretation of the report by Franke et al. [42] indicating a size exclusion limit of approximately 18 nm, whereas a more recent paper [40] indicates a nuclear pore size of 40 nm.

Although NP-induced cytotoxicity has been reported by several groups, many biomedical applications have been reported. Gold NPs conjugated to antibodies can be selectively targeted to cancer cells without significant binding to healthy cells [2, 43]. Gold nanospheres anticancer therapy by using their two-photon absorption of 800 nm laser light was reported by the same group [44]. Recently, Patra et al. [5] have developed a NP-based targeted drug delivery system (DDS) using an anti-epidermal growth factor receptor as a targeting agent, gemcitabine as the anti-cancer drug, and gold as the delivery vehicle in pancreatic cancer. They demonstrated that targeted DDS was much more effective to inhibit the proliferation of pancreatic cancer cells than its non-targeted counterpart. Kang et al. [4] proposed that AuNPs can be used alone as an anticancer therapeutic material if conjugated to the proper nuclear-targeting ligand such as the nuclear localization signal peptide sequence (NLS). NLS is known to associate with importin-protein in the cytoplasm after which translocation to the nucleus occurs. In this case, AuNPs induces DNA damage, causing cytokenesis arrest and apoptosis [4]. Our results strongly demonstrate that AuNP localization was observed in cell nuclei without any nanogold targeting signal.

Taken together, our results suggested that the three sizes of AuNPs have a capability of inducing DNA damage and subsequent events which can specially affect cellular functions leading to cell death. What kind of cell death AuNPs could induce: necrosis or apoptosis? No fragmentation of nuclei and/or cytoplasmic organelles indicating apoptosis process was visible. Results of Pan et al. [25] concluded to a size-dependent cytotoxicity, in that 1.4 nm particles trigger necrosis by oxidative stress and mitochondrial damage. Our TEM examinations demonstrated a swollen shape of nuclei which could lead to necrosis. That point requires attention because following necrosis, AuNPs could be externalized in the whole tissue and targeted toward other nuclei. In this way, necrosis could spread everywhere. The products released by necrosis process are highly inflammatory and could cause inflammation in the whole animal. AuNPs could be redistributed via the hemolymph as demonstrated by intravenous administration of AuNPs in mice [45, 46]. Nanoparticle exposure induces responses of biomarkers of defense such as MTs, involved in metal detoxification, GST produced in presence of xenobiotics, and SOD and CAT enzymes expressed in oxidative stress [47, 48]. Our parallel study [31] demonstrated that following exposure of S. plana to 5, 15, and 40 nm AuNPs, these biomarkers were responsive. As AuNPs are xenobiotics, GST activities highly increased. MT levels were higher in exposed than in control animals. It has been shown that MTs play an important role in metal detoxification in bivalves since they are responsible for the sequestration of metal ions [49]. MTs are also involved in the defense against oxidative stress [38]. The activities of CAT and SOD involved in the primary defenses were increased demonstrating the induction of an oxidative stress by AuNPs. It has been demonstrated that amine-coated AuNPs trigger MT overproduction and an oxidative stress in gills and visceral mass of the bivalve C. fluminea during a trophic contamination experiment [7].

The present study suggests morphological alterations of the nuclear membrane in experimental groups compared to controls. TEM observations in controls revealed a thin perinuclear space whereas in contaminated clams this space seemed to be enlarged. In the same way, nuclei shape appeared swollen and amount of condensed chromatin was

highly decreased. In addition, a more pronounced alteration was observed following 15 and 5 nm AuNP exposures. Such morphological features could indicate a disturbance of the nuclear membrane due to the induction of oxidative stress by reactive oxygen species which in excess cause protein, DNA, and membrane injury [50, 51]. However, in the absence of any significant increase of thiobarbituric acid-reactive substances, no lipid peroxidation was revealed in S. plana [31]. On the other hand, lipid peroxidation products were detected in digestive glands of M. edulis exposed to 5 nm AuNPs [11] whereas their previous investigation found no significant increase in tissues of mussels exposed to AuNPs at 13 nm [12].

Taken together, our results demonstrate that the presence of AuNPs is clearly corroborated to a morphological change in chromatin. To our knowledge, only few papers reported a nuclear localization following AuNPs exposure [24].

## Conclusions

We have shown that AuNPs at 5, 15, and 40 nm are able to penetrate within branchial and digestive epithelia of a benthic bivalve S. plana during a water-borne contamination experiment. Differences between the selective tissue bioaccumulation were observed. In gills, only few AuNPs were observed whereas in digestive glands they were numerous and located within the nuclei whatever the size (5, 15, and 40 nm). According to our previous study [31] demonstrating an increase in biomarker responses linked to oxidative stress, the present study suggests a potential cytotoxicity. Till now, most of the studies suggesting toxicity of AuNPs were based on in vitro experimentation. Our evaluation of toxicity in vivo suggest morphological disturbance of nuclear membrane and chromatin which could lead to a necrosis process. That points out the necessity to investigate the feasibility of minimizing the cytotoxicity of AuNPs before their use in various medical applications without any hazardous effects on human health. Moreover, our results demonstrate uptake and bioaccumulation of AuNPs from an aquatic ecosystem to a marine bivalve. These findings are of interest in a species which plays a major role in the coastal and estuarine food chain since recent reports have brought evidence for transfer of gold particles within a terrestrial food chain [52] and within an estuarine food chain [9]. Despite the doses tested in the present study are too high to be encountered in the environment, the fact that AuNPs may be accumulated within living organisms and the food chain, with potential toxicity at the level of cellular nuclei and chromatin indicates that the use of AuNPs must be developed in a precautionary manner to avoid environmental impacts. Till now, AuNPs were generally considered nontoxic like bulk gold, which is inert and biocompatible.

However, recent findings (this study and literature quoted therein) highlight that there is an urgent need to better understand their nanotoxicity.

**Acknowledgments** This work was supported by a post-doctoral scholarship from the Fondation Franco-Chinoise pour la Science des Applications (FFCSA), China Scholarship Council (CSC), and the Région Pays de la Loire. We kindly thank Prof. Jacques Taxi (UM 74, Université Pierre et Marie Curie, Paris) for helpful discussions and Hélène Terisse (Institut des Matériaux, Nantes) for Zetapotential analyzer. We also acknowledge INSERM U791 for facilities to TEM access. The authors thank the NanoReTox program (part of the EC FP7/2007-2013) for providing contaminated bivalves.

# References

1. Baptista P, Pereira E, Eaton P, Doria G, Miranda A, Gomes I, Quaresma P, Franco R (2008) Gold nanoparticles for the development of clinical diagnosis methods. Anal Bioanal Chem 391:943–950

2. El-Sayed IH, Huang X, El-Sayed MA (2006) Selective laser photothermal therapy of epithelial carcinoma using anti-EGFR antibody conjugated gold nanoparticles. Cancer Lett 239:129–135

3. Panyala NR, Pena-Méndez EM, Havel J (2009) Gold and nanogold in medicine: overview, toxicology and perspectives. J Appl Biomed 7:75–91

4. Kang B, Mackey MA, EL-Sayed MA (2010) Nuclear targeting of gold nanoparticles in cancer cells induces DNA damage, causing cytokinesis arrest and apoptosis. J Am Soc 132:1517–1519

5. Patra CR, Bhattacharya R, Mukhopadhyay D (2010) Fabrication of gold nanoparticles for targeted therapy in pancreatic cancer. Adv Drug Deliv Rev 62:346–361

6. Klaine SJ, Alvarez PJJ, Batley GE, Fernandez TF, Handy RD, Lyon DY, Mahendra S, McLaughlin MJ, Lead JR (2008) Nanomaterials in the environment: behavior, fate, bioavailability, and effects. Environ Toxicol Chem 27:1825–1851

7. Renault S, Baudrimont M, Mesmer-Dudon N, Gonzales P, Mornet S, Brisson A (2008) Impact of gold nanoparticle exposure on two freshwater species a phytoplanctonic alga (*Scenedesmus subspicatus*) and benthic bivalve (*Corbicula fluminea*). Gold Bulletin 41(2):116–126

8. Farkas J, Christian P, Urrea JAG, Roos N, Hassellôv M, Tollefsen KE, Thomas KV (2009) Effects of silver and gold particles on rainbow trout (*Oncorhynchus mykiss*) hepatocytes. Aquat Toxicol 96:44–52

9. Ferry JL, Craig P, Hexel C, Sisco P, Frey R, Pennington PL, Fulton MH, Scott IG, Decho AW, Kashiwada S, Murphy CJ, Shaw TJ (2009) Transfer of gold nanoparticles from the water column to the estuarine food web. Nature Nanotech 4:441–444

10. Ward JE, Kach DJ (2009) Marine aggregates facilitate ingestion of nanoparticles by suspension-feeding bivalves. Mar Environ Res 68:137–142

11. Tedesco S, Doyle H, Blasco J, Redmond G (2010) Oxidative stress and toxicity of gold nanoparticles in *Mytilus edulis*. Aquat Toxicol 100:178–186

12. Tedesco S, Doyle H, Blasco J, Redmond G, Sheehan D (2010) Exposure of the blue mussel, *Mytilus edulis*, to gold nanoparticles

13. Lewinsky N, Colvin V, Drezek R (2008) Cytotoxicity of nanoparticles. Small 4:26–49

14. Levy R, Shaheen U, Cesbron Y, Sée V (2010) Gold nanoparticles delivery in mammalian live cells: critical review. Nano Reviews 1:4889

15. Murphy CJ, Gole AM, Stone JW, Sisco PN, Alkilany AM, Goldsmith EC, Baxter SC (2008) Gold nanoparticles in biology: beyond toxicity to cellular imaging. Accounts Chem Res 41:1721–1730

16. Alkilany AM, Murphy CJ (2010) Toxicity and cellular uptake of gold nanoparticles: what we have learned so far? J Nanopart Res 12:2313–2333

17. Khlebtsov N, Dykman L (2010) Biodistribution and toxicity of engineered gold nanoparticles: a review of in vitro and in vivo studies. Chem Soc Rev 40:1647–1671

18. Nativo P, Prior IA, Brust M (2008) Uptake and intracellular fate of surface-modified gold nanoparticles. Acs Nano 8:1639–1644

19. Li JJ, Zou L, Hartono D, Ong CN, Bay BH, Lanry Yung LY (2008) Gold nanoparticles induce oxidative damage in lung fibroblasts in vitro. Adv Mater 20:138–142

20. Stelzer R, Hutz RJ (2009) Gold nanoparticles enter rat ovarian granulosa cells and subcellular organelles, and alter in-vitro estrogen accumulation. J Reprod Dev 55:685–690

21. Chitrani BD, Ghazani AA, Chan WCW (2006) Determining the size and shape dependence of gold nanoparticle uptake into mammalian cells. Nano Lett 6:662–668

22. Pernodet N, Fang X, Sun Y, Bakhtina A, Ramakrishnan A, Sokolov J, Ulman A, Rafailovich M (2006) Adverse effects of citrate/gold nanoparticles on human dermal fibroblasts. Small 6:766–773

23. Tsoli M, Kuhn H, Brandau W, Esche H, Schmid G (2005) Cellular uptake and toxicity of Au55 clusters. Small 1:841–844

24. Panessa-Warren BJ, Warren JB, Maye MM, Van der Lelie D, Gang O, Wong S, Ghebrehiwet B, Tortora GT, Misewich JA (2008) Human epithelial cell processing of carbon and gold nanoparticles. Int J Nanotechnol 5:55–91

25. Pan YS, Neuss LA, Fischler M, Wen F, Simon U, Schmid G, Brandau W, Jahen-Dechent W (2007) Size-dependent cytotoxicity of gold nanoparticles. Small 11:1941–1949

26. Patra HK, Banerjee S, Chaudhuri U, Lahiri P, Dasgupta AK (2007) Cell selective response to gold nanoparticles. Nanomedicine 3:111–119

27. Lanone S, Boczkowski J (2006) Biomedical applications and potential health risks of nanomaterials: molecular mechanisms. Curr Mol Med 6:651–663

28. Tedesco S, Doyle H, Redmond G, Sheehan D (2008) Gold nanoparticles and oxidative stress in *Mytilus edulis*. Mar Environ Res 66:131–133

29. Cho WS, Cho MJ, Jeong J, Choi M, Cho HY, Han BS, Kim SH, Kim HO, Lim YT, Chung BH, Jeong J (2009) Acute toxicity and pharmacokinetics of 13 nm-sized PEG-coated gold nanoparticlesToxicol. Appl Pharmacol 236:16–24

30. Young IS, Woodside JV (2001) Antioxidants in health and disease J Clin Pathol 54:176–186

31. Pan JF, Buffet PE, Poirier L, Amiard-Triquet C, Gilliland D, Guibbolini M, Christine C, Roméo M, Mouneyrac C (2012) Size dependent bioaccumulation and ecotoxicity of gold nanoparticles in an endobenthic invertebrate: the tellinid clam *Scrobicularia plana*. Environ Poll 168:37–43

32. Bonnard M, Roméo M, Amiard-Triquet C (2009) Effects of copper on the burrowing behavior of estuarine ans costal invertebrates, the polychaete *Nereis diversicolor* and the bivalve *Scrobicularia plana*. Hum Ecol Risk Assess 15:11–26

33. Solé M, Kopecka-Pilarczyk J, Blasco J (2009) Pollution biomarkers in two estuarine invertebrates, *Nereis diversicolor* and

and the pro-oxidant menadione. Comp Biochem Physiol C: Pharmacol Toxicol 151:167–174

*Scrobicularia plana*, from a Marsh ecosystem in SW Spain. Environ Int 35:523–531

34. Turkevich J, Stevenson PC, Hillier JA (1951) Study of the nucleation and growth processes in the synthesis of colloidal gold. Discuss Faraday Soc 11:55–59

35. RNO (2006) Surveillance de la qualité du milieu marin Ministère de l'environnement et Institut français de recherche pour l'exploitation de la mer (Ifremer), Paris et Nantes

36. Hull MS, Chaurand P, Rose J, Auffan M, Bottero JY, Jones JC, SchultzIR VPJ (2011) Filter-feeding bivalves store and biodeposit colloidally stable gold nanoparticles. Environ Sci Technol 45:6592–6599

37. Hughes RN (1969) A study of feeding in *Scrobicularia plana*. J Mar Biol Assoc UK 49:805–823

38. Viarengo A, Burlando B, Cavaletto M, Marchi B, Ponzano E, Blasco J (1999) Role of metallothionein against oxidative stress in the mussel *Mytilus galloprovincialis*. AMJ Physiol Reg Integr Comp Physiol 277:R1612–1619

39. Peckys DB, de Jonge N (2011) Visualizing gold nanoparticle uptake in live cells with liquid scanning transmission electron miscroscopy. Nano Lett 11:1733–1738

40. Jovanovic-Talisman T, Tetenbaum-Novatt J, McKenney AS, Zilman A, Peter R, Rout MP, Chait BT (2009) Artificial nanopores that mimic the transport selectivity of the nuclear pore complex. Nature 457:1023–1027

41. Payne DW, Thorpe NA (1993) Carbohydrate digestion in the bivalve *Scrobicularia plana* (da costa). Comp Biochem Physiol 104B:499–503

42. Franke WW, Scheer U, Krohne G, Jarasch ED (1981) The nuclear envelope and the architecture of the nuclear periphery. J Cell Biol 1(91):39s–50s

43. Huang X, El-Sayed EH, Quian W, El-Sayed MA (2006) Cancer cell imaging and photothermal therapy in the near-infrared region by using gold nanorods. J Am Chem Soc 128:2115–2120

44. Jain PK, Huang X, El-Sayed HI, El-Sayed MA (2008) Noble metals on the nanoscale: optical and photothermal properties and some applications in imaging, sensing, biology, and medicine. Acc Chem Res 41:1578–1586

45. De Jong WH, Hagens WL, Krysteck P, Burger MC, Sips AJAM, Geertsma RE (2008) Particle size-dependent organ distribution of gold nanoparticles after intravenous administration. Biomaterials 29:1912–1919

46. Sonavane G, Tomoda K, Makino K (2008) Biodistribution of colloidal gold nanoparticles after intravenous administration: effect of particle size. Colloids Surf B: Biointerf 66:274–280

47. Van der Oost R, Porte-Visa C, Van den Brink NW (2005) Ecotoxicological testing of marine and freshwater ecosystems. In: Munawar M (ed) P J Den Besten. Taylor and Francis, Boca Raton, pp 87–152

48. Marquis BJ, Love SA, Braun KL, Haynes CL (2009) Analytical Methods to Assess Nanoparticle Toxicity Analyst 134:425–439

49. Amiard JC, Amiard-Triquet A, Barka S, Pellerin S, Rainbow PS (2006) Metallothioneins in aquatic invertebrates: their role in metal detoxification and their use as biomarkers. Aquat Toxicol 76:160–202

50. Nel A, Xia T, Mädler L, Li N (2006) Toxic potential of materials at the nanolevel. Science 311:622–627

51. Aillon KL, Xie Y, El-Gendy N, Berkland CJ, Forres ML (2009) Effects of nanomaterial physicochemical properties on in vivo toxicity. Adv Drug Deliv Rev 61:457–466

52. Judy JD, Unrine JM, Bertsch PM (2011) Evidence for biomagnification of gold nanoparticles within a terrestrial food chain. Environ Sci Technol 45:776–781

# 18

# Green biosynthesis of gold nanometre scale plates using the leaf extracts from an indigenous Australian plant *Eucalyptus macrocarpa*

**Gérrard Eddy Jai Poinern · Peter Chapman · Xuan Le ·
Derek Fawcett**

**Abstract** In this preliminary study, we demonstrate an environmentally friendly process for the green synthesis of gold nanometre scale particles using the leaf extract from an indigenous Australian plant *Eucalyptus macrocarpa* as both the stabilising agent and the reducing agent. The synthesis process is straightforward, clean and non-toxic. It also has the advantages of being performed at room temperature and does not need complex processing equipment. Formation of the gold nanometre sized particles was confirmed and characterised by UV-visible spectroscopy, X-ray diffraction, transmission electron microscopy and field emission scanning electron microscopy. The antibacterial activity of the synthesised gold particles was also quantified using the sensitivity method of Kirby–Bauer.

**Keywords** Gold · Nano-particles · Green synthesis ·
Antibacterial · *Eucalyptus macrocarpa*

## Introduction

Metallic nanometre scale particles, in particular colloidal gold nanoparticles (Au NPs), have attracted considerable interest in many fields such as medicine, biotechnology, materials science, photonics and electronics [1–4]. The size, shape and surface morphology can have a profound influence on the chemical, physical, optical and electronic properties of nano-materials [5, 6]. This is indeed the case for a noble metal such as Au, where the metal exhibits a strong surface plasmon resonance when exposed to electromagnetic radiation [7]. Until recently, producing nanoparticles involved expensive chemical and physical processes that often used toxic materials with potential hazards such as environmental toxicity, cytotoxicity and carcinogenicity [8]. An attractive alternative to the traditional manufacturing techniques used for the production of nanoparticles involves using a green environmentally friendly technology based on biological systems such as plants [9, 10], bacteria [11, 12], fungus [13, 14] and similar organisms [15, 16]. Synthesising nanoparticles via biological systems offers a clean, nontoxic and environmentally friendly method with the potential to deliver a wide variety of nanoparticle types, sizes, shapes and morphologies. Out of the several biological systems mentioned above, synthesising nanoparticles via leaf extracts from plants is a relatively straight forward technique. The technique does not need any special culture preparation or isolation techniques that are normally required for bacteria and fungi based synthesis techniques.

Au NPs have attracted significant interest over the last decade as a potential platform for a number of biomedical applications such as biosensors [17], clinical chemistry [18], fluorescent labelling for immunoassays [19], tumour destruction via heating (hyperthermia) [20], targeted delivery of therapeutic drugs and genetic substances [21] and as antibacterial drugs [22, 23]. The cells of most living organisms have dimensions in the range of 10 to 50 µm, while the cells' internal organelles and other sub-cellular structures are all in the sub-micron size range. The difference in size between a typical cell and a nanometre scale particle can range from a hundred times up to ten thousand times. This significant difference in size range between the cell and a nanometre scale particle gives the NP the potential to biophysically interact with biological molecules both inside and on the surface of cell [3]. Recent studies have shown that membrane

G. E. J. Poinern (✉) · X. Le · D. Fawcett
Murdoch Applied Nanotechnology Research Group,
Department of Physics, Energy Studies, and Nanotechnology,
School of Engineering and Energy, Murdoch University,
Murdoch, WA 6150, Australia
e-mail: g.poinern@murdoch.edu.au

P. Chapman
Department of Chemistry, Curtin University of Technology,
Bentley, WA 6102, Australia

damage and toxicity can result from the biosorption and cellular uptake of NPs by bacteria [24, 25]. However, the processes behind NP inhibition of bacterial growth are not fully understood, but some studies have strongly suggested that the size, shape and surface modifications could influence the antibacterial properties of the NPs [25, 26].

Chrysotherapy is a branch of medicine that uses Au as a medicinal compound in the treatment of certain diseases such as rheumatoid arthritis. In addition, Au NPs have also been successfully used as a carrier platform for the delivery of double-stranded DNA in gene gun technology [27]. Another feature of Au NPs is their ability to passively accumulate in tumours, where their good optical and chemical properties can be used in thermal treatment therapies [28, 29]. Moreover, recent studies have shown that cancer drugs bonded to the surface of Au NPs can be more effectively targeted to tumours, thus improving delivery and minimising treatment durations and side effects of anticancer drugs [3, 30–33]. Au NPs have also attracted significant scientific interest as a new class of biomedical materials, since they can act as both antibacterial and antifungal agents capable of interacting with microorganisms to produce cellular damage [33]. The ability to induce cellular damage and the ultimate death of various strains of bacteria and yeast microorganisms has many significant health benefits. For example, both bacterial and fungal species have the ability to develop immunity against commonly used antibiotic with time. Therefore, new and more effective antimicrobial agents are needed to fight antibiotic resistant strains of microorganisms. Au NPs have the potential to be an effective antimicrobial agent, especially if they can be synthesised via a green chemistry route that ensures a clean, nontoxic and environmentally friendly method for producing NPs.

The southwest corner of the Australian continent is a global biodiversity hot spot and is also the home of the exquisite *Eucalyptus macrocarpa* [34]. The plant is also known as the *Rose of the West* or *the Mottlecah* and is easily recognised by its beautiful silvery foliage and red flowers as presented in Fig. 1a. And because of its attractive floral arrangement, the plant has been successfully propagated throughout the coastal

region of Western Australian. The silvery grey appearance of the plants leaves are the result of nanometre scale features formed on the surface of the leaf by epicuticular waxes. These waxes give the leaves their remarkable wetting and self-cleaning properties, which enhances the plants survival in its arid climate [35].

In this paper, we report for the first time the green synthesis of stable Au NPs by the direct reduction of $AuCl_4^-$ ions via Mottlecah leaf extracts without using conventional stabilising ligands. The advantages of using this approach include: (1) the leaf extract acts as both stabilising agent and reducing agent during the synthesis process; (2) the aqueous synthesis process is environmental friendly and produces no toxic waste; and (3) the technique is simple, straight forward and does not require specialised equipment. Furthermore, the antibacterial activity of both the synthesised Au NPs and the leaf extracts were tested against *Bacillus subtilis* and *Escherichia coli* using the Kirby–Bauer sensitivity method [36].

### Materials and methods

Chemicals

The $AuCl_4^-$ ions were produced by dissolving high purity Au wire (99.99 %) in a solution of aqua regia. Both $HNO_3$ and HCl were purchased from Sigma-Aldrich (Castle Hill, NSW, Australia) and used without further purification. Milli-Q® water was used throughout all synthesis procedures involving aqueous solutions and was produced by a Barnstead Ultrapure Water System D11931 (Thermo Scientific Dubuque IA 18.3 MΩ cm$^{-1}$).

Leaf material and preparation of leaf extract

The *E. macrocarpa* leaves were collected from several locations around the Murdoch University campus in Perth, Western Australia. A wide selection of Mottlecah leaves, ranging

**Fig. 1 a** *Eucalyptus macrocarpa* with its distinctive silvery foliage and prominent red flowers, **b** droplets consisting of: (*i*) filtered solution containing Au NPs, (*ii*) solution s1 composed of $AuCl_4^-$ and leaf extract solution (1:1), (*iii*) pure $AuCl_4^-$ solution and (*iv*) solution of raw leaf extract

from young to mature leaves, was harvested from various locations on each plant. Generally, five locations were selected (top, north, south, east and west), and on average, ten leaves were taken from each of the locations. Both the adaxial and abaxial sides of the leaf were examined, with only healthy leaves free from damage being harvested. The leaves were washed several times with Milli-Q® water to remove any dust or debris. After cleaning, 10 g of Mottlecah leaves were finely cut into small strips and added to a 100 mL solution of Milli-Q® water. The aqueous mixture was then poured into the blending bowl of an IKA® T25 Digital Ultra Turrax® homogenizer. The mixture was homogenised at 5,000 rpm for 10 min at a room temperature of 24 °C. At the end of this time, the solution was filtered using a Hirsch funnel to remove leaf debris. This was followed by two further filtrations using a 0.22 μm Millex® (33 mm diameter) syringe filter unit. At the end of the three-step filtration procedure, the leaf extract was placed in a glass vial ready for the synthesis of Au NPs.

## Removal of epicuticular waxes

To determine the influence of epicuticular wax in the leaf extract, it was necessary to prepare a leaf extract without wax. This was achieved by first selecting several pre-cleaned leaves (discussed above) and then using a solution of chloroform to wash each leaf. Both the adaxial and abaxial sides of the leaf were washed to remove all surface waxes. The washing procedure consisted of slowly pouring a 50 mL solution of chloroform over the entire inclined leaf collecting the run off in a small beaker, each side being done in turn. During the 30 s washing period, the chloroform efficiently removed the epicuticular waxes from the leaf surface. Subsequent scanning electron microscope (SEM) analysis revealed that the waxes had indeed been removed from the leaf surface leaving the underlining cutin undamaged.

## Synthesis of gold nanoparticles

The Au NPs used as the control were synthesised by first adding a 1.0 mL solution of 1 mM $AuCl_4^-$ to a 10 mL solution of Milli-Q® water, while the solution was stirred vigorously. This was followed by adding a 1.0 mL solution of 1 mM sodium citrate (stabilising and capping agent) to the aqueous solution at room temperature (24 °C). The reduction of the Au nanoparticles was initiated by the addition of a 1.0 mL solution of 0.01 M sodium borohydride to the aqueous solution, while the solution was stirred. The reduction process was allowed to proceed at room temperature (24 °C).

The biological reduction of a 1.0 mL solution of 10 mM aqueous $AuCl_4^-$ ion solution was investigated using three solutions with varying amounts of leaf extract without wax and with wax. The quantities of leaf extract used to make up the solutions consisted of 1 mL for $g_1$, 2 mL for $g_2$ and 3 mL

for $g_3$. Once the $AuCl_4^-$ ion solution was added to each quantity of leaf extract, the solutions were then vigorously stirred for 1 min. The reduction process was allowed to proceed at room temperature (24 °C), with stable Au NPs being produced within 1 h.

## Leaf droplets

A clean Mottlecah leaf was selected. The leaf was laid flat in the horizontal plane to prevent the various droplet types from rolling off the surface. Then using a fluid-specific clean glass pipette fitted with a rubber bulb, four individual droplets were placed onto the leaf surface. The droplets consisting of: (1) filtered solution containing Au NPs, (2) solution s1 composed of $AuCl_4^-$ and leaf extract solution, (3) pure $AuCl_4^-$ solution, and (4) solution of raw leaf extract. The progress of the reduction process was photographed over a 2-h period using a Canon EDS 600 D digital camera (Canon Inc, Tokyo Japan fitted with macro lens EF 100 mm 1:2:8 USM). A typical photograph taken after 1 h is presented in Fig. 1b.

## Characterisation of biologically reduced Au nanoparticles

All samples were examined and characterised using five advanced analysis techniques. These included: UV-visible spectrum analysis, X-ray diffraction spectroscopy (XRD), energy dispersive X-ray spectroscopy (EDAX), transmission electron microscopy (TEM) and field emission scanning electron microscopy (FESEM).

### UV-visible spectrum analysis

A series of samples were prepared. The first set consisted of three controls: (1) Milli-Q® water, (2) pure $AuCl_4^-$ solution and (3) pre-filtered pure leaf extract (filtered twice, each time using a new Whatman 0.22 μm syringe filter). The test solutions consisted of the three Au colloids $g_1$, $g_2$ and $g_3$. The UV-visible spectra of each of the samples was then measured using a Varian Cary 50 series UV-visible spectrophotometer version 3, over a spectral range from 200 to 1,100 nm, with a 1 nm resolution over the first hour at room temperature of 24 °C.

### XRD spectroscopy

After the end of each reduction procedure, samples for XRD examination were extracted from each glass vial using a clean glass pipette fitted with a rubber bulb. Then two to three drops of each sample were dispersed over the surface of a specific glass microscope slide. Then each glass slide was then dried under vacuum for a period of 4 h. At the end of this time, the dried samples were then characterised using XRD spectroscopy. The XRD spectra were recorded at room temperature (22 °C), using a Bruker D8 series diffractometer (Cu $K_\alpha$=1.5406 Å radiation source) operating at 40 kV and 30 mA. The diffraction patterns were collected over a $2\theta$ range from 15 to 80°, with an incremental step size of 0.04° using flat plane geometry. The acquisition time was 2 s. The powder XRD spectrum was used to identify the size of the Au particles and their crystalline structure. The particle size was calculated using the Debye–Scherrer equation (Eq. 2) from the respective XRD patterns and estimated from both TEM and FESEM images.

*EDAX spectroscopy*

Each sample for EDAX examination was initially deposited onto a thin mica strip using a glass pipette, the mica strip was attached to a SEM stub using carbon tape. The samples were then dried under vacuum overnight. The following day, all samples were sputter-coated with a 3-nm layer of platinum. The samples were then examined using an Oxford Instruments EDS X-ray detector (EDAX) and Oxford Instruments energy dispersive X-ray detectors (EDS). The electron back-scatter diffraction was used during the analysis, and the EDS aperture was set to 60 μm and operated at 20 kV.

*TEM*

The size and morphology of the Au NPs was investigated using TEM. Sample preparation consisted of filtering the suspensions two times, each time using a new Whatman 0.22 μm syringe filter. After filtration, a single drop from each sample was deposited onto its respective carbon-coated copper TEM grid using a micropipette and then allowed to slowly dry over a 24-h period. After sample preparation, a bright field TEM study was carried out using a Phillips CM-100 electron microscope (Phillips Corporation Eindhoven, The Netherlands) operating at 80 kV.

*FESEM*

Each sample for FESEM examination was initially deposited onto a thin mica strip using a glass pipette; the mica strip was attached to a SEM stub using carbon tape. The samples were then dried under vacuum overnight. The following day, all samples were sputter-coated with a 3-nm layer of platinum. The particle size and morphological features of the samples were investigated using a high resolution FESEM (Zeiss Neon 40EsB FIBSEM) at 5 kV, with a 30-μm aperture operating under a pressure of $1\times10^{-10}$ Torr.

Microbial sensitivity of Au nanoparticles

The antibacterial activity of the biologically reduced Au NPs was investigated using the sensitivity method of Kirby–Bauer [36]. In this technique, the test media was composed of Difco nutrient agar (23 g/1,000 mL), agar (5 g/1,000 mL), yeast extract (4 g/1,000 mL) and distilled water (1,000 mL). The components were thoroughly mixed via heating and agitation to completely dissolve the components. The medium was then autoclaved at 121 °C for 15 min. At the end of the autoclaving period, a 20-mL volume of the media was added to each Petri dish (90 mm diameter) and then allowed to solidify at room temperature before being stored at 4 °C for future use.

Two bacteria, namely *E. coli* and *B. subtilis* were used to investigate the antibacterial activity of the Au NPs. In each case, the respective bacteria were cultured in a media composed of Difco nutrient broth (8 g/1,000 mL), yeast extract (4 g/1,000 mL) and distilled water (1,000 mL). The nutrient medium was then autoclaved at 121 °C for 15 min. After autoclaving, the nutrient medium was ready for culturing bacteria; in the first stage, each respective bacteria was added to a 10-mL sample of the nutrient medium and then incubated overnight at 28 °C while being agitated. In the second stage, the cultures were then added to a 100-mL solution of nutrient media and further incubated overnight at 28 °C, while being agitated. The following day, a 1-mL sample of the culture medium was pipette onto the surface of the nutrient containing Petri dish. The flooded dish is then carefully tilted in various orientations, until the entire nutrient surface was covered. Once the nutrient surface was covered, the excess culture medium was carefully removed using a pipette. Then, the cover was placed on each Petri dish, inverted and left to dry for 45 min.

While the Petri dishes were drying, 25 μg/mL solutions of as prepared Au NPs (Au NPs control and leaf extract solution (with and without wax) and leaf extract + Au NPs solution $g_1$ (with and without wax)) were transferred onto a sterile 6-mm diameter Whatman AA (2017–006) sample disc using a pipette, and then allowed to dry for 20 min. After drying, sterile forceps were used to place the sample disc onto the centre of the inoculated Petri dish. After all the samples were placed in their respective inoculated dishes, the covers were fitted and then the dishes were incubated overnight at 28 °C. After incubation, the diameter of the bacterial inhibition zone of each sample was measured using a standard scale ruler with an accuracy of±0.5 mm, with all measurements being rounded up to the next whole millimetre. All sample experiments were carried out in triplicate and the mean inhibition zone diameters recorded.

In addition, optical microscopy was used throughout the studies to examine the bacterial inhibition zone. An Olympus BX51 compound microscope (Olympus Optical Co. Ltd., Tokyo, Japan) was used for all optical studies, and photographs were taken using the DP 70 camera attachment.

## Results and discussions

This preliminary study reports the green synthesis of Au NPs by the biological reduction of aqueous $AuCl_4^-$ using the leaf extract from *E. macrocarpa* (Mottlecah). The water soluble ingredients present in the leaf extract were found to be both a highly effective reducing agent and an efficient stabilising agent. The untreated leaf extract was found to be acidic, with a pH of 5 for leaf extract without wax and a pH of 5.5 with wax. The formation of the Au NPs could be easily monitored by the change in colour of the reactive mixture. The synthesis of Au NPs begins with the addition of the light yellow $AuCl_4^-$ aqueous solution (pH=1) to a solution of raw leaf extract forming the reactive mixture, which typically had a pH of 2. At this point, the reactive mixture is subjected to 10 min of homogenization. During homogenization, the prevailing process is Au metal nucleation, and the mixture changes from a light yellow colour to a pale brown colour due to the excitation of surface plasmon vibrations of the forming Au NPs. After the initial homogenization treatment, the reaction mixture was then allowed to stand, and over the next 60 min further reduction of $Au^{III}$ took place. During this period, the colour of the complex slowly darkens to a deeper brown colour, indicating the increasing numbers of Au nuclei steadily forming. The formation of Au NPs was confirmed by both UV-visible spectrum analysis and EDAX spectroscopy analysis for all three solutions ($g_1$, $g_2$ and $g_3$). UV-visible spectrum analysis for the solutions indicated that the maximum absorbance occurred at 570 nm, which is similar to the maximum absorbance of 573 nm reported by Philip using the leaf extract from *Hibiscus rosa sinensis* and the maximum absorbance of 560 nm reported by Singh et al. using the leaf extract of ginger (*Zingiber officinale*) [9, 37]. A representative UV-visible spectrum analysis for solution $g_1$ (1:1) is presented in Fig. 2a, while the corresponding EDAX spectroscopy analysis presented in Fig. 2b clearly indicates the presence of metallic Au. The remaining two solutions ($g_2$ and $g_3$) gave similar UV-visible and EDAX results. The net result of these two analysis techniques confirmed the formation of Au NPs and can be summarised in Eq. 1.

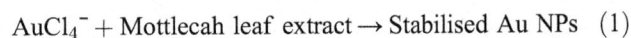

$$AuCl_4^- + \text{Mottlecah leaf extract} \rightarrow \text{Stabilised Au NPs} \quad (1)$$

TEM images taken after 60 min of biological reduction reveal that products present in the aggregate are NPs, with the main product being spherical particles ranging in size from 20 to 80 nm. However, coexisting with the spheres are Au nanometre-sized crystalline shapes (equilateral or truncated triangular, pentagon and hexagonal), all ranging in size from 50 to 100 nm, see Fig. 3a and b for representative images of the particle aggregates. This is unlike the result for Au NPs synthesised using sodium borohydride, which only produced spherical particles with a mean diameter of 30 nm. During biological reduction, NPs are produced during the first 60 min, after this initial period, nanometre scale crystalline shapes steadily grow into micrometre scale plates and spheres. After 120 min, micrometre-sized particles are the dominant particles found in the aggregate. Figure 4(a) presents a representative aggregate of micrometre-sized particles and plates. Triangular and hexagonal crystalline shapes are interspersed within the aggregate and are characterised by their smooth plate-like structure, which was also seen in the TEM images presented in Fig. 3a and b. Also presented in Fig. 4 are two colourised FESEM images (b) and (c), which highlight the morphology of the hexagonal and truncated triangular Au plates. The side lengths of the truncated triangular-shaped plates can generally reach a length of 6 μm, while the side lengths of the hexagonal-shaped plates are typically around 4 μm. The thickness of both plate types was found to vary from 300 to 600 nm. The aggregate was also found to have spherical particles with diameters ranging size from 200 nm up to 1.5 μm.

To investigate the crystalline nature of the Au NPs and nanometre scale platelets, dried Au powder samples were investigated using XRD spectroscopy. The Au crystalline phases present were found to be consistent with phases incorporated in the International Centre for Diffraction Data databases. A typical XRD pattern of a dried Au powder sample is presented in Fig. 3c. Inspection of Fig. 3c reveals the presence of intense peaks located at 38.3, 44.5, 64.5 and 77.9°, which correspond to the main (hkl) indices for Au: (111), (200), (220) and (311). The Bragg peaks indicate that the synthesised Au NP based powder was composed of pure crystalline metallic Au consisting of an fcc lattice structure. The intense peak at (111) indicates that this plane was the predominant orientation, and this was confirmed by evaluating the peak intensity ratio of the (200) and (111) peaks, which gave a value of 0.18. The value of 0.18 is much lower than the conventional bulk intensity ratio of 0.52 normally associated with Au, and clearly indicates that the (111) plane is indeed the predominant orientation found in the synthesised Au powders. The dominance of the (111) plane resulting from the biosynthesis of Au NPs has also been reported by Philip using *H. rosa sinensis* [9] and Zhang et al. using chloroplasts [38]. The crystalline size, $t_{(hkl)}$, of Au NPs was calculated from the XRD pattern using the Debye–Scherrer equation presented in Eq. 2 below:

$$t_{(hkl)} = \frac{0.9\lambda}{B\cos\theta_{(hkl)}} \quad (2)$$

where, $\lambda$ is the wavelength of the monochromatic X-ray beam, $B$ is the full width at half maximum (FWHM) of the peak at the maximum intensity, $\theta_{(hkl)}$ is the peak diffraction angle that satisfies Bragg's law for the (h k l) plane, and $t_{(hkl)}$

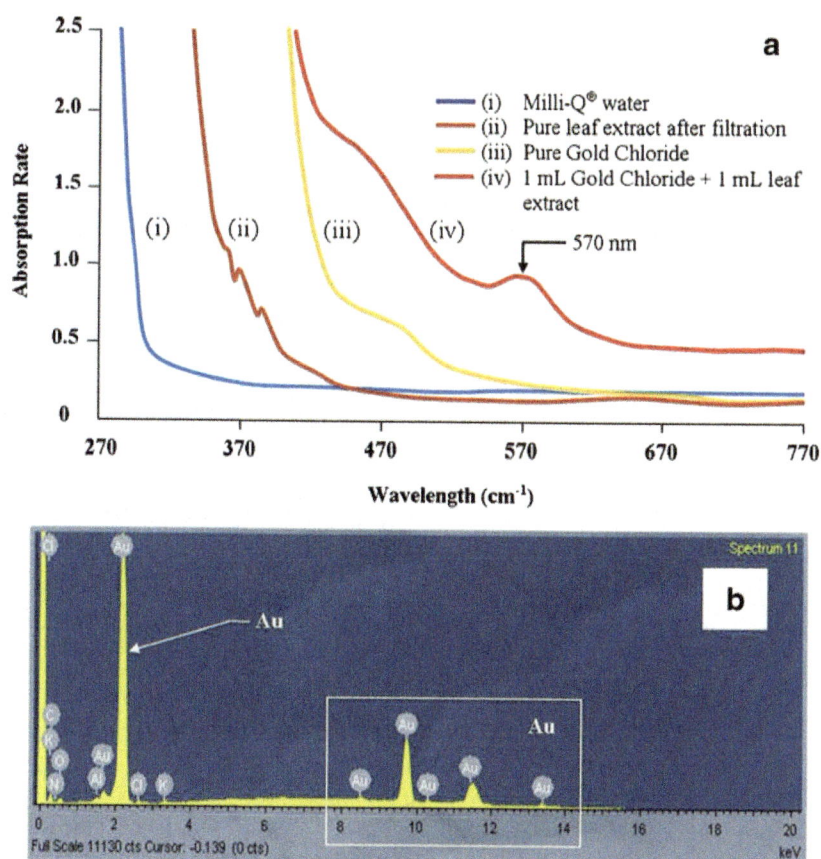

Fig. 2 **a** A representative UV-visible spectrophotometer analysis of Au NPs (g₁) synthesised using fresh *Eucalyptus macrocarpa* leaves, **b** corresponding EDAX spectra of sample

is the crystallite size. An estimate of the mean crystallite size using the FWHM of the (111) peak was calculated to be 84 nm. The TEM analysis revealed a particle size range from 20 to 100 nm, with the aggregation containing not only spherical-shaped particles but also truncated triangular, pentagon and hexagonal-shaped crystalline particles, see Fig. 3a and b.

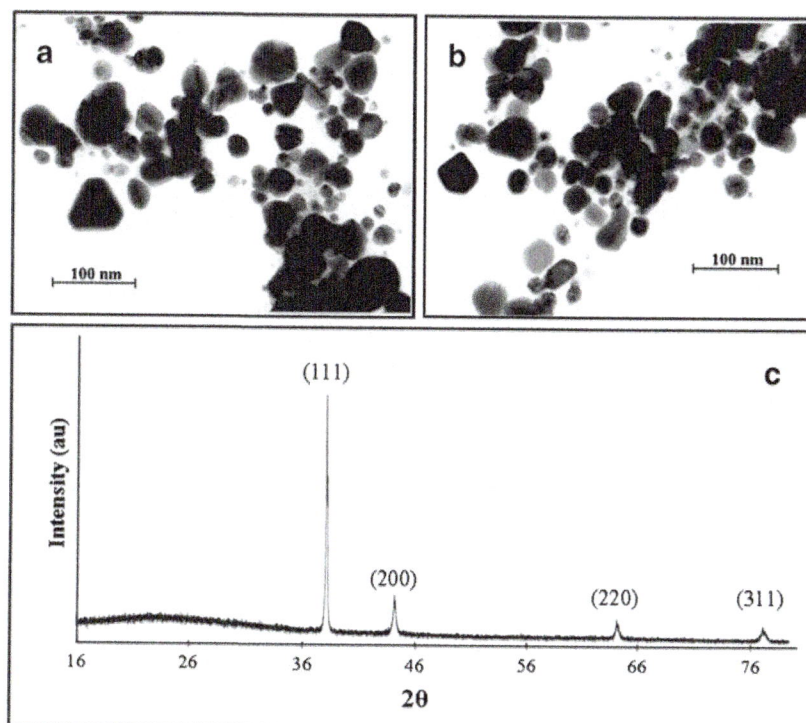

Fig. 3 **a** and **b** Representative TEM micrographs of Au NPs synthesised using leaf extracts with wax (*LW*) from *Eucalyptus macrocarpa* leaves, **b** a typical XRD pattern of Au NPs synthesised from leaf extracts

**Fig. 4** **a** A representative FESEM image of an aggregate containing Au particles and plates synthesised in leaf extract with wax (*LW*) over periods greater than 1 h; **b** and **c** colourised FESEM images highlighting truncated triangular and hexagonal Au plates

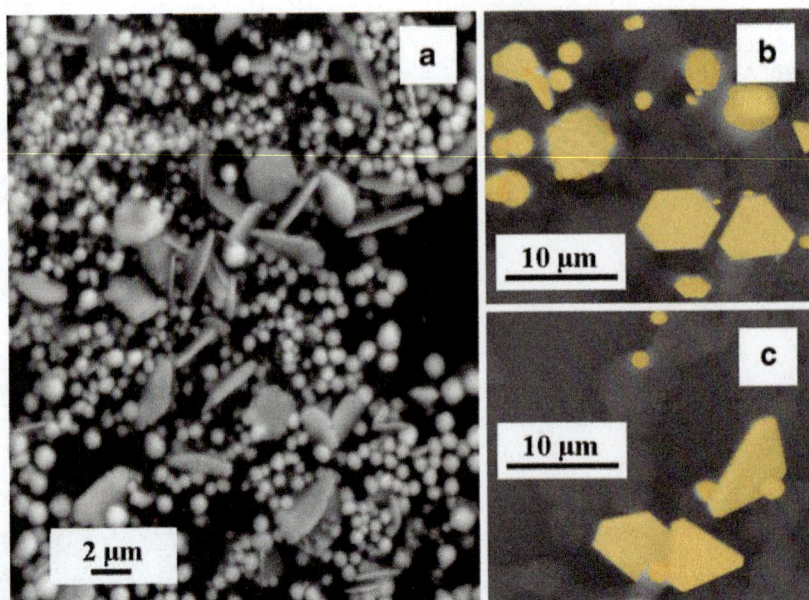

A possible mechanism to explain the plate-like growth and morphology of particles during reduction in the leaf extract could be the result of competitive growth between crystallographic surfaces, which results in a preferential attachment in a particular orientation [39]. An alternative mechanism proposed by Wang et al. arises from the initial spontaneous self-assembly of particles along a particular crystallographic orientation. This is followed by other Au particles coalescing at the planar interface, which effectively reduces their surface energy and contributes to the growth of the particle along this particular orientation [40]. In this study, analysis of the XRD data clearly indicates the dominance of the (111) orientation and confirms preferential particle growth along (111) crystal plane. This explains the inherent anisotropy of the atomically flat surface of the Au crystalline-shaped plates produced initially in the nanometre range. Subsequent growth in the micrometre range also produces the atomically flat surfaces, which are characteristic of the plate-like particles synthesised using *E. macrocarpa* leaf extracts.

**Fig. 5** **a** Zone of inhibition produced by colloid $g_1$ [Au NPs+leaf extract (with wax)] for *Escherichia coli* ~ diameter 29 mm, **b** microscopic view of the zone of inhibition boundary, **c** antibacterial activity of Au NPs against *Bacillus subtilis* and **d** antibacterial activity of Au NPs against *Escherichia coli*

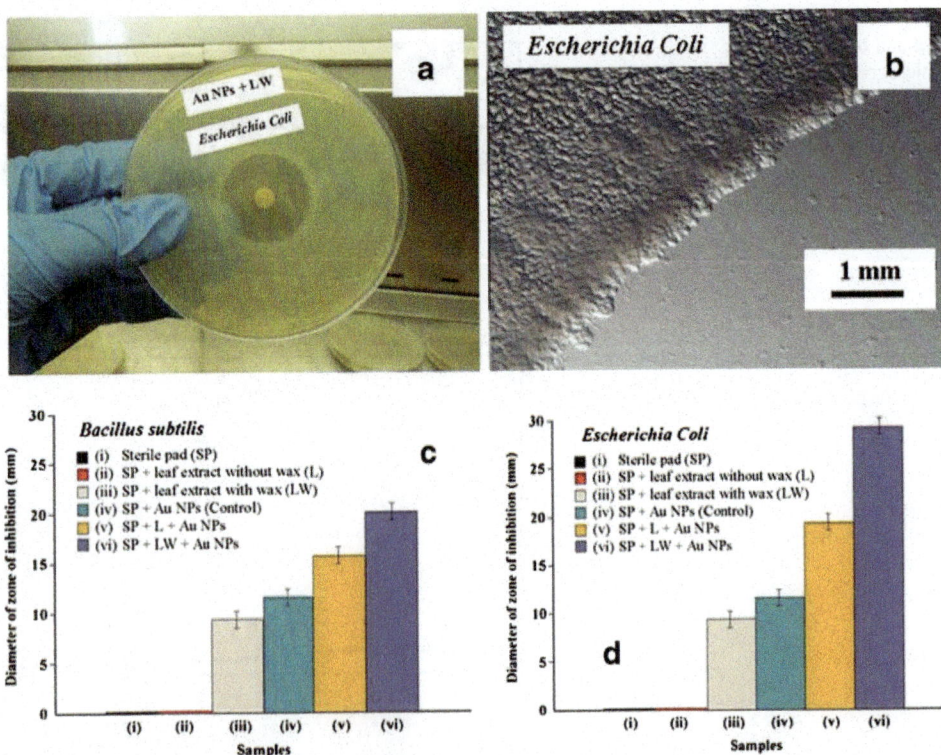

The antibacterial potential of the Au NPs synthesised using *E. macrocarpa* leaf extracts, the leaf extract itself and a combination of both (synergistic effects) were tested against *B. subtilis* (gram-positive) and *E. coli* (gram-negative). The leaf extract was initially tested against the two bacteria because earlier studies by Murata et al. were able to show an antibacterial effect of a compound (macrocarpal A) isolated from the leaf extract [41]. Further studies on the *E. macrocarpa* leaves by Yamakoshi et al. revealed the presence of a further six antibacterial compounds (macrocarpal B to G) [42]. Their study also determined the structure of the macrocarpal compounds from both XRD and NMR analysis. The analysis revealed that the compounds were composed of phloroglucinol dialdehyde diterpene derivatives. At this stage, it is not known if any of these compounds are directly involved in the biological reduction of the Au NPs, but there is clear evidence that these compounds are giving the leaf extract its antibacterial properties. For example, Petri dishes with only a sterile pad (control) and dishes with a sterile pad treated with leaf extract (without wax) recorded no zone of inhibition. However, dishes with a sterile pad treated with leaf extract (with wax) recorded a zone of inhibition of 9 mm for both *B. subtilis* and *E. coli*. The results are graphically presented in Fig. 5c (*B. subtilis*) and d (*E. coli*).

In the case of the Au NPs, the first tested was the Au NP (control), which was synthesised using sodium borohydride as the reducing agent and sodium citrate as the stabilising and capping agent. The Petri dishes containing a sterile pad treated with Au NPs (control) recorded a zone of inhibition of 12 mm for both *B. subtilis* and *E. coli*. The results of the zone of inhibition measurements indicate that both bacteria are resistant to the Au NP control. In the next step, Au NPs synthesised using the leaf extract were tested against the two bacteria. The Petri dishes containing a sterile pad treated with Au NPs+leaf extract (without wax) recorded a zone of inhibition of 16 mm for *B. subtilis* and 19 mm for *E. coli*. Since the leaf extract without wax was found to have no zone of inhibition (no antimicrobial effect), the larger zones of inhibition recorded for both bacteria are the result of the bio-reduced Au particles. This result indicates that both bacteria are more sensitive to the bio-reduced Au particle. And in the final test set, Petri dishes containing a sterile pad treated with Au NPs+leaf extract (with wax) recorded a zone of inhibition of 20 mm for *B. subtilis* and 29 mm for *E. coli*. The results of the Au NPs (control) and Au NPs synthesised using leaf extract are graphically presented in Fig. 5c (*B. subtilis*) and d (*E. coli*).

Inspection of both Fig. 5c and d reveals that the size of the diameters obtained for the zone of inhibition are much larger for Au NPs+leaf extract (with wax) than those without leaf extract. It is also evident that there is a synergistic effect between the Au NPs and the leaf extract. This is dramatically seen for Au NPs+leaf extract (with wax) against *E. coli* (zone of inhibition is 29 mm), which was significantly greater than the 19 mm for Au NPs+leaf extract (without wax). The

results also show that Au NPs+leaf extract (with wax) was much more effective against the gram-negative *E. coli* (diameter 29 mm) than the gram-positive *B. subtilis* (diameter 20 mm). A photograph of the 29-mm zone of inhibition for *E. coli* resulting from Au NPs+leaf extract (with wax) is presented in Fig. 5a, while an enlarged microscopic view of the interface between *E. coli* and the zone of inhibition is presented in Fig. 5b. The results of the antibacterial study reveal that both *B. subtilis* (gram-positive) and *E. coli* (gram-negative) were sensitive to Au NPs+leaf extract (with wax). The leaf extract without wax proved to have no antibacterial effect. However, leaf extract with wax produced a small zone of inhibition (9 mm) for both bacteria, indicating they were both resistant to the leaf extract. The results indicate that the antibacterial properties contained within the leaf are resident in the waxes, and when the waxes are combined with Au NPs, a significant synergistic effect was produced. Combining the inherent properties of the leaf wax with those of Au NPs has proven to be beneficial in producing an effective antibacterial agent against both *B. subtilis* and *E. coli*.

## Conclusion

A straightforward, clean and environmentally friendly method of biologically synthesising Au NPs was demonstrated using the leaf extracts from an indigenous Australian plant *E. macrocarpa*. The leaf extracts acted as both the stabilising agent and the reducing agent, with the process occurring at room temperature. The biological reduction of $AuCl_4^-$ produced an aggregate of NPs consisting of spherical particles ranging in size from 20 to 80 nm and nanometre-sized crystalline shapes (equilateral or truncated triangular, pentagon and hexagonal), ranging in size from 50 to 100 nm. Longer reduction times allowed the NPs to grow into micrometre-sized particles of varying size and morphology. The sensitivity method of Kirby–Bauer was used to establish a synergistic effect between the Au NPs and the waxes contained in the leaf extracts. And when combined, the Au NPs and leaf extracts (with wax) created an effective antibacterial agent against both *B. subtilis* and *E. coli*.

**Acknowledgments** This work was partly supported by the Western Australian Nanochemistry Research Institute (WANRI). The authors would also like to thank Mr Ravi Krishna Brundavanam for his assistance with the XRD measurements and Mrs. Monaliben Shah, Mrs. Sridevi Brundavanam and Dr Michelle Buttery for their assistance in the antibacterial studies.

**Conflict of interest** The authors claim no conflict of interest in this work.

# References

1. Sperling RA, Gil PR, Zhang F, Zanella M, Parak WJ (2008) Biological applications of gold nanoparticles. Chem Soc Rev 37:1896–1908

2. Sykora D, Kasicka V, Miksik I, Rezanka P, Zaruba K, Matejka P, Kral V (2010) Applications of gold nanoparticles in separation sciences. J Sep Sci 30:372–387

3. Cai W, Gao T, Hong H, Sun J (2008) Applications of gold nanoparticles in cancer nanotechnology. Nanotechnol Sci Appl 1: 17–32

4. Pankhurst QA, Connolly J, Jones SK, Dobson J (2003) Applications of magnetic nanoparticles in biomedicine. J Phys D: Appl Phys 36:R167–R181

5. Yan H, Park SH, Finkelstein G, Reif JH, LaBean TH (2003) DNA-templated self-assembly of protein arrays and highly conductive nanowires. Science 301:1882–1884

6. Keren K, Berman RS, Buchstab E, Sivan U, Braun E (2003) DNA-templated carbon nanotube field-effect transistor. Science 302: 1380–1382

7. Ghosh SK, Pal T (2007) Interparticle coupling effect on the surface plasmon resonance of gold nanoparticles: from theory to application. Chem Rev 107:4797–4862

8. Ai J, Biazar E, Jafarpour M et al (2011) Nanotoxicology and nanoparticle safety in biomedical designs. Int J Nanomedicine 6:1117–1127

9. Philip D (2010) Green synthesis of gold and silver nanoparticles using *Hibiscus rosa sinensis*. Physica E 42:1417–1424

10. Kumar P, Singh P, Kumari K et al (2011) A green approach for the synthesis of gold nanotriangles using aqueous leaf extract of *Callistemon viminalis*. Mater Lett 65:595–597

11. Lengke M, Southam G (2006) Bioaccumulation of gold by sulphate-reducing bacteria cultured in the presence of gold (I)-thiosulfate complex. Acta 70(14):3646–3661

12. Joerger TK, Joerger R, Olsson E, Granqvist CG (2001) Bacteria as workers in the living factor: metal-accumulating bacteria and their potential for materials science. Trends Biotechnol 19(1):15–20

13. Ahmad A, Senapati S, Khan MI et al (2003) Intracellular synthesis of gold nanoparticles by a novel alkalotolerant actinomycete Rhodococcus species. Nanotechnology 14:824–828

14. Kuber C, Souza SF (2006) Extracellular biosynthesis of silver nanoparticles using the fungus *Aspergillus fumigates*. Colloids Surf B 47:160–164

15. Ahmad A, Senapati S, Khan MI et al (2003) Extracellular biosynthesis of monodisperse gold nanoparticles by a novel extremophilic actinomycete Thermomonospora species. Langmuir 19:3550–3553

16. Simkist K, Wilbur KM (1989) Mulluscs-epithelial control of matrix and minerals. In: Simkist K, Wilbur KM (eds) Biomineralisation, cell biology and mineral deposition. Academic, New York, pp 231–256

17. Kreibig U, Vollmer M (1995) Optical Properties of Metal Clusters. Springer, Berlin

18. Liu X, Dai Q, Austin L, Coutts J et al (2008) A one-step homogeneous immunoassay for cancer biomarker detection using gold nanoparticle probes coupled with dynamic light scattering. J Am Chem Soc 130:2780–2782

19. Chan WCW, Nie SM (1998) Quantum dot bioconjugates for ultrasensitive nonisotopic detection. Science 281:2016–2018

20. Huang X, Jian PK, El-Sayed IH et al (2006) Determination of the minimum temperature required for selective photothermal destruction of cancer cells with the use of immune-targeted gold nanoparticles. Photochem Photobiol 82:412–417

21. Paciotti GF, Mayer L, Weinreich D et al (2004) Colloidal gold: a novel nanoparticle vector for tumour directed drug delivery. Drug Deliv 11:169–183

22. Sondi I, Salopek-Sondi B (2004) Silver nanoparticles as antimicrobial agent: a case study on *E. coli* as a model for gram-negative bacteria. J Colloid Interface Sci 275:177–182

23. Hsiao M, Chen S, Shieh D, Yeh C (2006) One-pot synthesis of hollow Au3Cu1 spherical-like and biomineral botallackite Cu2 (OH) 3Cl flowerlike architectures exhibiting antimicrobial activity. J Phys Chem B 110:205–210

24. Priester J, Stoimenov P, Mielke R, Webb S et al (2009) Effects of soluble cadmium salts versus CdSe quantum dots on the growth of planktonic *Pseudomonas aeruginosa*. Environ Sci Technol 43:2589–2594

25. Brayner R, Ferrari-Iliou R, Brivois N, Djediat S, Benedetti M, Fiévet F (2006) Toxicological impact studies based on *Escherichia coli* bacteria in ultrafine ZnO nanoparticles colloidal medium. Nano Lett 6:866–870

26. Simon-Deckers A, Loo S, Mayne-L'hermite M, Herlin-Boime N, Menguy N, Reynaud C, Gouget B, Carrie M (2009) Size-, composition- and shape-dependent toxicological impact of metal oxide nanoparticles and carbon nanotubes toward bacteria. Environ Sci Technol 43:8423–8429

27. Niemeyer CM (2001) Nanoparticles, proteins and nucleic acids: biotechnology meets materials science. Angew Chem Int Ed 40:4128–4158

28. Hirsch LR et al (2003) Nanoshell-mediated near-infrared thermal therapy of tumours under magnetic resonance guidance. Proc Natl Acad Sci U S A 100:13549–13554

29. Zheng Y, Sache L (2009) Gold nanoparticles enhance DNA damage induced by anti-cancer drugs and radiation. Radiat Res 172:114–119

30. Cheng Y, Samia AC, Li J et al (2010) Delivery and efficacy of a cancer drug as a function of the bond to the gold nanoparticle surface. Langmuir 26:2248–2255

31. Thomas M, Klibanov AM (2003) Conjugation to gold nanoparticles enhances polyethylenimine's transfer of plasmid DNA into mammalian cells. Proc Natl Acad Sci U S A 100:9138–9143

32. Mukherjee P, Bhattacharya R, Bone N et al (2007) Potential therapeutic application of gold nanoparticles in B-chronic lymphocytic leukaemia (BCLL): enhancing apoptosis. J Nanobiotechnol.

33. Hernandez-Sierra JF, Ruiz F, Pena DC et al (2008) Antimicrobial sensitivity of *Streptococcus mutans* to nanoparticles of silver, zinc oxide and gold. Nanomedicine NBM 4:237–240

34. http://eol/pages/635316/overview Encyclopaedia of Life: Eucalyptus Macrocarpa. Last accessed January 21 2013.

35. Poinern GEJ, Le XT, Fawcett D (2011) Super-hydrophobic nature of nanostructures on an indigenous Australian eucalyptus plant and its potential application. Nanotechnol Sci Appl 4:113–121

36. Jorgensen JH, Turnidge JD (2007) Susceptibility test methods: dilution and disc diffusion methods. In: Murray PR, Baron EJ et al (eds) Manual of clinical microbiology, 9th edn. ASM Press, Washington DC, pp 1152–1172

37. Singh C, Sharma V, Naik PK et al (2011) A green biogenic approach for synthesis of gold and silver nanoparticles using *Zingiber officinale*. Dig J Nanomater Biostruct 6(2):535–542

38. Zhang YX, Zheng J, Gao G et al (2011) Biosynthesis of gold nanoparticles using chloroplasts. Int J Nanomedicine 6:2899–2906

39. Alexandrides P (2011) Gold nanoparticle synthesis, morphology control and stabilisation facilitated by functional polymers. Chem Eng Technol 34(1):15–28

40. Wang L, Chen X, Zhan J et al (2005) Synthesis of gold nano and microplates in hexagonal liquid crystals. J Phys Chem B 109:3189–3194

41. Murata M, Yamakoshi et al (1990) Macrocarpal A, a novel antibacterial compound from *Eucalyptus macrocarpa*. Agric Biol Chem 54(12):3221–3226

42. Yamakoshi Y, Murata M, Shimizu A et al (1992) Isolation and characterization of macrocarpals B-G Antibacterial compounds from *Eucalyptus macrocarpa*. Biosci Biotechnol Biochem 56(10): 1570–1576

# 19

# Study of gold nanorods–protein interaction by localized surface plasmon resonance spectroscopy

Néné Thioune · Nathalie Lidgi-Guigui · Maximilien Cottat ·
Ana-Maria Gabudean · Monica Focsan · Henri-Michel Benoist ·
Simion Astilean · Marc Lamy de la Chapelle

**Abstract** In this paper, gold nanorods' (GNRs) interaction with different proteins (i.e. carbonic anhydrase, lysozyme, ovalbumin and bovine serum albumin (BSA)) at physiological pH is investigated using localized surface plasmon resonance (LSPR) spectroscopy. We observe that the incubation of these proteins at different concentrations with cetyltrimethylammonium bromide-capped GNRs of three aspect ratios induces dramatic changes in the extinction spectra of the nanoparticles. In particular, we correlate the position and shape of the longitudinal LSPR peaks to the ability of the proteins to specifically interact with GNRs' surface. The different types of behaviour observed are explained by the exposed molecular surface area of the proteins' cysteine residues as modelled on the basis of their respective X-ray crystallographic data structures. Cysteine is the only amino acid that exhibits an SH group that is well known to have a strong affinity to gold. The presence and the accessibility of such a residue may explain the protein binding to GNRs. The isoelectric point of the proteins is also an important characteristic to take into account, as the electrostatic strength between GNRs and protein explains some of the cases where aggregates are formed.

N. Thioune · N. Lidgi-Guigui (✉) · M. Cottat · M. L. de la Chapelle
Laboratoire CSPBAT UMR CNRS 7244 UFR SMBH, Université Paris 13, Sorbonne Paris Cité, 74 rue Marcel Cachin, 93017 Bobigny, France
e-mail: nathalie.lidgi-guigui@univ-paris13.fr

A.-M. Gabudean · M. Focsan · S. Astilean
Nanobiophotonics and Laser Microspectroscopy Center, Interdisciplinary Research Institute on Bio-Nano-Sciences, Faculty of Physics, Babes-Bolyai University, T. Laurian 42, 400271 Cluj-Napoca, Romania

N. Thioune · H.-M. Benoist
Département d'Odontologie de Dakar, Université Cheikh Anta Diop (UCAD) BP 5005, Dakar, Senegal

**Keywords** Proteins · LSPR · Gold nanorods (GNRs) · isoelectric point · Solvant accessible area

## Introduction

During the last decades when nanosciences have made their revolution, gold nanoparticles (GNPs) have attracted particular interests for many reasons. Among them, two have drawn our attention in the present study. First, their genuine optical properties make them of prime choice for many applications especially in medicine and biology [1, 2]. Second, they are highly stable in many liquid or gaseous environments, which makes them very versatile. GNPs provide high contrast in cellular and tissue imaging using confocal reflectance microscopy [3] or dark-field imaging [4]. They have been used in photothermal therapy of cancer [5]. Their optical relevance is mainly due to the localized surface plasmon resonance (LSPR) emerging in the visible spectral range for GNPs with size between ~2 to ~200 nm. Moreover, LSPR can be finely tuned by changing the GNPs size, environment or shape. The literature is rich in articles describing protocols to synthesize a wide range of GNPs of different shapes and geometries such as spherical particles, nanorods, nanoshells, nanostars and nanocages [6, 7]. Interestingly, a non-spherical nanoparticle will present several resonance characteristics of its shape [8, 9]. Gold nanorods (GNRs) are in this aspect, highly studied because their LSPR can be tuned on a wider wavelength scale. The aspect ratio of GNRs is also easily deduced from their absorption spectra. The LSPR sensitivity to the nanoparticle close environment makes GNPs good candidates for sensor use.

GNRs are more and more often used in biology and for biomedical applications. Reports are found on their use in biosensing, bio-imaging, photothermal therapy or even in theranostic [10–13]. The common point of all these works is the necessity to attach the appropriate biomolecules to the GNRs, in order for it to accomplish the task it was designed for. For this, several routs can be considered, namely ligand exchange, electrostatic interaction, biofunctional linkage and surface coating. All these techniques are described in detail in [14, 15]. Giving the increasing interest of scientist from all disciplines for working with these tremendous nanostructures in a bio-environment, it is of crucial importance to understand how they directly interact with biomolecules.

In the present study, cetyltrimethylammonium bromide (CTAB)-capped GNRs were synthesized using the seed-mediated growth method in order to study their direct interaction with four different proteins (i.e. carbonic anhydrase, lysozyme, ovalbumin and bovine serum albumin (BSA)). GNRs represent a unique class of metallic nanostructures with two well-defined LSPR bands in the visible/near infrared spectral range corresponding to electronic oscillations along the short and long axis of nanorods. In a previous work [16], we already showed that the position of these two LSPR bands is not dependent on the molecular weight of the proteins interacting with the GNRs surface. In this context, we are interested in investigating the interaction of the above mentioned proteins at physiological pH with GNRs as function of their concentrations using LSPR spectroscopy.

## Experimental

### Reagents

Tetrachloroauric(III) acid (HAuCl$_4$), CTAB and ascorbic acid were purchased from Aldrich. Sodium borohydride (NaBH$_4$, 99 %) and silver nitrate (AgNO$_3$) were obtained from Merck. Lysozyme from chicken egg white (ref. 62970), carbonic anhydrase from bovine erythrocytes (86.5 %, ref. c3934) and albumin from bovine serum (BSA ref. a8806, 96 %) were purchased from Sigma-Aldrich. Ovalbumin (ref. W377206) has been provided by GE Healthcare. All proteins were dissolved in distilled water at concentrations ranging from $10^{-10}$ to $10^{-4}$ M. Table 1 gives the main characteristics of the proteins used.

**Table 1** Proteins characteristics

| Protein | Molecular weight (kDa) | Isoelectric point (IEP) | pH | Number of accessible sulphur |
|---|---|---|---|---|
| BSA | 66.5 | 4.7 | 7±0.5 | 2 |
| Ovalbumin | 43 | 4.6 | 7±0.5 | 1 |
| Carbonic anhydrase | 29 | 5.9 | 7±0.5 | 0 |
| Lysozyme | 14.6 | 11.35 | 7±0.5 | 1 |

### Synthesis of GNRs

GNRs were synthesized utilizing the seed-mediated growth method described in details in [16]. The UV–visible spectra of each set of GNRs are presented on Fig. 1. The shift of the spectra is explained by the fact that the three solutions contain GNRs with different aspect ratios.

### Preparation of proteins/GNRs conjugates

All protein solutions were prepared in distilled water at pH=7±0.5. Six concentrations of protein solutions were prepared (100 pM, 1 nM, 10 nM, 100 nM, 1 μM, 10 μM and 100 μM) except for BSA for which just 1 nM, 100 nM and 10 μM solutions were prepared. Then, 250 μl of pure GNR solutions were mixed with 400 μl of the protein solution at different concentration values. The samples were incubated at room temperature for 20 min before the measurement.

### Instrumentation

The interaction between CTAB-capped GNRs and the different proteins was characterized by extinction spectroscopy. Spectra were recorded using a Kontron Uvikon 941 spectrophotometer in the range of 400 to 800 nm with 1 nm step and 1 nm resolution. The extinction spectra were fitted using the "multipeak fit" tool on Origin® with a Lorentzian type.

The GNRs morphology was analysed with conventional transmission electron microscopy (TEM) using a JEOL 100 U type TEM microscope operated at 100 kV accelerating voltage. The ζ potential of the colloidal solutions was measured using a particle analyser (Nano ZS90 Zetasizer, Malvern Instruments) equipped with a He–Ne laser (633 nm, 5 mW) and a measurement angle of 90°. Each sample was measured three times, and the mean value was reported.

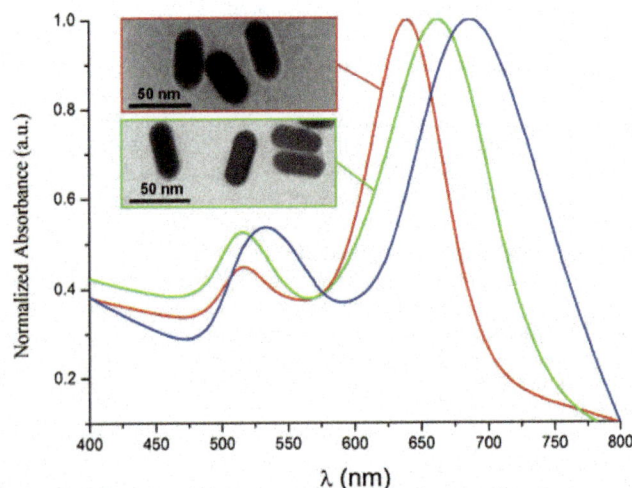

**Fig. 1** Normalized LSPR spectra of the three types of synthesized GNRs

The solvent molecular surface area (MSA) of the proteins was estimated using a solvent probe of 1.4 Å; numerical results were obtained using GetArea [17] and graphically confirmed using Jmol (http://jmol.sourceforge.net).

## Results

Figure 1 illustrates the normalized extinction spectra of the as-synthesized GNRs. The spectral position of transversal LSPR band and the longitudinal LSPR bands at 639, 662 and 686 nm correspond to GNRs with aspect ratios of 2.1 (average length/width, 46/22), 2.52 (average length/width, 43/17 nm) and 2.8 (average length/width, 42/15 nm), respectively, as given by TEM images (see inset in Fig. 1). In the following, data will be presented showing the interaction of proteins with one set of the GNR. All the experiences were repeated at least once with another set of GNR, and the results were reproducible. Herein, we will mainly focus on the study of the longitudinal LSPR (i.e. width and position) to explain the mechanism of GNRs/proteins interaction.

Figure 2 presents the modifications of the longitudinal LSPR peak of GNRs in the presence of different concentrations of proteins. Upon incubation of GNRs with carbonic anhydrase (Fig. 2a) and lysozyme (Fig. 2b) protein solutions, the positions of the longitudinal LSPR peaks vary as function of protein concentration. Moreover, we observe that the width of the bands is constant, indicating a good dispersion of the nanoparticles in the solution. On the opposite, dramatic changes of the longitudinal LSPR bands were observed after incubation of the colloidal solution with ovalbumin (Fig. 2c) and BSA (Fig. 2d) solutions. In the case of ovalbumin (Fig. 2c), the protein seems to interact with GNRs' surface in the same manner as carbonic anhydrase or lysozyme except at 100 nM and 1 μM. At these concentrations, the intensity of plasmonic band suddenly decreases together with its broadening, clearly suggesting the aggregation of nanoparticles. Similarly, this phenomenon was also induced by BSA protein at the concentrations of 1 and 100 nM (Fig. 2d). Taking into account that the spectral position of the LSPR band is strongly dependent on the dielectric function of GNRs' local environment, the modification of its position together with its broadening indicate the extent of individual or aggregated nanoparticles in colloidal solution.

In order to study these spectral modifications, we have fitted the longitudinal LSP band with Lorentzian curves. As

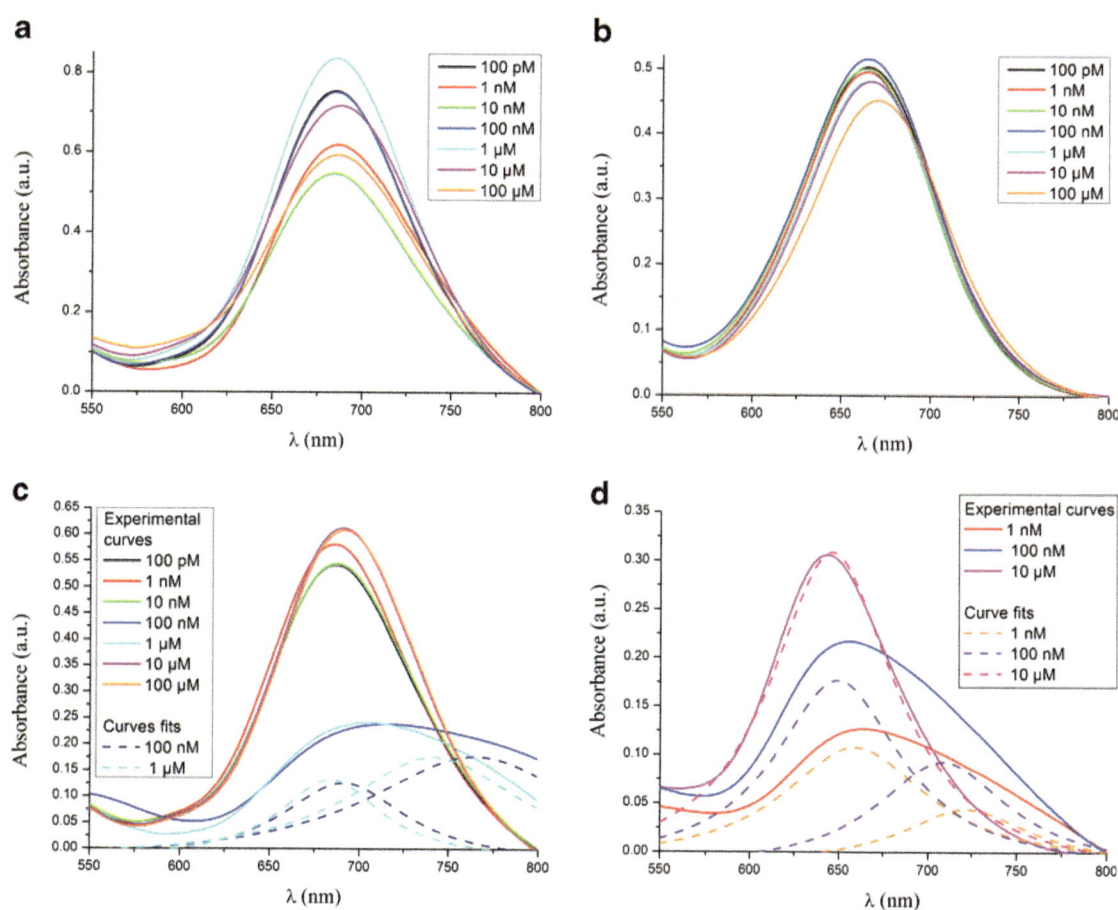

**Fig. 2** Longitudinal LSPR band of synthesized GNRs in the absence and presence of: **a** carbonic anhydrase, **b** lysozyme, **c** ovalbumin and **d** BSA of different concentrations

it reveals on Fig. 2c, d, the widest LSP were fitted with two Lorentzian when only one was needed for the unaltered LSP. The experimental recorded curve reflects a composite plasmonic resonances of (1) GNRs interacting with the protein as mentioned above (individual mode) and (2) protein-induced aggregation of GNRs (aggregation mode).

To understand the mechanism underlying these two modes, Fig. 3 was plotted. For this, the shift between the highest LSPR (aggregation mode if it exists, otherwise the normal mode) and the pure GNRs is measured and then plotted. Giving the resolution of the spectrometer, there is no observable shift for carbonic anhydrase. On the opposite, the GNRs' individual mode in presence of lysozyme and ovalbumin proteins, respectively, displays a red shift for a high enough concentration (~10 nM), indicating an increase of the dielectric constant of the GNRs surrounding medium. In other words, a material

with higher dielectric constant than CTAB has replaced or at least attached to the surfactant molecules. Obviously, this material can only be the proteins.

The extent of this red shift varies in a specific way. It is constant or slightly decreases as the protein concentrations increase until it reaches a minimum value at 10 nM for lysozyme and 1 nM for ovalbumin. For higher concentration, the extent of the shift increases again. We believe that this behaviour highlights some aspect of the GNRs/protein interactions as it is discussed in the next section.

## Discussion

We explain the behaviour of the proteins toward GNRs by exploiting the available data on their electric charges and structures. The electric charge of the proteins at the working pH was obtained from their isoelectric point (Table 1). The electrostatic interaction between proteins and GNRs was studied, taking into account that the GNRs are capped with CTAB, which makes them positively charged (the $\zeta$ potential was measured to be +20 mV).

The structure of the proteins was gathered from the RCSB Protein Data Bank[©]. The structure of the proteins studied here was obtained from X-ray crystallographic data [18–21]. The cysteine (Cys) residue is the centre of our interest since it is the only amino acid which owns a thiol group. In chemistry, thiols are compounds which are very famous to demonstrate strong affinity to gold. In the presence of thiols, strong sulphur–gold bonds are formed, which are renowned to be strong covalent bonds. Previous studies [22, 23] on GNPs/protein interactions have shown that the binding of protein to a very small GNP depends firstly on the existence of a Cys inside the protein and secondly, on the accessibility of the thiol group of the Cys residue. Herein, the sulphur accessibility is estimated via the notion of MSA [24]. This is defined as the contact surface area of a solvent probe sphere that rolls along the surface of the protein. Using the software GetArea, we have calculated for each Cys residue, its MSA and the surface energy of the corresponding sulphur. The idea here is to measure the availability of the sulphur to bind to the GNRs. The same approach was followed using Jmol. Figure 4 shows the MSA (red dots) for a 1.4-Å sphere rolled on the protein surface.

In particular, carbonic anhydrase has only one Cys residue which is located inside its volume, making the sulphur non-accessible for binding with GNRs. Thus, the only existing interaction between carbonic anhydrase and GNRs is through electrostatic forces. As the carbonic anhydrase is negatively charged at the working pH of 7, it can only bind to positively charged structures, here, the CTAB-capped GNRs. We suppose that carbonic anhydrase can only weakly adsorb on the CTAB double layer and not on the GNRs' surface. This explains that no shift is observed for carbonic anhydrase on

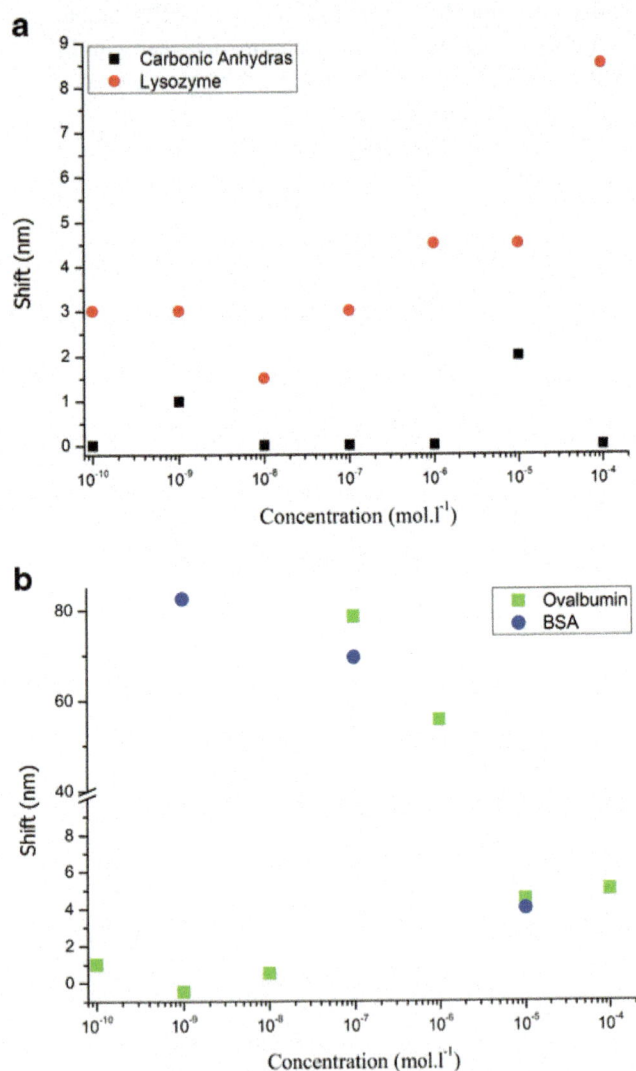

**Fig. 3** Longitudinal LSPR peak shift versus protein concentration. The longitudinal LSPR shift is calculated by comparing the position of the LSPR band of pure GNRs solution with the position of the corresponding band of conjugates

**Fig. 4** CPK view of **a** carbonic anhydrase, **b** lysozyme, **c** ovalbumin and **d** BSA. The *red dots* materialized the MSA; the Cys residues are coloured in *green* and the sulphur atoms in *yellow*

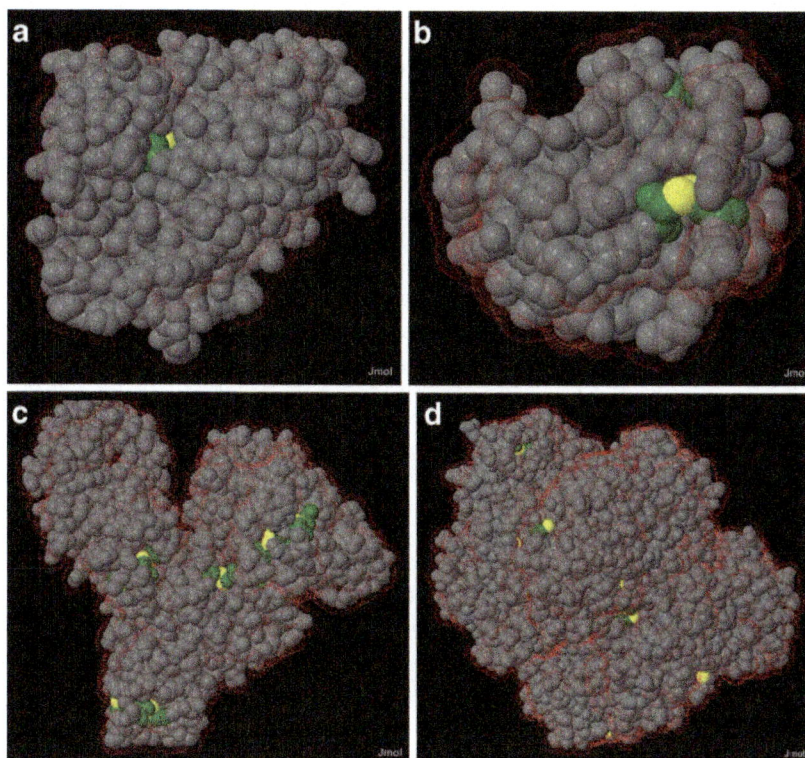

Fig. 3a: the GNRs environment does not change and so its longitudinal LSPR. However, small deviations can occur, that are probably due to the imperfection of the CTAB capping. Especially, it is known that the CTAB double layer is sparse at the GNRs extremities.

As mentioned above, in the case of lysozyme and ovalbumin, the behaviour of individual mode is quite different. For both proteins, a red shift is observed for all the concentrations. If the proteins were just electrostatically adsorbed to the CTAB double layer, we would observe no shift just as for carbonic anhydrase. Thus, we can conclude that lysozyme and ovalbumin replace CTAB. We also observe for both proteins that the red shift extent first decreases until a specific concentration before it increases again. The CTAB bilayer surrounding the GNRs is most probably in competition with the protein. As shown in Table 2 and Fig. 4, lysozyme and ovalbumin have one accessible thiol. It is known that the functionalization of CTAB-capped GNR with thiolated molecules is trivially made by the sulphur–gold interaction that overcomes the weak binding of CTAB to gold [14]. The consequence of the addition of lysozyme (respectively ovalbumin) is the binding of the protein to gold GNR and the removal of CTAB. We observe here that a minimum value of the concentration has to be reached before proteins replace most of the CTAB (when the red shift extent increases).

If the first part of the LSPR, shift is similar for lysozyme and ovalbumin; the latter shows two main specificities. First of all, in the case of ovalbumin, the lowest red shift of the LSPR band

occurs for a lower concentration than for lysozyme. Contrary to lysozyme, ovalbumin is negatively charged in the experimental conditions (see Table 1). As the GNRs are strongly positively charged, the electrostatic attraction forces are added to the thiol attraction. The result is an easier attachment of the proteins to the GNRs.

The second point to rise is the aggregation taking place at 100 nM and 1 μM (Fig. 3b). Once again, we explain this behaviour by the surface charge of ovalbumin. Electrostatic attraction forces act on several GNRs at the same time allowing the formation of aggregates. This is why the ovalbumin/GNRs ratio has to be very well controlled to avoid charge neutralization and hence aggregation. Many studies have already shown that a stabilized attachment of proteins

**Table 2** Summary of the cysteine (Cys) residues properties in carbonic anhydrase, lysozyme, ovalbumin, BSA and IgG

| Protein | Number of Cys residues | Number of Cys residues with MSA >20 Å | Number of SH groups with surface energy >15 mcal mol$^{-1}$ |
|---|---|---|---|
| Carbonic anhydrase | 1 | 0 | 0 |
| Lysozyme | 8 | 1 | 1 |
| Ovalbumin | 23 | 1 | 1 |
| BSA | 35 | 5 | 2 |
| IgG | 34 | 5 | 2 |

to GNPs thanks to electrostatic interaction required an optimization of both the pH and the concentration of the solutions [14, 25, 26].

The case of BSA has in common with ovalbumin that it is negatively charge at pH 7 and thus is electrostatically attracted to the CTAB-capped GNRs. Nevertheless, BSA has also the characteristic of containing quite a lot of Cys residues (see Table 2). Among them, only five are solvent accessible, and among these five, two sulphur have enough surface energy to bind to gold surface. Not only BSA has twice more accessible sulphur than lysozyme or ovalbumin but these two sulphurs (in Cys 200 and Cys 559) are located on opposite sides of the protein (Fig. 4d). This way, one BSA can bind to two different GNRs. This is how very strong aggregates are formed.

The longitudinal LSPR is becoming closer to the same band of pure GNRs as the BSA concentration increases. The origin of this can be the fact that a strong excess of BSA in the presence of CTAB is known to form aggregates of proteins [27]. The concentration of BSA actually interacting with GNRs is thus much lower than the nominal concentration. Instead of increasing, the number of BSA interacting with the GNRs is decreasing, explaining the reduction of the longitudinal LSPR shift.

## Conclusion

In this article, we have investigated the interaction of GNRs with carbonic anhydrase, lysozyme, ovalbumin, and BSA regarding the proteins concentration using LSPR spectroscopy. In a previous study [16], we have already demonstrated that the longitudinal LSPR shift is the most significant for the understanding of GNRs/protein interaction. In the same reference, we established the fact that the molecular weight of the proteins is not a relevant parameter to characterize this interaction. In the present letter, we have considered the protein concentration, isoelectric point and the MSA to explain the experimental results. The conclusion is that the number of accessible sulphurs is of crucial importance to explain the shape of the longitudinal LSPR shift evolution with the concentration. If there are too many, like in the case of BSA, strong aggregates are formed. If there is only one, the evolution of the longitudinal LSPR with the concentration presents two parts: the first one decreases and the second one increases. Then, the isoelectric point gives insight of the electric charge of the protein and explains some of the GNRs/proteins aggregation cases. Carbonic anhydrase which has no accessible sulphur and is negatively charged barely binds to GNRs. In comparison, ovalbumin has one accessible sulphur and is negatively charged; it binds to GNRs and in some conditions can even form electrostatic aggregates.

Liu et al. [28] conducted a similar study in 2012 using dynamic light scattering (DLS) instead of extinction spectroscopy. They come to the same conclusion as us, concerning the interaction between GNRs and BSA or human serum albumin, and they show that immunoglobulin (IgG and IgA) follow the same behaviour as ovalbumin. However, they fail to explain the difference of behaviour between these proteins. When looking at the available sulphur in IgG or IgA (Table 2) [29], we found that the three of them are in the same location (bottom of Fab and top of Fc). Steric interaction most probably cancels the duplicity of thiols, so they behave as there was only one accessible sulphur. This is in complete agreement with the explanation developed in our study.

**Acknowledgments** This work was supported by the Programme Hubert Curien Brancusi, by the ANCS, project number PN II Capacitati/Brancusi, 489/2011 and by the Nanoantenna European project (FP7-HEALTH-F5-2009 241818)

## References

1. El-Sayed IH, Huang X, El-Sayed M (2005) Surface plasmon resonance scattering and absorption of anti-EGFR antibody conjugated gold nanoparticles in cancer diagnostics: applications in oral cancer. Nano letters 5:829–834.
2. Aslan K, Lakowicz JR, Geddes CD (2005) Plasmon light scattering in biology and medicine: new sensing approaches, visions and perspectives. Current opinion in chemical biology 9:538–544.
3. Aaron J, Nitin N, Travis K et al (2007) Plasmon resonance coupling of metal nanoparticles for molecular imaging of carcinogenesis in vivo. Journal of biomedical optics 12:034007.
4. Liu GL, Yin Y, Kunchakarra S et al (2006) A nanoplasmonic molecular ruler for measuring nuclease activity and DNA footprinting. Nature nanotechnology 1:47–52.
5. Li W, Cai X, Kim C et al (2011) Gold nanocages covered with thermally-responsive polymers for controlled release by high-intensity focused ultrasound. Nanoscale 3:1724–1730.
6. Chen J, Saeki F, Wiley BJ et al (2005) Gold nanocages: bioconjugation and their potential use as optical imaging contrast agents. Nano letters 5:473–477.
7. Geddes CD (ed) (2012) Reviews in Plasmonics 2010. Springer, New York.
8. Hu M, Chen J, Li Z-Y et al (2006) Gold nanostructures: engineering their plasmonic properties for biomedical applications. Chemical Society reviews 35:1084–1094.
9. Sun Y, Xia Y (2003) Gold and silver nanoparticles: a class of chromophores with colors tunable in the range from 400 to 750 nm. Analyst 128:686.

10. Chen C-D, Cheng S-F, Chau L-K, Wang CRC (2007) Sensing capability of the localized surface plasmon resonance of gold nanorods. Biosensors & bioelectronics 22:926–932.

11. Yong K-T, Swihart MT, Ding H, Prasad PN (2009) Preparation of gold nanoparticles and their applications in anisotropic nanoparticle synthesis and bioimaging. Plasmonics 4:79–93.

12. Huang X, El-Sayed IH, Qian W, El-Sayed MA (2006) Cancer cell imaging and photothermal therapy in the near-infrared region by using gold nanorods. Journal of the American Chemical Society 128:2115–2120.

13. Young JK, Figueroa ER, Drezek RA (2012) Tunable nanostructures as photothermal theranostic agents. Annals of biomedical engineering 40:438–459.

14. Huang X, Neretina S, El-Sayed MA (2009) Gold nanorods: from synthesis and properties to biological and biomedical applications. Adv Mater 21:4880–4910.

15. Joshi PP, Yoon SJ, Hardin WG et al (2013) Conjugation of antibodies to gold nanorods through Fc. Synthesis and Molecular Specific Imaging. Bioconjugate chemistry, Portion.

16. Cottat M, Thioune N, Gabudean A-M et al (2012) Localized surface plasmon resonance (lspr) biosensor for the protein detection. Plasmonics 8:699–704.

17. Fraczkiewicz R, Braun W (1998) Exact and efficient analytical calculation of the accessible surface areas and their gradients for macromolecules. J Comput Chem 19(3):319–333

18. Diamond R (1974) Real-space refinement of the structure of hen egg-white lysozyme. J Mol Biol 82:371–391

19. Eriksson AE, Jones TA, Liljas A (1988) Refined structure of human carbonic anhydrase II at 2.0 A resolution. Proteins 4:274–282.

20. Stein PE, Leslie AGW, Finch JT, Carrell RW (1991) Crystal structure of uncleaved ovalbumin at 1·95 Å resolution. J Mol Biol 221:941–959

21. Bhattacharya A, Grüne T, Curry S (2000) Crystallographic analysis reveals common modes of binding of medium and long-chain fatty acids to human serum albumin. Journal of molecular biology 303:721.

22. Lidgi-Guigui N, Leung C, Palmer RE (2008) Weak precursor state binding of protein molecules to size-selected gold nanoclusters on surfaces. Surf Sci 602:1006–1009.

23. Prisco U, Leung C, Xirouchaki C et al (2005) Residue-specific immobilization of protein molecules by size-selected clusters. Journal of the Royal Society, Interface / the Royal Society 2:169–175.

24. Richards FM (1977) Areas, volumes, packing and protein structure. Annual review of biophysics and bioengineering 6:151–176.

25. Pissuwan D, Valenzuela SM, Killingsworth MC et al (2007) Targeted destruction of murine macrophage cells with bioconjugated gold nanorods. J Nanoparticle Res 9:1109–1124.

26. Geoghegan WD, Ackerman GA (1977) Adsorption of horseradish peroxidase, ovomucoid and anti-immunoglobulin to colloidal gold for the indirect detection of concanavalin A, wheat germ agglutinin and goat anti-human immunoglobulin G on cell surfaces at the electron microscopic level: a new method, theory and application. Journal of Histochemistry & Cytochemistry 25:1187–1200. doi:10.

27. Sharma A, Pasha JM, Deep S (2010) Effect of the sugar and polyol additives on the aggregation kinetics of BSA in the presence of N-cetyl-N, N, N-trimethyl ammonium bromide. Journal of colloid and interface science 350:240–248.

28. Liu H, Pierre-Pierre N, Huo Q (2012) Dynamic light scattering for gold nanorod size characterization and study of nanorod–protein interactions. Gold Bull 45:187–195.

29. Harris LJ, Larson SB, Hasel KW, Mcpherson A (1997) Refined structure of an intact IgG2a monoclonal antibody. Biochemistry 36(7):1581–97

# 20

# Self-assembled ferrite nanodots on multifunctional Au nanoparticles

Yukiko Yasukawa · Xiaoxi Liu · Akimitsu Morisako

**Abstract** The fabrication of magnetic oxide nanodots was studied without the use of conventional lithographic techniques and patterning masks. Self-assembled Au nanoparticles with an average size of approximately 17 nm were formed via simple sputtering and used as the underlayer for the magnetic oxide film. Subsequently, hexagonal ferrite, $SrFe_{12}O_{19}$, was sputtered on the Au nanoparticles, resulting in an $SrFe_{12}O_{19}/Au$ sample. Self-assembled $SrFe_{12}O_{19}$ nanodots were obtained with an average size of 40–50 nm. The morphology of the Au nanoparticle underlayer acted as a template for the $SrFe_{12}O_{19}$ film, such that the self-assembled $SrFe_{12}O_{19}$ nanodots were formed. In addition, the fabrication of the $SrFe_{12}O_{19}$ film on the Au nanoparticles induced the down-sizing of the magnetic domain structures of $SrFe_{12}O_{19}$ to the nanoscale. Importantly, although the nanodots showed nanometric magnetic domains, a sufficient magnetization magnitude in the $SrFe_{12}O_{19}$ nanodots was revealed. Furthermore, the $SrFe_{12}O_{19}$ nanodot fabrication area was ~8.5 $cm^2$, thereby the current technique can be applied to the development of future functional magnetic nanodots.

**Keywords** Sputtering · Self-assembly · Au nanoparticles · Hexaferrite · Magnetic nanodots

## Introduction

Processing of magnetic nanostructures is currently crucial in the field of magnetic engineering, such as for magnetic hyperthermia [1], drug delivery systems [2], environmental purification [3], and magnetic recording systems [4]. For instance, with respect to magnetic recording systems, the immediate development of ultrahigh density magnetic recording media is indispensable because of the explosive worldwide increase in digital information. Patterned media [5–7], in which data are stored in a physically isolated single magnetic domain, have the potential for application in ultrahigh-density magnetic recording systems offering magnetic bit areal densities above $2 \times 10^{12}$ bits/in.. On the basis of these requirements, nanofabrication techniques for magnetic films and subsequent magnetic nanostructures are the latest key topics.

Hexagonal structural ferrite, $SrFe_{12}O_{19}$, is known to exhibit a large anisotropy constant ($K_1$) value of $3.5 \times 10^6$ erg/$cm^3$ [8] and is widely used as a ferrite magnet. $SrFe_{12}O_{19}$ also shows sufficient chemical stability, corrosion resistance, and superior wear resistance; these characteristics make $SrFe_{12}O_{19}$ suitable for use in functional magnetic nanoparticles. The combined feature of superior wear resistance and a high $K_1$ value is one of the important factors in the fabrication of magnetic nanostructures. Sufficient magnetic properties can be expected, even after nanofabrication of $SrFe_{12}O_{19}$, due to the large $K_1$ value, while the wear resistance easily enables fabrication of $SrFe_{12}O_{19}$ at the nanoscale. In spite of these advantages, there are only a few reports on the processing of hexagonal ferrites that are ten to several tens of nanometers in size. Therefore, it is worth studying the fabrication of $SrFe_{12}O_{19}$ on the several tens of nanometers or slightly larger scale to investigate its potential in future applications such as functional magnetic nanodevices.

A wide variety of techniques have been reported for the preparation of magnetic nanostructures, such as imprint lithography [9–11], ion irradiation [12, 13], focused ion beam lithography [14, 15], electron beam lithography [6, 16, 17], and a combination of lithographic techniques and electrochemical deposition of magnetic materials [17]. However, structural and magnetic damage are unavoidable issues in lithography. In addition, the fabrication areas are limited, and many experimental procedures are required when using lithographic-based techniques. On the other hand, nanofabrication based on the self-assembly technique is a promising method for attaining magnetic nanostructures, because this approach involves

Y. Yasukawa (✉) · X. Liu · A. Morisako
Department of Computer Science and Engineering, Faculty of Engineering, Shinshu University, 4-17-1 Wakasato, Nagano 380-8553, Japan
e-mail: yasukawa@shinshu-u.ac.jp

inexpensive equipment, simple procedures, large fabrication areas, and there is no need for patterning masks. Furthermore, structural and magnetic damage does not occur. In fact, nanofabrication of magnetic films using the self-assembly technique has recently been widely reported [18, 19].

In this study, $SrFe_{12}O_{19}$ was selected as the target material. $SrFe_{12}O_{19}$ exhibits perpendicular magnetic anisotropy when the $SrFe_{12}O_{19}$ (00$l$) planes are stacked along the $c$-axis direction. The lattice parameter, $a$, of $SrFe_{12}O_{19}$ is 5.89 Å (JCPDS card; 33–1340). Meanwhile, the interatomic distance, $d$, of Au (111) calculated based on the JCPDS card (04–0784) is 5.77 Å. Because the misfit ratio between $a[SrFe_{12}O_{19}$ (00$l$)] and $d[Au(111)]$ is ~2.1 %, heteroepitaxial growth between $SrFe_{12}O_{19}$ (00$l$) and Au (111) could be expected. In fact, $SrFe_{12}O_{19}$ (00$l$) was successfully grown on Au (111) and its magnetic properties were studied in our previous works [20–23]. However, the physical sources for magnetic properties, e. g., magnetic domain structures and nanometric crystal structures of the $SrFe_{12}O_{19}$ (00$l$)/Au (111) samples, were not studied in detail.

On the basis of these backgrounds, the structuring of $SrFe_{12}O_{19}$ nanodots was studied using the self-assembly technique. First, fabrication of self-assembled Au nanoparticles on a substrate was achieved via sputtering. These Au nanoparticles were then used as an underlayer for the $SrFe_{12}O_{19}$ film. The Au nanoparticle underlayer was utilized to downsize the $SrFe_{12}O_{19}$ film in order to form $SrFe_{12}O_{19}$ nanodot structures because the morphology of the upper layer ($SrFe_{12}O_{19}$) must reflect the structure of the underlayer (self-assembled Au nanoparticles). The nanostructures of the $SrFe_{12}O_{19}$/Au samples were carefully observed, and correlations between the $SrFe_{12}O_{19}$ nanostructures and the macroscopic magnetic properties of $SrFe_{12}O_{19}$ in the $SrFe_{12}O_{19}$/Au samples were also discussed.

## Experimental

Samples were prepared on a $SiO_2$/Si substrate using a DC magnetron sputtering system of our own making. The vacuum chamber of sputtering machine was evacuated to reach the pressure of $0.8$–$1.4 \times 10^{-6}$ Torr. In order to form the self-assembled Au nanoparticles, Au sputtering was performed in an Ar atmosphere at a sputtering pressure of $2.0 \times 10^{-3}$ Torr with a deposition power density of ~0.49 W/cm$^2$ for 1 min at a substrate temperature ($T_S$) of 100 °C. Subsequently, $SrFe_{12}O_{19}$ was deposited by DC magnetron sputtering on Au in a mixture of Ar and $O_2$ gasses at a fixed partial-pressure ratio of $P_{Ar}$: $P_{O2}$=99:1. The $SrFe_{12}O_{19}$ sputtering was carried out at sputtering pressure of $2.0 \times 10^{-3}$ Torr with ~0.99 W/cm$^2$ for 1 h at a $T_S$ of 475 °C using a sintered off-stoichiometric ceramic target ($SrFe_{11}O_x$, $\phi$ 50.8 mm with a thickness of 5 mm; Kojundo Chemical Laboratory). The crystallization of $SrFe_{12}O_{19}$ could be

possible at $T_S$=475 °C, but it should be noted that this temperature is close to the lowest limit for the crystallization of $SrFe_{12}O_{19}$ [20, 22]. Therefore, a post-annealing was carried out for the sufficient crystallization of $SrFe_{12}O_{19}$. The sample was annealed in a furnace under an air atmosphere at 750 °C for 1 h after $SrFe_{12}O_{19}$ deposition.

An X-ray diffractometer (XRD; Rigaku SmartLab) was used to identify the obtained material phases. The local nanostructures of the samples were studied using field-emission scanning electron microscopy (FE-SEM; Hitachi SU8000) and transmission electron microscopy (TEM; JEOL JEM-2010). Atomic force microscopy (AFM; Veeco Innova) was used to acquire the morphologies of the samples. The magnetic domains and magnetic properties were evaluated by magnetic force microscopy (MFM; Veeco Innova) and vibrating sample magnetometer (VSM; Tamagawa Factory Co. Ltd. custom made), respectively, at room temperature.

## Results and discussion

It was possible to control the morphology of Au and change it from a film (Fig. 1a) to self-assembled nanoparticles (Fig. 1b) using an in situ heat treatment of the substrate at 100 °C during sputtering. The total energy ($E_{total}$) of this system can be classified into (1) the surface energy ($\gamma$) of the Au film, (2) the interfacial energy between the substrate and the Au film, (3) the $\gamma$ of the substrate, and (4) the elastic strain energy of the Au film. Energies (1) and (3) originate from the surface curvatures, whereas (2) is generated by the wettability of the Au and the substrate. Furthermore, the misfit between the Au film and the substrate induces (4). The summation of (1)–(4) generates $E_{total}$ in this system. Through in situ heating of the substrate during the sputtering, the Au adatoms diffuse on the substrate. The driving force for the formation of the Au nanoparticles from the Au film is a reduction of $E_{total}$ by altering the morphology of Au. Consequently, Au nanoparticles were obtained. Figure 1c shows the XRD pattern for the self-assembled nanoparticles (Fig. 1b). It was found that Au exhibits a (111) orientation. The $\gamma$ of $fcc$ structural metals including Au is known to be $\gamma(111)<\gamma(100)<\gamma(110)$ in general [24]. Therefore, Au is oriented toward the direction of the lowest $\gamma$, (111). The diameter (size) and distribution of the Au nanoparticles shown in Fig. 1b were evaluated by SEM as summarized in Fig. 1d. The average size of the self-assembled Au nanoparticles was approximately 17 nm with a standard deviation value of ~7.6. It was reported that the size of Au nanoparticles was 32 nm under the close experimental parameters to the current study [22]. We achieved finer and more homogeneous size of Au nanoparticles in the current case. The areal density of Au nanoparticles calculated from SEM image was ~$1.2 \times 10^{12}$ particles/in., which is

**Fig. 1** Top view SEM images of **a** initial Au film and **b** self-assembled Au nanoparticles. The sputtering time for both samples was 1 min. **c** XRD diagram of **b**. **d** Histogram of Au nanoparticle size in **b**

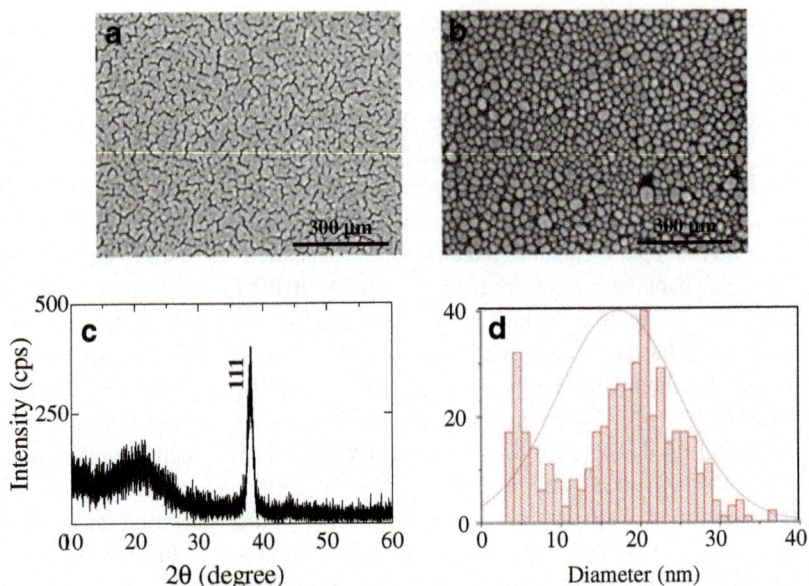

close to the density required for the future ultrahigh density magnetic recording media.

Figure 2a–h show cross-sectional TEM images of the Au nanoparticles (Fig. 2a and b), $SrFe_{12}O_{19}$ (Fig. 2e and f), and $SrFe_{12}O_{19}/Au$ (Fig. 2c, d, g, and h). Figure 2b, d, f, and h are high magnification images of Fig. 2a, c, e, and g, respectively. Completely isolated, island-like elliptical Au nanoparticles were obtained (Fig. 2a and b). The height of the nanodots was ~10 nm, which is the same as the thickness of the original Au film (Fig. 1a) as revealed in the TEM cross-sectional observations (data not shown). For the $SrFe_{12}O_{19}/Au$ sample (Fig. 2c and d), lattice fringes on the Au nanoparticles were observed (Fig. 2d). It is reasonable to consider that the lattice fringes originate from crystalline $SrFe_{12}O_{19}$, such that the crystallization of $SrFe_{12}O_{19}$ would be accelerated on the Au nanoparticles. Figure 2e and f show $SrFe_{12}O_{19}$ prepared directly on the substrate with subsequent post-annealing at 750 °C for 1 h in air without the Au underlayer. Well-oriented lattice fringes of $SrFe_{12}O_{19}$ are observed, but note the creation of a ~7.2-nm thick

amorphous layer between the crystalline $SrFe_{12}O_{19}$ and $SiO_2$ layers indicated by the arrows (Fig. 2f). This amorphous phase may be due to an interdiffusion layer consisting of $SrFe_{12}O_{19}$ and the substrate—namely, a mixture of Sr, Fe, Si, and O. Figure 2g and h show $SrFe_{12}O_{19}/Au$ after post annealing of the samples shown in Fig. 2c and d, respectively, at 750 °C for 1 h in air. Similar to the case of Fig. 2d, the lattice fringes of $SrFe_{12}O_{19}$ were observed on the Au nanoparticle (Fig. 2h).

To gain detailed information on the creation of the crystalline $SrFe_{12}O_{19}$ nanodots using the present fabrication technique, Fourier transformations of the obtained TEM image (Fig. 3a) for the $SrFe_{12}O_{19}/Au$ sample were carried out (Fig. 3b–g). The sample was similar to those of Fig. 2c and d. The transformed areas are indicated as rectangles in Fig. 3a. No diffraction spots were observed in Fig. 3b, d, and g, whereas spots were seen in Fig. 3c and f. These Fourier transformed spots suggest the existence of crystalline $SrFe_{12}O_{19}$. By comparing Fig. 3a, c, and f, the lattice fringes of $SrFe_{12}O_{19}$ were determined to be created on the

**Fig. 2** Cross-sectional TEM images for **a** Au nanoparticles corresponding to those in Fig. 1b and **b** enlarged image. **c** and **d** $SrFe_{12}O_{19}$ deposited on the Au nanoparticles at $T_S$=475 °C. **e** $SrFe_{12}O_{19}$ prepared directly on the substrate at $T_S$=475 °C with post annealing at 750 °C for 1 h in air. **f** Magnified image of **e**. **g** and **h** Effects of post annealing on samples **c** and **d** at 750 °C for 1 h in air

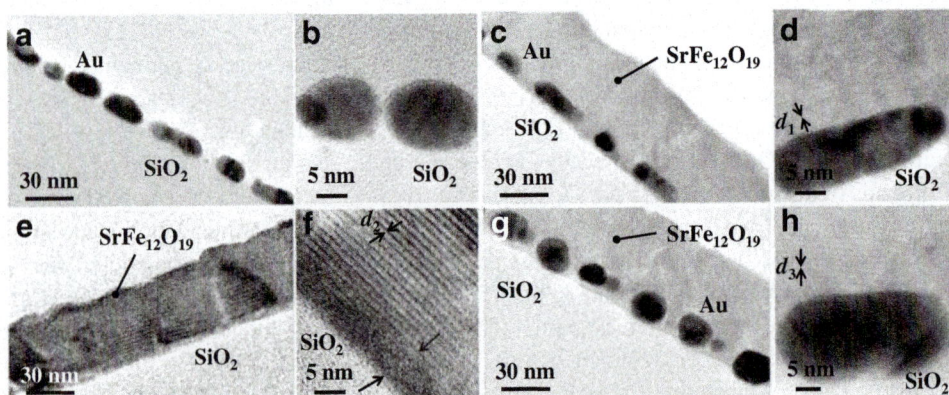

**Fig. 3** **a** Cross-sectional TEM and **b**–**g** local Fourier transformation images of $SrFe_{12}O_{19}$ on Au nanoparticles prepared at $T_S=475$ °C. **b, d,** and **g** Transformed images of $SrFe_{12}O_{19}$ at the interspaces of the Au nanoparticles. **c** and **f** Fourier transformed images of $SrFe_{12}O_{19}$ on the Au nanoparticles. **e** Transformed image of the $SiO_2$/Si substrate. Transformed areas are indicated as *rectangles*

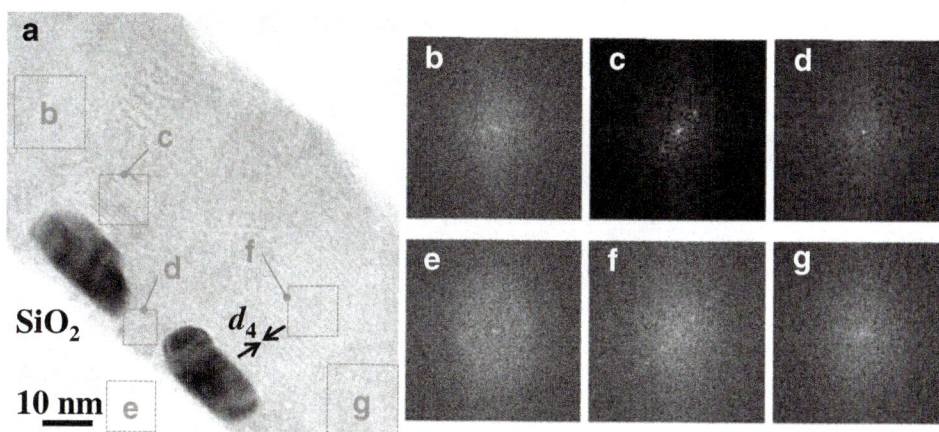

Au nanoparticles. On the other hand, the amorphous Sr–Fe–O seen in Fig. 3b, d, and g corresponds to the interspaces of the Au nanoparticles, i.e., the $SiO_2$ substrate portions. From these results, it was concluded that the crystalline $SrFe_{12}O_{19}$ formed preferentially on the Au nanoparticles rather than on the $SiO_2$ substrate. Promotion of $SrFe_{12}O_{19}$ crystallization on the Au nanoparticles is probably due to the relatively lower crystallization temperature of $SrFe_{12}O_{19}$ on Au than that on $SiO_2$. The coexistence of the crystalline $SrFe_{12}O_{19}$ and amorphous Sr–Fe–O sections in the same $SrFe_{12}O_{19}$/Au phase was then experimentally clarified. Figure 3e shows the Fourier-transformed image of the $SiO_2$ substrate area. Because $SiO_2$ is amorphous, no diffraction spots were observed, and the result is thus similar to those seen in Fig. 3b, d, and g.

Notably, the local nanostructures discussed above were well reflected in the macroscopic crystal structures of $SrFe_{12}O_{19}$. Single-phase $SrFe_{12}O_{19}$ can be seen in Fig. 4a, which shows the diffraction lines from the (00$l$) planes except the (107) plane, which is the strongest in the $SrFe_{12}O_{19}$ powder system. In Fig. 4b, the preferential (00$l$) orientation of $SrFe_{12}O_{19}$ coincident with the Au (111) orientation was indexed. The XRD measurements indicated that the Au nanoparticle underlayers enhanced the (00$l$) orientation of $SrFe_{12}O_{19}$. Taking into account the present XRD results, it is reasonable to denote the spacing of the

lattice planes ($d_1$–$d_4$) indicated in Figs. 2d ($d_1$), 2f ($d_2$), 2h ($d_3$), and 3a ($d_4$) as $SrFe_{12}O_{19}$ (00$l$) planes and/or the (107) plane. We estimated the $d_1$–$d_4$ values from TEM images, and the representative $SrFe_{12}O_{19}$ plane for each $d$ value was determined. The experimentally obtained $d$ values ($d_1$–$d_4$) and those reported in the JCPDS card (No. 33–1340) were then compared. The differences ($\Delta$) between the experimentally determined $d$ values and those of the JCPDS card were evaluated as $(d_{experiment}-d_{JCPDS})/d_{JCPDS}\times100$. The results are tabulated (Table 1). For the $SrFe_{12}O_{19}$/Au samples ($d_1$, $d_3$, and $d_4$), the (004) and (008) planes were indexed, suggesting that the Au underlayers promoted the (00$l$) orientation of $SrFe_{12}O_{19}$, which is essential for the perpendicular magnetic anisotropy of $SrFe_{12}O_{19}$. Meanwhile, the (107) plane was identified from the $d_2$ value for the $SrFe_{12}O_{19}$ sample without the Au underlayer. The $SrFe_{12}O_{19}$ (107) component deteriorates the perpendicular magnetic anisotropy of the overall $SrFe_{12}O_{19}$.

Figure 5a and d, respectively, show the surface morphologies of the $SrFe_{12}O_{19}$ and $SrFe_{12}O_{19}$/Au samples. The samples seen in Fig. 5a–c and d–f are the same as those shown in Figs. 2e, f, and 4a and Figs. 2g, h, and 4b, respectively. For the $SrFe_{12}O_{19}$ monolayer, the surface roughness is large and exhibits abnormal growth of grains, which are indicated by the highlighted contrasts in the image. In the meantime, Fig. 5d shows the self-assembled $SrFe_{12}O_{19}$ nanodot structures. The estimated size of the $SrFe_{12}O_{19}$ nanodots was found to be 40–50 nm. The

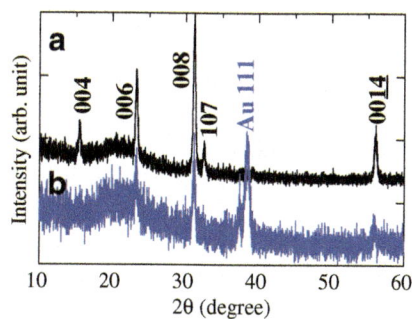

**Fig. 4** XRD patterns of **a** $SrFe_{12}O_{19}$ and **b** $SrFe_{12}O_{19}$/Au samples. Both samples are the same as those shown in Fig. 2e and f and Fig. 2g and h, respectively

**Table 1** Calculated $d$ values based on the TEM observations and corresponding $SrFe_{12}O_{19}$ crystallographic planes. Reported $d$ values from the JCPDS card are also listed. $\Delta$ represents the difference in the experimentally determined and JCPDS $d$ values

|  | $d_{experiment}$ (Å) | Possible plane | $d_{JCPDS}$ (Å) | $\Delta$ (%) |
|---|---|---|---|---|
| $d_1$ | 2.857 | (008) | 2.878 | −0.73 |
| $d_2$ | 2.681 | (107) | 2.765 | −3.04 |
| $d_3$ | 5.739 | (004) | 5.753 | −0.24 |
| $d_4$ | 5.499 | (004) | 5.753 | −4.42 |

**Fig. 5** Nanostructures and magnetic properties of the $SrFe_{12}O_{19}$ (**a**–**c**) and $SrFe_{12}O_{19}/$Au (**d**–**f**) samples shown in Fig. 2e and f and Fig. 2g and h, respectively. **a** and **d** are AFM images, while **b** and **e** are the corresponding MFM images. The magnetic hysteresis loops of both samples are shown in **c** and **f**. Two different directions of magnetic field were applied to the samples, i.e., perpendicular ($\perp$) to the sample surface and parallel (//) in relation to the sample surface, respectively

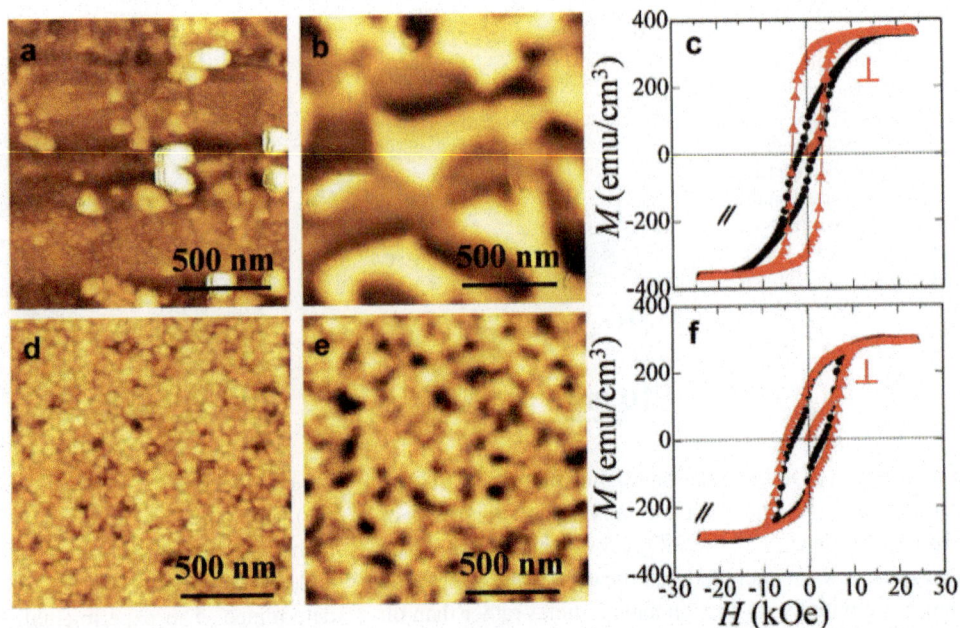

downsizing of the $SrFe_{12}O_{19}$ film to form the self-assembled structures was demonstrated by using the self-assembled Au nanoparticle underlayers. The Au nanoparticles are presumed to hinder the continuous growth of $SrFe_{12}O_{19}$ grains. Figure 5b and e show the magnetic domain structures of both samples (the observation areas for Fig. 5b and e are the same as those in Fig. 5a and d, respectively). The bright and dark contrasts indicate the direction of the perpendicular magnetization components of the magnetic domains. That is, perpendicularly upward or downward. Microscale magnetic domains with various shapes were found in $SrFe_{12}O_{19}$ (Fig. 5b), while nanoscale magnetic domains were seen in $SrFe_{12}O_{19}/$Au (Fig. 5e). By comparing the AFM and MFM images, a single $SrFe_{12}O_{19}$ nanomagnetic domain (Fig. 5e) was determined to be slightly larger than a single $SrFe_{12}O_{19}$ nanodot (Fig. 5d). However, it is possible that the resolution of the MFM image is lower than that of the AFM image, because the MFM measurements detect the leakage magnetic fields from the sample surface. Thus, equivalent discussions should not be made regarding the sizes of the magnetic domains and nanodots. The downsizing of the magnetic domains to the nanoscale was actualized through the downsizing of the $SrFe_{12}O_{19}$ grains.

The magnetic hysteresis loops of both samples were acquired at room temperature (Fig. 5c and f). Judging from the magnetic behavior, it was concluded that the $SrFe_{12}O_{19}$ monolayer exhibits perpendicular magnetic anisotropy. The (00*l*) crystallographic orientation of $SrFe_{12}O_{19}$ is essential for perpendicular magnetic anisotropy. In the case of $SrFe_{12}O_{19}/$Au sample, however, isotropic magnetic behavior contributed from out-of-plane and in-plane magnetization components was observed. This result suggests an imperfection in the (00*l*) orientation of $SrFe_{12}O_{19}$ in

$SrFe_{12}O_{19}/$Au, which could be attributed to the fluctuation of Au orientation. From Fig. 1c, there might be a trace of Au (200) in the vicinity of $2\theta=44°$, while the existence of Au (200) was not clear from Fig. 4b. The magnetic characteristics for $SrFe_{12}O_{19}$ and $SrFe_{12}O_{19}/$Au did not agree with the results of the TEM and XRD studies. This issue is still an open question in the present study. The saturation magnetization ($M_S$) value of the $SrFe_{12}O_{19}$ sample was 360 emu/cm³, which is comparable to that for bulk $SrFe_{12}O_{19}$ [25]. For the $SrFe_{12}O_{19}/$Au sample, the $M_S$ value was smaller than that of the $SrFe_{12}O_{19}$ at 290 emu/cm³. The reduction of the $M_S$ value for $SrFe_{12}O_{19}/$Au is believed to be reasonable, because the volume fraction of crystalline $SrFe_{12}O_{19}$ in the $SrFe_{12}O_{19}/$Au sample is smaller than that in the $SrFe_{12}O_{19}$ monolayer due to the coexistence of crystalline $SrFe_{12}O_{19}$ and amorphous Sr–Fe–O. Nanofabricated magnetic materials usually show a suppression of magnetic properties attributed to the thermal fluctuations of the small magnetic crystallites; however, serious deterioration of the magnetic characteristics of the $SrFe_{12}O_{19}$ nanodots in $SrFe_{12}O_{19}/$Au was suppressed exhibiting a tolerable magnitude of magnetization, even after nanostructuring of $SrFe_{12}O_{19}$. Furthermore, the magnetic reversal behaviors are discussed herein. An abrupt magnetic reversal was observed in the $SrFe_{12}O_{19}$ sample with a perpendicular coercivity value of 3.4 kOe (Fig. 5c), while the hysteresis loops exhibited gradual slope in the $SrFe_{12}O_{19}/$Au sample (Fig. 5f). The perpendicular coercivity of $SrFe_{12}O_{19}/$Au was 4.7 kOe. We consider that the gradual magnetic switching and increase in the coercivity of $SrFe_{12}O_{19}/$Au sample can be attributed to two reasons; (1) the coexistence of magnetic $SrFe_{12}O_{19}$ and nonmagnetic Sr–Fe–O sections in the same phase and (2) variations in the magnetic anisotropy direction (*c*-axis direction) of $SrFe_{12}O_{19}$

grains. Because the isotropic magnetic hysteresis loops indicated the existence of various anisotropy directions of $SrFe_{12}O_{19}$. In the case of (1), the strength of the intergranular interaction between the magnetic $SrFe_{12}O_{19}$ grains is weaker than that of $SrFe_{12}O_{19}$ monolayer, because the magnetic $SrFe_{12}O_{19}$ grains are separated by the nonmagnetic Sr–Fe–O amorphous sections (Fig. 3a–g). As a result, each $SrFe_{12}O_{19}$ grain independently behaves as an isolated grain. The coercivity value could be different from grain to grain, such that the magnetization of each $SrFe_{12}O_{19}$ grain reverses at various strengths of magnetic field. Therefore, the overall magnetic reversal occurred gradually and overall perpendicular coercivity increased. Concerning (2), the magnetizations of $SrFe_{12}O_{19}$ grains reverse in a wide range of magnetic fields due to the fluctuations of magnetic anisotropy direction. Based on the considerations (1) and (2), the gradual magnetic switching behaviors and increase in the perpendicular coercivity were generated.

The next subject in our study is more precise control of $SrFe_{12}O_{19}$ nanodot intervals. It is essential to control the order of the self-assembled Au nanoparticles used as the underlayer. On the basis of our primary experiments, RF magnetron sputtering with in situ heating of the substrate at a moderate temperature was effective to improve the periodicities of the Au nanoparticles. The size of the Au nanoparticles was also markedly reduced to as small as $\leq 4$ nm. Such Au nanoparticle underlayers will enable the formation of precisely ordered $SrFe_{12}O_{19}$ nanodots at the $\sim 10$-nm level.

## Conclusions

Spontaneously organized $SrFe_{12}O_{19}$ nanodot structures with a size of 40–50 nm were obtained via plasma-aided nanofabrication. In the present study, the self-assembled Au nanoparticles with the areal density as high as $\sim 1.2 \times 10^{12}$ particles/in. were utilized as underlayers for $SrFe_{12}O_{19}$ and played the role of a template. The site-preferential crystallization of $SrFe_{12}O_{19}$ on the Au nanoparticles was experimentally confirmed. The downsizing of the magnetic domains to the nanometer level was evidenced along with the downsizing of the $SrFe_{12}O_{19}$ grains, preserving the magnetic properties by using the Au nanoparticle underlayers. The effects of the Au nanoparticle underlayers can be summarized as (1) acceleration of the crystallization of $SrFe_{12}O_{19}$, (2) nanoscaling of $SrFe_{12}O_{19}$, and (3) a tendency to improve the (00$l$) orientation of $SrFe_{12}O_{19}$, although this effect cannot be perfectly concluded.

The current fabricated area of the $SrFe_{12}O_{19}$ nanodots is $\sim 8.5$ cm$^2$, which is much larger than that obtained using conventional lithography. The self-assembled $SrFe_{12}O_{19}$ nanostructures were obtained using a simple sputtering method, such that the present technique has significant potential for the manufacture of magnetic nanodots. In the future, we expect to realize a patterned magnetic recording $SrFe_{12}O_{19}$ medium consisting of coexisting nanocrystalline $SrFe_{12}O_{19}$ (magnetic) regions and nanoamorphous Sr–Fe–O (nonmagnetic) regions in the same phase at regular nanointervals.

**Acknowledgment** This work was financially supported by a Grant-in-Aid for Young Scientists (B) No. 23760280 from the Japan Society for the Promotion of Science (JSPS).

## References

1. Perigo EA, Silva SC, de Sousa EMB, Freitas AA, Cohen R, Nagamine LCCM, Takiishi H, Landgraf FJG (2012) Properties of nanoparticles prepared from NdFeB-based compound for magnetic hyperthermia application. Nanotechnol 23:175704/1–175704/10
2. Anirudhan TS, Sandeep S (2012) Synthesis, characterization, cellular uptake and cytotoxicity of a multi-functional magnetic nanocomposite for the targeted delivery and controlled release of doxorubicin to cancer cells. J Mater Chem 22:12888–12899
3. Mink JE, Rojas JP, Logan BE, Hussain MM (2012) Vertically grown multiwalled carbon nanotube anode and nickel silicide integrated high performance microsized (1.25 μL) microbial fuel cell. Nano Lett 12:791–795
4. Asghar G, Nasir S, Awan MS, Tariq GH, Anis-ur-Rehman M (2012) Anomalous behavior of chemically synthesized magnetoplumbite strontium ferrite nano particles. Key Eng Mater 510–511:330–334
5. Lambert SE, Sanders IL, Patlach AM, Krounbi MT, Hetzler SR (1991) Beyond discrete tracks: other aspects of patterned media. J Appl Phys 69:4724–4726
6. Hellwig O, Bosworth JK, Dobisz E, Kercher D, Hauet T, Zeltzer G, Risner-Jamtgaard JD, Yaney D, Ruiz R (2010) Bit patterned media based on block copolymer directed assembly with narrow magnetic switching field distribution. Appl Phys Lett 96:052511/1–052511/3
7. Grobis MK, Hellwig O, Hauet T, Dobisz E, Albrecht TR (2011) High-density bit patterned media: magnetic design and recording performances. IEEE Trans Magn 47:6–10
8. Chikazumi S, Ohta K, Adachi K, Tsuya N, Ishikawa Y (2006) Handbook for magnetic materials. Asakurashoten, Tokyo (in Japanese)
9. Dong Q, Li G, Ho CL, Faisal M, Leung CW, Pong PWT, Liu K, Tang BZ, Manners I, Wong WY (2012) A polyferroplatinyne precursor for the rapid fabrication of $L1_0$-FePt-type bit patterned media by nanoimprint lithography. Adv Mater 24:1034–1040
10. Schmid GM, Miller M, Brooks C, Khusnatdinov N, La Brake D, Resnick DJ, Sreenivasan SV, Gauzner G, Lee K, Kuo D, Weller D, Yang XM (2009) Step and flash imprint lithography for manufacturing patterned media. J Vac Sci Technol B 27:573–580
11. Yang XM, Xu Y, Seiler C, Wan L, Xiao S (2008) Toward 1 Tdot/in.$^2$ nanoimprint lithography for magnetic patterned media: opportunities and challenges. J Vac Sci Technol B 26:2604–2610
12. Chappert C, Bernas H, Ferré J, Kottler V, Jamet JP, Chen Y, Cambril E, Devolder T, Rousseaux F, Mathet V, Launois H (1998) Planar patterned magnetic media obtained by ion irradiation. Science 280:1919–1922

13. Suharyadi E, Oshima D, Kato T, Iwata S (2011) Switching field distribution of planar-patterned CrPt$_3$ nanodots fabricated by ion irradiation. J Appl Phys 109:07B771/1–07B771/3

14. Rettner CT, Best ME, Terris BD (2001) Patterning of granular magnetic media with a focused ion beam to produce single-domain islands at >140 Gbit/in$^2$. IEEE Trans Magn 37:1649–1651

15. Adam JP, Jamet JP, Ferre J, Mougin A, Rohart S, Weil R, Bourhis E, Gierak J (2010) Magnetization reversal in Pt/Co(0.5 nm)/Pt nano-platelets patterned by focused ion beam lithography. Nanotechnol 21:445302/1–445302/5

16. Yang JKW, Chen Y, Huang T, Duan H, Thiyagarajah N, Hui HK, Leong SH, Ng V (2011) Fabrication and characterization of bit-patterned media beyond 1.5 Tbit/in$^2$. Nanotechnol 22:385301/1–385301/6

17. Sohn JS, Lee D, Cho E, Kim HS, Lee BK, Lee MB, Suh SJ (2009) The fabrication of Co–Pt electro-deposited bit patterned media with nanoimprint lithography. Nanotechnol 20:025302/1–025302/5

18. Zhou Q, Heard PJ, Schwarzacher W (2011) Fabrication and magnetic properties of patterned NiFeMo films electrodeposited in self-assembled nanosphere templates. J Appl Phys 109:054313/1–054313/4

19. Albrecht M, Makarov D (2012) Magnetic films on nanoparticle arrays. Open Surf Sci J 4:42–54

20. Kaewrawang A, Ishida G, Liu X, Morisako A (2008) Epitaxial growth of SrM(00$l$) film on Au(111). IEEE Trans Magn 44:2899–2902

21. Kaewrawang A, Ghasemi A, Liu X, Morisako A (2010) Fabrication, crystallographic and magnetic properties of SrM perpendicular films on Au nanoparticle arrays. J Alloys Compd 492:44–47

22. Kaewrawang A, Ghasemi A, Liu X, Morisako A (2010) Self-assembled strontium ferrite dot array on Au underlayer. J Magn Magn Mater 322:2043–2046

23. Kaewrawang A, Ghasemi A, Liu X, Morisako A (2010) A simple method toward high density SrM dot arrays. J Magn Soc Jpn 34:277–280

24. Yoshida S (1990) Thin films. Baifukan, Tokyo (in Japanese)

25. Ramamurthy Acharya B, Krishnan R, Prasad S, Venkataramani N, Ajan A, Shringi SN (1994) Sputter deposited strontium ferrite films with $c$-axis oriented normal to the film plane. Appl Phys Lett 64:1579–1581

# 21

# In vivo uptake and cellular distribution of gold nanoshells in a preclinical model of xenografted human renal cancer

Mariana Pannerec-Varna · Philippe Ratajczak · Guilhem Bousquet · Irmine Ferreira · Christophe Leboeuf · Raphaël Boisgard · Guillaume Gapihan · Jérôme Verine · Bruno Palpant · Emmanuel Bossy · Eric Doris · Joel Poupon · Emmanuel Fort · Anne Janin

**Abstract** Large-sized gold nanoparticles, promising imaging and therapeutic tools in human cancer, need long-term studies evaluating tissue bio-distribution in blood, organs and tumor. In a preclinical model of mouse xenografted with human renal cancer, we analysed the bio-distribution of a single dose (160 μg/kg) intravenously injected of poly-ethylene glycol (PEG)ylated gold nanoshells (~150 nm), in blood, normal and tumoral tissues. Using inductively coupled plasma mass spectrometry (ICP-MS), dark field and electron microscopy, we performed a sequential study of nanoshell uptake and distribution in the tumor. We also studied microscopically the organs most sensitive to efficient anticancer drugs to detect a possible long-term toxicity. Gold quantities significantly decreased in blood between early and late time points, whereas they significantly increased in liver and spleen. In addition, gold nanoshells did not induce any tissue damage, such as necrosis, inflammatory infiltrate or fibrosis in mouse liver, spleen, kidney or bone marrow after 6 months. In human renal cancer xenografts, ICP-MS showed an early decrease of gold, with 1-week stability before decrease at Day 15. Dark field microscopy showed gold particles within the vessel lumen 5 to 30 min after nanoshell injection, while 24 h later, gold particle distribution was mainly intracellular. Electron microscopy identified nanoshells within blood vessels at 5 and 30 min, within endothelial cells at 3 and 6 h and within cytoplasms of macrophages in the tumoral tissue after 24 h. In conclusion, no toxicity was observed in mice 6 months after administration of PEGylated gold nanoshells and the distribution kinetics progressed from intravascular flow at 30 min to intratumoral cells 24 h later.

M. Pannerec-Varna (✉) · P. Ratajczak · G. Bousquet · I. Ferreira · C. Leboeuf · G. Gapihan · J. Verine · A. Janin (✉)
Université Paris Diderot, Sorbonne Paris Cité, UMR-S 728, 75010 Paris, France
e-mail: mariannavarna@yahoo.fr
e-mail: anne.janin728@gmail.com

M. Pannerec-Varna · P. Ratajczak · G. Bousquet · I. Ferreira · C. Leboeuf · G. Gapihan · J. Verine · A. Janin
INSERM, U728, 75010 Paris, France

R. Boisgard
Commissariat d'Energie Atomique et aux Energies Alternatives, SHFJ, INSERM, U1023, 91401 Orsay, France

B. Palpant
Laboratoire de Photonique Quantique et Moléculaire, Ecole Centrale Paris, Grande Voie des Vignes, 92295 Châtenay-Malabry Cedex, France

E. Bossy · E. Fort
Institut Langevin, ESPCI ParisTech, 75238 Paris, France

E. Doris
Service de Chimie Bioorganique et de Marquage, CEA, iBiTecS, 91191 Gif-sur-Yvette, France

J. Poupon
Laboratoire de Toxicologie biologique, AP-HP-Hôpital Lariboisière, Paris 75010, France

J. Verine · A. Janin
Laboratoire de Pathologie, AP-HP-Hôpital Saint-Louis, Paris 75010, France

**Keywords** Gold nanoshells · Mouse xenograft · Human renal cancer · Sequential study · Tissue biodistribution · Long-term tolerance

# Introduction

Gold nanoparticles are now synthesised into a large variety of forms. Only some of them are suitable for in vivo hyperthermia. A recent comparative study of small-sized gold nanoparticles including nanorods, nanocages and nanohexapods demonstrated the value of nanohexapods for in vivo photothermal destruction using [18]F-fluorodesoxyglucose positron emission tomography/computed tomography [1]. Large-sized gold nanoparticles (NPs) such as nanoshells about 150 nm in diameter are also promising tools for imaging and therapeutic approaches in cancer [2].

In preclinical models, the spherical particles of silica–gold nanoshells have shown optical properties for imaging and photothermal targeted therapy [3, 4]. In addition, a polyethylene glycol (PEG) coating reduces the adsorption of blood serum proteins to the nanoparticles [1]. Liu et al. [5] also demonstrated that PEGylation is a key factor that governs the fate of gold nanoparticles in the animal organisms and their accumulation in target organs. PEGylated gold nanoshells enter more easily into organs and tissues [6]. Encouraging results showed, at the early checkpoint of 7 days after injection, the absence of toxicity of PEGylated small-sized gold nanoparticle in liver spleen, kidney, heart and lung [1].

In the field of cancer, the possibility to link antibodies to nanoshells using PEG enables the use of these biocompatible nanoshells for the delivery of drugs to targeted tumor cells. The goal of such innovative treatment is double to enhance the drug effect on targeted tumor cells and to reduce the toxic effect of the anticancer drug on normal organs. A prerequisite for the use of these biocompatible nanoparticles in oncology daily practice is to check in vivo their tolerance in the long-term and on the organs that are the most sensitive to efficient anticancer drugs, particularly the liver, bone marrow, kidney, heart and lung [7, 8].

To perform this study, we used human tumor xenografts in immunodeficient mice. This in vivo model enables sequential analyses in the different physiological compartments represented by blood, normal and tumoral tissues [9], [10]. In this preclinical model, we studied human renal cancer, since this highly vascularised cancer frequently develops haematogenous metastases and resists to anticancer drugs at non-toxic therapeutic doses [11]. We performed a sequential analysis of uptake and distribution of large-sized PEGylated gold nanoshells and focussed the study on their intracellular distribution in the xenografted human cancers, as well as on their long-term tolerance in normal mouse organs.

# Experimental section

## Nanoparticle characteristics

Gold nanoshells composed of a ~20-nm Au shell around a ~130-nm silica core and PEGylated (MW=5,000) were purchased from Nanospectra Biosciences (Houston, TX, USA). The nanoparticles were characterised using transmission electron microscopy (TEM) and the absorbance spectrum was evaluated with a Genesys 10S UV-VIS spectrophotometer (Thermo Scientifique, France).

Dynamic light scattering (DLS) and zeta potential measurements were carried out with a Nanosizer ZS90 instrument (MALVERN) at 25 °C. Hydrodynamic diameters were obtained using the cumulant method.

PEG density is estimated at ~10–15 pmol/cm$^2$ (about 13, 000 molecules/particle) (Nanospectra Biosciences, TX, USA).

## Human renal cancers and mice

For the xenografts, human renal cancer cell carcinoma samples were taken from surgical pieces for Tumorothèque of Hopital Saint Louis by pathologists, after the tumoral tissue necessary to establish the diagnostic had been taken. Informed consent for the use of the tumor sample for research was obtained for each patient. The study was approved by the University Board Ethics Committee and conducted in accordance with the Declaration of Helsinki.

The animal study was approved by the Ethics Committee for animal experimental studies of the University Institute board. NMRI/nude 7-week-old female mice purchased from Janvier (R. Janvier, France) were maintained in specific pathogen-free animal housing.

Ten cubic millimeters of human renal cancer were grafted subcutaneously on mice, under xylasin (10 mg/kg body weight) and ketamin (100 mg/kg body weight) anaesthesia. The follow-up was performed daily and included a clinical score and assessment of tumor growth. Xenografted tumors were measured in two perpendicular diameters with a calliper every day and calculated as $V = L \times l^2/2$, $L$ being the largest diameter (length), $l$ the smallest (width) [12]. Daily, mice were weighed and assessed for behavioural changes and all data were recorded using the FileMakerPro software.

## Pharmacokinetics in mice

At different time points and on different organs, we quantified the amount of gold and also assessed a possible toxicity in nude mice and in mice xenografted with human renal cancer.

*Gold nanoshells administration*

In normal nude mice, gold nanoshells were intravenously injected with 100 μL NPs ($3 \times 10^9$ NP/mL) ($n = 5$) or with PBS as control ($n = 3$). In nude mice, xenografted with human renal cancer gold nanoshells were intravenously injected with 100 μL NPs ($3 \times 10^9$ NP/mL) ($n = 5$) or with PBS as controls ($n = 3$). After injection, mice were weighed and assessed daily for behavioural changes and all data were registered using the FileMakerPro software.

*Sequential blood and tissue sampling*

For all mice, the whole blood, heart, kidney, lung, liver, spleen, ovary, adrenal gland, lymph node and brain were systematically sampled. For xenografted mice, a sample from the tumor was also taken. The time points for sampling for normal nude mice were 1, 7 and 15 days and 1, 2, 3 and 6 months. The time points for sampling for xenografted nude mice were 5 and 30 min, 2 and 6 h and 1, 7 and 15 days to respect the ethic rules regarding maximal tumor growth. All tissue samples were cut into three parts: one part was snap-frozen for inductively coupled plasma mass spectrometry (ICP-MS) analysis, another part was glutaraldehyde-fixed for electron microscopy and a third was formalin-fixed and paraffin-embedded for dark field microscopy analysis and histological analysis.

*Spectrometric analyses in blood and tissue samples*

To quantify the gold content in blood and tissue samples at the different time points described above, ICP-MS was performed. Fresh samples were weighed. After drying at 80 °C overnight, they were weighed again and digested in nitric acid/HCl (Sigma). Remaining minerals were analysed using ICP-MS (Elan DRCe, Perkin Elmer, Les Ulis, France). Samples were nebulized in an argon plasma (6000–8000 °C). The ions formed by nebulization were extracted from the plasma, introduced into a mass spectrometer and separated according to their mass on charge ratio (m/z). Gold content was measured at mass 197, a level at which no interference occurs. The detection limit was 1 ng/L.

Gold nanoshell distribution and assessment of toxicity in the different mouse organs

In the different tissue samples systematically taken during autopsies of mice performed at different time points from 1 day to 6 months, histological analyses were performed on hematoxilin eosin staining to assess whether tissue damages linked to toxicity could be detected. In the organs from mice euthanised early (between 1 and 15 days), particular attention was paid to the following features: acute hepatotoxicity with hepatocyte necrosis and inflammatory infiltrate in the portal areas; hypoplasia or complete aplasia of the bone marrow; acute nephrotoxicity with endothelial damage, fibrin deposit in the glomerular areas and necrosis of epithelial tubular cells; cardiotoxicity with endocardial damage. In the organs from mice euthanised at late time points (1, 2, 3 and 6 months), particular attention was paid to the following features of long-term toxicity: fibrosis and inflammatory infiltrate in the liver; glomerular sclerosis and interstitial fibrosis with inflammatory infiltrate in the kidney; diffuse interstitial fibrosis and inflammatory infiltrate in the lung; and possible bone marrow fibrosis or brain damage.

Assessement of gold nanoshell distribution in xenografted human renal cancer

The distribution of gold nanoshells in xenografted human renal cancer was assessed using dark field microscopy and TEM to analyse the different types of cells within the tumor.

*Dark- field microscopy*

Five-micromillimeter-thick tissue sections from xenografted human renal cancer were studied using a microscope (Olympus AX, Tokyo, Japan) equipped for both bright field and dark field microscopy. Using bright field, we could focus our analyses on tumoral cells and avoid areas of necrosis. Using dark field on the tumor area identified with bright field microscopy, we could assess the distribution and the relative density of the nanoparticles. Images were captured at ×40 magnification, on the same tumor areas successively analysed using bright field and dark field microscopy.

*Transmission electron microscopy*

After fixation in 2 % glutaraldehyde in caccodylate buffer, the samples of human renal cancers were embedded in Epon resin. Ultrathin sections were observed using a Hitachi HF-2000 transmission electron microscope (Hitachi, Tokyo, Japan). TEM analyses focussed on the presence of gold nanoshells in the different types of cells observed in the human renal cell cancer: tumoral epithelial cells, vascular endothelial, smooth muscle and pericyte cells, cells from the inflammatory infiltrate including macrophages, and fibroblasts and fibrocytes from the conjunctive stroma.

Statistics

All data were expressed as mean results ± SEM (standard error of mean). The paired Student's *t* test was used for statistical analysis. A *p* value of less than 0.05 was taken to indicate statistical significance.

## Results and discussion

We studied here in vivo uptake and cellular distribution of gold nanoshells in a preclinical model of xenografted human renal cancer.

For this study, we chose gold–silica nanoshells with a size between 130 and 150 nm because (i) these large nanoshells are suitable for imaging and hyperthermia in vivo [13] and (ii) gold is well tolerated and does not induce toxicity in human beings after intravenous or intra-articular injection [14].

We studied gold–silica nanoshells coated with PEG because PEG polymers form a hydrophilic layer and sterically block the electrostatic or hydrophobic interactions with opsonins [15]. These plasma proteins are less adsorbed on nanoshell surfaces, reducing their aggregation and their clearance by phagocytic cells. At a temperature of 25 °C, the zeta potential of the solution of PEGylated nanoshells used in our study was −25 mV for pH 5.6. The dynamic light scattering of gold nanoshells showed a good dispersion of gold nanoshells (supplementary Fig. S1A). The hydrodynamic size of PEGylated gold nanoshells in water was 185 nm (PDI 0.104). This was concordant with the result obtained in transmission electron microscopy (supplementary Fig. S1B). To determine the optimal dose of nanoshell to be injected in mice, we performed in vitro cytotoxicity test (MTT) using three types of cell lines, one normal human endothelial cell line (HUVEC) and two human cancer cell lines (CAKI 1, 786-0). We checked these cells lines for 24 h using different concentrations of gold nanoshells: $1.5 \times 10^9$ NP/mL, $3 \times 10^9$ NP/mL, $15 \times 10^9$ NP/mL, $3 \times 10^{10}$ NP/mL. The concentration of $3 \times 10^{10}$ NP/mL induced toxicity on the three cell lines, the concentration of $15 \times 10^9$ NP/mL induced toxicity on the tumor cell but not on the endothelial cell line. The concentration of $3 \times 10^9$ NP/mL was the highest concentration that did not induce any sign of toxicity on any of the three cell lines tested. This dose is in the range of doses previously tested by other teams, i.e between 10 µg/kg [16] and 8,000 µg/kg [17]. These previous studies focussed on early cytotoxic effect while we analysed in vivo both early and late toxic effects in our preclinical model. The late effect was assessed 6 months after injection because 6 months is a time point commonly used in clinics to assess chronic toxicity.

We analysed the uptake of these nanoshells from blood flow using ICP-MS because it is a sensitive method that enables detection of small quantities of gold [18]. As expected, we observed a decrease from Day 1 (381±111 ng/g) to Day 180 (1 ±1 ng/g), a time point when quantities of gold were nearly undetectable (Table 1). The uptake of gold spherical nanoparticles of 100, 200 and 250 nm, has been studied after intravenous injection in mice [19] and in rats [20]. The quantity of gold detected in blood by ICP-MS 24 h after injection was higher for 100-nm than for 250-nm gold nanoparticles in the experiments performed in rats [20] (Table 2), whereas no gold was detected in blood in the experiments performed in mice receiving 100-nm nanoparticles[19].

We systematically studied the biodistribution of PEGylated gold–silica nanoshells in organs of nude mice at different time points: Day 1, 15, 30, 60, 90 and 180. Using ICP-MS, we found a large amount of gold in the liver and spleen 24 h after injection (9,738±816 ng/g and 9,200±730 ng/g, respectively; Table 1). In the liver, the peak accumulation was observed at Day 30 (35,872±2,492 ng/g) with a significantly lower level when compared to Day 180 (3,723±376 ng/g) ($p < 0.005$). In the spleen, we also found a large amount of gold at Day 1 (9,200±730 ng/g). However, in the spleen the ICP-MS performed at different time points showed a progressive accumulation with a significant higher level when gold quantities at Day 1 (9,200±739 ng/g) were compared to gold quantities at Day 180 (33,857±2,564 ng/g) ($p < 0.005$). Dark field analyses performed on liver and spleen tissue sections at 3 and 6 months after nanoparticle injections showed a large accumulation of gold nanoparticles in the two organs.

**Table 1** Gold quantification in organs of normal non-xenografted nude mouse (ng/g of tissue) from Day 1 to Day 180

| Tissues | Day 1 | Day 15 | Day 30 | Day 60 | Day 90 | Day 180 |
|---|---|---|---|---|---|---|
| Blood | 381 (±111) | 197 (±50) | 147 (47) | 99 (±28) | 43 (±12) | 1 (±1) |
| Liver | 9,738 (±816) | 16,615 (±800) | 35,872 (±2,492) | 16,601 (±765) | 8,187 (±534) | 3,723 (±376) |
| Bone marrow | 13 (±8) | 70 (±23) | 57 (±13) | 20 (±5) | 19 (±9) | 19 (±5) |
| Spleen | 9,200 (±730) | 13,511 (±2,360) | 19,398 (±635) | 21,406 (±1,956) | 27,654 (±300) | 33,857 (±2,564) |
| Kidney | 395 (±19) | 309 (±34) | 161 (±45) | 126 (±36) | 29 (±15) | 61 (±20) |
| Lung | 55 (±21) | 43 (±13) | 54 (±12) | 15 (±12) | 21 (±12) | 17 (±12) |
| Adrenal gland | 306 (±12) | 76 (±20) | 53 (±15) | 56 (19) | 21 (±6) | 47 (±6) |
| Lymph node | 135 (±27) | 239 (±73) | 208 (±77) | 255 (±75) | 146 (±66) | 120 (±25) |
| Ovary | 46 (±14) | 32 (±13) | 12 (±6) | 15 (±6) | 10 (±8) | 32 (±27) |
| Heart | 14 (±5) | 17 (±7) | 13 (±5) | 18 (±6) | 15 (±7) | 8 (±3) |
| Brain | 12 (±4) | 5 (±5) | 1 (±1) | 11 (±2) | 5 (±3) | 7 (±4) |

**Table 2** Comparative assessment of early and late toxic effects in our study and eight previously published studies

| | Nanoparticles | | Administration | | | | Protocol | | | Toxic effect | |
| --- | Ref | Size (nm) | PEG | IV/IP | Dose | Injection number | Animal | Methods of analysis | Time of analysis | Early | Late (73 months) |
| --- | --- | --- | --- | --- | --- | --- | --- | --- | --- | --- | --- |
| | Our study | 130–150 | Yes | IV | 160 µg/kg | 1 | Mouse | ICP-MS, H&E, TEM | Min 5, 30; Day 1, 7, 15, Months 1, 2, 3, 6 | No | No |
| | [20] | 10, 50, 100, 250 | No | IV | 80–120 µg/g | 1 | Rat | ICP-MS | Day 1 | No | No |
| | [3] | 150 | Yes | IV | 2,550–3,775 µg/g | 1, 3, 5 | Mouse | NAA | Day 1, 3, 5 | No | – |
| | [6] | 155 | Yes | IV | 2,340 µg/kg | 1 | Mouse | H&E, NAA | Day 1, 7, 28, 56, 182, 404 | No | No |
| | [9] | 10, 50, 100, 200 | No | IV | 1,000 µg/kg | 1 | Mouse | ICP-MS | Day 1 | No | – |
| | [29] | 40 | No | IV | 1,400–1,600 µg/kg | 1 | Mouse | ICP-MS, AMG stain | Day 1; Months 1, 3, 6 | No | No |
| | [17] | 3, 5, 8, 12, 17, 37, 50 | No | IP | 8,000 µg/kg | 1 | Mouse | H&E | Day 21 | Yes (liver, lung) | – |
| | [30] | 5, 25 | No | IV | 1,000 µg/kg | 1 | Rabbit | ICP-MS, TEM | Day 1 | No | – |
| | [16] | 20 | No | IV | 10 µg/kg | 1 | Rat | ICP-MS | Days 1, 7; Months 1, 2 | – | – |
| | [18] | 12, 5 | No | IP | 320, 1,600, 3,200 µg/kg | 8 | Mouse | ICP-MS, GF-AAS | Day 8 | No | – |

PEG poly-etylene glycol, IV intravenous, IP intraperitoneal, min minutes, M month, ICP-MS inductively coupled plasma-mass specrometry, H&E hematoxylin and eosin staining, NAA neutron activation analysis, TEM transmission eletron microscopy, GF-AAS graphite furnace atomic absorbtion, IHC immunohistochemistry

Comparison of the density of the nanoparticles at the two time points showed an increase in the spleen and a decrease in the liver, thus confirming the results obtained with ICP-MS (Fig. 1). The large accumulation of gold nanoparticles in the liver and spleen, also observed in mice for different size of spherical gold nanoparticles [20], could be linked to the fact that the capillaries in the spleen and liver are discontinuous [2] and lined by cells with phagocytic capacity. Phagocytic cells, particularly macrophages, are more numerous in the spleen than in the liver [21] and this fact could explain the discrepancy between the progressive diminution of gold nanoparticles in the liver and the progressive accumulation in the spleen. In addition, experimental data on mouse Kupffer cells, i.e. liver phagocytic cells, following intravenous injection of bacteriophage [22] demonstrated that after initial phagocytosis, there was an immediate and rapid decrease in the number of plaque-forming units (PFU) which could be recovered from both liver and spleen; however, as soon as 3 days after injection, the number of PFU which could be recovered from the spleen was greater than that found in the liver. At Day 5, there was as many as 50 times more PFU in the spleen than in the liver. In the present study, we assessed a concentration of gold nanoparticles ten times greater in the spleen than in the liver 6 months after injection.

We performed systematic pathological analyses to assess signs of toxicity at the different time points when the different organs of nude mice were analysed: Days 1, 15, 30, 60 and 180. We did not detect any signs of acute toxicity such as hypoplasia or complete aplasia of the bone marrow; acute nephrotoxicity with endothelial damage, fibrin deposit in the glomerular areas and necrosis of epithelial tubular cells; or cardiotoxicity with endocardial damage.

Regarding long-term toxicity, the greatest concentration of PEGylated gold silica nanoshells was found in the liver and spleen. Therefore, after dark field analysis, the same tissue sections from the liver and spleen 3 and 6 months after gold nanoshell injection were analysed using bright field microscopy to detect signs of toxicity. As shown in Fig. 1, no necrotic cell, no fibrosis and no inflammatory infiltrate were found. This is in accordance with the absence of any abnormalities in the clinical score used for the mouse follow-up. No similar study on long-term organ toxicity has been performed, as far as we know, in preclinical models.

In our study, we injected intravenously a single dose of 160 µg/kg gold nanoshells and we did not observe any toxicity either at early or long term. In three other studies using intravenous injection of gold nanoshells with a diameter larger than 100 nm, no toxicity was observed [13, 14], even in cases of repeated injections [3]. An early toxic effect on the liver and the lung was observed in only one mouse series, for gold particles of 8 to 37 nm of diameter injected intraperitoneally. As far as we know, only one previously published study focussed on long-term effect at 6 months [18]. It was different from our study by

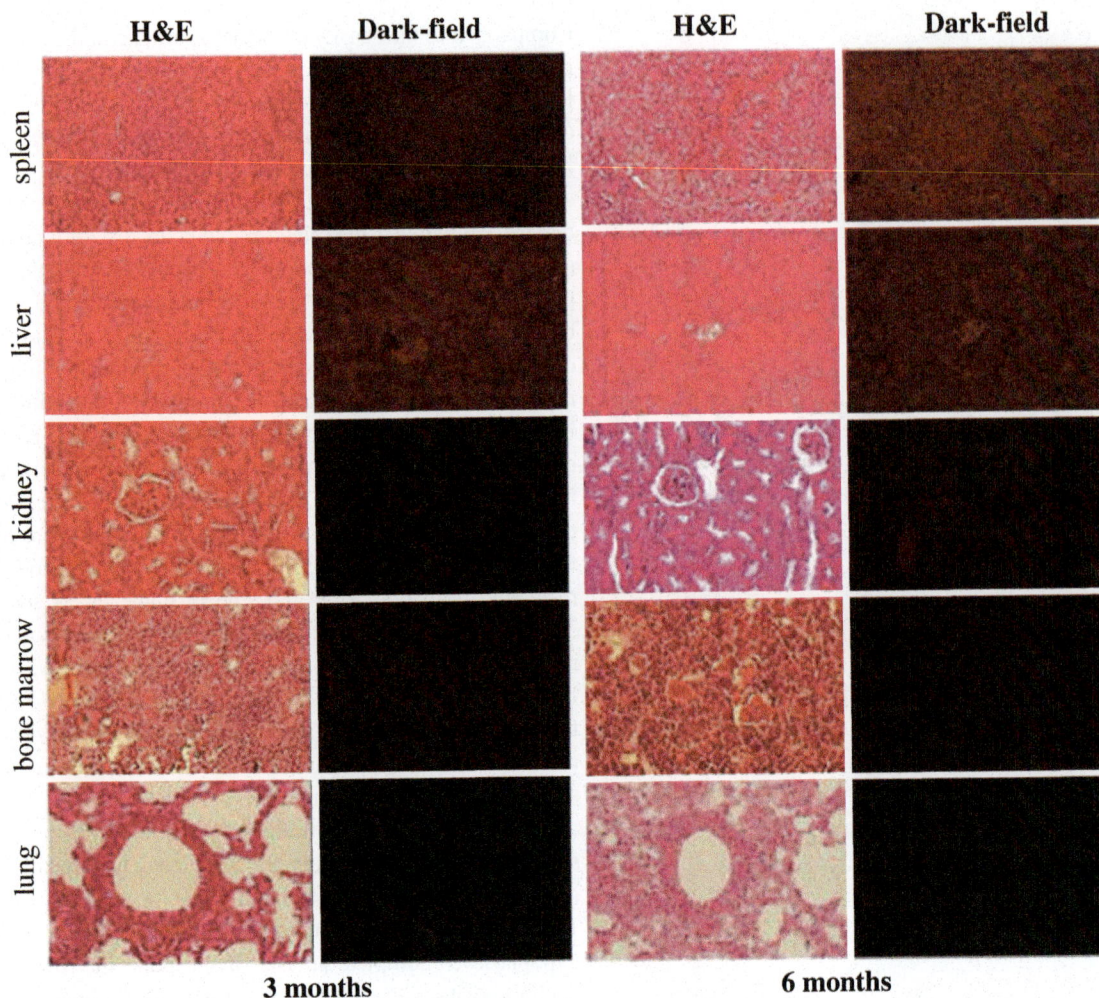

**Fig. 1** Long-term toxicity study. Data obtained at late time points in normal non-xenografted nude mouse. Systematic histological analysis (H&E staining), 3 and 6 months after injection, in the spleen, liver, bone marrow and lung, did not show any tissue damage linked to toxicity, such as necrosis, fibrin deposit in vascular areas, inflammatory infiltrate, fibrosis or bone marrow hypoplasia or aplasia. Dark field microscopy analysis on the same areas, at the same time points, shows a high intensity signal in the liver and spleen at 3 months followed at 6 months by a decrease in the liver and an increase in the spleen. A very weak signal is detected in the kidney at 3 and 6 months and no signal is observed in the bone marrow and lung at the same time points. Magnification ×400

the dose injected (1,400 vs 160 μg/kg) and the size of the nanoparticles (40 nm vs 150 nm) but no long-term toxicity was found, as in our study.

Apart from the liver and spleen, in the other organs analysed at different time points in normal nude mice, results of ICP-MS are also confirmed by dark field analyses, but the quantities of gold were much smaller than in the liver and spleen. The kidney is the organ where we detected the less low quantity of gold, with a decrease between Day 1 (395±19 ng/g) and Day 180 (61±20 ng/g). These data show that gold nanoparticles moved to the kidney, as expected, since the kidney physiological function is filtering the entire blood flow through the fenestrated endothelium of glomerular capillaries [23]. However, the urinary excretion was low and this was probably linked to the 5.5-nm size of the capillary endothelium pores [18, 24], far smaller than the 150-nm diameter of the gold silica nanoshells we had injected.

**Fig. 2** Uptake of gold particles. Data obtained at early time points in ▶ nude mouse xenografted with human renal cancer. **a** Induced coupled plasma-mass spectrometry (ICP-MS) for gold quantification (logarithmic representation) in the blood, liver, spleen, bone marrow, kidney and tumor at 5 and 30 min, 2 and 6 h and Day 1, 7 and 15. A significant decrease is observed in the blood, bone marrow and tumor when gold quantities are compared at 5 and 30 min, reflecting the rapid uptake of gold particles in highly vascularized areas, whereas an increase is observed in the liver and spleen when gold quantities are compared at 5 min and 15 days, reflecting the progressive uptake and stock of gold nanoparticles in these organs rich in phagocytic cells. **b** Distribution of gold nanoparticles in the xenografted human renal cancer at the same early time points. Dark field analysis shows a preferential distribution of gold nanoparticles in capillary wall areas at 2 h and in tumor cells at Day 1. Transmission electron microscopy shows an uptake of gold nanoparticles by endothelial cells at 6 h with a distribution in both endothelial and tumoral cells at Day 1, followed by an uptake and stock in phagocytic cell cytoplasm at Day 7. *$p < 0.05$

**a**

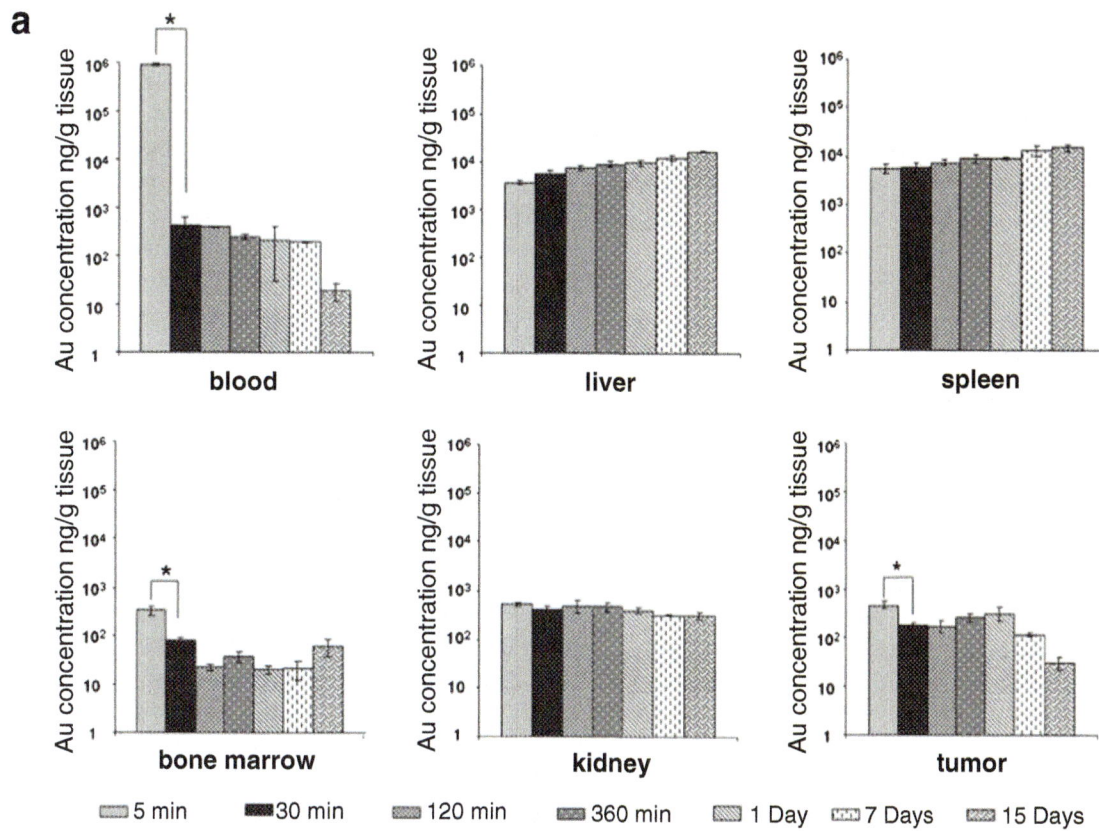

**b**

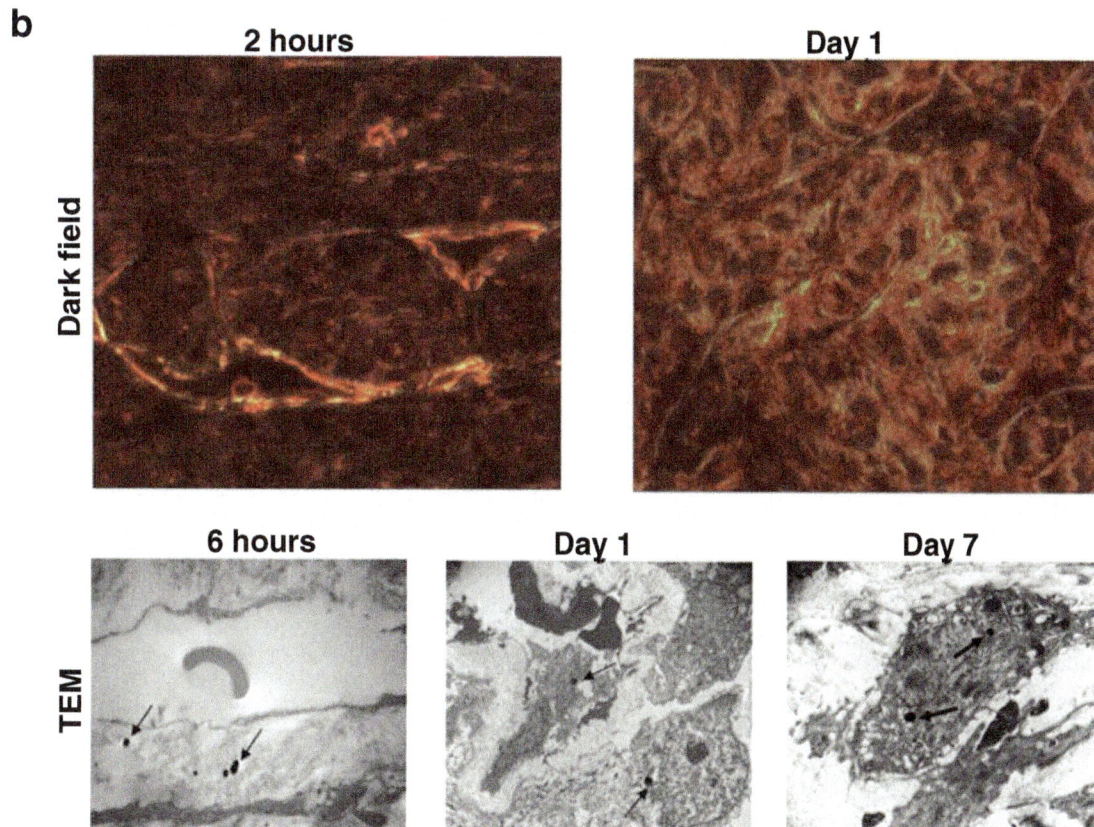

In the bone marrow (13±8 ng/g), lung (55±21 ng/g), adrenal gland (306±12 ng/g), lymph node (135±27 ng/g) and ovary (46 ±14 ng/g), very small gold quantities were detected at Day 1 and these quantities decreased over time (Table 1). In the brain, only traces of gold were found. This is in accordance with the clinical follow-up over 6 months since the mice did not show any behavioural abnormality, weight loss or mortality.

Regarding toxicity in the different organs studied, apart from the liver and spleen, pathological analyses of tissue sections of different time points did not show any signs of thrombosis, fibrin deposits or inflammatory infiltrate in the glomeruli at early time points and no glomerular sclerosis, interstitial fibrosis or inflammatory infiltrate in the kidney at late time points. There was no sign of hypoplasia or complete aplasia of the bone marrow. No interstitial fibrosis or inflammatory infiltrate was detected in the lung or heart, whether at early or late time points (Fig. 1). These data, provided by our preclinical model, are important because the treatment of cancer associating radiotherapy and poly-chemotherapy can induce vascular and interstitial damages in the kidney and heart [25], and life-threatening damages such as diffuse pulmonary fibrosis, and bone marrow aplasia. These severe complications linked to radiotherapy and poly-chemotherapy are systematically sought during follow-up of cancer patients under treatment and it is important to demonstrate that the uptake and the accumulation of large spherical gold silica nanoshells do not, on its own, induce any systemic organ damage.

This absence of toxicity, at early and late time points, on the organs most sensitive to efficient anticancer drugs enables further applications of gold nanoshells for local delivery of drugs on cells targeted through specific antibodies [26].

The study on xenografted mice was focussed on the tumor xenograft, but we also systematically studied the organs at different time points: 5 and 30 min, 2 and 6 h and 1, 7, 15 days (Fig. 2a).

When we compared the gold quantities detected by ICP-MS and gold nanoshell distribution in tissues using dark field analysis, we did not find any significant difference in the different organs of normal nude mice and of the xenografted nude mice. This implies that the engraftment of the human renal cancer did not change significantly the systemic distribution of large gold nanoshells after intravenous injection.

Regarding the xenografted human renal cancer, we successively studied the uptake of gold nanoshells from the blood and their distribution in the different cellular components of the tumor: tumor cells, microvessels, phagocytic interstitial cells.

When we compared the gold quantities in blood and xenograft at different time points, we found that in the two types of samples the larger amounts were found at the first time points (5 min), with a decrease at the successive time points, and significantly smaller gold quantities at Day 15 compared to the 5-min time point. These kinetic particularities were not found in the liver, spleen, bone marrow or kidney (Fig. 2a). These kinetic data in the blood and xenografted tumor could be linked to the fact that human renal cancer is a highly vascularised malignant tumor [11], and that the microvessels of malignant tumors are dystrophic with possible absence of pericytes, irregular thickness or absence of continuity of basement membrane [27]. In addition, experimental data in mice have demonstrated that gold nanoparticles are able to pass through large gaps in endothelial cells and accumulate into interstitial spaces [3]. This phenomenon is known as the enhanced permeability and retention effect (EPR effect) [28].

To precisely determine the distribution of large gold nanoshells in the different cellular components of the human renal cancer, we used two imaging methods: dark field and TEM.

Gold nanoshells could be detected in tumor sections using dark field microscopy as early as 5 min after injection. They remained located within the microvessels as long as 2 h (Fig. 2b). There was a striking distribution change between Day 1 and Day 15, with a distribution within cytoplasm of cells located outside the microvessel network. Transmission electron microscopy confirmed these results and enabled a more precise analysis at higher magnification. As early as 6 h after injection, gold nanoshells, easily detected because they are electron dense with a regular round shape and a constant diameter, were found in the sub-endothelial compartment, thus showing they had crossed over the microvessel endothelial wall. At Day 1, gold nanoshells could be detected both in the cytoplasm of microvessel endothelial cells and in the cytoplasm of tumor cells. After Day 7, gold nanoshells were mainly found within the cytoplasm of macrophages, easily identified because of their phagolysosomes. Rare nanoshells were detected into the cytoplasm of tumor cells at Day 15 (supplementary Fig. 2).

This original sequence not previously reported, using TEM to identify the different cell components of human renal cancer, opens new fields for medical applications using large PEGylated gold nanoshells by optimising the time for targeted hyperthermia or local drug delivery to tumor cells.

**Acknowledgments** We thank Tumorothèque Hopital Saint Louis; S Arien and K Pereira for the electron microscopy technique, N Sanson for the DLS and zeta potential measurements and A Swaine for the revision of the English language. Grants are from the ANR Golden Eye and Plan Cancer 2009-2013-projet GoldFever, ANR Ibisa.

**Conflict of interest** The authors declared no conflict of interest.

# References

1. Wang Y, Black KC, Luehmann H, Li W, Zhang Y, Cai X, Wan D, Liu SY, Li M, Kim P, Li ZY, Wang LV, Liu Y, Xia Y (2013) Comparison study of gold nanohexapods, nanorods, and nanocages for photothermal cancer treatment. ACS Nano 7:2068–2077

2. Papasani MR, Wang G, Hill RA (2012) Gold nanoparticles: the importance of physiological principles to devise strategies for targeted drug delivery. Nanomedicine 8:804–814

3. Puvanakrishnan P, Park J, Diagaradjane P, Schwartz JA, Coleman CL, Gill-Sharp KL, Sang KL, Payne JD, Krishnan S, Tunnell JW (2009) Near-infrared narrow-band imaging of gold/silica nanoshells in tumors. J Biomed Opt 14:024044

4. Day ES, Thompson PA, Zhang L, Lewinski NA, Ahmed N, Drezek RA, Blaney SM, West JL (2011) Nanoshell-mediated photothermal therapy improves survival in a murine glioma model. J Neurooncol 104:55–63

5. Liu H, Liu T, Wang H, Li L, Tan L, Fu C, Nie G, Chen D, Tang F (2013) Impact of PEGylation on the biological effects and light heat conversion efficiency of gold nanoshells on silica nanorattles. Biomaterials 34:6967–6975

6. Gad SC, Sharp KL, Montgomery C, Payne JD, Goodrich GP (2012) Evaluation of the toxicity of intravenous delivery of auroshell particles (gold-silica nanoshells). Int J Toxicol 31:584–594

7. Chatelut E, Delord JP, Canal P (2003) Toxicity patterns of cytotoxic drugs. Invest New Drugs 21:141–148

8. Keefe DM, Bateman EH (2012) Tumor control versus adverse events with targeted anticancer therapies. Nat Rev Clin Oncol 9:98–109

9. Taurin S, Nehoff H, Greish K (2012) Anticancer nanomedicine and tumor vascular permeability; Where is the missing link? J Control Release 164(3):265–75

10. Varna M, Ratajczak P, Ferreira I, Leboeuf C, Bousquet G, Janin A (2012) In vivo distribution of inorganic nanoparticles in preclinical models. J Biomater Nanobiotechnol 3:269–279

11. Huang D, Ding Y, Li Y, Luo WM, Zhang ZF, Snider J, Vandenbeldt K, Qian CN, Teh BT (2010) Sunitinib acts primarily on tumor endothelium rather than tumor cells to inhibit the growth of renal cell carcinoma. Cancer Res 70:1053–1062

12. Varna M, Lehmann-Che J, Turpin E, Marangoni E, El-Bouchtaoui M, Jeanne M, Grigoriu C, Ratajczak P, Leboeuf C, Plassa LF, Ferreira I, Poupon MF, Janin A, de The H, Bertheau P (2009) p53 dependent cell-cycle arrest triggered by chemotherapy in xenografted breast tumors. Int J Cancer 124:991–997

13. Bardhan R, Lal S, Joshi A, Halas NJ (2011) Theranostic nanoshells: from probe design to imaging and treatment of cancer. Acc Chem Res 44:936–946

14. Thakor AS, Jokerst J, Zavaleta C, Massoud TF, Gambhir SS (2011) Gold nanoparticles: a revival in precious metal administration to patients. Nano Lett 11:4029–4036

15. Kah JC, Wong KY, Neoh KG, Song JH, Fu JW, Mhaisalkar S, Olivo M, Sheppard CJ (2009) Critical parameters in the pegylation of gold nanoshells for biomedical applications: an in vitro macrophage study. J Drug Target 17:181–193

16. Balasubramanian SK, Jittiwat J, Manikandan J, Ong CN, Yu LE, Ong WY (2010) Biodistribution of gold nanoparticles and gene expression changes in the liver and spleen after intravenous administration in rats. Biomaterials 31:2034–2042

17. Chen YS, Hung YC, Liau I, Huang GS (2009) Assessment of the in vivo toxicity of gold nanoparticles. Nanoscale Res Lett 4:858–864

18. Lasagna-Reeves C, Gonzalez-Romero D, Barria MA, Olmedo I, Clos A, Sadagopa Ramanujam VM, Urayama A, Vergara L, Kogan MJ, Soto C (2010) Bioaccumulation and toxicity of gold nanoparticles after repeated administration in mice. Biochem Biophys Res Commun 393:649–655

19. Sonavane G, Tomoda K, Makino K (2008) Biodistribution of colloidal gold nanoparticles after intravenous administration: effect of particle size. Colloids Surf B: Biointerfaces 66:274–280

20. De Jong WH, Hagens WI, Krystek P, Burger MC, Sips AJ, Geertsma RE (2008) Particle size-dependent organ distribution of gold nanoparticles after intravenous administration. Biomaterials 29: 1912–1919

21. Nagayama S, Ogawara K, Fukuoka Y, Higaki K, Kimura T (2007) Time-dependent changes in opsonin amount associated on nanoparticles alter their hepatic uptake characteristics. Int J Pharm 342:215–221

22. Inchley CJ (1969) The actvity of mouse Kupffer cells following intravenous injection of T4 bacteriophage. Clin Exp Immunol 5: 173–187

23. Molema G, Aird WC (2012) Vascular heterogeneity in the kidney. Semin Nephrol 32:145–155

24. Alkilany AM, Murphy CJ (2010) Toxicity and cellular uptake of gold nanoparticles: what we have learned so far? J Nanopart Res 12:2313–2333

25. Khouri MG, Douglas PS, Mackey JR, Martin M, Scott JM, Scherrer-Crosbie M, Jones LW (2012) Cancer therapy-induced cardiac toxicity in early breast cancer: addressing the unresolved issues. Circulation 126:2749–2763

26. You J, Zhang R, Xiong C, Zhong M, Melancon M, Gupta S, Nick AM, Sood AK, Li C (2012) Effective photothermal chemotherapy using doxorubicin-loaded gold nanospheres that target EphB4 receptors in tumors. Cancer Res 72:4777–4786

27. Dvorak HF, Weaver VM, Tlsty TD, Bergers G (2011) Tumor microenvironment and progression. J Surg Oncol 103:468–474

28. Maeda H, Wu J, Sawa T, Matsumura Y, Hori K (2000) Tumor vascular permeability and the EPR effect in macromolecular therapeutics: a review. J Control Release 65:271–284

29. Sadauskas E, Danscher G, Stoltenberg M, Vogel U, Larsen A, Wallin H (2009) Protracted elimination of gold nanoparticles from mouse liver. Nanomedicine 5:162–169

30. Glazer ES, Zhu C, Hamir AN, Borne A, Thompson CS, Curley SA (2011) Biodistribution and acute toxicity of naked gold nanoparticles in a rabbit hepatic tumor model. Nanotoxicology 5:459–468

# 22

# One-pot synthesis of various Ag–Au bimetallic nanoparticles with tunable absorption properties at room temperature

Brett W. Boote · Hongsik Byun · Jun-Hyun Kim

**Abstract** This report describes the formation of gold-coated silver bimetallic nanoparticles prepared by the one-pot synthetic approach which involves the subsequent reduction of silver and gold ions at ambient conditions. The reduction of silver ions by excess L-ascorbic acid initially led to the formation of silver cores. This step was followed by the addition of gold ions into the preformed cores, resulting in the formation of silver-core gold-shell type bimetallic nanoparticles at room temperature. This process systematically allowed for the formation of various bimetallic nanoparticles which exhibited tunable absorption properties corresponding to the visible and near-IR regions. The thickness of the gold shells and the diameter of the silver-core nanoparticles were readily controlled; the morphological and structural properties of the resulting bimetallic nanoparticles were thoroughly analyzed by SEM/TEM, DLS, and UV–Vis spectrophotometry. The overall results demonstrated not only that these gold-coated silver nanoparticles were reliably prepared by our one-pot synthetic approach, but also that their optical properties were tunable in the visible and near-IR areas as a function of the core size and shell thickness.

**Keywords** Bimetallic nanoparticles · Gold · Silver · Near infrared · Core–shell

B. W. Boote · J.-H. Kim (✉)
Department of Chemistry, Illinois State University, Normal, IL 61790-4160, USA
e-mail: jkim5@ilstu.edu

H. Byun
Department of Chemical System Engineering, Keimyung University, Daegu 704-701, South Korea
e-mail: hsbyun@kmu.ac.kr

## Introduction

The preparation of metal nanoparticles in small sizes that absorb in the visible and near-IR spectral regions remains an ongoing challenge to colloidal science [1, 2]. Access to these broad absorption areas is especially important for solar energy based and/or biological applications because optically driven solar cells and therapies represent some of the most promising advances in the emerging field of renewable energy systems and nanomedicine [3, 4]. These developing nanotechnologies take advantage of the fact that there are not many chromophores in biological tissue that broadly absorb in the visible and near-IR regions [5, 6]. As such, metal nanoparticles are often preferred over organic-based systems because of the tunable and strong absorption bands from the visible to near-IR areas, as well as their extended stability in solutions and/or solid states [7–9].

Numerous studies have proposed strategies to construct or transform metal nanoparticles into diverse nanostructures (including prisms, disks, rods, and core–shells) that can tune the absorption properties [10–14]. In particular, the preparation of metal–metal core–shell nanoparticles possessing desired optical and electrical properties with a biocompatible nature have been extensively studied by the subsequent chemical reduction method using two or more metal ions [9, 13, 15–20]. Moreover, controlling the absorption properties of these nanoparticles on a sub-100-nm scale may bring additional advantages for practical use in biotechnology and photoinduced reactions due to their high surface-to-volume ratio. Unfortunately, most preparation methods for such nanoparticles often require high-reaction temperatures, multiple steps, and/or the proper use of unique stabilizing agents [21–26]. Here, we demonstrate a simple one-pot synthetic method that allows for the reliable preparation of stable core–shell type bimetallic nanoparticles with strong and tunable optical properties at ambient conditions. Specifically, two different sizes of Ag core nanoparticles (~15 and ~40 nm in diameter) were prepared at room temperature and used as cores for Au shells. Varying thicknesses of Au shells were grown on the Ag cores simply by adding selected amounts of HAuCl$_4$ in a potassium carbonate (K$_2$CO$_3$) solution to an aqueous solution of the preformed Ag cores in situ. The absorption wavelengths of these "core–shell type bimetallic nanoparticles" can be systematically tuned from the visible to the near-IR region by adjusting the molar ratio of the initial reagents. Importantly, these bimetallic nanoparticles are reliably prepared and can be modified further with substantially greater ease and more anisotropic shapes than the related gold-coated dielectric or metallic nanoshells/nanocubes [11, 13, 14, 20, 27–29]. Upon the formation of various bimetallic nanoparticles possessing a strong and broad absorption band, these nanoparticles may serve as interesting materials in the area of optically tunable devices, SERS enhancers, and biomedical applications [15, 30–34].

## Experimental section

*Materials* Nitric acid, hydrochloric acid, potassium hydroxide, isopropyl alcohol, methanol, ethanol, hexadecyltrimethylammonium bromide (CTAB, ≥99.0 %) (all from Fisher Scientific), potassium carbonate (≥99.0 %), sodium borohydride (~98 %), hydrogen tetrachloroaurate(III) hydrate (99.999 % trace metals basis), L-ascorbic acid (AsA, ≥99.0 %) (all from Aldrich), and silver nitrate (from Mallinckrodt) were used without purification from the indicated commercial suppliers. Deionized water was purified to a resistance of 18 MΩ (Nanopure Water System; Barnstead/Thermolyne) and filtered through a 0.2 μm membrane to remove impurities. All glassware was cleaned with an aqua regia solution, followed by treatment in a base bath, and then rinsed with pure water prior to use.

*Characterization methods* All nanoparticles were characterized by ultraviolet–visible (UV–Vis) spectroscopy for the absorption properties, by environmental scanning electron microscopy (SEM) and transmission electron microscopy (TEM) for the morphology and structure, and by dynamic light scattering (DLS) for the size distribution.

An FEI Quanta 450 instrument operating at 20 kV and Zeiss 10 TEM operating at an accelerating voltage of 80 kV were used to evaluate the general size distribution and the overall structure of the nanoparticles, respectively. All samples were deposited from solution onto silicon wafers (for SEM) and 300 mesh carbon-coated copper grids (for TEM) and then completely dried at room temperature overnight.

DLS on a ZetaPALS equipped with a particle analyzer (Brookhaven Instruments Corp., Holtsville, New York) with a 35 mW solid-state laser (90 and 15° angular measurements) was employed to examine the hydrodynamic diameter and polydispersity of the nanoparticles as well as zeta potentials. The diameters were collected at 20 °C from an average of five measurements over 100 s.

An Agilent 8453 UV–Vis spectrometer was used to characterize the absorption properties of the nanoparticles over the wavelength range of 200 to 1,100 nm. All samples were prepared in pure water and transferred to a quartz UV–Vis cell.

*Preparation of HAuCl$_4$ and AgNO$_3$ stock solution* All preparation methods involved the use of HAuCl$_4$ or AgNO$_3$ in a K$_2$CO$_3$ solution (the K–Au solution and K–Ag solution, respectively). Specifically, the preparation of the K–Au solution was as follows: 0.025 g of K$_2$CO$_3$ and 98 mL of water were added to a 150 mL Erlenmeyer flask. The solution was stirred for at least 15 min to completely dissolve the K$_2$CO$_3$, followed by the rapid addition of 2 mL of 1 wt.% HAuCl$_4$ H$_2$O solution. The color of the mixture changed from light yellow to colorless within 30 min. In the case of the K–Ag solution, 1 mL of 1 wt.% AgNO$_3$ was introduced to an aqueous 99 mL solution of K$_2$CO$_3$, which exhibited a color change from colorless to bright yellow in a few seconds. The final solutions were stored overnight in a refrigerator prior to use.

*Preparation of large Ag core–Au shell nanoparticles* Five milliliter of the prepared K–Ag solution was placed in a 20 mL glass vial containing a magnetic stirring bar. L-ascorbic acid (0.3 mL of 100 mM: 0.176 g/10 mL water) was quickly introduced to the yellow K–Ag solution, resulting in the formation of Ag nanoparticles with a greenish yellow color. Subsequently, varying amounts of K–Au solution were slowly (dropwise) or rapidly introduced into the mixture. The final solution was stirred for an additional 5 min and exhibited a reddish-yellow to brownish-blue color as a function of the concentration of the K–Au solution.

*Preparation of small Ag core–Au shell nanoparticles* Five milliliter of the prepared K–Ag solution (2.94 μmol) was placed in a 20 mL glass vial containing a magnetic stirring bar. NaBH$_4$ (0.02 mL of 3.2 mM: 1.2 mg/10 mL of water) was quickly added to the yellow K–Ag solution to form small Ag seed nanoparticles. Subsequently, L-ascorbic acid (0.3 mL of 100 mM) was added to the reaction mixture, resulting in the formation of small dark yellow Ag cores that were presumably prepared through the seed growth process. Varying amounts of the K–Au solution were then slowly (dropwise) introduced into the mixture. The final solution was stirred for an additional 5 min; it exhibited a reddish-yellow to brownish-blue color as a function of the concentration of the K–Au solution.

*Transformation of the Ag–Au core–shells into large uniform or anisotropic structures under light irradiation* Ten milliliter of preheated hexadecyltrimethylammonium bromide solution (CTAB, 10 mM, 0.0364 g in 10 mL water, ≥35 °C) was placed in a 15 mL polystyrene centrifuge tube. Subsequently, HAuCl$_4$ (0.2 mL of 1 wt.% solution) was added, and the tube was swirled several times, resulting in a homogeneous orange color. L-ascorbic acid (0.6 mL of 100 mM) was quickly added to the orange solution, which led to a colorless solution. Finally, an aliquot of preformed 0.4 M ratio bimetallic core–shell nanoparticles (0.1–0.5 mL) was added to the colorless solution; the tube was capped loosely and then placed under fluorescent light irradiation (a 35 W desk lamp at a distance of 5 cm providing 100 mW/cm$^2$, as measured by a handheld optical power meter; Newport Corp.) for 30 min. The final solution was centrifuged at 3,000 rpm twice, and the precipitates were resuspended in 3 mL of pure water. The final colors of the solutions were brownish-red under reflection and deep blue under transmission.

## Results and discussion

Various Ag–Au core–shell type bimetallic nanoparticles prepared by our developed one-pot synthetic approach largely cover from the visible to the near-IR range simply by adjusting the ratio of the initial silver and gold ions. The small Ag core nanoparticles ~15 nm diameter and large Ag core nanoparticles ~40 nm in diameter were prepared by the reduction of Ag ions (K–Ag) with the combination of NaBH$_4$ and AsA at ambient conditions. Subsequently, these core nanoparticles were coated with tunable Au layers through the introduction of varying amounts of the Au growth solution (K–Au) in the presence of a residual reducing agent (i.e., AsA). The systematic control of the diameters of these nanoparticles allowed for tunable absorption properties depending on the core size and the shell thickness. This entire process was systematically completed in situ one-pot at room temperature. Interestingly, the rate of adding the K–Au growth solution (slow vs. rapid) to the Ag seed nanoparticles slightly affected the reproducible formation, the surface morphology, and the absorption bands of the resulting nanoparticles. Based on previous studies, the reduction of Au ions in the presence of preformed Ag cores generally led to the formation of Ag-rich cores and Ag–Au alloy shells with compositions that were enriched in the gold component with the increasing Au mole fraction [18, 35, 36]. Although understanding the precise compositions of these bimetallic nanoparticles is important, the main goals of this research were to demonstrate a simple synthetic method for the preparation of various sub-100-nm bimetallic nanoparticles and the subsequent modification of these core–shells to possess tunable absorption properties. As such, these Ag core–Au shell type bimetallic nanoparticles were examined as a function of the Ag and Au molar ratio from 0.1 to 1.0 (K–Au/K–Ag). To the best of our knowledge, this is the first study to demonstrate the reliable formation of

the Ag–Au core–shell nanoparticles in the absence of a surfactant at room temperature via a one-pot synthetic approach.

Figures 1 and 2 show the SEM and TEM images of the relatively large and rough Ag core nanoparticles and the core–shell nanoparticles of varying thicknesses that were prepared by the subsequent reduction of Ag and Au ions in the presence of an excess reducing agent. The irregular and polydisperse Ag cores ~40 nm in diameter (~0.30 PDI by DLS) were initially formed by the reduction of Ag ions at a high pH (pH ~10) with excess AsA. This result is similar to previous studies that demonstrated the formation of rough and partially aggregated nanoparticles upon the reduction of metal ions with excess AsA in the absence of a surfactant [37–40]. After a short period of time (≤5 min), the K–Au solution was subsequently introduced either slowly (i.e., dropwise) or rapidly to the solution containing these preformed Ag core nanoparticles. This process allowed for the gradual coating of gold layers around the Ag cores. We noted that the nanoparticles prepared by rapid addition exhibited slightly rougher, more irregular structures (as well as partial aggregation) than those formed through the slow addition of the solution. The thickness of the shell was estimated by comparing the total diameter of the nanoparticles before and after the shell growth. The nanoparticles prepared by both methods at or below room temperature generally showed rough and irregular shapes. These results were comparable to the previous nanocube system, in which the rougher Au coating was often accomplished on preformed Ag nanocubes at a low temperature (i.e., 20 °C) rather than a high temperature (i.e., 100 °C) [25]. In addition to the low temperature, it is also important to remember that such rough core–shell type nanoparticles could also be affected by the presence of excess AsA, which could cause the formation of rough and partially aggregated nanoparticles [37, 39, 41, 42]. As the amount of the K–Au solution increased, the size of the core–shell nanoparticles systematically increased, probably due to the thicker coating of the Au layer. When excess K–Au solution (over a 1:1 ratio of K–Au to Ag seed solution) was introduced into the preformed Ag seed solution, the Au shell layer was too thick to clearly visualize the core–shell structures, even with TEM analysis. The highly reproducible and consistent formation of the Ag core–Au shell particles was observed when the molar ratio between the K–Au growth solution and the Ag seed solution was higher than 0.4 for the rapid addition and 0.3 for the slow addition of the solution, respectively. While newly formed small Au nanoparticles and the increased aggregation of the nanoparticles were often observed under the rapid addition condition (zeta potentials of the nanoparticles, from −25 to −35 mV, implied stable colloids in solution, but partially aggregated nanoparticles were often observed), the slow addition process generally allowed for the more reliable and gradual coating of Au layers around the preformed Ag core nanoparticles.

A UV–Vis spectrophotometer was employed to monitor the absorption patterns of the core–shell nanoparticles prepared by subsequent rapid or dropwise addition of varying amounts of the K–Au growth solution to the preformed Ag nanoparticles (Fig. 3a, b). The K–Ag solution initially showed a very weak absorption peak at ~400 nm (data not shown). Upon the addition of excess AsA, an intense and broad absorption band appeared at ~410 nm, indicating the formation of rough and polydisperse Ag nanoparticles. At a ratio of 0.1 of the K–Au growth solution to Ag cores, a notably decreased absorption intensity of the Ag cores occurred at ~410 nm and a new peak forms at ~500 nm under both conditions, which were presumably caused by the formation of Ag–Au alloy shells on Ag core nanoparticles. As the ratio of the K-Au solution to Ag seed solution increased, the gradual decrease of the strong Ag core peak at ~410 nm and new absorption bands at longer wavelengths between typical pure gold and silver nanoparticles with a shoulder peak clearly suggested the formation of Au–Ag alloy shells on Ag cores. It has been reported that a certain degree of alloying occurs during the reduction of Au ions in the presence of Ag seeds [13, 36]. At the ratio of 0.3, a peak centered at ~654 nm with a broad shoulder peak below

~500 nm still pointed to a Ag core on a Au–Ag alloy shell. When the ratio reached 0.5, the longest absorption band centered at ~700 nm without a notable shoulder peak below ~500 nm was observed, probably for the Ag-rich core and Au-rich shell nanoparticles. The absorption peaks were then slightly blue-shifted as the ratios of the K–Au solution to Ag core solution increased above a molar ratio of 0.5. A further increase of the ratio over 1.0 resulting in the slight red-shift of the absorption band again suggested the increase in the nanoparticle diameters [43, 44], which was consistent with the SEM images and DLS size measurements. Unlike the rapid addition method, slightly narrower absorption bands of the core–shell nanoparticles prepared by the slow addition process might suggest a smoother (less rough) surface of the Au layers on the Ag cores with fewer partial aggregations.

Interestingly, the absorption bands of the nanoparticles prepared by the slow addition remained similar, but the adsorption bands of the nanoparticles prepared by the rapid addition were gradually blue-shifted (shift of $\lambda_{max}$, 5–10 nm) and became narrow upon aging at room temperature for over 2 days. This observation might be explained by the possible surface restructuring process of the nanoparticles [39, 40].

Fig. 1 SEM/TEM images of a K–Ag cores and Ag–Au core-shells with b 0.3, c 0.5, and d 1.0 Au/Ag molar ratios (slow addition)

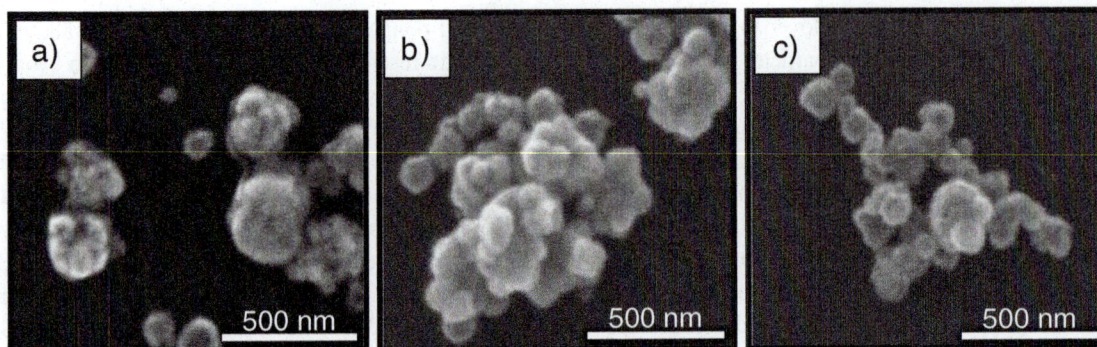

**Fig. 2** SEM images of Ag–Au core–shells with **a** 0.3, **b** 0.5, and **c** 1.0 Au/Ag molar ratios (rapid addition)

Based on our previous studies, partially aggregated gold nanoparticles with a rough surface prepared by excess AsA underwent a notable reshaping process upon aging at room temperature to form polydisperse gold nanoparticles with a smooth surface. This morphological evolution of the gold nanoparticles was able to be thoroughly examined by the significantly blue-shifted absorption maxima and overall absorption patterns [37, 39, 40]. Although we did not microscopically observe the differences in the surface roughness, the slight shift in the absorption bands for our core-shell nanoparticles as a function of time could explain our speculation for the change in the surface structures. In a separate study done by Moskovits and his group, the blue-shift of the absorption bands and less roughness of the nanoparticles might be affected by the slow reduction of the adsorbed Au ions on the surface of the nanoparticles as well [45]. As such, it was clear that the addition rate of the Au growth solution to the preformed Ag cores played an important role in the reproducible formation, the surface morphology, and the absorption property of the resulting core–shell type bimetallic nanoparticles. We are still investigating the main cause of

the slight shift of the absorption bands of the nanoparticles prepared by the rapid addition process.

An additional feature of our core–shell type bimetallic nanoparticles was that the nanoparticles did not exhibit severe pinholes or hollow interiors as well as any destruction (i.e., the Ag atoms were dissolved by $HAuCl_4$) regardless of the addition rate of the Au growth solution. The formation of Ag–Au core–shell nanoparticles is often accomplished via the galvanic replacement reaction when an Au growth solution is introduced to citrate-stabilized Ag core nanoparticles in the absence of reducing agents [13, 25]. The final core–shell nanoparticles readily possess pinhole/hollow structures due to the galvanic reaction of the Ag core nanoparticles by $AuCl_4^-$ ions. A similar reaction process was also observed by the Xia group during the formation of hollow Au nanocages on Ag core nanoparticles via the polyol method at high temperatures. As the amount of the Au growth solution increased, a dealloying process typically took place in the absence of reducing agents [25]. Upon this dealloying process, the broad absorption band of the nanoparticles flattened/disappeared due to the destruction of the core–shell

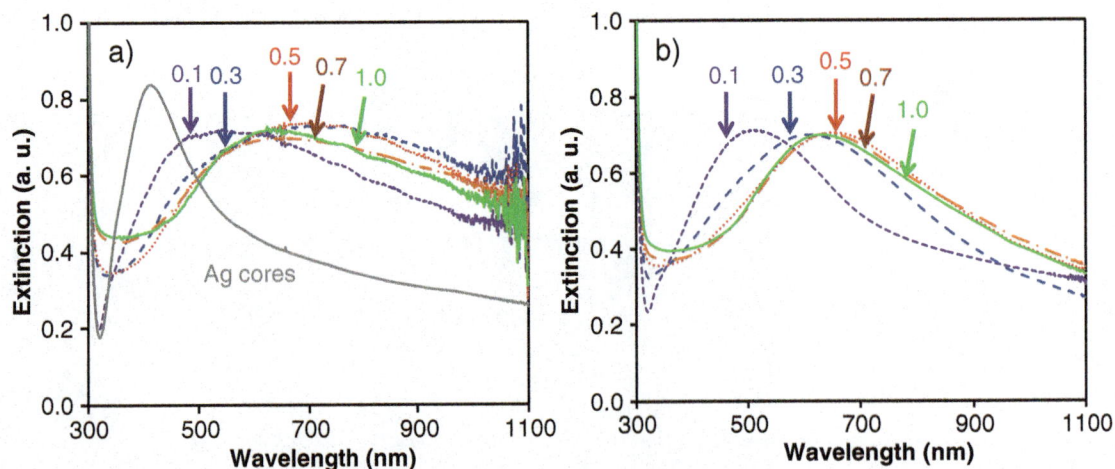

**Fig. 3** UV–Vis spectra of various Ag–Au core–shells noted by initial Au/Ag molar ratio using the **a** rapid addition of K–Au and **b** slow addition of K–Au

structures. However, since our core–shell nanoparticles were prepared in the presence of a high concentration of residual AsA, the reduction of Au ions on the preformed Ag core nanoparticles by AsA was speculated to be preferred over the galvanic reaction process because our core–shell nanoparticles were less likely to exhibit pin holes or hollow structures. Additionally, the formation of a thicker gold coating was easily achieved rather than the rapid dealloying of the nanoparticles with an increase of the K–Au solution. The maintenance of strong and broad absorption peaks even after the addition of excess K–Au solution (i.e., a two times greater concentration) strongly supported the preservation of the Ag–core and Au-rich shell type structures under our reaction conditions. In a separate study, Liz-Marzan and his group also proposed the favorable reduction of Au ions on preformed Ag seeds in the presence of AsA as a reducing agent [18, 46].

The slow (dropwise) addition of the K–Au solution onto Ag core nanoparticles yielded the reliable formation of a slightly smooth surface and less aggregation of the bimetallic nanoparticles. The small Ag seed nanoparticles (~2–3 nm in diameter) were initially formed by the reduction of Ag ions with a strong reducing agent (NaBH$_4$); this led to a bright yellow color. This was followed by the subsequent addition of excess AsA, resulting in the formation of polydisperse Ag nanoparticles ~15 nm in diameter (~0.35 PDI by DLS) at room temperature via the seed growth process. A selected amount of the K–Au solution was then slowly introduced to

these preformed Ag nanoparticles to prepare various Au shells on small Ag cores. Figure 4 shows the SEM and TEM images of the small Ag core nanoparticles and Ag–Au core–shell nanoparticles with varying shell thicknesses. While bare Ag core nanoparticles initially exhibited slightly irregular shapes with partial aggregation, the core-shell bi-metallic nanoparticles showed the systematic increase in the diameters of the nanoparticles, suggesting the successful growth of Au shells. The TEM image (Fig. 4b) again clearly shows two different contrasts throughout the nanoparticles coming from the phase boundary between the Au shell and Ag core. The apparent core–shell contrast consisting of darker outer shells and the lighter inner cores originated from the gold and silver elements because gold scatters more electrons than silver [47]. Such a contrast in the Ag–Au core–shell structures has also been observed in other studies as well [35, 47, 48], which were consistent with our results and confirmed the successful growth of Au layers on small Ag core nanoparticles with increased surface-to-volume ratios.

The absorption spectra of the small core–shell nanoparticles were collected before and after the growth of various thicknesses of the Au shells (Fig. 5). The introduction of an aliquot of NaBH$_4$ initially led to the formation of small Ag seeds possessing a very weak absorption peak at 390 nm; the subsequent addition of excess AsA resulted in the further reduction of residual Ag ions to complete the growth of Ag nanoparticles possessing an intense absorption band at 400 nm with a broad

**Fig. 4** SEM/TEM images of **a** 15 nm Ag cores and small Ag–Au core–shell particles with **b** 0.3, **c** 0.5, **d** 1.0 Au/Ag molar ratios

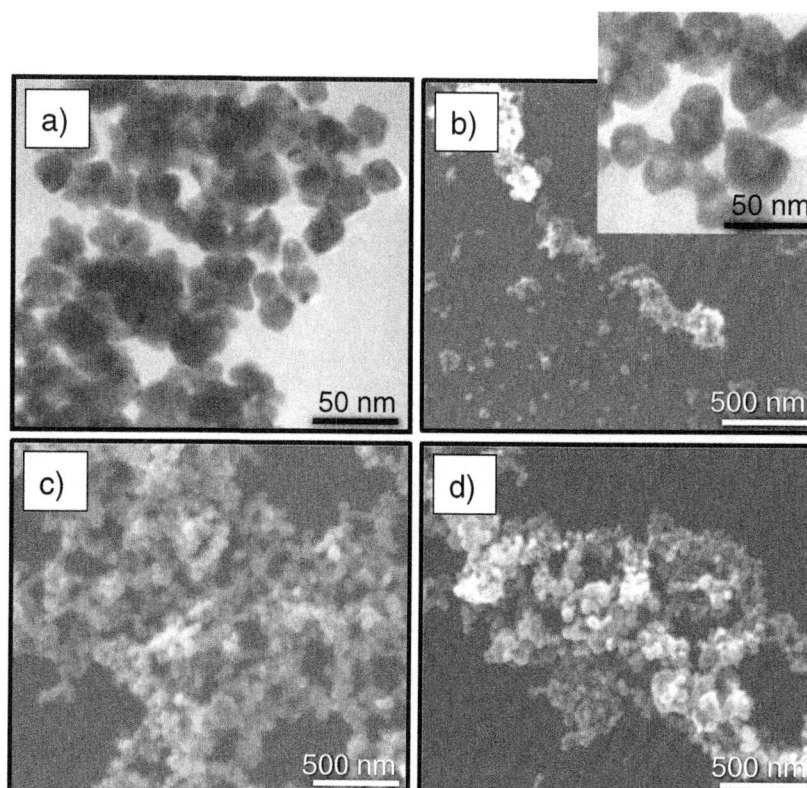

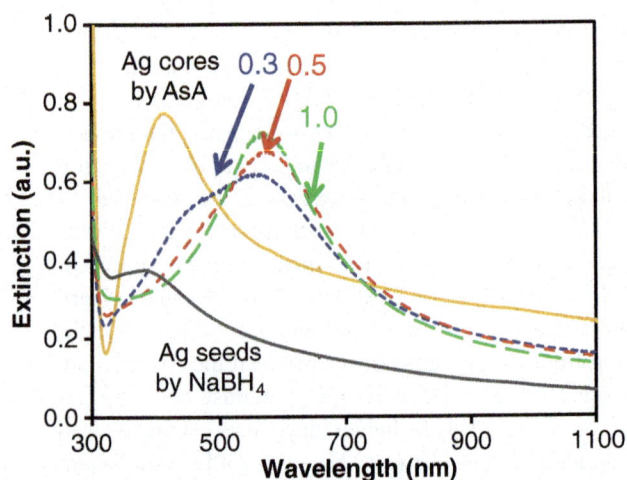

**Fig. 5** UV–Vis spectra of small Ag cores and various core–shells prepared by different Au/Ag molar ratios

shoulder peak in the visible range. This broad absorption pattern of the resulting nanoparticles suggested the formation of rough polydisperse nanoparticles, which was consistent with the microscopic images. Upon the slow addition of the K–Au solution to these preformed Ag nanoparticles, the gradual decrease of the Ag core peak at 400 nm and the appearance of new peaks at over 550 nm indicated the formation of Ag–Au alloy type core–shell nanoparticles. As a small shoulder peak below 500 nm appeared as a result of the use of a 0.5 M ratio of K–Au to K–Ag, the small Ag cores required slightly more of the K–Au solution to form a Au-rich shell. The maximum absorption bands of the core–shell nanoparticles prepared by greater than 0.5 M ratios were placed in the visible (550–600 nm) range regardless of the shell thickness due to the relatively small size of the cores [45, 49], which is comparable to the

previous studies done by the Henglein (~8 nm core) [23] and Srnova-Sloufova (~9 nm core) [35] groups. The Liz-Marzan group prepared multiple core–shell structures using 17 nm core nanoparticles [18], and these nanoparticles also possessed limited absorption bands in the visible areas. However, our nanoparticles possessing highly increased surface-to-volume ratios exhibited slightly broader absorption bands (400–700 nm) than those of the precedent core–shell nanoparticles (400–600 nm), perhaps due to the roughness of the shells and their high polydispersity.

In addition, the resulting nanoparticles were found to be stable in the absence of any surfactants that provided great potential for easy modifications and/or applications. As a proof-of-concept example, the pre-synthesized large core–shell particles were transformed into either large uniform or anisotropic structures to completely cover broad absorption bands across the visible and near-IR areas. This modification process was simply accomplished by the addition of an aliquot of the core–shell particles (1 and 5 % seed particles to growth solution) into a fixed growth solution containing HAuCl$_4$, AsA, and a surfactant (CTAB), then exposed to visible light irradiation for 30 min. Figure 6 shows the SEM images of the initial core–shell particles and the transformed large uniform and anisotropic nanoparticles. Interestingly, the structures of the final nanoparticles were highly affected by the initial ratios of the core–shell nanoparticles to the growth solution. While symmetric growth was favored with the use of a high concentration of the core–shell nanoparticles, anisotropic growth was observed with a low concentration of the nanoparticles. The large uniform nanoparticles (~200 nm in diameter) exhibited two distinctive peaks at 560 nm for multipole resonance and at 760 nm for the dipole plasmon band, these peaks were consistent with those of large bare gold nanoparticle systems [50]. The highly anisotropic nanoparticles have shown a much broader absorption band across visible to near-IR range that is comparable to the previous work as well [51]. More thorough study is underway to elicit the concentration-related structural information.

## Conclusions

The Ag–Au core–shell type bimetallic nanoparticles with varying sizes were reliably prepared by a very simple one-pot synthetic approach at ambient conditions. The thorough characterization of these nanoparticles by UV–Vis spectroscopy, SEM, DLS, and TEM collectively supported the reliable formation of various Ag–Au core–shells with tunable optical properties. Furthermore, the easy transformation of these nanoparticles into either large uniform or anisotropic structures under light irradiation can allow for their potential applications requiring strong optical properties in the visible and/or near-IR regions of the spectrum.

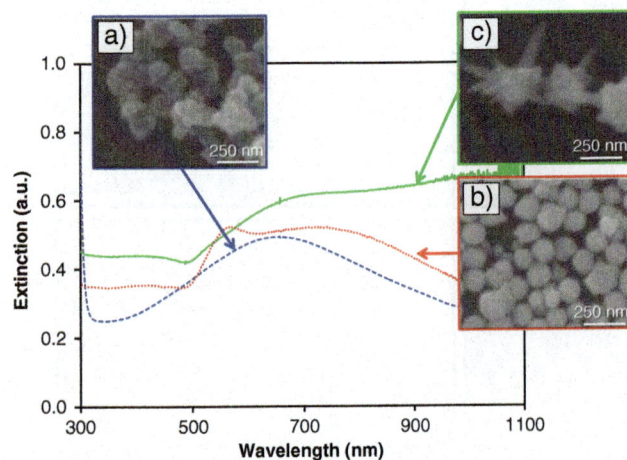

**Fig. 6** UV–Vis spectra and corresponding SEM images of the transformation of 0.4 M ratio core–shells into large uniform or anisotropic structures. Note that the final structures of the particles vary based on the core–shell concentration

**Acknowledgments** We gratefully acknowledge the financial support from the Cottrell College Science Award of Research Corporation and Illinois State University. This research is also supported by Korea Ministry of Environment as "The Eco-Innovation project (Global Top project, no. GT-SWS-11-01-0040-0)". In addition, we thank Dr. M. E. Cook for assistance with the SEM and TEM measurements.

# References

1. Daniel M-C, Astruc D (2004) Gold nanoparticles: assembly, supramolecular chemistry, quantum-size-related properties, and applications toward biology, catalysis, and nanotechnology. Chem Rev 104:293–346
2. El-Sayed MA (2004) Small is different: shape-, size-, and composition-dependent properties of some colloidal semiconductor nanocrystals. Acc Chem Res 37:326–333
3. Gobin AM, O'Neal DP, Watkins DM, Halas NJ, Drezek RA, West JL (2005) Near infrared laser-tissue welding using nanoshells as an exogenous absorber. Las Surg Med 37:123–129
4. Loo C, Lowery A, Halas N, West J, Drezek R (2005) Immunotargeted nanoshells for integrated cancer imaging and therapy. Nano Lett 5:709–711
5. Nie SR, Emroy SR (1997) Probing single molecules and single nanoparticles by surface-enhanced Raman scattering. Science 275:1102–1106
6. O'Neal DP, Hirsch LR, Halas NJ, Payne JD, West JL (2004) Photothermal tumor ablation in mice using near infrared-absorbing nanoparticles. Cancer Lett 109:171–176
7. Busbee BD, Obare SO, Murphy CJ (2003) An improved synthesis of high aspect ratio gold nanorods. Adv Mater 15:414–416
8. Wu H-Y, Huang W-L, Huang MH (2007) Direct high-yield synthesis of high aspect ratio gold nanorods. Crystal Growth & Design 7:831–835
9. Chaudhuri GR, Paria S (2012) Core/shell nanoparticles: classes, properties, synthesis mechanisms, characterization, and applications. Chem Rev 112:2373–2433
10. Averitt RD, Westcott SL, Halas NJ (1999) Linear optical properties of gold nanoshells. J Opt Soc Am B 16:1824–1832
11. Devi P, Badllescu S, Packlrlsamy M, Jeevanandam P (2010) Synthesis of gold-poly (dimethylsiloxane) nanocomposite through a polymer-mediated silver/gold galvanic replacement reaction. Gold Bull 43:307–315
12. Kim J-H, Bryan WW, Lee TR (2008) Preparation, characterization, and optical properties of gold, silver, and gold–silver alloy nanoshells having silica cores. Langmuir 24:11147–11152
13. Sun Y, Wiley B, Li Z-Y, Xia Y (2004) Synthesis and optical properties of nanorattles and multiple-walled nanoshells/nanotubes made of metal alloys. J Am Chem Soc 126:9399–9406
14. Wang W, Pang Y, Yan J, Wang G, Suo H, Zhao C, Xing S (2012) Facile synthesis of hollow urchin-like gold nanoparticles and their catalytic activity. Gold Bull 45:91–98
15. Liu X, Knauer M, Ivleva NP, Niessner R, Haisch C (2010) Synthesis of core–shell surface-enhanced Raman tags for bioimaging. Anal Chem 82:441–446
16. Mott D, Lee J, Thuy NTB, Aoki Y, Singh P, Maenosono S (2011) A study on the plasmonic properties of silver core gold shell nanoparticles: optical assessment of the particle structure. Japanese J App Phys 50:p065004–p065011
17. Pande S, Ghosh SK, Praharaj S, Panigrahi S, Basu S, Jana S, Pal A, Tsukuda T, Pal T (2007) Synthesis of normal and inverted gold–silver core–shell architectures in B-cyclodextrin and their applications in SERS. J Phys Chem C 111:10806–10813
18. Rodriguez-Gonzalez B, Burrows A, Watanabe M, Kielyb CJ, Liz-Marzan LM (2005) Multishell bimetallic AuAg nanoparticles: synthesis, structure, and optical properties. J Mater Chem 15:1775–1759
19. Vongsavat V, Vittur BM, Bryan WW, Kim J-H, Lee TR (2011) Ultrasmall hollow gold–silver nanoshells with extinctions strongly red-shifted to the near-infrared. ACS Appl Mater Interfaces 3:3616–3624
20. Xu W, Niu J, Shang H, Shen H, Ma L, Li LS (2013) Facile synthesis of AgAu alloy and core/shell nanocrystals by using Ag nanocrystals as seeds. Gold Bull 46:19–23
21. Anandan S, Grieser F, Ashokkumar M (2008) Sonochemical synthesis of Au–Ag core–shell bimetallic nanoparticles. J Phys Chem C 112:15102–15105
22. Chen Y-H, Nickel U (1993) Superadditive catalysis of homogeneous redox reactions with mixed silver–gold colloids. J Chem Soc Faraday Trans 89:2479–2485
23. Mulvaney P, Giersig M, Henglein A (1993) Electrochemistry of multilayer colloids: preparation and absorption spectrum of gold-coated silver particles. J Phys Chem 97:7061–7064
24. Radziuk D, Shchukin D, Mohwald H (2008) Sonochemical design of engineered gold–silver nanoparticles. J Phys Chem C 112:2462–2468
25. Sun Y, Xia Y (2004) Mechanistic study on the replacement reaction between silver nanostructures and chloroauric acid in aqueous medium. J Am Chem Soc 126:3892–3901
26. Treguer M, de Cointet C, Remita H, Khatouri J, Mostafavi M, Amblard J, Belloni J (1998) Dose rate effects on radiolytic synthesis of gold–silver bimetallic clusters in solution. J Phys Chem B 102:4310–4321
27. Brongersma ML (2003) Nanoshells: gifts in a gold wrapper. Nature Materials 2:296–297
28. Oldenburg SJ, Jackson JB, Westcott SL, Halas NJ (1999) Infrared extinction properties of gold nanoshells. Appl Phys Lett 75:2897–2899
29. Salgueirino-Maceira V, Caruso F, Liz-Marzan LM (2003) Coated colloids with tailored optical properties. J Phys Chem B 107:10990–10994
30. Chen J, Wang D, Xi J, Au L, Siekkinen A, Warsen A, Li Z-Y, Zhang H, Xia Y, Li X (2007) Immuno gold nanocages with tailored optical properties for targeted photothermal destruction of cancer cells. Nano Lett 7:1318–1322
31. Cui Y, Ren B, Yao J-L, Gu R-A, Tian Z-Q (2006) Synthesis of Ag core-Au-shell bimetallic nanoparticles for immunoassay based on surface-enhanced Raman spectroscopy. J Phys Chem B 110:4002–4006
32. Pavan Kumar GV, Shruthi S, Vibha B, ARB A, Kundu TK, Narayana C (2007) Hot spots in Ag core-Au shell nanoparticles potent for surface-enhanced Raman scattering studies of biomolecules. J Phys Chem C 111:4388–4392
33. Prevo BG, Esakoff SA, Mikhailovsky A, Zasadzinski JA (2008) Scalable routes to gold nanoshells with tunable sizes and response to near-Infrared pulsed-laser irradiation. Small 4:1183–1195
34. Stern JM, Stanfield J, Kabbani W, Hsieh J-T, Cadeddu JA (2008) Selective prostate cancer thermal ablation with laser activated gold nanoshells. J Urology 179:748–753
35. Srnova-Sloufova I, Lednicky F, Gemperle A, Gemperlova J (2000) Core–shell (Ag)Au bimetallic nanoparticles: analysis of transmission electron microscopy images. Langmuir 25:9928–9935
36. Srnova-Sloufova I, Vlckova B, Bastl Z, Hasslett TL (2004) Bimetallic (Ag)Au nanoparticles prepared by the seed growth method: two-dimensional assembling, characterization by energy dispersive X-ray analysis, X-ray photoelectron spectroscopy, and surface enhanced Raman spectroscopy, and proposed mechanism of growth. Langmuir 20:3407–3415

37. Kim J-H, Lavin BW (2011) Preparation of gold nanoparticle aggregates and their photothermal heating property. J Nanosci Nanotechnol 11:45–52

38. Kim J-H, Lavin BW, Boote BW, Pham JA (2012) Photothermally enhanced catalytic activity of partially aggregated gold nanoparticles. J Nanopart Res 14:p995–p1004

39. Kuo C-H, Huang MH (2005) Synthesis of branched gold nanoparticles by a seeding growth approach. Langmuir 21:2012–2016

40. Wu H-Y, Liu M, Huang MH (2006) Direct synthesis of branched gold nanocrystals and their transformation into spherical nanoparticles. J Phys Chem B 110:19291–19294

41. Andreescu D, Sau TK, Goia DV (2006) Stabilizer-free nanosized gold sols. J Colloid Interface Sci 298:742–751

42. Goia DV, Matijevic E (1998) Preparation of monodispersed metal particles. New J Chem 22:1203–1215

43. Haiss W, Thanh NT, Jenny Aveyard J, Fernig DG (2007) Determination of size and concentration of gold nanoparticles from UV–Vis spectra. Anal Chem 79:4215–4221

44. Kim J-H, Lavin BW, Burnett RD, Boote BW (2011) Controlled synthesis of gold nanoparticles by fluorescent light irradiation. Nanotechnology 22:p285602–p285607

45. Moskovits M, Srnová-Sloufová I, Vlcková B (2002) Bimetallic Ag–Au nanoparticles: extracting meaningful optical constants from the surface-plasmon extinction spectrum. J Chem Phys 116:10435–10446

46. Liz-Marzan LM (2006) Tailoring surface plasmons through the morphology and assembly of metal nanoparticles. Langmuir 22:32–41

47. Hutter E, Fendler JH (2002) Size quantized formation and self-assembly of gold encased silver nanoparticles. Chem Commun:378–379

48. Ramos M, Ferrer DA, Chianelli RR, Correa V, Serrano-Matos J, Flores S (2011) Synthesis of Ag-Au nanoparticles by galvanic replacement and their morphological studies by HRTEM and computational modeling. J Nanomater 2011:5 pages

49. Mie G (1908) Contributions to the optics of turbid media, particularly of colloidal metal solutions. Ann Phys 25:377–445

50. Rodriguez-Fernandez J, Perez-Juste J, Garcia de Abajo FJ, Liz-Marzan LM (2006) Seeded growth of submicron Au colloids with quadrupole plasmon resonance modes. Langmuir 22:7007–7010

51. Sanchez-Gaytan B, Park S-J (2010) Spiky gold nanoshells. Langmuir 26:19170–19174

# 23

# Optimizing the immobilization of gold nanoparticles on functionalized silicon surfaces: amine- vs thiol-terminated silane

Maroua Ben Haddada · Juliette Blanchard ·
Sandra Casale · Jean-Marc Krafft · Anne Vallée ·
Christophe Méthivier · Souhir Boujday

**Abstract** Immobilization of gold nanoparticles on planar surfaces is of great interest to many scientific communities; chemists, physicists, biologists, and the various communities working at the interfaces between these disciplines. Controlling the immobilization step, especially nanoparticles dispersion and coverage, is an important issue for all of these communities. We studied the parameters that can influence this interaction, starting with the nature of the terminal chemical function. Thus, we have carefully grafted silanes terminated by either amine or thiol groups starting from aminopropyltriethoxysilane (APTES) or mercaptopropyltriethoxysilane. We also changed the chain length for thiol-terminated layers through covalent grafting of mercaptoundecanoic acid (MUA) on APTES-modified layers, and the protocol of nanoparticles deposition to evaluate whether other factors must be taken into consideration to rationalize this interaction. The formed layers were characterized by X-ray photoelectron spectroscopy and gold nanoparticles deposition was monitored by scanning electron microscopy and surface-enhanced Raman scattering. We observed significant differences in terms of nanoparticles dispersion and density depending on the nature of the chemical layer on silicon. The use of ultrasounds during the deposition process was very efficient to limit aggregates formation. The optimal deposition procedures were obtained through the use of APTES and APTES/MUA functionalization. They were compared in terms of coverage, dispersion, and densities of isolated nanoparticles. The APTES/MUA surfaces clearly showed better results that may arise from both the longer chain and the dilution of thiol end groups.

**Keywords** Gold nanoparticles · Silicon surface · Silane grafting · Surface functionalization · XPS · Scanning electron microscopy

## Introduction

The fabrication of solid substrates with gold nanoparticles assemblies immobilized on their surface is currently the subject of growing interest because of the key role these substrates could play in the development of several devices [1–3]. The physical and chemical properties of these devices will depend not only on the size and shape of the gold nanoparticles but also on their spatial arrangement and on the nature of their interaction with the substrate surface [4–6]. Biosensors represent one of the areas for which the use of nanoparticles is booming. Indeed, biosensors are analytical tools whose effectiveness is highly dependent on the accuracy of the measurement and thus on its reproducibility [7, 8]. Input from gold nanoparticles in this area is twofold: on the one hand, they allow amplification of the signal transduction for many techniques and, on the other hand, they can provide surface nanostructuration [9, 10]. In both cases, achieving a densely packed layer and a regular arrangement is crucial to ensure an optimal amplification while preserving a reproducible and quantitative response of transduction techniques [11, 12].

M. Ben Haddada · J. Blanchard · S. Casale · J.-M. Krafft ·
A. Vallée · C. Méthivier · S. Boujday
UPMC Univ Paris 6, UMR CNRS 7197, Laboratoire de Réactivité de
Surface, 75005 Paris, France

M. Ben Haddada · J. Blanchard · S. Casale · J.-M. Krafft ·
A. Vallée · C. Méthivier · S. Boujday (✉)
Laboratoire de Réactivité de Surface, UMR CNRS 7197, Université
Pierre et Marie Curie-Paris VI, 4 Place Jussieu, 75252 Paris cedex 05,
France
e-mail: souhir.boujday@upmc.fr

Usually, the anchoring of gold nanoparticles on the solid substrates is carried out by using an intermediate layer of organic molecules grafted on the solid surface, whose terminal functional groups are selected for their electrostatic or chemical interactions with the nanoparticles [11–16]. These platforms should also be stable over time and the interaction of the nanoparticles with the surface should be strong enough to ensure that the nanoparticles remains attached to the surface during further functionalization and upon utilization. The molecules commonly used for the functionalization of oxidized silicon wafer or glass slides are organosilane with up to three hydrolysable groups (either $-OR$ or $-Cl$) that would react with the silanols groups from the substrate surface and (at least) a nonhydrolysable group bearing the terminal function responsible for the interaction with the gold nanoparticles [17]. Among the large variety of available organosilanes, aminopropyltriethoxysilane (APTES; $NH_2(CH_2)_3Si(OCH_2CH_3)_3$) is, by far, the most studied surface modifier (references [18–21] and references therein). APTES is relatively easy to handle thanks to its moderate reactivity. Its three hydrolysable ethoxy groups ensure a robust anchoring of the silane to the surface (silanization step), whereas the amine function of the aminopropyl group remains available for further reaction. Terminal amine functions are extensively used for surface functionalization because they easily react with acid, aldehyde, or thiocyanates through covalent bond. Moreover, amine groups have affinity to gold nanoparticles and are widely used to immobilize them through electrostatic interactions [4, 11, 14].

The numerous studies dealing with the functionalization of silica/oxidized silicon surface with APTES have shown that the silanization step is very sensitive toward experimental conditions and that the final surface state of the modified substrate depends strongly on temperature, presence of water, concentration of APTES, and the duration of the grafting step [18].

In this work, we studied the organic layer influence on the grafting of gold nanoparticles on functionalized silicon substrates. Our aim was to find the optimal experimental conditions for both an optimal density and dispersion of gold nanoparticles on silicon surfaces. For this purpose, we explored some of the parameters that can influence this interaction, starting with the nature of the terminal chemical function. In a first step, APTES was used to form amine-terminated silane layers. Then, starting from mercaptopropyltriethoxysilane (MPTES), thiol-terminated silane layers were constructed. We also changed the chain length for thiol-terminated layers through covalent grafting of mercaptoundecanoic acid (MUA) on APTES-modified layers. This latter functionalization procedure was successfully used by Kaminska et al. to immobilize CTAB-covered nanoparticles on silicon and reach a densely packed layer [22]. X-ray photoelectron spectroscopy (XPS) and contact angle measurements were used to characterize the formed layers and the protocol of nanoparticle deposition was

modified, particularly using ultrasonication during the deposition of the spherical gold nanoparticles. The grafting of nanoparticles was monitored by surface-enhanced Raman scattering (SERS). Finally, scanning electron microscopy (SEM) was used to compare gold nanoparticles coverage and dispersion for the considered surface functionalizations.

## Experimental section

### Materials

$N$-ethyl-$N'$-(3-(dimethylamino) propyl) carbodiimide hydrochloride (EDC; 98 %), sodium citrate (HOC(COONa) $(CH_2COONa)_2 \cdot 2H_2O$; 99 %), and gold(III)chloride trihydrate (HAuCl$_4$.H$_2$O; ≥99.9 %) were purchased from Sigma Aldrich. $N$-hydroxysuccinimide (NHS; 97 %), 11-mercaptoundecanoic acid (MUA; 95 %), (3-aminopropyl)triethoxysilane (APTES; 99 %) and (3-mercaptopropyl)trimethoxysilane (MPTES; 95 %) were purchased from Aldrich and ethylenediamine (EDA; ≥99.5 %) from Fluka. Silicon wafer <100> from Sigma Aldrich was cut into $1 \times 1$ cm$^2$ pieces. Sulfuric acid, 96 % (H$_2$SO$_4$) and hydrogen peroxide, 30 % in water (H$_2$O$_2$) were supplied by Carlo Erba. Toluene, acetone, and ethanol were purchased from Analar Normapur. MilliQ water (18 MΩ, Millipore, France) was used for the preparation of the solutions and for all rinses. All chemicals were reagent grade or higher and were used without further purification.

### Surface chemistry

First, silicon surfaces were cleaned following a procedure which includes several washing step, a treatment with a piranha solution, and finally a treatment with UV ozone. A detailed description of this procedure is given in reference [18].

The same experimental conditions were applied for silane grafting for the amine-terminated silane (APTES) and for the thiol-terminated one (MPTES): the surface oxidized and cleaned silicon wafer was immersed in a 50 mM solution of silane in anhydrous toluene at 75 °C for 24 h. After silanization step, the samples were washed twice, sonicated for 10 min in anhydrous toluene, dried under nitrogen, and heated at 90 °C for 2 h.

The MUA grafting was achieved after APTES deposition on silicon surfaces. First, MUA solution was activated using a mixture of EDC and NHS in ethanol during 90 min and then the APTES-modified silicon surfaces were placed in the activated solution of MUA for 90 min, then washed twice in ethanol and dried under nitrogen.

## Gold nanoparticles preparation and deposition

Citrate-stabilized gold nanoparticles (GNPs) were prepared according to the standard method developed by J. Turkevich and co-workers and refined by G. Frens. Details on solution storage and particles dispersion are given in references [11, 12].

Freshly synthesized Au NPs were then deposited, with no further dilution, on the modified silicon surfaces for 2 h either using a gentle agitation or using a sonication bath (Elma, 90 W, 45 kHz). Silicon substrates were then washed twice in water and dried under nitrogen.

## Characterization techniques

*TEM* Transmission electron microscopy measurements were performed using a JEOL JEM 1011 microscope operating at an accelerating voltage of 100 kV. The transmission electron microscopy (TEM) grids were prepared as follows: typically 1.5 mL of the solution was centrifuged at 10,000 rpm (equivalent to 11,200 relative centrifugal forces) for 10 min to precipitate the particles. The colorless supernatant was discarded. The heavy residue was redispersed in a suitable volume of deionized water depending on the quantity of the residue. Of this redispersed particle suspension, 2 μL was placed on a carbon-coated copper grid and dried at room temperature.

*SEM* Scanning electron microscopy images of the gold nanoparticles on the modified silicon wafers were obtained using a SEM FEG Hitachi SU-70 scanning electron microscope with a low voltage of 1 kV and distance of 1.9–2.3 mm; the secondary electron detector "in Lens" was used. The distribution of gold nanoparticles on the surfaces was established considering 1,000–1,500 particles.

*Contact angle measurements* Static water contact angles were measured at room temperature using the sessile drop method and image analysis of the drop profile. The instrument, which uses a CCD camera and an image analysis processor, was purchased from Krüss (Germany). The water (Milli-Q) droplet volume was 1 μL and the contact angle (θ) was measured 5 s after the drop was deposited on the sample. For each sample, the reported value is the average of the results obtained on three droplets and the overall accuracy in the measurements was better than ±5°.

*XPS* XPS analyses were performed with a PHOIBOS 100 X-ray photoelectron spectrometer from SPECS GmbH (Germany) with an monochromated Al Kα X-ray source (hν=1,486.6 eV), operating at $10^{-10}$ Torr or less, in a "fixed analyzer transmission" analysis mode with a 7×20 mm entrance slit, leading to a resolution of 0.1 eV. A 10 eV pass energy for the survey scan and 10 eV pass energy for the small regions were applied. The spectra were fitted using Casa XPS software (version 2.3.13, Casa Software, UK).

*SERS* Surface-enhanced Raman scattering spectra were recorded in the 500–1,900 $cm^{-1}$ range on a modular Raman spectrometer (Model HL5R of Kaiser Optical Systems, Inc.) equipped with a high-powered near-IR laser diode working at 785 nm. Before spectra acquisition, an optical microscope (Olympus; objective, ×50) was used to focus the laser beam. Measurements were carried using an objective ×100. The laser output power was 10 mW, which corresponds to ~1 mW on samples. For each spectrum, 30 acquisitions of 5 s were recorded to improve the signal-to-noise ratio. To ensure a representative characterization of surfaces, a minimum of three measurements were taken on different parts of the surface.

## Results and discussion

### Silane grafting and surface characterization

Gold nanoparticles are commonly immobilized on amine-terminated layers, through electrostatic interaction or on thiol-terminated layers through covalent bonds. These two terminations were compared herein through the use of APTES or MPTES for silanization. Adding MUA on the APTES layer would lead to a longer chain terminated also by a thiol function, and thus enabling us to establish the influence of an additional parameter, namely, the hydrocarbon chain length. The three chemical modification of silicon surfaces are shown in Fig. 1. In a first stage, the efficiency of silane binding and further chemical grafting was investigated through contact angle measurements and XPS analysis of the functionalized silicon surfaces.

Contact angle measurements were performed to investigate the hydrophilic character of the three functionalized silicon surfaces. The pictures obtained with the CCD camera and the corresponding angles are shown in Fig. 2.

The initial clean silicon surface (Fig. 2a) was very hydrophilic and the contact angle (<10°) could not be measured due to drop spreading. This result confirms the efficiency of the applied cleaning procedure to remove the organic contaminants from the surface [23]. Upon APTES grafting (Fig. 2b), the measured contact angle was 83°, indicating a hydrophobic layer. This result is in agreement with previously observed values [18], the hydrophobic character, in apparent contradiction with the hydrophilic terminal amine functions, is probably due to the folding of amine terminal function to form H bonds with free silanol groups [24]. For the MPTES functionalization, the contact angle was lower than for

**Fig. 1** Schematic description of
the three functionalized surfaces

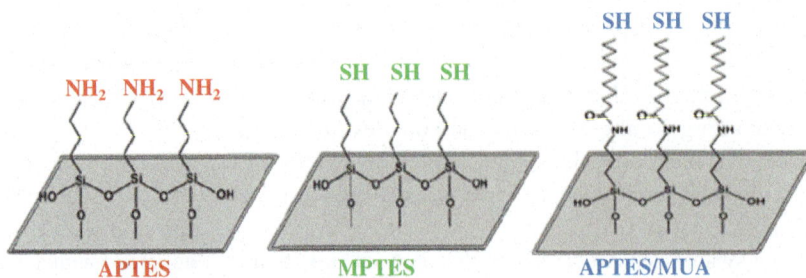

APTES-modified surfaces (~61°). This higher hydrophilicity may be due to the lower tendency of sulfur to form hydrogen bonds; which could induce less folding of the silane chains. But another possibility, that will be discussed hereafter, on XPS data basis, is the incomplete coverage of silicon surface by MPTES; this would indeed lower contact angle values as the measured value will include a contribution of the hydrophilic-bare silicon oxide. Finally, upon grafting MUA on the APTES-modified layer, an even lower contact angle was recorded, ~40°, suggesting a higher hydrophilic character of the resulting surfaces. The increase of hydrophilic character with chain length was observed likewise by Wasserman et al. upon assembling chloroalkylsilanes on silicon surfaces [25]. The most likely explanation is the increase of the layer crystallinity and organization due to the increase of Van der Waals interactions between alkyl chains [26]. This higher crystallinity would reduce the folding of the terminal groups, thus exhibiting the hydrophilic part of the grafted molecule on the upmost of the surface [27].

The modified silicon surfaces were also characterized by XPS analysis. The main XPS spectra are shown in Fig. 3. On all the survey spectra, in addition to the silicon and oxygen peaks arising from silicon wafers and the silica layer, carbon, nitrogen, and sulfur for APTES/MUA- and MPTES-modified surfaces were present. On the Si 2p peaks, shown in Fig. 3, two contributions were observed; the first one at ~99 eV, arises from bulk silicon wafer $Si^0$, and the second contribution at higher binding energy, ~103 eV is attributed to $Si^{IV}$ in both oxide layer and grafted silanes. The ratio $Si^{ox}/Si^0$ was calculated to compare silane coverage, assuming the silica layer was similar in all cases. For the three surfaces, the values were very close, 0.64, 0.62, and 0.56, for APTES, MPTES, and APTES/MUA, respectively, suggesting similar average coverages with silanes, and not enabling a precise comparison of the amounts of grafted silanes.

The N 1s peaks included two contributions attributed to amine groups [C–$NH_2$] (at 399.8 eV) and protonated amines C–$NH_3^+$ (at 401.6 eV). The shape of this peak was modified

**Fig. 2** Static water contact angles
pictures, and corresponding
angles; **a** cleaned (oxidized)
silicon wafer, **b** silicon-grafted
APTES, **c** silicon-grafted
MPTES, and **d** silicon-grafted
APTES/MUA

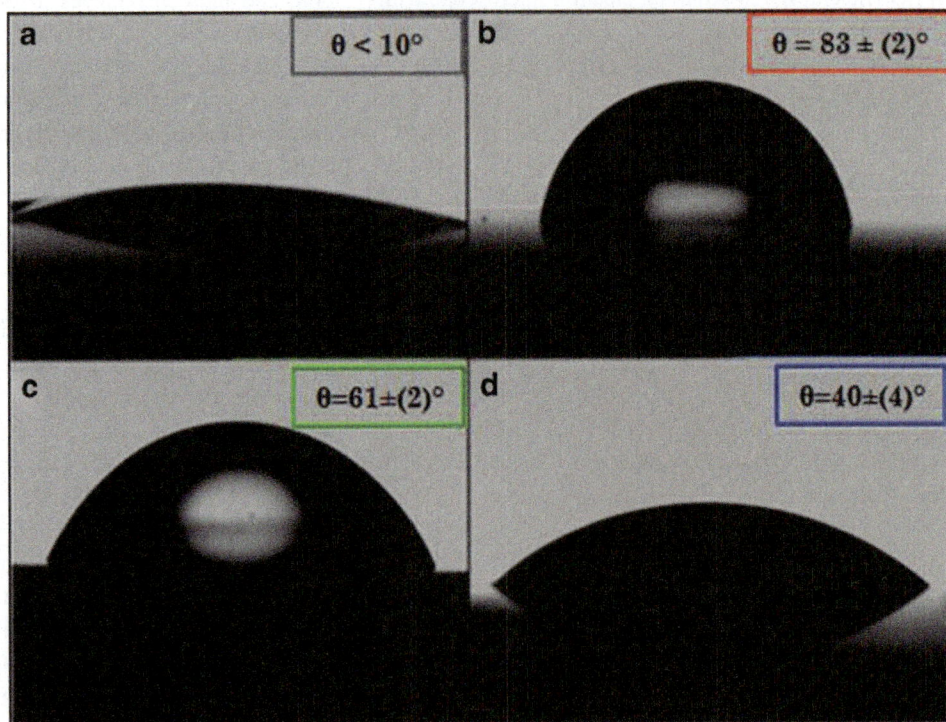

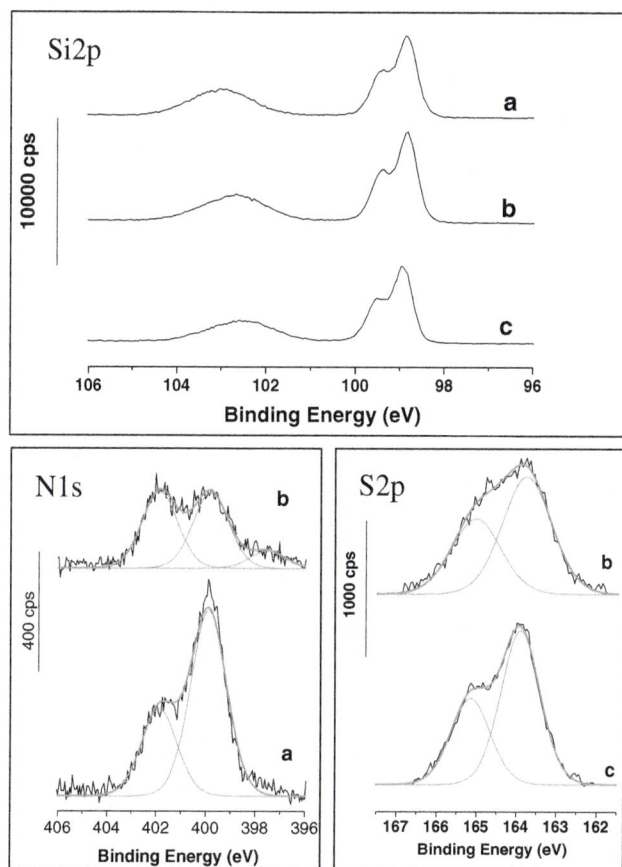

**Fig. 3** XP spectra Si 2p, N1s, and S2p signals after APTES (**a**), APTES/MUA (**b**), and MPTES (**c**) grafting on silicon surfaces

upon grafting MUA, but it is difficult to discriminate the contribution of amide nitrogen [C=O)–$N$H] from the amine one [28]. The S 2p spectra showed the S 2p doublet at 163.6 and 164.9 eV for S $2p_{3/2}$ and $2p_{1/2}$, respectively, corresponding to the sulfur in the SH groups.

For the APTES/MUA-modified surface, the ratio S/N calculated using S 2p and N 1s peak was 0.24. Thus, despite the

large excess of MUA in solution, only ca. 24 % of amine groups have reacted with the acid function of MUA. Therefore, thiol-end groups on these surfaces are "diluted" and four times less numerous than amine functions on APTES-modified surfaces. In addition, the integration of the S2p for both APTES/MUA and MPTES led to similar amounts of sulfur. Therefore, the use of MPTES would have led to a silane coverage four times lower than the use of APTES. This low value is not surprising as it was established than the terminal amine groups of APTES act as catalysts for silane reaction with silanols, thus enhancing the efficiency of grafting [24, 29]. We confirmed these observations by performing the MPTES grafting using a small amount of ethylenediamine as catalyst. The XPS results (see Electronic supplementary material (ESM) 1) showed an increase of sulfur amount by a factor of 3.7, indicating a silane cover close to that obtained for APTES.

### Gold nanoparticles deposition and characterization

Before interacting with the silane layers, the solution of gold nanoparticles was characterized by UV–visible spectroscopy. On this spectrum (see ESM), a narrow plasmon band was present at 525 nm, indicating a homogeneous particle size of ~13 nm [30, 31]. Gold nanoparticles size and distribution were also confirmed by TEM images (see ESM 1).

Figure 4 shows the Raman spectra of functionalized silicon surfaces before and after gold nanoparticles deposition. On the Raman spectrum of the functionalized silicon surface before the deposition of GNPs, the vibration bands of silanes could not be seen and only silicon signal was observed around 500 and 1,000 cm$^{-1}$ for the silica layer. After gold nanoparticles deposition, a SERS effect is expected [32, 33]. For the three functionalized surfaces, a Raman signal was observed, evidencing the grafting of gold nanoparticles [34, 35].

**Fig. 4** Raman spectra obtained for gold nanoparticles deposited on silicon surfaces modified by APTES/MUA (**a**), APTES (**b**), and MPTES (**c**). **d** Silane layers prior to gold nanoparticles deposition

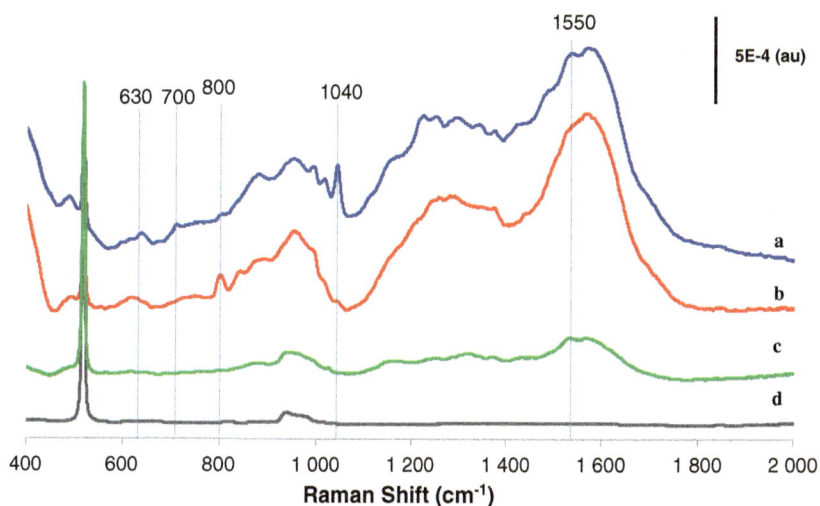

**Fig. 5** SEM images of silicon surfaces after nanoparticle deposition without (**a**) and with (**b** and **c**) sonication on APTES, MPTES, and APTES/MUA

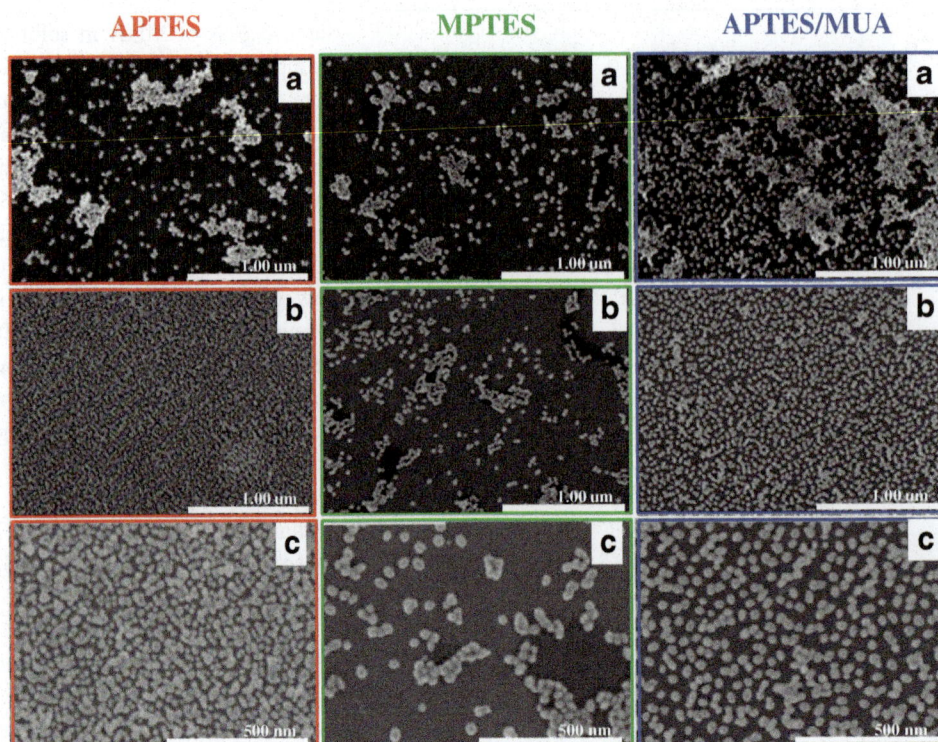

The main bands observed on SERS spectra arise from the characteristic bands of citrates and mainly of the carboxylate groups between 1,200 and 1,600 cm$^{-1}$ [11, 36]. The $\nu$ (C–C) arising from citrates and silane carbon chain formed massif of bands in the region 900–1,100 cm$^{-1}$. Some differences, related to the different underlayer compositions, were observed and highlighted on the figure. On APTES/MUA, bands were present corresponding to $\nu$ (C–S)$_G$ at 630 cm$^{-1}$ and $\nu$ (C–S)$_T$ at 700 cm$^{-1}$, which is absent from the APTES spectrum. The intensities of the bands at 800 and 1,040 cm$^{-1}$ arising from amine functions were modified upon MUA grafting on APTES and an additional band, amide II, appeared at 1,552 cm$^{-1}$ [37].

Considering the citrates bands, the intensity of the signal was dependent on the system: While APTES- and APTES/MUA-modified surfaces led to similar intensities; the MPTES-modified surfaces exhibited lower exaltation. This may be due to a lower coverage of GNPs as discussed below.

Figure 5 shows SEM images recorded after deposition of gold nanoparticles on APTES-, APTES/MUA-, and MPTES-modified silane surfaces. In Fig. 5a, the images obtained when nanoparticle deposition was carried out with a gentle agitation are shown. The highest coverage in gold nanoparticles was clearly observed on APTES/MUA-modified surfaces. However, for the three surface functionalization nanoparticles, aggregates are clearly observable with this deposition procedure.

When gold nanoparticles deposition was done under sonication, the largest nanoparticles aggregates were removed on both APTES and APTES/MUA surfaces. On these two systems, the dispersion was also clearly improved. The GNPs

coverage on APTES was considerably increased. On MPTES layers, no significant improvement was observed using the sonication bath. The same results were obtained upon depositing the nanoparticles on the MPTES catalyzed by ethanediamine with or without sonication (images in the ESM 1). The sulfur density at silicon surface is therefore not a determining factor for this system.

Despite the improved dispersion and the removal of the largest aggregates, a closer inspection of the SEM images at higher magnification (Fig. 5c) showed the presence of small aggregates especially on APTES layers. To compare efficiently the dispersion on APTES and APTES/MUA layers, particle distributions were quantified for these two functionalizations and are presented as histograms in Fig 6.

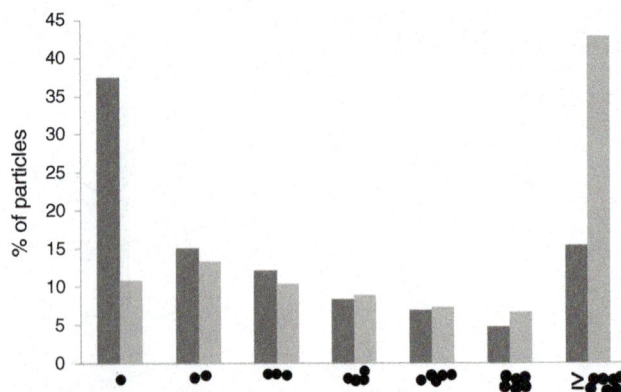

**Fig. 6** Gold nanoparticles distribution on APTES/MUA (*dark*) and on APTES (*bright*)-modified layers

The comparison of the aggregate size distribution for the two functionalized surfaces clearly showed that the APTES/MUA functionalization allowed a better dispersion of the gold nanoparticles with a large fraction of isolated particles whereas, on the APTES functionalized surface, the nanoparticles formed higher amount of large aggregates. This better dispersion of gold nanoparticles on the MUA/APTES layer may result from the dilution of terminal functions on this system, 24 % SH on MUA/APTES vs 100 % $NH_2$ on APTES. Indeed, similar behavior was observed using mixed thiol self-assembled monolayers to immobilize proteins: the dispersion was better when the terminal functions were diluted [38, 39]. The length of the alkyl chain induces also a better flexibility of the organic layer that probably favors nanoparticles binding compared to the MPTES layer [8].

## Conclusion

We studied the organic layer influence on the grafting of gold nanoparticles on functionalized silicon substrates. To this aim, we grafted silanes terminated by either amine or thiol groups starting from APTES or MPTES. A third layer, also terminated by a thiol function, was constructed by covalent attachment of MUA on APTES-modified layers. XPS characterizations confirmed the successful grafting of APTES and MPTES on silicon surfaces. Upon adding MUA to the APTES-modified layers, only 24 % of the amine functions reacted. Spherical gold nanoparticles were then deposited on these layers. SEM revealed that sonication during the deposition procedure limited the presence of aggregates on the surfaces. The higher coverage in gold nanoparticles was observed on APTES and APTES/MUA layers. SEM images also showed that gold nanoparticles dispersion was better on APTES/MUA than on APTES, with a larger fraction of isolated particles and few aggregates. This system likely benefits from both the dilution of terminal functions, 24 % SH, and the flexibility of the organic layer.

**Acknowledgments** This work financially supported by the Agence Nationale de la Recherche "NArBIoS" project (ANR-DFG program; grant number: ANR-11-INTB-1013). The authors are grateful to Claire-Marie Pradier for her interest in the work.

## References

1. Sardar R, Funston AM, Mulvaney P, Murray RW (2009) Langmuir 25:13840–13851
2. Hoa XD, Kirk AG, Tabrizian M (2007) Biosens Bioelectron 23:151–160
3. Daniel MC, Astruc D (2004) Chem Rev 104:293–346
4. Zheng J, Zhu Z, Chen H, Liu Z (2000) Langmuir 16:4409–4412
5. Haes AJ, Van Duyne RP (2004) Anal Bioanal Chem 379:920–930
6. Ghosh SK, Pal T (2007) Chem Rev 107:4797–4862
7. Luppa PB, Sokoll LJ, Chan DW (2001) Clin Chim Acta 314:1–26
8. Boujday S, Bantegnie A, Briand E, Marnet P-G, Salmain M, Pradier C-M (2008) J Phys Chem B 112:6708–6715
9. Hutter E, Fendler JH (2004) Adv Mater 16:1685–1706
10. Lalander CH, Zheng Y, Dhuey S, Cabrini S, Bach U (2010) ACS Nano 4:6153–6161
11. Morel A-L, Boujday S, Méthivier C, Krafft J-M, Pradier C-M (2011) Talanta 85:35–42
12. Morel A-L, Volmant R-M, Méthivier C, Krafft J-M, Boujday S, Pradier C-M (2010) Colloids Surf B Biointerfaces 81:304–312
13. Zhong Z, Patskovskyy S, Bouvrette P, Luong JHT, Gedanken A, Phys J (2004) Chem B 108:4046–4052
14. Aureau D, Varin Y, Roodenko K, Seitz O, Pluchery O, Chabal YJ (2009) J Phys Chem C 114:14180–14186
15. Khire VS, Lee TY, Bowman CN (2007) Macromolecules 40:5669–5677
16. McNally H, Janes DB, Kasibhatla B, Kubiak CP (2002) Superlattice Microst 31:239–245
17. Gooding JJ, Ciampi S (2011) Chem Soc Rev 40:2704–2718
18. Aissaoui N, Bergaoui L, Landoulsi J, Lambert JF, Boujday S (2012) Langmuir 28:656–665
19. Kim J, Cho J, Seidler PM, Kurland NE, Yadavalli VK (2010) Langmuir 26:2599–2608
20. Pasternack RM, Rivillon Amy S, Chabal YJ (2008) Langmuir 24:12963–12971
21. Howarter JA, Youngblood JP (2006) Langmuir 22:11142–11147
22. Kaminska A, Inya-Agha O, Forster RJ, Keyes TE (2008) Phys Chem Chem Phys 10:4172–4180
23. Mittal KL (1979) Surface contamination. In: Mittal KL (ed). Springer, New York, pp 3–45
24. Kanan SM, Tze WTY, Tripp CP (2002) Langmuir 18:6623–6627
25. Wasserman SR, Tao YT, Whitesides GM (1989) Langmuir 5:1074–1087
26. Love JC, Estroff LA, Kriebel JK, Nuzzo RG, Whitesides GM (2005) Chem Rev 105:1103–1169
27. Vallée A, Humblot V, Al Housseiny R, Boujday S, Pradier C-M (2013) Colloids Surf B Biointerfaces 109:136–142
28. Mercier D, Boujday S, Annabi C, Villanneau R, Pradier C-M, Proust A (2012) J Phys Chem C 116:13217–13224
29. Blitz JP, Murthy RSS, Leyden DE (1987) J Am Chem Soc 109:7141–7145
30. Turkevich J, Hillier J, Stevenson PC (1951) Discuss Faraday Soc 11:55
31. Seitz O, Chehimi MM, Cabet-Deliry E, Truong S, Felidj N, Perruchot C, Greaves SJ, Watts JF (2003) Coll Surf A: Physicochem Eng Aspects 218:225–239
32. Grabar KC, Freeman RG, Hommer MB, Natan MJ (1995) Anal Chem 67:735–743
33. Chumanov G, Sokolov K, Gregory BW, Cotton TM (1995) J Phys Chem 99:9466–9471
34. Moskovits M (2005) J Raman Spectrosc 36:485–496
35. Tian ZQ (2005) J Raman Spectrosc 36:466–470
36. Bantz KC, Nelson HD, Haynes CL (2012) J Phys Chem C 116:3585–3593
37. Richard Allen N (2001) Interpreting infrared, Raman, and nuclear magnetic resonance spectra. Academic, San Diego, pp 143–148
38. Briand E, Gu C, Boujday S, Salmain M, Herry JM, Pradier CM (2007) Surf Sci 601:3850–3855
39. Thébault P, Boujday S, Sénéchal H, Pradier C-M (2010) J Phys Chem B 114:10612–10619

# 24

# Influence of the cluster's size of random gold nanostructures on the fluorescence of single CdSe–CdS nanocrystals

**Damien Canneson · Stéphanie Buil · Xavier Quélin ·**
**Clémentine Javaux · Benoît Dubertret ·**
**Jean-Pierre Hermier**

**Abstract** It is well known that coupling a single emitter to metallic structures modifies drastically its fluorescence properties compared to single emitter in vacuum. Depending on various parameters such as the nature of the metal or the geometry of the metallic structure, quenching or intensity enhancement as well as radiative processes acceleration are obtained through the creation of new desexcitation channels. The use of metallic random structures gives the opportunity to magnify the effect of the coupling by strongly confined electromagnetic fields. A gold film at the percolation threshold is an interesting illustration of that effect. Here, we study the influence of the method used to realize these films through two different examples. First, we show that the mean size of the gold clusters constituting the film depends on the deposition method. Even if similar optical properties (in particular far-field absorption) are exhibited by the structures, crucial differences appear in the fluorescence of single emitters when coupled to the two kinds of random gold film. Especially, we focus our attention on the creation of desexcitation channels and show that they are cluster size dependent.

**Keywords** Nanocrystals · Metallic random structures · Surface plasmon resonances · Purcell effect

D. Canneson · S. Buil · X. Quélin (✉) · J.-P. Hermier
Groupe d'Étude de la Matière Condensée,
Université de Versailles-Saint-Quentin-en-Yvelines,
CNRS UMR8635, 45 avenue des États-Unis,
78035 Versailles, France
e-mail: quelin@physique.uvsq.fr

J.-P. Hermier
e-mail: jean-pierre.hermier@uvsq.fr

C. Javaux · B. Dubertret
Laboratoire de Physique et d'Étude des Matériaux,
CNRS UMR8213, ESPCI, 10 rue Vauquelin,
75005 Paris, France

J.-P. Hermier
Institut Universitaire de France, 103 boulevard
Saint-Michel, 75005 Paris, France

## Introduction

Colloidal core–shell nanocrystals (NCs) are promising single emitters. They are synthesized in solution and so are easy to use and to prepare at low cost. They are bright (high quantum yield (QY)) and photostable at room temperature, opening a wide variety of applications, such as quantum optics [1], biomedical applications [2, 3], or optoelectronics [4–6]. But for an optimal use of NCs, a fine tailoring of their fluorescence properties is needed.

It is well known [7] that the control of the fluorescence properties, such as polarization, emission diagram, or photoluminescence (PL) decay, can be achieved by modifying the electromagnetic (EM) surroundings of the emitter. For example, the modification of the PL decay can be achieved by the use of cavities. In the weak coupling regime, the modification of the PL decay can be quantified through the Purcell factor $\left( F_P = \frac{\tau_0}{\tau} \right)$ [8], where $\frac{1}{\tau_0}$ and $\frac{1}{\tau}$ are the decay rates of the emitter in free space and in the cavity, respectively.

Specific dielectric structures have been investigated, such as micro-disks [9], micro-pillars [10, 11], or other micro-cavities [12]. Dielectric cavities with very high quality factors ($Q$) can be designed and Purcell factors ($F_P$) greater than 10 have been obtained. In these cavities, $F_P$ is proportional to the ratio $\frac{Q}{V}$, where $V$ is the modal volume in the cavity. It is obvious that increasing $F_P$ can be obtained either by increasing $Q$ or decreasing $V$. The reduction of $V$ is limited by the diffraction: it is not possible to obtain, in cavities, smaller modal volume than the cube of the wavelength. Then, increasing $Q$ is the key to achieve high $F_P$. But this is meaningful only if the bandwidth of the emitter is narrower than $Q$. Otherwise, in the $F_P$ expression, $Q$ is replaced by the bandwidth of the emitter. This is a strong limitation for controlling the emitter's emission with dielectric cavities at room temperature, where emitters usually exhibit a large bandwidth.

Since the results obtained by Drexhage et al. [13], it is known that metallic structures can also be used to modify the fluorescence properties of an emitter. Two major advantages of the metallic structures are due to plasmon resonances: strong localizations of EM field and wide bandwidths. Thanks to the strong localization of the EM field, modal volumes smaller than the cube of the wavelength can easily be obtained and thanks to the spectral width of the plasmon resonances, it is possible to couple efficiently large bandwidth emitters or different kind of emitters to the same metallic structure.

However, a strong drawback of the metallic films is losses. Some metallic structures [14, 15] have been specifically designed to redirect the non-radiative plasmons. The parameters involved in these studies are the size of the metallic nanostructure and its distance and orientation with respect to the emitter [16, 17]. In these kind of studies, mastering the relative position of the emitter to the metallic cluster plays a crucial role. In this paper, we propose to use random gold films, allowing us to skirt this major difficulty. These random gold films have many advantages for the coupling with the fluorescence of the emitter. Due to a large number of plasmon resonances, the coupling can be efficient whatever the emission wavelength is. A wide variety of plasmons is also observed on these films, including radiative ones that can induce an efficient collection of the fluorescence. It will be shown in this article that randomness is not the only parameter controlling the coupling between emitters and metallic structures. The size of the random metallic clusters also plays a great role in the modification of the fluorescence properties.

## Samples and experimental setup

The random gold structures were elaborated using two different methods: thermal evaporation under ultrahigh vacuum ($10^{-9}$ torr, method 1) and radio frequency sputtering (method 2). For both methods, the films were obtained on the same kind of substrate and below the percolation threshold. In this concentration regime, the deposited layer exhibit a wide variety, in shapes and sizes, of unconnected gold clusters.

The local structure of the gold films was observed with an Atomic Force Microscope (AFM, Dimension 3100, Bruker AXS). Their optical properties (transmission, reflection, and absorption) were studied with a Perking Elmer LAMBDA 950 spectrophotometer.

The NCs have been synthesized following the method described in [18]. They are core–thick shell CdSe–CdS NCs, with a 5-nm-diameter core and a 10-nm-thick shell. The fluorescence emission is around 660 nm with a full width at half maximum of 30 nm.

The fluorescence of the NCs was analyzed with an inverted confocal microscope (IX 71, Olympus). The excitation was performed with a pulsed diode laser (LDH 485, PicoQuant, pulse duration $\sim$ 100 ps) emitting at 485 nm, with a repetition rate of 2.5 MHz for NCs deposited on glass coverslip (reference) and 40 MHz for NCs deposited on the gold films.

The fluorescence was collected through an air objective with numerical aperture (NA) of 0.95. An oil objective with NA of 1.4 was also used for the characterization on glass coverslip.

The optical signal is sent in a high-sensitivity Hanbury-Brown and Twiss (HB-T) setup (avalanche photodiodes PDM series, MPD, time resolution 50 ps). The signal is recorded by a PicoHarp 300 module (PicoQuant, time resolution 64 ps). For each experiment, the absolute time of arrival of the photons with close to 100 ps accuracy is given. Then the time evolution of the intensity, the correlation between photon pairs, and the PL decay are extracted.

## The CdSe–CdS nanocrystals

Unlike usual CdSe–ZnS NCs, for which it is impossible to grow a thick shell without defects, CdSe–CdS NCs can have shells as thick as 15 nm. Due to relative positions of the conduction and valence bands at the interface between CdSe and CdS, when an electron-hole (e-h) pair is created, the electron is delocalized in the whole structure of the nanocrystal while the hole remains confined in the core. This affects the fluorescence lifetime, increasing from $\sim$ 20 ns for CdSe–ZnS NCs to $\sim$ 70 ns for CdSe–CdS NCs. The blinking phenomenon, which is the main drawback of CdSe–ZnS NCs and consists on non-emitting periods due to ionization and Auger effect [19], is strongly reduced. More precisely, the ionized state of CdSe–CdS NCs is still a radiative state but with a reduced efficiency. The duration of the ionization periods does not exceed tens of milliseconds [20]. The thick shell also plays the role of a spacer. It prevents the quenching due to non-radiative processes occurring close to a metallic surface. Depositing directly these thick-shell NCs on gold films is then possible [21].

Typical fluorescence properties are depicted on Fig. 1. The record of the intensity versus time (Fig. 1a) shows the fluctuations of the intensity. The histogram of this trace exhibits a maximum value with a tail due to the remaining flickering. The most probable value will be used to compare intensities on gold films.

The PL decay is shown on Fig. 1b and is well fitted by a bi-exponential decay curve. The longest time decay of the bi-exponential corresponds to the neutral state of the NC and the shortest one is due to the ionized state of the NC. Details can be found in [22].

NCs were excited in the low power excitation regime (10 % of the saturation regime). In this regime, the probability to create e-h pairs is described by a Poisson statistics [20], and 90 % of the pulses will not generate an e-h pair, 9.5 % will generate one e-h pair, and only 0.5 % will generate more than one e-h pair. The fraction of multi-excitons is small enough to be neglected for future intensity comparisons.

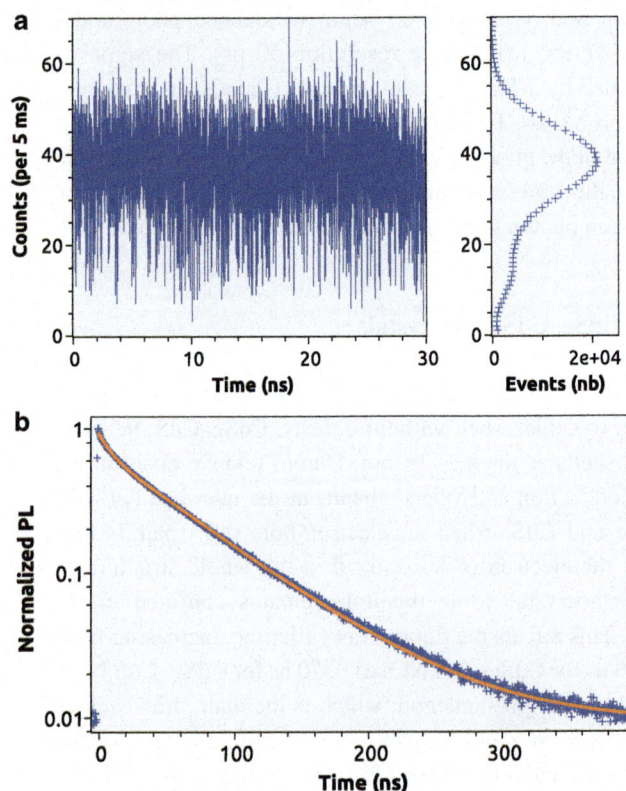

**Fig. 1** Properties of the fluorescence of a single NC on reference substrate (glass coverslip): **a** intensity versus time of the fluorescence and its histogram, and **b** PL decay, fitted by a bi-exponential decay curve ($\tau_{slow} = 70$ ns, $\tau_{fast} = 15$ ns)

## The gold structures

The two methods used to elaborate the gold films lead to structures with similarities and differences which can be observed through the local structure and the far-field spectrum (Fig. 2). The topographic images (Fig. 2a, b) show a similar general structure of the films, i.e., non-connected random clusters of widely different shapes, but either the mean lateral dimensions or the height of the clusters are smaller, roughly by a factor two, for the films elaborated by method 2 (Fig. 2b) compared to those elaborated by method 1 (Fig. 2a). The mass thickness ($\sim 7.5$ nm) is very close for the two films. Despite the difference in size of the clusters, the same quantity of metal is deposited on the films. To the end of the study, the structures elaborated by method 1 will be called *big* structures and the structures elaborated by method 2 will be called *small* structures.

Far-field optical spectrum are depicted on Figs. 2c, d. The general behavior of absorption, reflection, and transmission are nearly the same. The values of the reflection and the transmission are slightly different, but the absorption is the same for both structures with a plateau covering a part of the visible and extending to the infrared. This plateau corresponds to a wide range of plasmon resonances due to the distribution of metallic cluster sizes and shapes.

To evaluate the disorder, it is common to calculate the fractality of the perimeter of the clusters [23]. The perimeter $P_c$ of a cluster is said to be fractal when $P_c \propto S^{\frac{D}{2}}$, where $S$ is the surface of the cluster and $D$, the fractal dimension, is a non-integer number. Here, for the big structures, $D = 1.88$. For the small ones, $D \simeq 1.9$. For the two films, $D$ is nearly the same and very close to the well-known Hausdorff value for these kinds of structure [23].

Due to the disorder of these metallic structures, strong localizations and enhancements of the EM fields have been observed [24]. These localizations of energy are usually called *hot spots*. Their position, number, and intensity depend on the excitation parameters. The higher the wavelength, the stronger the enhancement and localization.

## Coupling NCs to the gold structures

NCs are directly deposited on the gold nanostructures without any spacer. Only their fluorescence is coupled to the plasmon resonances, but not the excitation laser which is outside the absorption plateau. The main modification induced by the coupling between plasmons and NCs emission is a strong acceleration of the fluorescence decay (Fig. 3a). Due to this strong acceleration, the lifetime of the ionized state becomes too short to be correctly extracted with our setup. Consequently, we only focus on the modifications of the neutral state's lifetime. The acceleration is

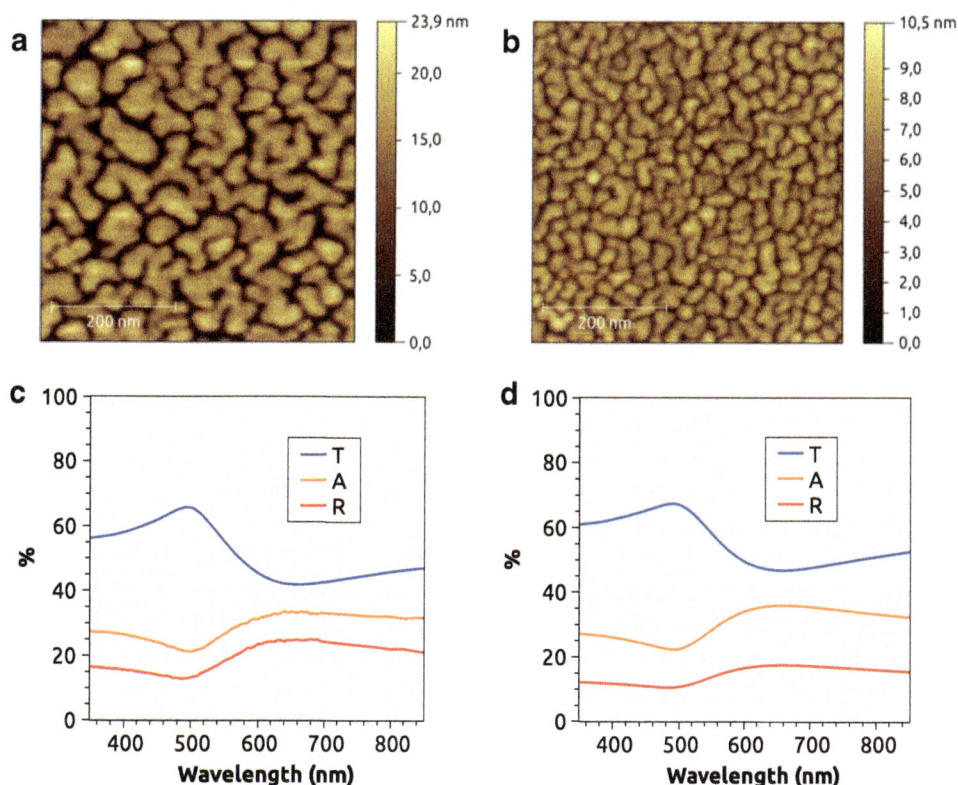

**Fig. 2** AFM topography (**a**, **b**) and far-field spectra (**c**, **d**) of the random gold structures deposited by thermal evaporation under ultrahigh vacuum (**a**, **c**) and radio frequency sputtering (**b**, **d**)

due to the opening of new desexcitation channels that can be radiative or non-radiative. The intensity of the fluorescence is then strongly modified. Figure 3a shows these modifications on both structures, where the intensity is plotted versus $F_P$ for a large set of NCs.

On Fig. 3a, a strong increase of the decay rate is observed. For each set of data, the range of $F_P$ values is wide, with a factor of 10 between the less and the most accelerated NCs. For the big structures, $F_P$ ranges from 6 to 60. For the small structures, $F_P$ ranges from 12 to 120. This wide distribution is attributed to the random distribution of

positions and intensities of the *hot spots* on the surface of the film. Comparing the two sets of data, it is clear that the emission is more accelerated on the small structures: the $F_P$ distribution is centered on higher values on the small structures than on the big ones but with a comparable width.

The strong modification of the collected fluorescence intensity is observed on Fig. 3a. The intensity of fluorescence collected by the setup for NCs on gold structures is normalized by the intensity collected on glass coverslip. The different repetition rates of the laser used in both cases has been taken into account. The intensity is far lower on

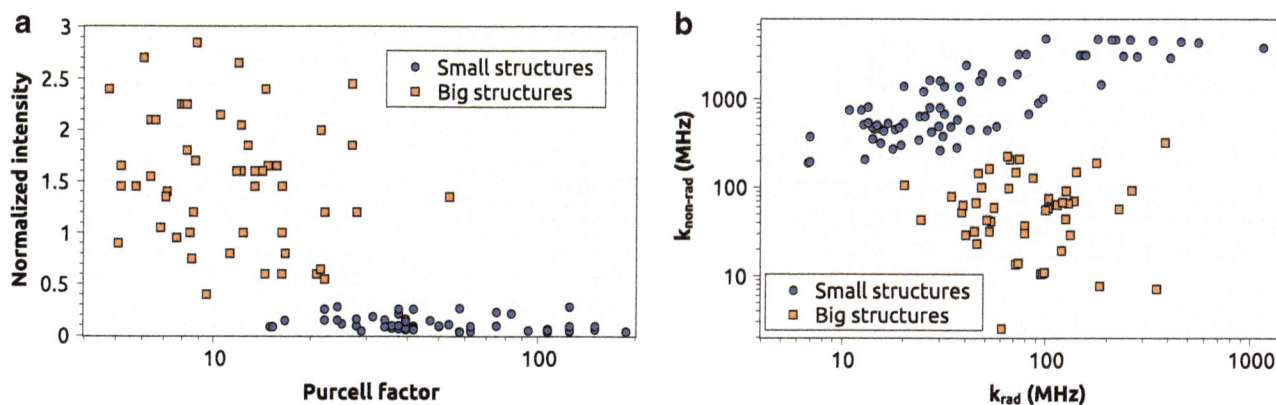

**Fig. 3** Modification of the single NCs fluorescence properties due to the coupling with the gold films: **a** the normalized intensity $\left(\frac{I}{I_{\text{glass}}}\right)$ versus the Purcell factor $\left(\frac{\tau_0}{\tau}\right)$ for NCs deposited on big (*orange*

*squares*) and small (*blue circles*) structures and **b** radiative against non-radiative decay rates both for the coupling on big (*orange squares*) and small (*blue circles*) structures

small structures and the ratio of signal-to-noise is too low to keep the same excitation power than on big structures. A higher excitation, still remaining in a low power excitation regime, has been needed but intensity results have been re-normalized. A wide range of intensities can be obtained, up to three times and down to 0.05 times the reference intensity on glass coverslip. Here again, this distribution is due to the wide variety of plasmons existing on this type of structures. The intensities are clearly higher for the big structures, with values ranging from 0.4 to 2.9 times the reference on glass coverslip. For the small structures, most of the fluorescence is lost in non-radiative channels as the intensity is always lower than 0.4 times that on the glass coverslip.

A common set of $F_P$ values can be found on both structures, from 12 to 60, but even for this set, intensities are far lower on small structures. This means that the desexcitation channels on the two structures are completely different. The number of non-radiative channels is more important on the small structures. The calculation of the number of radiative and non-radiative channels is done as follows. First, we select the intensity corresponding to the neutral state, giving us $F_{coll}$. Then, we extract the fluorescence lifetime, giving us the total decay rate $\tau = \frac{1}{k_{tot}}$. Using the two well-known equations

$$k_{tot} = k_{rad} + k_{non-rad}$$

and

$$F_{coll} = \frac{k_{rad}}{k_{rad} + k_{non-rad}},$$

we can deduce the values of $k_{rad}$ and $k_{non-rad}$, respectively (details can be found in [25]). This procedure has been applied on the experimental data, and the results are plotted on Fig. 3b.

A wide range of radiative and non-radiative decay rates is obtained. For both structures, the range of radiative decay rates is nearly the same, slightly wider for the small structures. The radiative channels are comparable on both structures. On the contrary, the value of the non-radiative decay rate is always higher for the small structures. The distribution is still wide, clearly higher for small structures. Non-radiative channels are much more likely than radiative ones. This explains the data shown on Fig. 3a. For the small structures, desexcitation is mostly due to non-radiative processes. The acceleration of PL decay is higher on small structures due to a higher total number of decay channels. Then, higher $F_P$s on the small structures correspond to lower intensities.

The size of the initial clusters seems to play a crucial role. This could be in agreement with results obtained for the coupling of single emitter with gold nanospheres [26]. The smallest is the sphere, the strongest is the coupling, but also the lowest is the collection. Nevertheless, one has to bear in mind that the metallic structures studied in this article are much more complex than single nanoparticles. For a more complete understanding, one has to take into account the distance between the clusters, the fact that the emitter inter-acts with many of them and the roughness of the film [27]. The influence of long range interaction processes might also be considered.

However, our results could explain crucial differences obtained by several groups concerning the coupling of sin-gle emitters with random structures [28–30]: it strongly depends on the size of the gold clusters and the method used to obtain the random gold films.

## Conclusion

All the above results show that the desexcitation channels are really different on both structures even if the two films are at the vicinity of the percolation threshold. A higher number of desexcitation channels are present on the small structures, but they are mainly non-radiative. It results in strongly accelerated decay processes with very low col-lected intensity. Fastening the desexcitation processes on small structures is mainly due to non-radiative processes. On the contrary, on the big structures, the total number of desexcitation channels is lower, but the ratio between radia-tive and non-radiative ones is more balanced. The radiative processes are slightly less accelerated, but the collection of the fluorescence intensity is far higher. The coupling then realizes a compromise between the fastening of recom-bination process and the collection efficiency. Our results finally show that the coupling between a single emitter and a random gold film near the percolation threshold depends strongly on the microscopic structure of the film, both in terms of Purcell effect and far-field emission.

**Acknowledgments**  This work has been supported by the Region Ile-de-France in the framework of du DIM "des atomes froids aux nanosciences" by Agence Nationale de la Recherche (under grants QDOTICS ANR-12-BS-008), and by the Institut Universitaire de France.

## References

1. Brokmann X, Messin G, Desbiolles P, Giacobino E, Dahan M, Hermier J-P (2004) Colloidal CdSe/ZnS quantum dots as single-photon sources. New J Phys 6:99.
2. Michalet X, Pinaud FF, Bentolila LA, Tsay JM, Doose S, Li JJ, Sundaresan G, Wu AM, Gambhir SS, Weiss S (2005) Quan-tum dots for live cells, in vivo imaging, and diagnostics. Science 307:538–544.
3. Tomczak N, Jańczewski D, Dorokhin D, Han M-Y, Vancso G J (2012) Enabling biomedical research with designer quantum dots. Nanotechnol Regen Med, Methods Mol Biol 811:245–265.

4. Talapin DV, Lee JS, Kovalenko MV, Shevchenko EV (2010) Prospects of colloidal nanocrystals for electronic and optoelectronic applications. Chem Rev 110:389–458.

5. Sargent EH (2012) Colloidal quantum dot solar cells. Nat Photonics 6:133–135.

6. Shirasaki Y, Supran GJ, Bawendi MG, Bulović V (2013) Emergence of colloidal quantum-dot light-emitting technologies. Nat Photonics 7:13–23.

7. Lakowicz JR (2006) Principles of fluorescence spectroscopy. Springer, New York

8. Purcell EM (1946) Spontaneous emission probabilities at radio frequencies. Phys Rev 69:681.

9. Gayral B, Gérard JM, Sermage B, Lemaître A, Dupuis C (2001) Time-resolved probing of the Purcell effect for InAs quantum boxes in GaAs microdisks. Appl Phys Lett 78:2828.

10. Gérard JM, Sermage B, Gayral B, Legrand B, Costard E, Thierry-Mieg V (1998) Enhanced spontaneous emission by quantum boxes in a monolithic optical microcavity. Phys Rev Lett 81:1110–1113.

11. Kahl M, Thomay T, Kohnle V, Beha K, Merlein J, Hagner M, Halm A, Ziegler J, Nann T, Fedutik Y, Woggon U, Artemyev M, Pérez-Willard F, Leitenstorfer A, Bratschitsch R (2007) Colloidal quantum dots in all-dielectric high-Q pillar microcavities. Nano Lett 7:2897–2900.

12. Goldberg D, Menon VM (2013) Enhanced amplified spontaneous emission from colloidal quantum dots in all-dielectric monolithic microcavities. Appl Phys Lett 102(081119).

13. Drexhage KH, Kuhn H, Schäfer FP (1968) Variation of the fluorescence decay time of a molecule in front of a mirror. Ber Bunsenges Phys Chem 72:329.

14. Song J-H, Atay T, Shi S, Urabe H, Nurmikko AV (2005) Large enhancement of fluorescence efficiency from CdSe/ZnS quantum dots induced by resonant coupling to spatially controlled surface plasmons. Nano Lett 5:1557–1561.

15. Belacel C, Habert B, Bigourdan F, Marquier F, Hugonin J-P, Michaelis de Vasconcellos S, Lafosse X, Coolen L, Schwob C, Javaux C, Dubertret B, Greffet J-J, Senellart P, Maitre A (2013) Controlling spontaneous emission with plasmonic optical patch antennas. Nano Lett 13:1516–1521.

16. Pfeiffer M, Lindfors K, Wolpert C, Atkinson P, Benyoucef M, Rastelli A, Schmidt OG, Giessen H, Lippitz M (2010) Enhancing the optical excitation efficiency of a single self-assembled quantum dot with a plasmonic nanoantenna. Nano Lett 10:4555–4558.

17. Manjavacas A, García de Abajo F J, Nordlande P (2011) Quantum plexcitonics: strongly interacting plasmons and excitons. Nano Lett 11:2318–2323.

18. Mahler B, Spinicelli P, Buil S, Quélin X, Hermier J-P, Dubertret B (2008) Towards non-blinking colloidal quantum dots. Nat. Mater 7:659–664.

19. Efros AL, Rosen M (1997) Random telegraph signal in the photoluminescence intensity of a single quantum dot. Phys Rev Lett 78:1110–1113.

20. Spinicelli P, Buil S, Quélin X, Mahler B, Dubertret B, Hermier J-P (2009) Bright and grey states in CdSe–CdS nanocrystals exhibiting strongly reduced blinking. Phys Rev Lett 102(136801).

21. Canneson D, Mallek-Zouari I, Buil S, Quélin X, Javaux C, Mahler B, Dubertret B, Hermier J-P (2011) Strong Purcell effect observed in single thick-shell CdSe/CdS nanocrystals coupled to localized surface plasmons. Phys Rev B 84(245423).

22. Javaux C, Mahler B, Dubertret B, Shabaev A, Rodina AV, Efros AL, Yakovlev DR, Liu F, Bayer M, Camps G, Biadala L, Buil S, Quélin X, Hermier J-P (2013) Thermal activation of non-radiative Auger recombination in charged colloidal nanocrystals. Nat Nano 8:206–212.

23. Mandelbrot B (1983) The fractal geometry of nature. Freeman, New York

24. Buil S, Aubineau J, Laverdant J, Quélin X (2006) Local field intensity enhancements on gold semicontinuous films investigated with an aperture nearfield optical microscope in collection mode. J Appl Phys 100(063530).

25. Brokmann X, Coolen L, Hermier J-P, Dahan M (2005) Emission properties of single CdSe/ZnS quantum dots close to a dielectric interface. Chem Phys 318:91.

26. Mertens H, Koenderink AK, Polman A (2007) Plasmon-enhanced luminescence near noble-metal nanospheres: comparison of exact theory and an improved Gersten and Nitzan model. Phys Rev B 76(115123).

27. Biehs S-A, Greffet J-J (2011) Statistical properties of spontaneous emission from atoms near a rough surface. Phys Rev A 84(052902).

28. Cazé A, Pierrat R, Carminati R (2012) Radiative and non-radiative local density of states on disordered plasmonic films. Photonics Nanostruct Fundam Appl 10:339–344.

29. Canneson D, Mallek-Zouari I, Buil S, Quélin X, Javaux C, Dubertret B, Hermier J-P (2012) Enhancing the fluorescence of individual thick shell CdSe/CdS nanocrystals by coupling to gold structures. New J Phys 14(063035).

30. Shimizu KT, Woo WK, Fisher BR, Eisler HJ, Bawendi MG (2002) Surface-enhanced emission from single semiconductor nanocrystals. Phys Rev Lett 89(117401).

# 25

# Stimuli-responsive gold nanohybrids: chemical synthesis and electrostatic directed assembly on surfaces by AFM nanoxerography

**Stéphane Lemonier · Pierre Moutet · Wissam Moussa ·
Mathias Destarac · Laurence Ressier ·
Jean-Daniel Marty**

**Abstract** Stimuli-responsive nanohybrids based on gold nanoparticles coated with poly(acrylic acid) were synthesized. Their intrinsic properties (i.e., zeta potential and hydrodynamic diameter) were easily adjusted through the control of pH. This allows modulating the intensity of the electrophoretic forces exerted by charge patterns on the nanohybrids during the development step in AFM nanoxerography and controlling the directed assembly of nanohybrids onto charged patterns.

**Keywords** Nanohybrids · AFM nanoxerography ·
pH responsive · Directed assembly

## Introduction

In recent years, there has been a considerable interest in the development of metal- and semiconductor-based colloidal nanoparticles (NPs). This interest arise first from their unique optical, electronic, and catalytic properties and also because nanometer-sized structures are appropriate for interfacing with biomacromolecules (proteins, DNA…) and probing intracellular environments [1]. These inorganic nanoparticles are usually coated with an organic or inorganic layer that provides solubility, long-term colloidal stability, and functionalization. These coated NPs, also called "nanohybrids", adopt some characteristics from the components that compose them; synergistic effects can also produce properties not present in any of the parts. Hence, a tremendous amount of studies has been performed, either to control the size and shape of the inorganic core and/or to choose the chemical structure of the stabilizing ligand shell. This allows tailoring NPs with specific properties for applications based on electron transfer (electronics, catalysis, electrochemistry, photochemistry) or lock–key interaction (recognition, sensorial operations,…) [2]. Among them, those which are able to respond to an external stimulus are referred to as "stimuli responsive." Different kinds of stimuli have been described in the literature: (1) physical stimuli: temperature, ionic strength, polarity of solvent, photochemical irradiation, electric or magnetic field, mechanical stress,…; (2) chemical stimuli: pH, specific complexation of ions, redox reactions,…; and (3) biochemical stimuli: DNA hybridization, enzyme/substrate complexes…. These responses are manifested as dramatic changes in the material properties, dimensions, structure, and interactions and may lead to their rearrangement or changes in their aggregation state. This could be taken in good account in an increasing number of applications. For instance, catalytic activity of nanohybrids dispersed in solution can be controlled by applying such a stimulus in order to induce either a change in the organic shell diffusion properties of the nanohybrids or a reversible aggregation useful for the recovering of the catalyst [3]. In biological applications, external stimuli allowed for instance the control of drug delivery for small drugs encapsulated inside those nanohybrids [1].

Additionally, during the past decade, a fascinating new challenge involving the spatially controlled arrangement of

S. Lemonier · W. Moussa · J.-D. Marty (✉)
Laboratoire IMRCP, CNRS UMR 5623, Université de Toulouse,
118, route de Narbonne, 31062 Toulouse, France
e-mail: marty@chimie.ups-tlse.fr

P. Moutet · L. Ressier (✉)
LPCNO, INSA-UPS-CNRS, Université de Toulouse, 135 avenue de
Rangueil, 31077 Toulouse Cedex 4, France
e-mail: laurence.ressier@insa-toulouse.fr

M. Destarac
Laboratoire HFA, CNRS UMR 5069, Université de Toulouse, 118,
route de Narbonne, 31062 Toulouse, France

NPs, into two- or three-dimensional architectures has emerged. Depending on their final structure, these assemblies exhibit many interesting optical, magnetic, or conducting properties, paving the way to novel materials and applications [4–6]. Until now, different approaches based on the use of self-assembled systems [7–10], template effect [11, 12], or application of an external field (magnetic, electric, rheological,...) have been used to reach such a goal [13–15]. However, in order to successfully exploit NP assembly in the many foreseen applications and to ensure efficient scaling-up, a high level of direction and control is required. This necessitates having access to technology allowing formation of nanohybrid building blocks with controlled morphology and properties and to a controlled assembling process. In that context, the use of stimuli-responsive systems could be of special interest either to facilitate the assembly process or to modify the spatial arrangement of the assembled nanohybrids [16]. For instance, Mirkin and colleagues have obtained temperature-controlled organization of gold NPs by modifying them with DNA strands [17, 18]. Liquid crystalline thermotropic or lyotropic hybrid materials whose organization strongly depends on temperature or concentration have been also extensively studied [19–22].

Since the pioneering work of G. Decher on the assembly of organic compounds of opposite charges [12], the use of ionic interactions to direct the assembly of hybrid systems has attracted much attention. Hence, we have previously demonstrated that gold NPs covered with anionic or catanionic compounds can be selectively deposited on simple or complex pattern, bearing the opposite charge using AFM nanoxerography [3, 23]. We aim here to demonstrate and understand in which extent deposition of nanohybrids can be controlled by finely tuning the nanohybrid charge. For this, pH-responsive nanohybrids based on gold nanoparticles coated with poly(acrylic acid) (PAA) were synthesized. Fine adjustment of the zeta potential of these nanohybrids through pH was achieved to study the key parameters leading to the electrostatic directed assembly of gold nanohybrids on surfaces by AFM nanoxerography.

## Experimental section

*Reagents* Tetrachloroauric acid trihydrate (HAuCl$_4$·3H$_2$O), sodium citrate (Na$_3$C$_6$H$_5$O$_7$), acrylic acid (99 %), and sodium hydroxide (NaOH) were purchased from Aldrich Fine Chemicals and were used without further purification. The $O$-ethyl-$S$-(1-methoxycarbonyl) ethyldithiocarbonate MADIX agent (Rhodixan A1) was obtained from Rhodia and used as received. The 4,4′-azobis(4-cyanovaleric acid) (ACVA) initiator (>98 %) was purchased from Janssen Chimica and recrystallized from ethanol before use. Absolute ethanol (AnalaR normapur VWR) was used as received. Using a Purite device, water was purified through a filter and ion exchange resin (resistivity≈16 MΩ cm).

*Characterization of polymer samples* Average number molecular weights ($M_n$) and dispersities ($Đ$) were determined by size exclusion chromatography on an apparatus comprising a Varian ProStar 325 UV detector (dual wavelength analysis) and a Waters 410 refractive index detector, using two Shodex K-805L columns (8×300 mm, 13 μm). For PAA, a 0.1-mol L$^{-1}$ NaNO$_3$ aqueous solution containing 100 ppm of NaN$_3$ was used as eluent, and the system was calibrated with narrow sodium poly(acrylate) standards ranging from 1,250 to 193,800 g mol$^{-1}$. $^1$H NMR spectra were recorded on a Bruker ARX 400 at 400.13 MHz.

Dynamic light scattering (DLS) and zeta potential measurements were carried out at 25 °C with a Malvern Instrument Nano-ZS equipped with a He–Ne laser ($\lambda =$ 633 nm). Samples were introduced into cells (pathway, 10 mm) after filtration through 0.45-μm PTFE micro-filters. The correlation function was analyzed via the nonnegative least square (NNLS) algorithm to obtain the distribution of diffusion coefficients ($D$) of the solutes, and then, the apparent equivalent hydrodynamic diameter ($<D_h>$) was determined using the Stokes–Einstein equation. Mean diameter values were obtained from three different runs. Standard deviations were evaluated from diameter distribution and were equal to 5 nm for all samples. For zeta potential measurements, zeta potential were extracted from mobility values using the Smoluchowski model.

*Transmission electron microscopy* A drop of the aqueous dispersions was placed on a carbon-coated copper transmission electron microscopy (TEM) grid and left to dry under air. For samples needing negative staining, the TEM grids were successively placed on a drop of the NP dispersion for 2 min, on a drop of an aqueous solution of uranyl acetate (2 wt%, 2 min), and finally, on a drop of distilled water, after which the grids were then air-dried before introduction into the electron microscope. The samples were observed with a HITACHI HU12 microscope operating at 70 kV. Size distribution was determined by manual counting on ca. 150 particles, using the WCIF Image J software.

*UV–visible spectroscopy measurements* An Analytik Jena diode array spectrometer (Specord 600) or a BMG Labtech diode array spectrometer (Spectrostar nano) equipped with a temperature control system was used for UV–visible absorption spectra recording (optical path length, 1 cm).

*Polymer synthesis* The typical procedure for synthesizing the RAFT/MADIX polymers of the study were as follows, with AA taken as an example: 7 mg of ACVA, 43 mg of Rhodixan A1, 5 g of acrylic acid, 4 g of ethanol, and 12.5 g of water were placed in a two-neck round-bottomed flask equipped with a magnetic stirrer and a reflux condenser. The solution was then degassed for 15 min by bubbling argon. It was then heated at 70 °C during 4 h, keeping a slow stream of argon in the reactor. After this period of time, the solution was cooled down to ambient temperature, and the polymer was analyzed. AA conversion was >99 % ($^1$H NMR in D$_2$O). Dispersity value was determined by size exclusion chromatography in water equal to 1.7.

*Formation of Au NPs with sodium citrate* Spherical gold nanoparticles were prepared according to the following procedure: to 2 mL of a $1.10^{-2}$-mol L$^{-1}$ HAuCl$_4$ solution, 18 mL of distilled water was added [24]. This solution was boiled under stirring. Then 2 mL of a $38.8 \ 10^{-3}$-mol L$^{-1}$ sodium citrate solution was added under reflux. The NP dispersion turned from yellow to dark, then red color in 10 min. The solution was then cooled down to room temperature.

*Coating of preformed NPs and modification of pH* Typically, for a final polymer concentration of $5·10^{-4}$ wt%, 0.75 mL of a $10^{-3}$ wt% aqueous stock solution of PAA was added to 0.75 mL of aqueous gold NPs stock solution to yield coated NPs ([Au(0)]$_{final}$=$4.5 \ 10^{-4}$ mol L$^{-1}$). Nanohybrid dispersions were then centrifuged to remove excess of unbound organic species. Modification of pH values was performed by addition of small quantities of NaOH or HCl solutions.

*AFM nanoxerography* Charge writing was carried out into 100-nm thin films of 996,000 molecular weight poly(methyl methacrylate) (PMMA) spin-coated on $p$-doped ($10^{16}$ cm$^{-3}$) silicon wafers. It was performed in air under ambient conditions, using a Multimode 8 atomic force microscope from Bruker. Charges were injected into the PMMA thin film by applying voltage pulses with an external generator to a conductive AFM tip. The $z$-feedback was adjusted to control the tip-sample distance during charge writing. The pulse length and the frequency of voltage pulses were fixed at 1 ms (long value compared to the cantilever oscillation period) and 50 Hz, respectively, while the pulse amplitude was fixed at ±80 V. The tip velocity was fixed at 10 μm/s for all experiments. These specific writing conditions were chosen regarding previous works because of their reliability and reproducibility, and this especially at such high voltages without tip and/or sample damaging [25]. Surface potential mappings of these charge patterns was carried out by the AFM-based electric technique of amplitude modulation Kelvin force microscopy (KFM), using a lift height of 20 nm and a scan rate of 0.5 Hz.

After charge writing, a two-stage development was performed [23]: a 30-μL drop of colloidal dispersion with desired pH was deposited on charge patterns for 30 s. pH adaptation from the original pH 5 solution was performed only few seconds prior to the drop in order to prevent agglomeration of the gold nanohybrids. After a 30-s rinsing in pure ethanol, delicate drying of the samples was achieved using a nitrogen steam.

The resulting nanohybrid assemblies were characterized by AFM topographical observations in tapping mode.

## Results and discussion

### Polymer synthesis

The precise control of the macromolecular characteristics is essential to better understand the role of the polymer in NP stabilization and surface modification. In the last years, RAFT/ MADIX polymerization appeared as a method of choice to prepare well-defined water-soluble polymers whose molecular weights can be predicted from reactants stoichiometry [2, 26]. Moreover, this technique is highly tolerant to functional groups usually present in hydrophilic monomers, and it can be carried out in water or polar solvents. PAA is a well-known pH-responsive polymer, which can be easily prepared by RAFT/ MADIX. Thus, PAA with a low molecular weight was synthesized by RAFT/MADIX polymerization mediated by the $O$-ethyl-$S$-(1-methoxycarbonyl)ethyl dithiocarbonate MADIX agent (Rhodixan A1). The conditions of reactions were chosen so that the targeted $M_n$ was 10,000 g mol$^{-1}$ [3]. Indeed, Schneider and Decher showed that the chain length is a crucial parameter for simultaneously avoiding flocculation while obtaining good rates of polyion deposition at the particle surface. Longer polymer chains will bridge and aggregate particles, whereas smaller ones will limit bridging flocculation [27]. Acrylic acid was polymerized in water at 70 °C with a minimum of ethanol to solubilize the hydrophobic Rhodixan A1, using 4′-azobis(4-cyanovaleric acid) as initiator. A monomer conversion greater than 99 % was achieved after 4 h of reaction. At this stage, the MADIX agent had fully reacted as shown by high-performance liquid chromatography (HPLC).

The $M_n$ values of PAA estimated from SEC performed in DMF (with poly(methyl methacrylate) calibration) matched well those predetermined by the initial concentrations of both monomer and Rhodixan A1.

### Synthesis and characterization of Au NPs modified with PAA

Citrate-stabilized gold nanoparticles were obtained through the Turkevitch reaction and the reduction of a HAuCl$_4$·3H$_2$O by sodium citrate [24]. Figure 1a, b shows TEM images and corresponding particle size histogram of the synthesized NPs. The NPs were approximately spherical and well dispersed. The mean particle diameter was 15.6 nm with a standard deviation of 4.0 nm. The nanoparticle dispersions obtained via this procedure exhibited a broad surface plasmon resonance absorption band around 520 nm, resulting in a pink–red color of the solutions. Negatively charge citrate species at the surface of the nanoparticles induced a short-term stability of the Au NPs in water. Moreover, addition of HCl on those NPs induced their irreversible aggregation. Therefore, in order to obtain NPs with a tunable zeta potential, those pristine NPs were subsequently covered with PAA (see "Experimental section") using a "grafting to" approach. PAA have been often reported to act as efficient steric, electrostatic, and also electrosteric barriers against destabilization of aqueous dispersions of inorganic particles. Interaction of PAA with Au NPs was first evidenced by small changes in absorbance spectra. Indeed wavelength at maximum absorbance shifted from 522 for pristine Au NPs to 520 nm after modification with PAA polymer (see Figure S1 in electronic SUPPORTING INFORMATION). Moreover, Fig. 1c displays a typical image obtained after negative staining from Au@PAA$_{10k}$ dispersions. No bare NPs were observed. The nanohybrids clearly showed to have a core-shell morphology, the dark cores corresponding to the electron-dense Au atom embedded into a circular brighter thin polymer shell (lower than 2 nm in the dry state). The data collected confirm the adsorption of PAA on the surface of Au NPs. This adsorption may occur from interactions of either main carboxylic functions or dithiocarbonate terminal functions [2, 28].

The modifications of properties induced by a change of pH were then studied. In solution, PAA has good solubility at high pH values; however, it has poor solubility and even collapses, entangling together to precipitate from solution at low pH values (pH<4) in its acidic form. Although PAA has been extensively used to stabilize Au NPs, its pH-induced properties

have not been thoroughly investigated [3]. When a PAA polymer layer is physically or chemically attached to the gold NP surface, not only a steric stability is acquired but also the properties of the Au NPs can be controlled to some extent in response to one of these small changes in the environmental parameters like pH. Due to their interaction with the inorganic surfaces, the mobility of PAA chains is restricted. For polymer segments close to the anchoring points, relatively higher energy input will be required to undergo transitions [2]. Nevertheless, far from the anchoring point, more space and free volume are available, providing energetically and spatially favorable conditions for rearrangement. Hence, any change of pH value will induce a modification of the ionic charges present on the surface of Au@PAA nanohybrids. The resulting modification of the hydrodynamic diameter and zeta potential of Au@PAA nanohybrids as a function of pH is depicted in Fig. 2a, b. At pH 2, nanohybrids have a zeta potential equal to −5±2 mV. Increasing the pH value up to 7.5 induces a decrease in zeta potential to −55±2 mV due to the deprotonation of the carboxylic groups of PAA. However, complete ionization on polyelectrolytes is certainly difficult to reach due to electrostatic effects exerted by other adjacent ionized groups, especially when considering this adsorbed polymer systems. We then studied the evolution of hydrodynamic diameter as a function of pH value. For pH values above 4, the hydrodynamic diameter was mainly affected by hybridization level of the nanohybrids. Hence, at pH 4.5, a hydrodynamic diameter of, i.e., 13±5 nm was observed in good agreement with the expected values from structural parameters of the nanohybrids. Upon ionization, the coiled chains extended responding to the electrostatic repulsions of the generated negative charges. As a result, the hydrodynamic diameter of nanohybrids slightly increased (up to 20 nm) at higher pH values. For low pH values (pH<4), NPs were covered by PAA in its carboxylic form. The lowering of repulsive interactions and the possibility of hydrogen bonding induced the formation of larger aggregates, which ultimately results in complete flocculation–decantation of the nanohybrids after a few hours. In order to avoid agglomeration of the gold nanohybrids during AFM nanoxerography (see "AFM nanoxerography"), pH adaptation from an original pH 5 solution was performed only few seconds prior to the nanohybrid assembly. In addition, the chemical modifications of the polymer shell resulted in the modification of the spectral properties of the nanohybrids as clearly observed in the UV–visible spectra. Apart from a decrease in intensity with time for low pH values (pH<3, see Figure S1 in electronic SUPPORTING INFORMATION), a slight decrease of lambda max was also observed by decreasing pH (Fig. 2c). Those nanohybrids were easily redispersed by increasing the pH to its initial value.

Thus, those nanohybrids present a negative effective surface charge that could be easily adjusted by controlling the pH value of the colloidal aqueous dispersion.

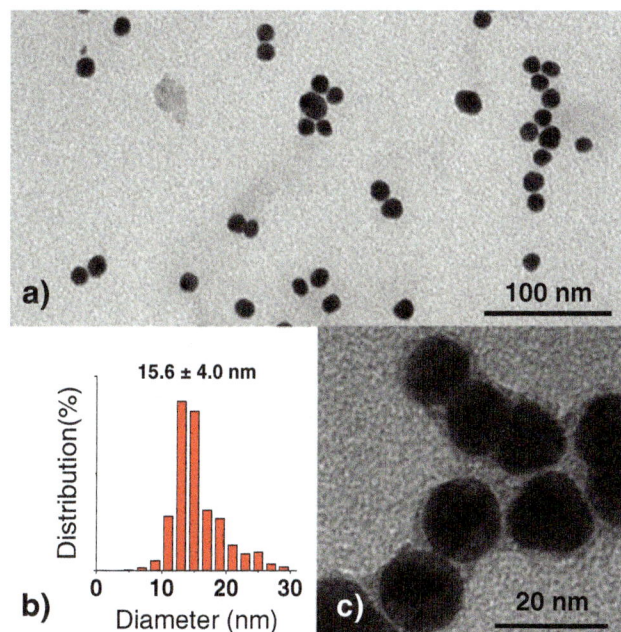

Fig. 1 a TEM image and b corresponding mean diameter distribution of Au NPs obtained by the Turkevitch method ; c TEM image of Au NPs subsequently covered with PAA polymer and negatively stained

**a)**

**b)**

**c)**

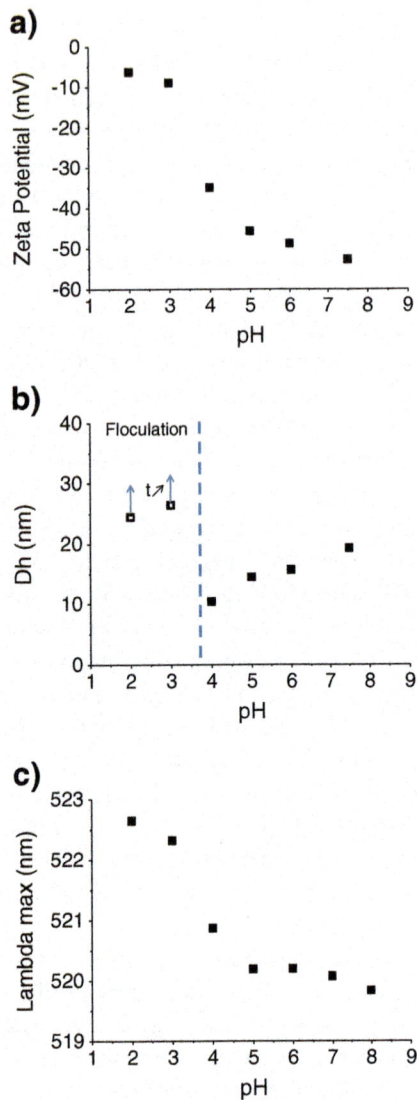

**Fig. 2** Variation of zeta potential (**a**), hydrodynamic diameter Dh (**b**), and lambda max (**c**) as a function of pH for Au@PAA$_{10k}$ nanohybrids ([Au(0)]=4.5.10$^{-4}$ mol L$^{-1}$, [polymer]=5·10$^{-4}$ wt%)

## AFM nanoxerography

Over the past few years, nanoxerography by atomic force microscopy (AFM) has emerged as a versatile method for colloid assembly, directly from the bulk liquid phase onto solid templates [3, 23, 29–32]. This technique is a nanoscale adaptation of the industrial printing process of xerography. It uses the strong electric fields generated by charge patterns written into electret thin films to trap any charged and/or polarizable colloidal nanoparticles via electrostatic interactions. It requires neither expensive cleanroom nor vacuum equipment.

In this work, we used the protocol of AFM nanoxerography that we developed in a previous work [23] to assemble Au nanohybrids from their aqueous dispersions

(Fig. 3). The first step consists of writing charge patterns of desired polarity and geometry into a PMMA thin film through a polarized AFM tip. The second step consists of incubation of a drop of nanohybrid aqueous dispersion on the electrostatically patterned substrates followed by immersion into pure ethanol. The samples are finally dried under nitrogen flow in order to remove any traces of solvent.

During this development step, two electrostatic forces, generated by the local electric field $E$ created by charge patterns, act on the nanohybrids: the electrophoretic forces and the dielectrophoretic forces.

The electrophoretic forces describe the Coulomb interaction between the effective charge of the nanohybrids dispersed in solution and the electric field $E$ generated by the charge patterns. They can be expressed by the Eq. (1):

$$F_{EP} = Q_{ef}E \qquad (1)$$

where $Q_{ef}$ is the effective charge of the nanohybrids dispersed in solution. These forces cause nanohybrids carrying an effective surface charge to be attracted toward the charge patterns of opposite polarity and repelled away from the charge patterns of same polarity. The dielectrophoretic forces arise from the interaction of the nonuniform electric $E$ generated by the charge patterns and the dipole induced by $E$ in the nanohybrids by distortion of the electrical double layer around them. For a spherical nanohybrid of radius $R$, in a suspending solvent with a relative permittivity $\varepsilon_s$, the dielectrophoretic forces can be expressed by the Eq. (2) [33]:

$$F_{DEP} = 2\pi R^3 \varepsilon_s \varepsilon_0 \, \mathrm{Re}[K] \nabla E^2 \qquad (2)$$

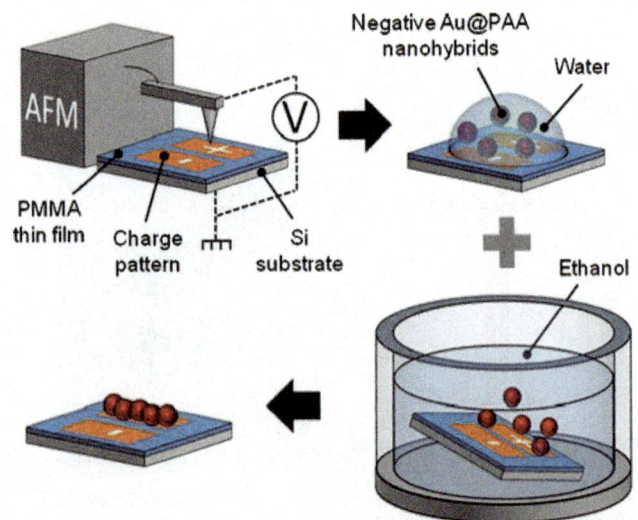

**Fig. 3** Principle of AFM nanoxerography

where Re $[K]$ is the real part of the Clausius–Mossotti factor $K$, given by:

$$K = \frac{\varepsilon_{NH}^* - \varepsilon_S^*}{\varepsilon_{NH}^* + 2\varepsilon_S^*} \quad (3)$$

with $\varepsilon_{NH}^*$ and $\varepsilon_S^*$ the complex permittivities of the nanohybrids and the suspending solvent, respectively. The dielectrophoretic forces are independent of the effective charge of the nanohybrids. They cause nanohybrids which are more polarizable than their solvent to be attracted toward the charge patterns of both polarities.

Figure 4a presents a typical KFM image of two $5 \times 10$-μm rectangular charge patterns of opposite polarity written by AFM in a 100-nm PMMA thin film. As shown on the KFM cross sections, their surface potential is about ±7 V. Figure 2b shows typical AFM topographical images of the resulting directed assembly of gold nanohybrids on charge patterns similar to those presented in Fig. 2a, for various pH of the colloidal dispersions.

At pH 5, these observations revealed that gold nanohybrids were massively grafted on the positively charged rectangle into a compact monolayer and strongly repelled from the negative one (a depletion zone is observed on the negatively charged pattern). In this case, the nanohybrids, presenting a strong negative zeta potential of (−50 mV) (Fig. 4b), were driven by the predominant electrophoretic forces. These forces attracted the negatively charged nanohybrids toward the charge pattern of opposite polarity and repelled them from the charge pattern of same polarity. The dielectrophoretic forces, attractive for charge patterns of both polarities since the nanohybrids are more polarizable than the aqueous solvent, were weak compared to the electrophoretic forces. They contributed for a small part to the directed assembly of nanohybrids on the positive pattern but were not sufficient to counterbalance the repulsive electrophoretic forces on the negative charge pattern.

At pH 4, nanohybrids were selectively assembled on the positive rectangle with a lower density, without any repulsion by the negative one. In that case, the combination of the attractive electrophoretic and dielectrophoretic forces on nanohybrids resulted in their directed assembly on the positive pattern, even if the electrophoretic forces were weaker than in the previous case due to the smaller zeta potential in absolute

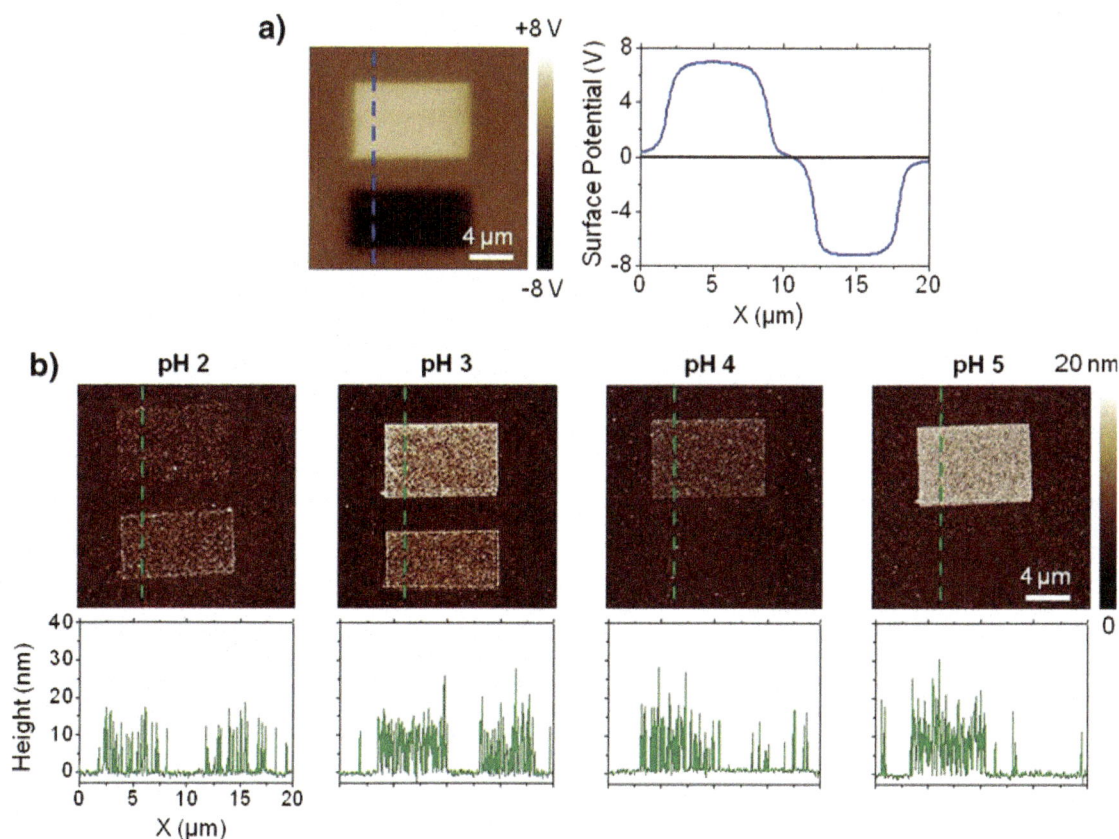

Fig. 4 a KFM surface potential image and associated section of rectangular charge patterns (the *bright pattern* is positively charged, the *dark one* is negatively charged) written by AFM in a 100-nm PMMA thin film, b AFM topographical images and associated sections of the resulting directed assembly of Au@PAA nanohybrids on charge patterns similar to those presented in (a), for different pHs of the colloidal dispersions N.B: from repeated experiments, the density of nanoparticles could not be regarded as significantly different on pattern of opposite charge at pH 2

value (−35 mV) of nanohybrids at this pH. However, the repulsive electrophoretic forces generated by the negative charge pattern were not strong enough in that case to overcome the attractive dielectrophoretic and nonselective forces acting on nanohybrids, resulting into a nonspecific absorption of nanohybrids on the negatively charged pattern.

At pH 3 and pH 2, selective assembly of nanohybrids occurred on both positive and negative charge patterns with similar nanohybrid density, indicating that the directed assembly of nanohybrids was only governed by the attractive dielectrophoresis forces. The effective charge of nanohybrids was too weak at these pH (zeta potential below −10 mV) to induce significant electrophoretic forces. For all pH, it is to note that the directed assemblies of gold nanohybrids were dense enough to be visible by optical microscopy. Very few particles were adsorbed outside the charge patterns.

These results clearly demonstrate that the fine tuning of the effective charge of Au@PAA nanohybrids through pH modulation of the colloidal dispersion allows controlling the intensity of the electrophoretic forces exerted by charge patterns on the nanohybrids during the development step. This offers the opportunity to drive the directed assembly of nanohybrids onto positively charged patterns only or on charge patterns of both polarities.

## Conclusion

Stimuli-responsive nanohybrids based on gold nanoparticles coated with PAA were synthesized. Their intrinsic properties (i.e., zeta potential and hydrodynamic diameter) were easily adjusted through the control of pH. This allows modulating the intensity of the electrophoretic forces exerted by charge patterns on the nanohybrids during the development step in AFM nanoxerography and controlling the directed assembly of nanohybrids onto charged patterns.

**Acknowledgments**   The authors thank the CMEAB for TEM facilities and Juliette Fitremann and Etienne Palleau for fruitful discussions.

## References

1.  Beija M, Salvayre R, Lauth-de Viguerie N, Marty JD (2012) Colloidal systems for drug delivery: from design to therapy. Trends Biotechnol 30(9):485–496.
2.  Beija M, Marty JD, Destarac M (2011) RAFT/MADIX polymers for the preparation of polymer/inorganic nanohybrids. Prog Polym Sci 36(7):845–886.
3.  Beija M, Palleau E, Sistach S, Zhao XG, Ressier L, Mingotaud C, Destarac M, Marty JD (2010) Control of the catalytic properties and directed assembly on surfaces of MADIX/RAFT polymer-coated gold nanoparticles by tuning polymeric shell charge. J Mater Chem 20(42):9433–9442.
4.  Cheon J, Park JI, Choi JS, Jun YW, Kim S, Kim MG, Kim YM, Kim YJ (2006) Magnetic superlattices and their nanoscale phase transition effects. Proc Natl Acad Sci U S A 103(9):3023–3027.
5.  Collier CP, Vossmeyer T, Heath JR (1998) Nanocrystal superlattices. Annu Rev Phys Chem 49:371–404.
6.  Zhuang JQ, Shaller AD, Lynch J, Wu HM, Chen O, Li ADQ, Cao YC (2009) Cylindrical superparticles from semiconductor nanorods. J Am Chem Soc 131(17):6084.
7.  Klein J, Kumacheva E, Mahalu D, Perahia D, Fetters LJ (1994) Reduction of frictional forces between solid-surfaces bearing polymer brushes. Nature 370(6491):634–636.
8.  McMillan RA, Paavola CD, Howard J, Chan SL, Zaluzec NJ, Trent JD (2002) Ordered nanoparticle arrays formed on engineered chaperonin protein templates. Nat Mater 1(4):247–252.
9.  Nie ZH, Fava D, Kumacheva E, Zou S, Walker GC, Rubinstein M (2007) Self-assembly of metal-polymer analogues of amphiphilic triblock copolymers. Nat Mater 6(8):609–614.
10. Park S, Lim JH, Chung SW, Mirkin CA (2004) Self-assembly of mesoscopic metal-polymer amphiphiles. Science 303(5656):348–351.
11. Gavioli L, Cavaliere E, Agnoli S, Barcaro G, Fortunelli A, Granozzi G (2011) Template-assisted assembly of transition metal nanoparticles on oxide ultrathin films. Prog Surf Sci 86(3–4):59–81.
12. Decher G (1997) Fuzzy nanoassemblies: toward layered polymeric multicomposites. Science 277(5330):1232–1237.
13. Gao Y, Tang ZY (2011) Design and application of inorganic nanoparticle superstructures: current status and future challenges. Small 7(15):2133–2146.
14. Mann S (2009) Self-assembly and transformation of hybrid nano-objects and nanostructures under equilibrium and non-equilibrium conditions. Nat Mater 8(10):781–792.
15. Min YJ, Akbulut M, Kristiansen K, Golan Y, Israelachvili J (2008) The role of interparticle and external forces in nanoparticle assembly. Nat Mater 7(7):527–538.
16. Karg M, Hellweg T, Mulvaney P (2011) Self-assembly of tunable nanocrystal superlattices using poly-(NIPAM) spacers. Adv Funct Mater 21(24):4668–4676.
17. Macfarlane RJ, Lee B, Jones MR, Harris N, Schatz GC, Mirkin CA (2011) Nanoparticle superlattice engineering with DNA. Science 334(6053):204–208.
18. Rosi NL, Mirkin CA (2005) Nanostructures in biodiagnostics. Chem Rev 105(4):1547–1562.
19. Hegmann T, Qi H, Marx VM (2007) Nanoparticles in liquid crystals: synthesis, self-assembly, defect formation and potential applications. J Inorg Organomet Polym Mater 17(3):483–508.
20. Mitov M, Bourgerette C, de Guerville F (2004) Fingerprint patterning of solid nanoparticles embedded in a cholesteric liquid crystal. J Physics-Condens Mat 16(19):S1981–S1988.
21. Saliba S, Davidson P, Imperor-Clerc M, Mingotaud C, Kahn ML, Marty JD (2011) Facile direct synthesis of ZnO nanoparticles within lyotropic liquid crystals: towards organized hybrid materials. J Mater Chem 21(45):18191–18194.

22. Saliba S, Mingotaud C, Kahn ML, Marty JD (2013) Liquid crystalline thermotropic and lyotropic nanohybrids. Nanoscale 5(15):6641–6.

23. Palleau E, Sangeetha NM, Viau G, Marty JD, Ressier L (2011) Coulomb force directed single and binary assembly of nanoparticles from aqueous dispersions by AFM nanoxerography. Acs Nano 5(5): 4228–4235.

24. Rahme K, Gauffre F, Marty JD, Payre B, Mingotaud C (2007) A systematic study of the stabilization in water of gold nanoparticles by poly(ethylene oxide)-poly(propylene oxide)-poly(ethylene oxide) triblock copolymers. J Phys Chem C 111(20):7273–7279.

25. Ressier L, Le Nader V (2008) Electrostatic nanopatterning of PMMA by AFM charge writing for directed nano-assembly. Nanotechnology 19(13):135301.

26. Boyer C, Stenzel MH, Davis TP (2011) Building nanostructures using RAFT polymerization. J Polym Sci Polym Chem 49(3):551–595.

27. Schneider G, Decher G (2008) Functional core/shell nanoparticles via layer-by-layer assembly. investigation of the experimental parameters for controlling particle aggregation and for enhancing dispersion stability. Langmuir 24(5):1778–1789.

28. Glaria A, Beija M, Bordes R, Destarac M, Marty JD (2013) Understanding the role of ω-end groups and molecular weight in the interaction of PNIPAM with gold surfaces. Chem Mater 25: 1465–2004.

29. Palleau E, Sangeetha NM, Ressier L (2011) Quantification of the electrostatic forces involved in the directed assembly of colloidal nanoparticles by AFM nanoxerography. Nanotechnology 22(32): 325603.

30. Ressier L, Palleau E, Garcia C, Viau G, Viallet B (2009) How to control AFM nanoxerography for the templated monolayered assembly of 2 nm colloidal gold nanoparticles. Ieee T Nanotechnol 8(4):487–491.

31. Seemann L, Stemmer A, Naujoks N (2007) Selective deposition of functionalized nano-objects by nanoxerography. Microelectron Eng 84(5–8):1423–1426.

32. Tzeng SD, Lin KJ, Hu JC, Chen LJ, Gwo S (2006) Templated self-assembly of colloidal nanoparticles controlled by electrostatic nanopatterning on a Si3N4/SiO2/Si electret. Adv Mater 18(9):1147.

33. Jones TB (1995) Electromechanics of particles. Cambridge University Press, Cambridge

# 26

# Control of selective silicate glass coloration by gold metallic nanoparticles: structural investigation, growth mechanisms, and plasmon resonance modelization

N. Pellerin · J-P. Blondeau · S. Noui · M. Allix · S. Ory · O. Veron · D. De Sousa Meneses · D. Massiot

**Abstract** Soda lime silicate oxide glasses are studied to perform coloration thanks to gold nanoparticles' crystallization. This precipitation is conducted by chemical reduction of gold ions with stannous or antimony oxides as reducing agents. A control of the rendered coloration between blank to red shades has been obtained using $Sb_2O_3$ agent and appropriate thermal treatments. The glasses remain colorless while heating up to 450 °C. Structural glasses evolution is studied by MAS NMR spectroscopy of $^{29}Si$ and $^{23}Na$ nuclei to investigate the silicate network polymerization change and the modification of sodium/oxygen bond length versus nucleation state and growth of Au nanoparticles. A clear decrease of the $Q^2$ species part is observed with nanoparticles growth confirmed by the evolution of chemical shift for $^{23}Na$ resonance. A slight network polymerization is then showed independently of the only thermal treatment. This structural change could be induced by the antimony oxidation and change towards higher coordinations. Finally, the glasses chemical durability has been studied by leaching tests and shows lower alteration for colored glass. The optical spectroscopy applied to colored glasses has given rise to plasmon resonance phenomena at around 600 nm which is the typical surface plasmon resonance of gold for a refractive medium index of 1.5, with a shift of the resonance towards the higher wavelengths with increasing thermal treatment temperature. This shift is modelized by Drude and MIE approaches and confirms the trend observed by UV-visible measurement with an increasing absorption at the SPR correlated to a typical Ostwald growth mechanism according to the increase of the annealing temperature.

**Keywords** Silicate glass–ceramic · Gold nanoparticles · Plasmonic effect · Drude and Mie Model · Chemical durability · MAS NMR spectroscopy · Structure

## Introduction

Cadmium sulfoselenide is currently used as pigment for red glass making [1, 2]. According to the cadmium toxicity, another common way to elaborate ruby glasses is the addition of gold (or cupper) and reducing agents as stannous oxide or antimony oxides to promote the gold reduction, according to the method delivered by the cultural heritage whose first historical trace is at the fourth century [1, 3, 4]. Some thermal treatments are then carried out for gold nanoparticles nucleation and growth associated to the color striking. The gold ruby glass is also still used punctually in luxury bottling, for example by Guerlain or Christian Dior for their perfumes, thanks to metallic dispersions of copper, silver, and gold to obtain colored decorative glass. Nonlinear optical properties of Au nanoparticles constitute also a large field of interest considering chemical and biological functionalities and optical and sensing applications [5].

The peculiar optical properties observed in the visible spectral range are explained by the surface plasmon resonance originating an absorption band usually around 530 nm (for gold) produced by these metal particles in the nanometer size range. However, the colors are variable, depending in part on the concentration and sizes of the gold particles dispersed throughout the glass and in another part of the processing parameters like thermal treatments. Typically, about 0.001–0.1 wt% gold is required to provide coloration in silicate

N. Pellerin · J.-P. Blondeau · S. Noui · M. Allix · S. Ory · O. Veron · D. De Sousa Meneses · D. Massiot
CNRS, UPR3079 CEMHTI, 1D avenue de la recherche scientifique, 45071, Orléans Cedex 2, France

N. Pellerin (✉) · J.-P. Blondeau · S. Noui · M. Allix · S. Ory · O. Veron · D. Massiot
Université d'Orléans, Faculté des Sciences, Avenue du parc floral, BP 6749, 45067, Orléans Cedex 2, France
e-mail: nadia.pellerin@univ-orleans.fr

D. De Sousa Meneses
Institut de Chimie, CNRS, 3 rue Michel-Ange, 75794, Paris cedex 16, France

glasses, and colors are variable between yellow for gold particles less than 5 nm, pink for particles around 10 nm, purple to red between 10 and 20 nm, and finally deep purple hue are observed for gold nanoparticles of 20–50 nm [6].

The glass structure can also be an important parameter if different redox states or coordination sites can exist for the chemical elements responsible of the coloration or if the network fragility can prevent or improve the nanoparticles growth. Wilk and collaborators have developed an example in acetate glasses where gold nanoparticles size can attain large dimensions in the fragile lithium-rich lead–lithium acetate glasses and liquids thanks to their flexibility in the contrary of sodium–potassium–calcium acetate strong samples where the gold nanoparticles are about 10 nm [6]. In the same way, Vosburgh and Doremus have obtained great changes in the kinetics of gold nanoparticles growth in boro-alumino silicate glasses of various viscosities [7].

The goal of this study is to elaborate glasses with a controlled coloration thanks to gold nanoparticles. This constitutes a challenge taking into account the parameters number to optimize the distribution, the size, and the form of the particles and taking into account the difficulties to characterize these nanoparticles. In the present work, conditions are established to get silicate glasses colored or colorless, thanks to reducing agents and specific thermal treatments controlling the color appearance. The nanoparticles formation is studied from optical absorption and couple to Drude and MIE modelization. Moreover, a structural approach is developed, thanks to nuclear magnetic resonance (NMR) spectroscopy to correlate the structure change and the nanoparticles growth.

## Material and method

### Glass elaboration

Silicate oxide glasses have been synthesized from the base molar composition 70 %$SiO_2$–10 %CaO–20 %$Na_2O$ called "base glass." Reducing agents SnO and/or $Sb_2O_3$ have been added for 1 to 2 wt%, and gold has been introduced as gold chloride (AuCl) for a rate between 0.1 and 0.4 wt%. In the following, glasses would be identified with the nomenclature "reducer wt% Au wt%–T," T referring to the temperature of the last thermal treatment following the quench.

A solid phase mixing of high purity powders is prepared from $SiO_2$ (Acros Organics, France, Ultra Pure), $CaCO_3$ (Acros Organics, France, 99 %), $Na_2CO_3$ (Acros Organics, France, 99.95 %), $Sb_2O_3$ (Chempur, Deutschland, 99.9 %), and AuCl (Alpha, 99.9 %) to obtain 30 g of glass. Oxides are put in a 10 % RhPt crucible of 10 $cm^3$ of volume and melted under air in an electric glass furnace. A temperature of 1, 350 °C is applied during 3 h to assure the melting and homogenization, and then the melt is casted in vitreous carbon

crucible (Sigradur®) to obtain glass cylinder of 15 mm high and 10 mm of diameter. Finally, each vitreous sample is annealed at 450°C during 16 h to relax the residual constraints. It is important to follow a precise and equal protocol as variation of cooling rate modifies the glass structure, and consequently, generates broadening of $^{29}Si$ NMR spectra [8]. Finally, the cylinder is cut to form slides of 1 mm thickness that are optically polished.

### DSC analysis

The glass transition temperatures (Tg) have been measured by differential scanning calorimetry (DSC, SETARAM multi-HTC) on a small amount of glass powder (about 0.7 g) in a platinum crucible. The powder is heated in argon with a rate of 10°C/min up to 1,400 °C. Accuracy obtained for Tg determination is ±2 °C.

### TEM study

Samples for transmission electron microscopy (TEM) were prepared from crushed powder, dispersed in absolute ethanol, and deposited onto a holey carbon film supported by a copper grid. Bright or dark field images and electron diffraction patterns were carried out with a TEM (Philips CM20) operating at 200 kV and equipped with an EDX probe.

### UV visible measurements

UV visible measurements have been performed with a dual beam spectrophotometer Jasco V530 in transmission mode in the range 200 to 1,000 nm in order to follow the glass transmission evolution and particularly to detect the surface plasmon resonance appearance and evolution during the annealing treatment.

### Infrared spectroscopy measurements

The measurements were performed with a homemade spectrometer able to measure reflectance and emittance spectra. The device is built around two spectrometers, an air-purged Bruker 70 and a Bruker 80 v working under vacuum. The system is equipped with a set of beam splitters and detectors allowing to acquire spectra in the whole infrared range that is from 50 to 12,500 $cm^{-1}$. A $CO_2$ laser is used to heat the sample at temperatures up to 2,500 K. The laser beam (11 mm of diameter) is divided into two parts by a beam splitter which allows to heat both sides of the sample ensuring a good axial temperature homogeneity. Furthermore, the acquisition of the sample flux is limited to a small area 2 mm of diameter to avoid radial gradients.

NMR spectroscopy

The $^{29}$Si magic angle spinning (MAS) NMR experiments are performed on a Bruker Avance WB 300 MHz (field of 7 T) operating at 59.63 MHz with a 4 mm Bruker MAS probe and ZrO$_2$ rotor. The excitation pulse duration (for a $\pi/2$ pulse angle) was 5.5 µs for a recycle time of 900 s, 192 scans accumulation, and a spinning rate of 10 kHz. The chemical shift of $^{29}$Si spectra is referenced to tetra-methyl-silane at 0 ppm.

In the case of $^{23}$Na (I=3/2), MAS NMR spectra are collected on Bruker Avance 750 MHz (17.6 T) spectrometer using a 2.5 mm rotor. Spectra are acquired with a $\pi/18$ pulse to ensure a quantitative excitation of the central transition and a recycle time of 1 s for a spinning rate of 30 kHz. Chemical shifts are reported with respect to NaCl aqueous solution (0.1 mol/L).

The calculation of NMR spectra was carried out using DM-FIT software [9] from Gaussian-Lorentzian lines in the case of silicon and sodium, the quadruolar coupling constant being very low at this field for $^{23}$Na nuclei.

Leaching experiment and ESEM analysis

The alteration behavior of the colored glasses has been studied in deionized water (pH 7.6) by static leaching experiments at 90 °C in an oven. The experiments are driven in Teflon® vessel of 50 ml, on monoliths, during 44 days. Two samples have been studied, a colorless glass Sb1Au0.1-450 treated at 450 °C during 16 h and a red colored glass Sb2Au0.2-550 treated at 550 °C during 16 h.

After leaching, the cationic concentrations have been determined by inductively coupled plasma–atomic emission spectroscopy (ICP-AES) (Thermo). Three aliquots of 2 ml are taken off during the experiment for each sample. The normalized mass loss NL(X) (in grams per square meter) is given by [10]:

$$NL(X) = \frac{[X]}{f_m(X) \times \frac{S}{V}}$$

Where [X] is the element concentration in the solution (grams per cubic meter), $f_m(X)$ is the mass rate of the element X in the glass (without unity), S/V the ratio of the glass surface area to the solution volume (per meter).

The calculation of the normalized mass losses allows comparing the releasing of the various elements each other in the same matrix or between different glasses. An accuracy of 2 % is retained for Si, Na, and Ca normalized mass losses, 4 % for Sb, and 10 % for Au. The S/V ratio is of the order of 10 m$^{-1}$ for both samples. At the end of the test, the samples are rinsed and dried for observations by environmental scanning electron microscope (ESEM Philips XL40) and qualitative chemical analysis performed by energy dispersive X-ray spectroscopy (EDX). The retained accuracy is 1 % for metals.

**Results and discussion**

The base glass composition 70 %SiO$_2$–10 %CaO–20 %Na$_2$O has allowed to test the red color appearance according to the reducing agents choice and to the thermal treatments performed. In compounds formed with a low amount of gold (0.1 wt%), the stannous oxide is a very efficient reducer for gold cations, and red glass is obtained as soon as the glass quench for 1 wt% of SnO. On the contrary, replacing tin to antimony, glass is kept colorless at the elaboration and after the first thermal treatment at 450 °C performed for the relaxation of the residual constraints. In presence of the both reducer oxides (1 wt% each), the glass is also conserved blank up to the first thermal treatment at 450 °C.

Using antimony, the red coloration appears toward 530 °C, covering only the central part of the slide for a thermal treatment of 16 h (Fig. 1). The increase of the temperature involves a change of the color towards the violet and the extension of the colored region up to the total volume. At a temperature of 590 °C, we have observed the pink color first appearance after 15 mn for a very small area just in the middle of the plate and the extension of the colored area with the duration of the temperature stage. After 1 h, the coloration covers half of the surface, and the total-volume is treated after 5 h. From 3 h, some regions change towards purple coloration. At 650 °C, around 100 °C above the Tg, the glass becomes blue.

A few dispersed nanoparticles have been observed by TEM in the glass heat treated at 590 °C for 3 h 30 min (Fig. 2). The observed particles are very small (5–10 nm mean size) but crystallized as proved by both the electron diffraction pattern matching the cubic structure of gold (4.08 Å, Fm-3 m) and the dark field image. In a sample treated at the same temperature during 16 h, we observe larger spherical nanoparticles with a characteristic size of around 15 nm, along with small ones (3–7 nm). The large ones are unambiguously analyzed by EDX as gold. The dark field images prove the crystalline nature of both size nanoparticles. At 650 °C, in a blue colored glass, the observed particles attain 35 nm in mean but are less spherical. TEM analyses have also been performed on a colorless glass treated at 450 °C. At this stage, no gold nanoparticles have been detected. Furthermore, no demixtion is present in the glasses whatever the stage of the process.

The base-glass structure has been analyzed thanks to MAS NMR spectroscopy. $^{29}$Si NMR spectrum shows a broad signal in the region −110, −70 ppm attributed to Q$^n$ species (n: number of bridging oxygen BO) [11] (Fig. 3). The calculation of the spectra is carried out from three lines (Gaussian-

**Fig. 1** Gold samples coloration according to annealing temperature

Gold samples coloration according to annealing temperature

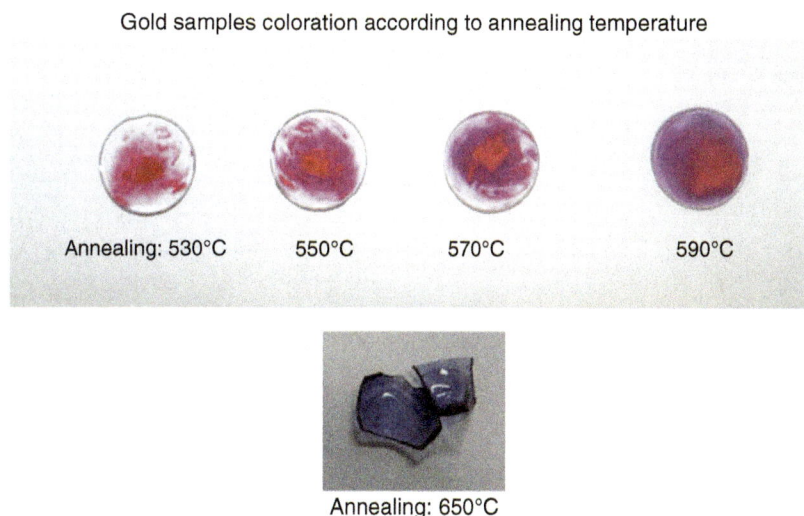

Annealing: 530°C    550°C    570°C    590°C

Annealing: 650°C

lorentzian type) assigned to $Q^2$, $Q^3$, and $Q^4$ species. The major specie is $Q^3$ (78 %) in agreement with the large part of modifier cations sodium and calcium in the glass. A basic calculation leads to a mean ratio NBO/Si=0.86 nonbridging oxygen per silicon, considering the glass composition (one NBO for one $Na^+$, 2 NBO for one $Ca^{2+}$), that can be compared to the ratio NBO/Si=0.88 deduced of the NMR calculation ($Q^3$: 1 NBO/Si, $Q^2$: 2 NBO/Si) (Table 1). The glasses incorporating reducer agents and gold have been analyzed at successive stages of the process. At first, it is important to remark that the NMR spectra changes are low. Figure 4 compares the signals for colored glass obtained after various thermal treatments of the same duration (16 h) to the corresponding colorless glass treated at 450 °C and to the base glass. The signal of the colorless glass Sb1Au0.1-450 is clearly shifted towards the low fields according to the base glass. This is associated to a slight decrease of the Tg from 546 °C for the base glass to 538 °C for the colorless. The spectrum calculation shows a consistent increase of the $Q^2$ part (Table 1). The colored glass Sb1Au0.1-590 offers a slightly modified signal with a shift

towards the lower frequencies according to the corresponding colorless glass. This tendency is verified with the glass annealed at 650 °C. The spectra calculation highlights a decrease of the $Q^2$ specie intensity with the annealing temperature increase. A decrease of the Tg is then observed.

To evaluate the network change with temperature, a further study has been performed on glasses Sb2Au0 without gold and treated in the same conditions (Fig. 4b). We observe a very slight change of the spectra in the temperature range 450–650 °C and an increase of Tg with the temperature annealing. The network polymerization observed in glasses with gold is then relative to the reducing and nanoparticles crystallization processes. The comparison of the spectra for the glasses Sb1Au0.1-450 and Sb2Au0-450 treated at the same temperature shows surprisingly a larger polymerization degree for the second glass with higher antimony content and a lower Tg. This observation shows the change of antimony structural part in these two glasses consequently to the oxidizing state.

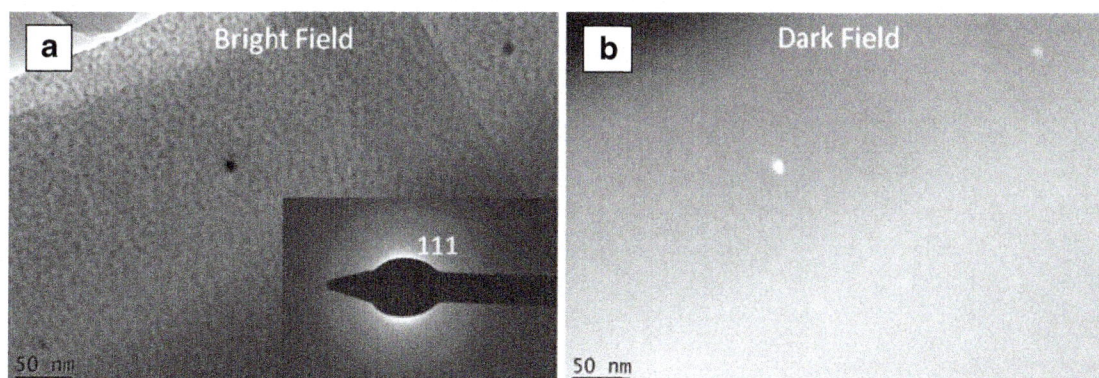

**Fig. 2** TEM patterns of a colored glass (0.4 wt% AuCl) treated at 590 °C for 3 h 30 min. **a** Bright field mode image showing the presence of nanoparticles. The electron diffraction pattern is embedded. **b** Corresponding dark field mode image showing the crystallinity of the nanoparticles

**Fig. 3** $^{29}$Si MAS NMR spectrum
for the base glass calculation
from three $Q^n$ species

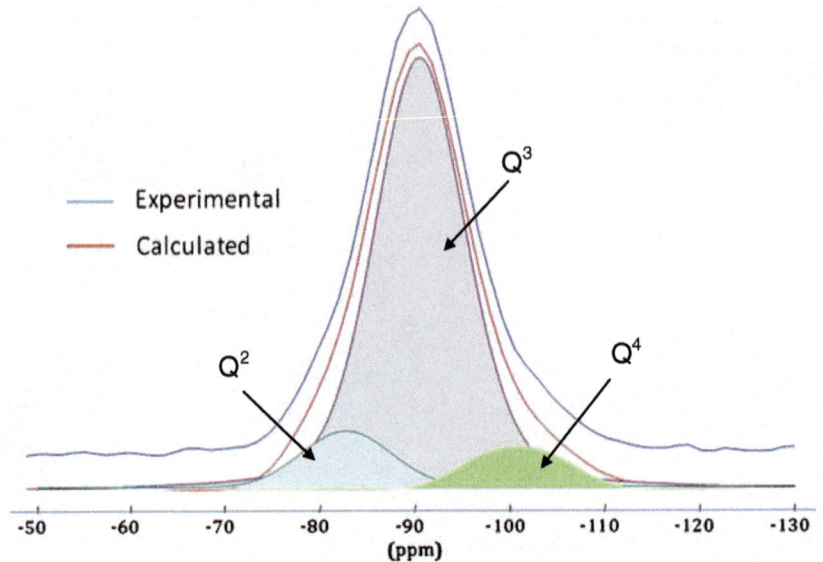

A similar approach has been driven by $^{23}$Na MAS NMR. The resonance is observed towards 4 ppm that is an usual value for the sodium acting as modifier network cation [12]. The spectra comparison shows (Fig. 5) a shift towards the low fields for the line corresponding to the colorless glass Sb1Au0.1-450 according to the base glass. According to Stebbins and collaborators [13, 14], a shift of $^{23}$Na line towards the higher frequencies indicates that the mean bond length Na$^+$–O is reduced, so Na$^+$ coordination is also lower and then the part of NBO in the Na$^+$ neighboring increases. A silicate network depolymerization is then consistent with this $^{23}$Na line shift. For the colored glass Sb1Au0.1-590, Fig. 5 shows that the signal is on the contrary slightly shifted towards the lower chemical shifts (around 0.5 ppm). This tendency is confirmed by the glass Sb1Au0.1-650. The thermal treatments could then induce a noticeable decrease of the NBO part in agreement with the

conclusions concerning $^{29}$Si spectra. To separate the effect of the temperature and the effect of nanoparticles crystallization, the glasses series without gold (Sb2Au0) have also been studied by $^{23}$Na NMR. Here, the spectra are undistinguishable for all thermal treatments. So, we can conclude that the nucleation and growth process for gold nanoparticles induces a slight network polymerization.

The study of chemical durability of glasses has been carried out from leaching tests comparing a non-colored glass (NCG) Sb1Au0.1-450 and a colored Sb2Au0.2-550. The monitoring of leachates pH during time (Table 2) shows an increase relatively similar for the both glasses, independently of the composition differences. This rise indicates the ion-exchange reactions in which modifier cations are replaced by protons and released of the matrix. The pH values of around 10 obtained at the end of the experiment are favorable to hydrolysis phenomenon [15].

**Table 1** NMR parameters: chemical shift $\delta$ (ppm), width $\Delta\delta$ (ppm), and intensity $A$ (%) of the lines deduced from $^{29}$Si NMR spectra calculation and glass transition temperatures Tg (°C)

| Glass | Tg (°C) | $Q^4$ $\delta$ (ppm)/$\Delta\delta$ (ppm)/$A$ (%) | $Q^3$ $\delta$ (ppm)/$\Delta\delta$ (ppm)/$A$ (%) | $Q^2$ $\delta$ (ppm)/$\Delta\delta$ (ppm)/$A$ (%) |
|---|---|---|---|---|
| Base glass-/ | 546 | −101.7/12.5/*16* | −90.8/11.4/*78* | −83.1/11.0/*5* |
| Sb1Au0.1–450 (blank) | 538 | −100.2/12.0/*14* | −89.8/10.3/*71* | −82.2/10.7/*15* |
| Sb1Au0.1–590 (red) | 531 | −100.1/12.8/*17* | −90.0/11.0/*71* | −82.0/11.0/*12* |
| Sb1Au0.1–650 (blue) | 527 | −101.1/12.5/*17* | −89.8/11.4/*77* | −79.5/9.8/*6* |
| Sb2Au0-/ | 539 | −101.4/13.5/*20* | −90.2/11.4/*74* | −80.9/9.8/*6* |
| Sb2Au0–450 | 529 | −100.3/13.5/*17* | −89.7/11.5/*77* | −80.0/10.0/*6* |
| Sb2Au0–590 | 536 | −100.7/12.3/*15* | −89.9/11.5/*77* | −79.7/11.0/*8* |
| Sb2Au0–650 | 540 | −101.0/13.0/*15* | −89.9/11.5/*76* | −80.2/10.5/*9* |

**Fig. 4** $^{29}$Si MAS NMR spectra for **a** base glass compared to a colorless and two colored glasses Sb1Au0.1, respectively, treated at 450, 590, and 650 °C. **b** Glasses without gold nonannealed and treated at 450, 590, and 650 °C

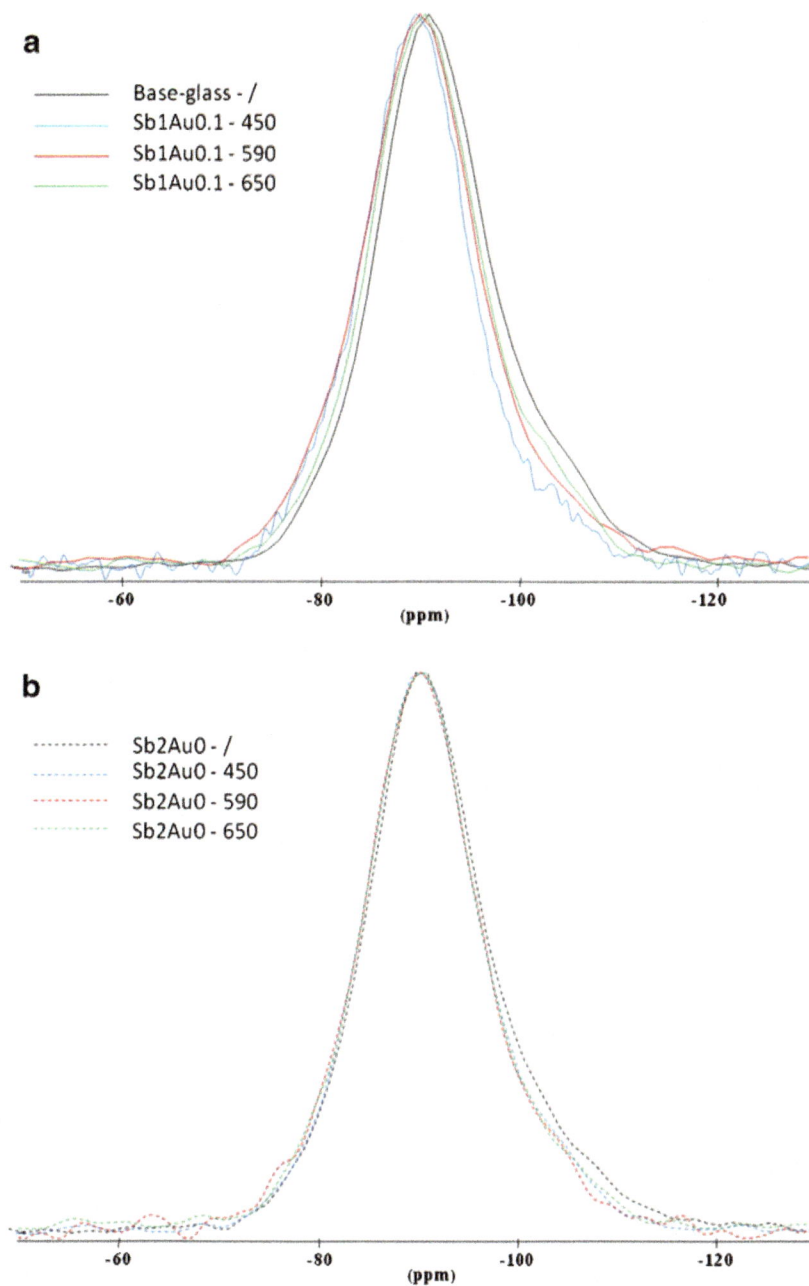

The presence here of large cations can improve the network alteration because they leave behind larger voids providing larger opening into which water can diffuse. According to the calculations of normalized mass losses thanks to ICP-AES measurements, the modifiers cations are strongly released in the solution, following the order Ca>Sb>Na at the beginning (Fig. 6, Table 2). For all cations, the dissolution rate tends to be reduced with leaching time to a lesser extent for sodium according to its strong diffusion ability. Gold is very few released in particular in colored glasses that can be associated to the nanoparticles presence. We observe also a slight decreasing of mass loss for sodium and calcium cations in colored glass compared to colorless, but the normalized mass losses behavior for calcium is complex with a decrease, especially in the case of the colored glass and a slow down followed by a releasing resumption in the colorless glass. The leached glasses' surface has been analyzed by ESEM. An alteration film (a few microns depth) commonly named "gel" is formed in the two cases on large areas (Fig. 7b) and regions with some crystals in spherules have also been observed for the NCG glass

**Fig. 5** $^{23}$Na MAS NMR spectra
for base glass compared to a
colorless glass (450 °C), a red
colored glass treated at 590 °C,
and a blue glass (650 °C)

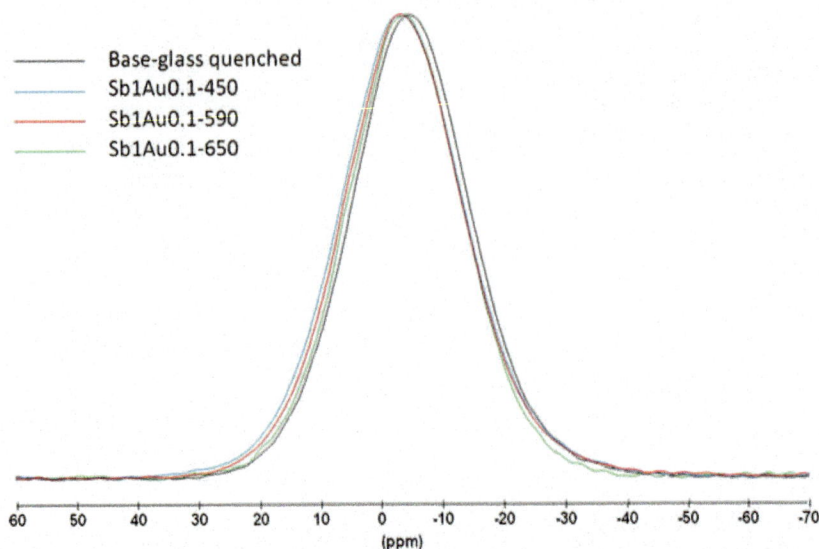

(Fig. 7a). The qualitative mean gel composition has been analyzed by EDX:

CG : 85 mol.% $SiO_2$–13 %mol.CaO–2 mol.% $Na_2O$
NCG : 83 mol.% $SiO_2$–16 %mol.CaO–1 mol.% $Na_2O$

The gel composition is strongly depleted in sodium according to the glass composition in agreement with its "tracer element" nature that is little retained in the condensed products [16]. In another part, the gel layer is slightly enriched in calcium in the NCG sample that can be due to the presence of crystallites enriched in calcium. Then, the gel formation can explain the reduction of cations releasing rate with time, according to its diffusion barrier role and retention properties [17]. The decrease of the calcium mass loss for the colored glass implies some condensation or precipitation of hydrolyzed elements phenomena. The crystallization observed for

**Table 2** Normalized mass losses calculated from ICP-AES for static leaching experiments on colored glass Sb2Au0.2-550 (CG) and non-colored glass Sb1Au0.1-450 (NCG) in g/m² versus time of leaching in hours and pH measurements of leachates during leaching test

| Glass | Time (h) | pH | Si | Ca | Na | Sb | Au |
|-------|----------|-----|-------|------|------|------|-----|
|       | 0        | 7.6 | 0.03  | 6.4  | 1.0  | 0.3  | 0   |
| CG    | 24       | 8.8 | 3.46  | 17.8 | 6.4  | 10.7 | 0.4 |
|       | 576      | 9.9 | 25.55 | 8.8  | 46.6 | 38.9 | 1.4 |
|       | 1056     | 9.8 | 26.83 | 4.1  | 77   | 51.8 | 0.5 |
|       | 0        | 7.6 | 0.03  | 6.4  | 1.0  | 0.3  | 0   |
| NCG   | 24       | 8.8 | 3.76  | 28.2 | 9.5  | 14.5 | 4.7 |
|       | 576      | 10.5| 22.75 | 28.0 | 55.5 | 37.9 | 1.9 |
|       | 1056     | 10.2| 30.84 | 64   | 84   | 50.6 | 1.8 |

the non-colored glass can explain the dissolution resumption in the case of calcium for the colorless glass. Some authors have shown the important influence of pH in the basic range on the crystallization phenomena and the reduction of the protective properties of the gel that can be induced [16]. This difference of behavior between the two glasses could result in the structure change involves by the growth of gold nanoparticles and highlighted by the NMR structural study. Angeli and coworkers have shown how the glass structure and composition influence the gel properties especially in basic media [18].

UV visible measurements acquired in transmission mode (Fig. 8) clearly show the typical plasmon resonance of gold nanoparticles located around 600 nm with an increase of the maximum of the absorption with the increasing temperature until 600 °C (Fig. 9b). The annealed sample at 650 °C shows a drastically move of the resonance towards the higher wavelengths and a decrease of the absorption. Coloration evolution of the annealed samples (Fig. 1) leads to purple coloration with a color expansion when increasing temperature until a blue coloration for the sample annealed at 650 °C. These hue changes are correlated to the evolution of the SPR position (Fig. 9a). For a better comprehension of the UV visible measurements evolution, we have developed Drude and Mie Modelization.

The Drude model allows from known experimental data such as those of "Palik" to consider the interband electronic transitions and intraband for the determination of the contribution of the electrons participating in the absorption effect due to the surface plasmon resonance [19]. The dielectric function $\varepsilon(\omega)$ of the noble metal can be written as:

$$\varepsilon = \varepsilon_1 + i\varepsilon_2$$

**Fig. 6** Comparison of the normalized mass losses during leaching for cations in a colored glass (CG) Sb2Au0.2-550 and colorless glass (NCG) Sb1Au0.1-450

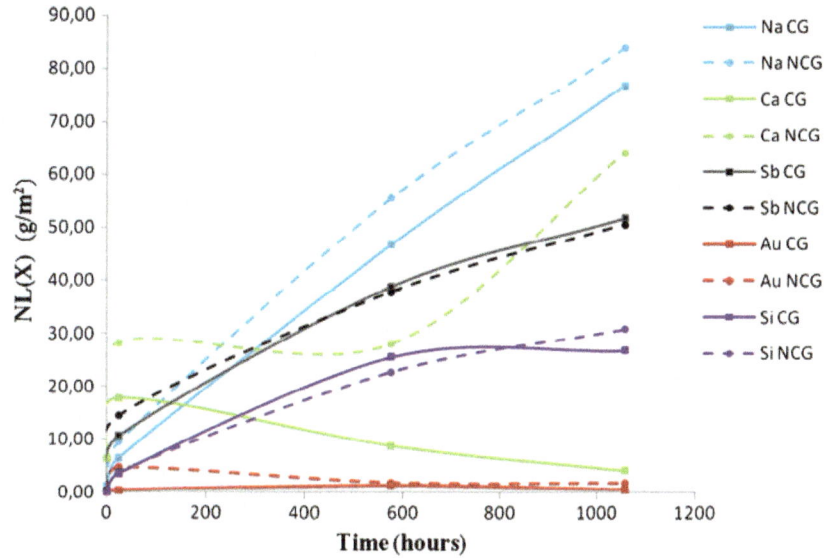

Where $\varepsilon_1$ and $\varepsilon_2$ are, respectively, the real and imaginary parts of the dielectric function. In the noble metals, the contributions of s and d electrons of $\varepsilon(\omega)$ (respectively the band and interband contributions) can be separated:

$$\varepsilon(\omega) = 1 + \chi^s(\omega) + \chi^d(\omega)$$

With:

$$\varepsilon^s(\omega) = 1 + \chi^s(\omega)$$

$$\varepsilon^d(\omega) = 1 + \chi^d(\omega)$$

$$\varepsilon(\omega) = \varepsilon^s(\omega) - 1 + \varepsilon^d(\omega)$$

Where $\chi^s(\omega)$ is the Drude part of the dielectric susceptibility and $\chi^d(\omega)$ the interband part (electrons). Consequently, the dielectric function is given by:

$$\varepsilon^s(\omega) = 1 - \frac{\omega_p^2}{\omega[\omega + i\Gamma(R)]}$$

In many theoretical models using Mie theory, a size effect is introduced by the radius parameter $R$ of a particle in the dielectric function associated with conduction electrons:

$$\Gamma(R) = \Gamma(\infty) + \frac{Av_F}{R}$$

Where $\Gamma(\infty)$ (collision coefficient) and $v_F$ (Fermi velocity) are given in Table 3 for different noble metals, and A=1 is an arbitrary parameter depending on the model. $\omega_p$ is the bulk plasmon frequency Drude given by:

$$\omega_p = \sqrt{\frac{ne^2}{m_e\varepsilon_0}}$$

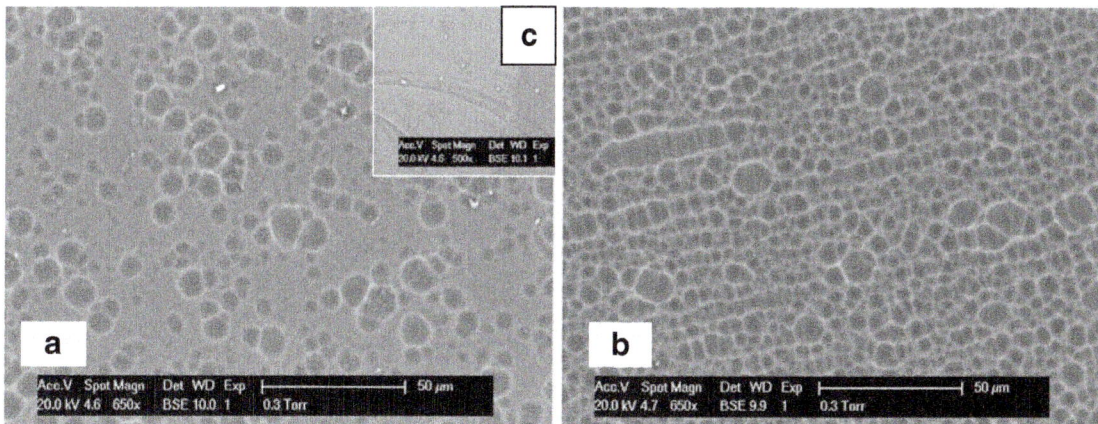

**Fig. 7** ESEM micrographs obtained in BSE mode of leached glass surface. **a** Non-colored glass. **b** Colored glass. As comparison, **c** ESEM micrograph of the non-colored glass where the alteration film has been removed

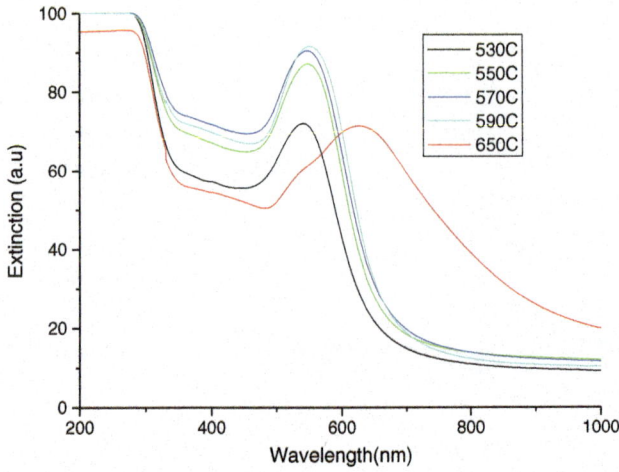

**Fig. 8** Extinction spectra versus wavelength according to annealing temperature (T) for Sb1Au0.1-T glasses

**Table 3** Main characteristics of gold

| Metal | Au |
|---|---|
| Electronic structure | $[Xe]5d^{10}6s^1$ |
| Interband Value (eV) | 1.85 |
| $r_s$ (WS) u.a. | 3.01 |
| $m_e$ u.a. | 1.01 |
| $\hbar\omega_p$ (eV) | 8.98 |
| $\Gamma(\infty)$ (eV) | 0.14 |
| $v_F$ ($10^8$ cm/s) | 1.38 |

Where $n$ is the electron density of the s electrons, $e$ the elementary electric charge, $m_e$ their effective mass, and $\varepsilon_0$ the vacuum permittivity.

$$\frac{1}{n} = \frac{4\pi r_s^3}{3}$$

Where $r_s$ is the Wigner-Seitz radius (WS). It remains to include the interband contribution from data "Palik." Knowing the imaginary part of the interband contribution and taking into account the interband threshold (minimum energy for which the transition occurs more), the determination of the interband contribution is made by the Kramers-Kronig relationship with $P$ representing the main part of the integral. The integration is performed to values up to 9,000 eV.

$$\varepsilon_{re}^d(\omega) = 1 + \frac{2}{\pi}P\int_{\omega_{IB}}^{\infty}\frac{\Omega\varepsilon_{im}^d(\Omega)}{\Omega^2-\omega^2}d\Omega$$

We use the "focus" software developed by Meneses [20] including interpolation type Fritsch-Carlson in order to rebuild the real part. Necessary relations are:

$$\varepsilon_1(\omega) = \varepsilon_1^{(d)}(\omega) - \frac{\omega_p^2}{\omega^2 + \Gamma^2}$$

$$\varepsilon_2(\omega) = \varepsilon_2^{(d)}(\omega) + \frac{\Gamma\omega_p^2}{\omega(\omega^2 + \Gamma^2)}$$

According to Mie's theory [21], we consider a metal sphere of radius $R$ exposed to an external electromagnetic field of wavelength $\lambda$. If the radius is in the nanometer range, we use the quasi-static approach and the effects of delays can be neglected ($R/\lambda \ll 1$). Thus, in the dipole approximation, the absorption cross-section of a metal sphere embedded in a matrix is written:

$$C_{abs}(\omega) = \frac{\omega}{c\varepsilon_0\varepsilon_m^{1/2}}\mathrm{Im}[\alpha(\omega)]$$

Where $\alpha(\omega)$ is the dynamic polarizability of the particle, $c$ is the speed of light, and $\varepsilon_m$ is the dielectric function of the matrix. In the case of a homogeneous sphere of dielectric function $\varepsilon$, $\alpha(\omega)$ is written:

$$\alpha(\omega) = 3V\varepsilon_0\varepsilon_m\frac{\varepsilon-\varepsilon_m}{\varepsilon + 2\varepsilon_m}$$

With $V = \frac{4\pi}{3}R^3$

**Fig. 9 a** and **b** SPR position evolution and maximum extinction value according to annealing temperature

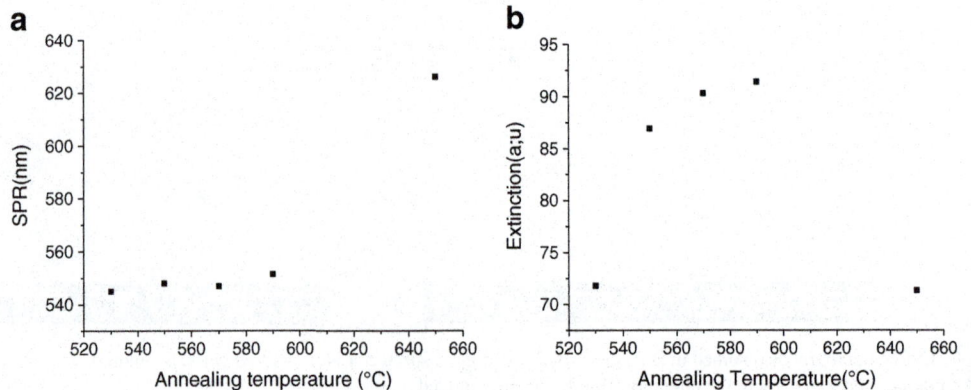

The contribution of the diffusion is given by (in the case of a spherical particle):

$$C_{diff}(\omega) = \frac{3V^2\omega^4}{2\pi c^4}\varepsilon_m^2\left(\frac{(\varepsilon_1-\varepsilon_m)^2 + \varepsilon_2^2}{(\varepsilon_1 + 2\varepsilon_m)^2 + \varepsilon_2^2}\right)$$

The sum of absorption and diffusion cross-section corresponds to the total extinction with:

$$C_{abs} + C_{diff} = C_{ext}$$

As seen in Fig. 10 which gives absorption and diffusion cross-section according to increasing particles size for a refractive medium index of 1.5, we can notice that absorption dominates for particles under 80 nm and diffusion cross-section becomes effective above this size value. We can then conclude from the TEM observations which give a particle estimation between 15 to 35 nm, that absorption is in our case the dominant phenomenon for these temperature ranges below 600 °C. To conclude, according to the modelization and coupled to the UV visible and TEM measurements, the increase of the absorption at the SPR with the annealing temperature is attributed to nanoparticles growth certainly governed by an Ostwald process.

For the particular case of the annealed sample at 650 °C, we observe a broadening of the absorption curve and a red shift of the SPR. Two hypotheses can be proposed to explain this result. The first one comes from the TEM observations which evidence some non-spherical particles leading to a nonhomogeneous field over the extension of the particle and consequently a decay of the SPR. Moreover, we observe for this sample a drastic increase in the FWHM of the transmission curve which is linked to a broadening of the size particles distribution.

The second hypothesis is illustrated in Fig. 11 where the absorption cross-section is plotted for an increasing refractive medium index and a particle size of 8 nm where absorption dominates and lead to a red shift of the SPR. Infrared measurements in reflectivity mode have been driven according to an annealing temperature of the Sb1Au0.1 glass sample from 440 to 1,150 K and for wavenumber from 150 to 2,750 cm$^{-1}$. These measurements are plotted in Fig. 12 according to the annealing temperature and illustrate the increase of the medium refractive index which is visible above 2,500 cm$^{-1}$. However, these two associated hypothesis can explain the behavior of the 650 °C annealed sample.

Gold is known for its low solubility in glasses. Experiments by $^{197}$Au Mössbauer have proved the Au$^+$ state of gold in the quenched colorless glass and the presence of nondissolved metallic gold at this stage [3]. Complementary experiments by $^{119}$Sn Mössbauer spectroscopy are consistent with mainly Sn$^{2+}$ component in the quenched colorless glass and Sn$^{4+}$ in

the red glass [3]. Schreiber and collaborators have established a relative electromotive force series (E' values) of redox couples in soda–lime–silicate melts at 1,400 °C through an indirect procedure by experimentally measuring the equilibrium redox ratios of the individual elements as a function of the imposed oxygen fugacity. E' is deduced of the equation $\log(X)=(n/4)(-\log fo_2)+E'$, where $X$ is the ratio of the concentrations of the element in the reduced state to the oxidized state, $n$ is the number of electrons transferred in the redox couple, and $fo_2$ is the imposed oxygen fugacity [22]. Thanks to this method the authors proposed the following values ($\pm0.3$ units) for E': Au$^{3+}$/Au$^0$>3.6, Sn$^{+4}$/Sn$^{+2}$: −4.9 (estimation), Sb$^{5+}$/Sb$^{3+}$: +0.3 [23]. Otherwise, they show the good correlation between the relative electromotive force series established in the melt and the series of standard reduction potentials in aqueous solution. No value has been proposed for the redox couple Au$^+$/Au$^0$, but considering the correspondence between the two series (melt/aqueous solution), we can also suppose a positive value for this couple, superior or close to the estimated value given for the couple Au$^{3+}$/Au$^0$ by Schreiber et al.

Stannous and antimony oxides are then convenient reducing agents with a stronger power for Sn$^{2+}$ compared to Sb$^{3+}$ following the probable redox equations:

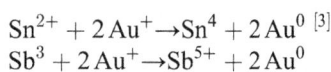

$$Sn^{2+} + 2Au^+ \rightarrow Sn^4 + 2Au^0 \text{ [3]}$$
$$Sb^3 + 2Au^+ \rightarrow Sb^{5+} + 2Au^0$$

However, considering the gold nanoparticles redox potential, it is now well established that the values vary with the nuclearity (atoms number), and that clusters (a few atoms) present a redox potential much more negative than the bulk metal, but this value increases with the nanoparticle formation [24]. Then, the redox mechanism implied during the nucleation and growth process is complex and certainly requires to consider a row of redox couples Au$^+$/Au$_n^0$ according to the gold aggregates size ($n$).

Tin has been found to speed the formation of the metallic nanoparticles by a catalyst role during annealing and an alloy could be formed between tin and gold [3]. In this case, highly dispersed tin seems to provide the presence of many condensation nuclei and the low total amount of gold will then prevent the formation of large gold particles (few nanometers). In the same way, nanocrystalline phases Cu$_y$Sb$_{2-x}$(O, OH)$_{6-7}$ have been highlighted with copper and Cu$_2$O phases in ruby colored cuprous oxide glasses [25]. In our case, no alloy has been detected between gold and antimony and the stronger releasing of Sb compared to Au in our leaching tests favors the hypothesis of alloy absence. However, antimony is most likely involved in the gold nucleation. The TEM observations of two nanoparticles size populations are in agreement with an Ostwald ripening mechanism for particles growth. The color appearance is easier to control using antimony as reducing agent, and colorless glass can be prepared in this case for

Gold: $\omega_p = 9.03\text{eV}$, $\varepsilon_m = 2.25$, $A = 1$     — $C_{abs}$   — $C_{scat}$

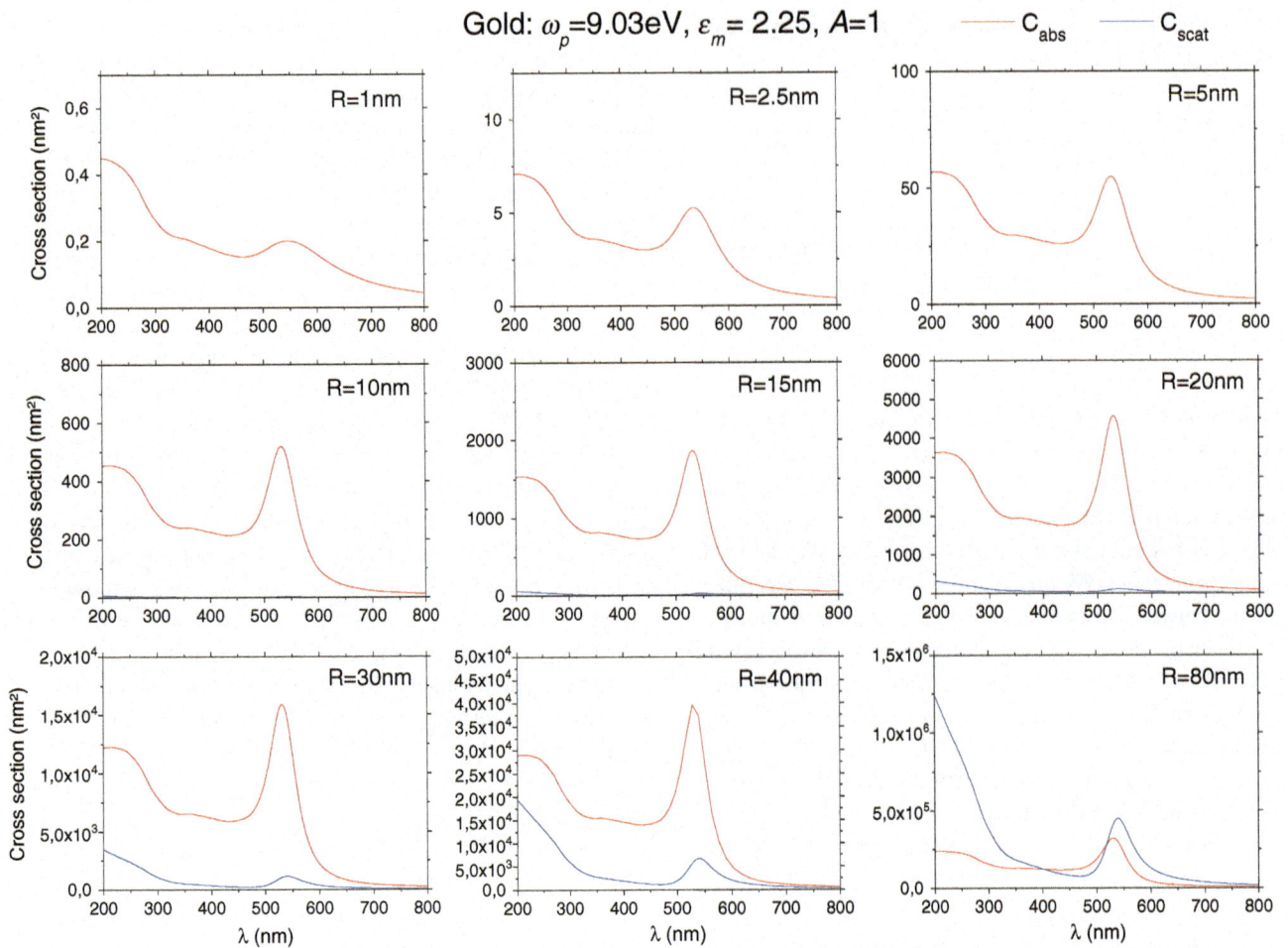

**Fig. 10** Theoretical absorption and diffusion cross-section for gold nanoparticles between 1 to 80 nm

annealing up to 450–500 °C. Local treatments by laser or in a thermal gradient furnace could then be applied for decorative functions or specific marking. First promising results have been obtained by $CO_2$ laser treatment, with local coloration without additional annealing.

As highlighted by the optical measurements (Fig. 9), the coloration process is largely favored by the temperature because it contributes to the species diffusion but also because in this temperature range close or upper the Tg, it contributes to change the network, slightly reducing the rigidity. Then, small particles (15 nm) are observed with treatments up to Tg, but nonspherical larger particles have been obtained with thermal

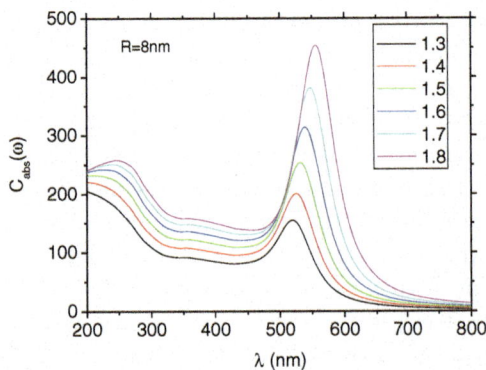

**Fig. 11** Absorption cross-section for increasing the value of the medium refractive index from 1.3 to 1.8

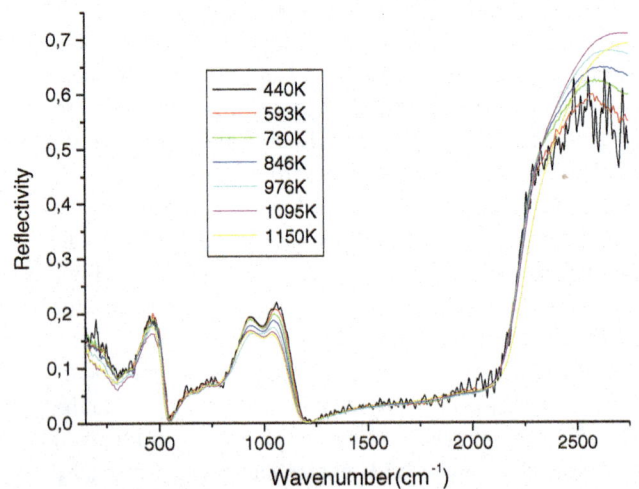

**Fig. 12** Infrared reflectivity measurement of the Sb1Au0.1 glass according to the annealing temperature

treatment above Tg (particles of 35 nm for the treatment 110 °C above Tg). These larger particles are certainly responsible of the SPR red shift promote by a nonconstant electric field over his axis associated to refractive index change due to glass structural change for treatment above Tg. This medium refractive index increases versus the glass temperature has been evidenced by in situ Fourier transform infrared spectroscopy measurements in reflectivity mode.

NMR results show clearly a silicon network change during coloration process with a polymerization tendency. Taking into account the very low gold content, we suppose that these observations result essentially of antimony part. In crystals, $Sb_2O_3$ exists in a cubic form (senarmontite) and an orthorhombic form (valentinite—more stable form at high temperature) with $SbO_3$ trigonal pyramids forming chains in valentinite [26]. The orthorhombic cervantite $Sb_2O_4$ crystals incorporate $Sb^{3+}$ in $Sb^{3+}O_4$ pseudo trigonal bipyramids and $Sb^{5+}$ in $Sb^{5+}O_6$ octahedra [27]. $Sb_2O_5$ antimony oxide is found in monoclinic structure consisting exclusively in $Sb^{5+}O_6$ octahedra. In glasses, there are few studies concerning the structural part of antimony, but authors are often agree to consider the former role of $Sb^{3+}$ according to its field strength. The studies on amorphous systems $Sb_2O_3$, $(1-x) B_2O_3-x Sb_2O_3$ and $(1-x) SiO_2-x Sb_2O_3$ conclude to a former role of antimony with the presence of mainly $Sb^{3+}O_3$ trigonal pyramids (as in valentinite) but also in a low amount of higher coordinated antimony $Sb^{3+}O_4$ pseudo trigonal bipyramid units and $Sb^{5+}O_6$ octahedra [28]. The Mössbauer spectroscopy has proved the $Sb^{5+}$ presence in these glasses in low amount [29]. The structural part of $SbO_6$ octahedra is not clear, but the two structures $SbO_4$ and $SbO_6$ could be paired and serve to connect the network. Also, Wood et al. show the increase of silicon network polymerization when low amounts of antimony are added in an alumino borosilicate glass [30].

During coloration process, $Sb^{3+}$ is expected to oxidize in $Sb^{5+}$ and the annealing treatments lead to the gold nanoparticles nucleation and growth, inducing slight changes in linkages and atomic arrangements. Here, the network polymerization change could be connected to the relative amount of $Sb^{3+}O_4$ and $Sb^{5+}O_6$ units. NMR analyses for $^{29}Si$ and $^{23}Na$ provide these changes attesting a slight polymerization development with a decrease of $Q^2$ units after treatments at higher temperature, but $^{23}Na$ resonance shift could also evidence of a charge balance part of sodium in the antimony environment. Mössbauer spectroscopy studies are in progress to attest and quantify the $Sb^{5+}$ valence.

## Conclusion

Soda lime silicate glasses have been used as a matrix for chemical reduction of gold and nanoparticles crystallization to obtain pink to purple "ruby" glasses type. The reducing chemical agent choice and thermal treatments allow a good control of the coloration appearance and consequently to possible selective volume and surface coloration with local laser heating. The mechanisms have been characterized by complementary analyses. Optical UV-visible analyses and Drude and Mie modelization performed show the distinct part for absorption and diffusion according to the particle size. Absorption is dominant and increase in the range size 15 to 35 nm obtained with annealing carried out between 590 and 650 °C and correlated to Ostwald mechanisms growth. Larger and less spherical particles obtained with treatments well above Tg are associated to the red shift of the SPR with a typical blue coloration, larger size distribution, and change in the refractive medium index evidenced by infrared spectroscopy measurements. The structure analysis by NMR spectroscopy gives indirect information of gold nanoparticles formation thanks to the silicon network changes. Coloration induces a slight increasing of the network polymerization. The signal is very sensitive to antimony presence that is suspected to adopt a tetrahedrally and octahedrally coordinated environment with oxygen. The chemical durability is also slightly increased after coloration with a very low releasing of gold for the colored glass.

**Acknowledgments** We wish to acknowledge the "Region Centre" (France) and the "Cosmetic Valley" hub for their financial support to this work. Furthermore, the authors would like to thank Emmanuel Véron (CNRS - CEMHTI Orléans) for very fruitful discussions and ESEM analysis.

## References

1. Torun Bring (2006) *Red glass coloration: a colorimetric and structural study.* Stockholm, Sweden doctoral thesis in Chemistry Stockholm, Växjö University Sweden
2. Apte SK, Kale BB, Sonawane RS, Naik SD, Bodhale SS, Daset BK (2006) Homogeneous growth of CdS/CdSSe nanoparticles in glass matrix. Mater Lett 60(4):499–503
3. Haslbeck S, Martinek K-P, Stievano L, Wagner FE (2005) Formation of gold nanoparticles in gold ruby glass: the influence of tin. Hyperfine Interact 165:89–94
4. Lafait J, Berthier S, Andraud C, Reillon V, Boulenguez J (2009) Physical colors in cultural heritage: surface plasmons in glass. CR Physique 10:649–659
5. Gonella F, Mazzoldi P (2000) Metal nanocluster composite glasses. In: Nalwa HS (ed) Handbook of Nanostructured Materials and Nanotechnology. Academic, San Diego, p 81
6. Wilk NR Jr, Schreiber HD (1998) Optical properties of gold in acetate glasses. J Non-Cryst Solids 239:192–196
7. Vosburgh J, Doremus RH (2004) Optical absorption spectra of gold nano-clusters in potassium borosilicate glass. J Non-Cryst Solids 349:309–314

8. Sato RK, McMillan PF, Dennison P, Dupree R (1991) High resolution $^{27}$Al and $^{29}$Si MAS NMR investigation of SiO$_2$-Al$_2$O$_3$ glasses. J Phys Chem 95:4483–4489

9. Massiot D, Fayon F, Capron M, King I, Le Calvé S, Alonso B, Durand J-O, Bujoli B, Gan Z, Hoatson G (2002) Modelling one- and two-dimensional solid-state NMR spectra. Magn Reson Chem 40:70–76

10. Jégou C, Gin S, Larché F (2000) Alteration kinetics of a simplified nuclear glass in an aqueous medium: effects of solution chemistry and of protective gel on diminishing the alteration rate. J Nucl Mater 280:216–229

11. Lippmaa E, Maegi M, Samoson A, Engelhardt G, Grimmer A-R (1980) Structural studies of silicates by solid-state high-resolution silicon $^{29}$Si NMR. J Am Chem Soc 102:4889–4893

12. Bunker BC, Tallant DR, Kirkpatrick RJ, Turner GL (1990) Multinuclear nuclear magnetic resonance and Raman investigation of sodium borosilicate glass structures. Phys Chem Glasses 31(1):30–41

13. Stebbins J-F (1998) Cation sites in mixed-alkali oxide glasses: correlations of NMR chemical shift data with site size and bond distance. Sol Stat Ion 112:137–141

14. Stebbins J-F, Zhao P, Kroeker S (2000) Non-bridging oxygens in borate glasses: characterization by $^{11}$B and $^{17}$O MAS and 3QMAS NMR. Solid State NMR 16:9–19

15. Bunker BC (1994) Molecular mechanisms for corrosion of silica and silicate glasses. J Non-Cryst Solids 179:300–308

16. Vernaz EY (2002) Estimating the lifetime of R7T7 glass in various media. CR Physique 3:813–825

17. Gin S (2000) Protective effect of the alteration gel : a key mechanism in the long-term behavior of nuclear waste glass. Mater Res Soc XXIV 663:207–215

18. Angeli F, Gaillard M, Jollivet P, Charpentier T (2006) Influence of glass composition and alteration solution on leached silicate glass structure: a solid-state NMR investigation. Geochim Cosmochim Acta 70:2577–2590

19. Palik ED (1998) Handbook of Optical Constants of Solids, vol III. Academic Press

20. Focus Web Site. Available from: <http://www.crmhti.cnrs-orleans.fr/pot/software/focus.html>.

21. Bohren CF, Huffman DR (2004) Absorption and Scattering of Light By Small Particles. WILEY-VCH Verlag GmbH&Co.

22. Schreiber HD, Coolbaugh MT (1995) Solvatation of redox ions in glass-forming melts. J Non-Cryst Solids 181:225–230

23. Schreiber HD, Wilk NR Jr, Schreiber CW (1999) A comprehensive electromotive force series of redox couples in soda-lime-silicate glass. J Non-Cryst Solids 253:68–75

24. Belloni J (2006) Nucleation, growth and properties of nanoclusters studied by radiation chemistry, application to catalysis. Catal Today 113:141–156

25. Som T, Karmakar B (2011) One-step synthesis and properties of monolithic photoluminescent ruby colored cuprous oxide antimony oxide glass nanocomposites. J Alloys Compounds 509:4999–5007

26. Svensson C (1974) The crystal structure of orthorhombic antimony trioxide Sb$_2$O$_3$. Acta Crystallogr B30:458–561

27. Orosel D, Balog P, Liu H, Qian J, Jansen M (2005) Sb$_2$O$_4$ at high pressures and high temperatures. Journal Solid State Chem 178:2602–2607

28. Orman RG (2010) Characterization of novel antimony (III) oxide-containing glasses. Doctoral Thesis in physics department, University of Warwick, UK

29. Holland D, Hannon AC, Smith ME, Johnson CE, Thomas MF, Beesley AM (2004) The role of Sb(5+) in the structure of Sb(2)O(3)-B(2)O(3) binary glasses - an NMR and Mössbauer spectroscopy study. Solid State Nucl Magn Reson 26(3-4):172–179

30. Wood JG, Prakabar S, Mueller KT, Pantano CG (2004) The effects of antimony oxide on the structure of alkaline—earth alumino borosilicate glasses. J Non-Cryst Solids 349:276–284

# 27

# Shape-dependent electrocatalytic activity of free gold nanoparticles toward glucose oxidation

**Seydou Hebié · K. Boniface Kokoh · Karine Servat ·
Teko W. Napporn**

**Abstract** The synthesis of shape and size-controlled free gold nanoparticles (AuNPs) was achieved by wet chemical methods. The UV–vis spectroscopy measurements and transmission electron microscopy characterizations confirmed the fine distribution in size and shape of the AuNPs. The zeta potential measurements permitted the evaluation of the stability of the AuNPs suspension. For the first time, the shape dependence on the electrocatalytic activity of these NPs is thoroughly investigated. The underpotential deposition (UPD) of lead reveals that their crystallographic facets are affected by their shape and growth process. Moreover, the glucose oxidation reaction strongly depends on the shape of AuNPs. Indeed, the gold nanocuboids (GNCs) and the spherical gold nanoparticles (GNSs) are significantly more active than the gold nanorods (GNRs) followed by the polyhedrons (GNPs). The oxidation process occurs at low potential for GNCs whereas the current densities are slightly higher for GNSs electrodes. Most importantly, the control of the shape and structure of nanomaterials is of high technological interest because of the strong correlation between these parameters and their optical, electrical and electrocatalytic properties.

**Keywords** Gold nanoparticles · Size · Shape · Lead · UPD · Electrocatalysis · Glucose oxidation

## Introduction

Gold nanoparticles continue to emerge in many scientific fields because of their unique physicochemical properties such as optical, magnetic, electronic and catalytic properties compared to those of the bulk material. These unusual properties make them able to offer a wide range of applications [1–3]. Since these properties strongly depend on the shape, the size, the crystallinity and the surface structure of nanoparticles, significant progresses in synthesis methods emerged in recent years for controlling the particles' morphology [4–10]. Among the mentioned synthesis approaches in the literature, the template method, the electrochemical method, the microwave rapid heating, the laser ablation, and the seed-mediated growth methods have been undertaken to yield gold nanoparticles with uniform sizes and morphologies [5, 9, 11–14]. The size and shape control in a wet chemical process has the advantages of low cost and easy scale-up over methods. Within the seed-mediated growth methods, hexadecyltrimethylammonium bromide (CTAB) has been commonly used as a capping agent. It is known as a structure-directing molecule that interacts differently with gold facets in the order $(100) \approx (110) > (111)$ [15]. The preferential adsorption of CTAB on the gold surface allows the preparation of nanoparticles with specific shape and surface crystallographic orientation. It is found that some metallic cations added to the growth solution, such as $Cu^{2+}$ or $Ag^+$, are effective agents that favor the control of the NPs' shape [14, 16]. In fact, $Cu^{2+}$ ions are adsorbed on (111) facets, selectively retarding their growth rate and preferentially leading to a particle growth along (100) facets [16]. The role of $Ag^+$ is similar but promotes the rod-shaped growth [14]. The possibility of controlling the crystallography on the surface of AuNPs has been investigated on gold nanomaterials since the electrocatalytic reactions are structure sensitive [17, 18]. On the one hand, the electrochemical reactivity of a molecule depends on the facility of the surface electrons of the particles to promote a

S. Hebié · K. B. Kokoh · K. Servat · T. W. Napporn (✉)
IC2MP, UMR 7285 CNRS, Université de Poitiers, 4, rue Michel Brunet B-27, BP 633, 86022 Poitiers Cedex, France
e-mail: teko.napporn@univ-poitiers.fr

catalytic process. Therefore, the (111), (100) and (110) reflection planes of a face-centered cubic (fcc) metal such as noble metals have different surface atom densities, electronic structures and chemical reactivity. Thereby, an effective control of the nanocrystal preparation with different shapes becomes essential. On the other hand, a modeling approach by Nørskov et al. [19] showed that small gold nanoparticles can exhibit unexpected catalytic activity toward the carbon monoxide oxidation.

Moreover, gold nanoparticles with high electrocatalytic activities play crucial roles in the performance of biofuel cell design and electrochemical sensors [20]. In the literature, the electrocatalytic oxidation of glucose in alkaline medium was studied on Cu, Ag–Au, Au–Pt, Ag, Pt and Au materials [21–31]. In alkaline and neutral media, gold appeared as an excellent electrocatalyst for several saturated oxygenated organic compounds due to its catalytic activity at low potentials compared to the other catalysts in the same conditions [22, 25, 30]. In addition, Vuyyuru et al. [32] showed that Au has a better catalytic activity than Pt, Pd, Ru and Rh for biomass-derived 5-hydroxymethylfurfural at pH 13.

The electrooxidation of glucose has been investigated on gold single crystal electrodes [33, 34] and also on the anisotropic gold octahedral, nanobelts, and nanoplates whose edges exhibited either (110) or (111) facets [26, 31]. For the oxidation of glucose on gold material in alkaline medium, it was shown that the (110) surface is more active than that of (111) [26, 31, 34]. Therefore, the size and the surface structure of the gold nanoparticles are important parameters for investigating glucose oxidation. The surface sensitive reactions, such as underpotential deposition (UPD) of adatoms, i.e. lead on gold, are a useful tool for characterizing its surface structure distribution [17, 18]. More recently, Hebié et al. [28] showed that the contribution of (100) facets on gold nanorods enhances the electroactivity toward the glucose oxidation.

The present investigation aims to establish a relationship between the shape, the surface structure, and the electrocatalytic activity of gold nanoparticles. Therefore, different shape-controlled gold nanoparticles were synthesized. The UPD of lead is considered as the main electrochemical characterization technique. Even if the shape of the AuNPs is considered as an important parameter, the surface reactivity of such particles is discussed in terms of structure involved in the chemical reaction process.

## Experimental section

*Reagents, apparatus and measurements* Hydrogen tetrachloroaurate (III) trihydrate (HAuCl$_4$·3H$_2$O, 99.9 %), sodium borohydride (NaBH$_4$, 99 %) and D-glucose (>99 %) were obtained from Sigma-Aldrich. In addition, ascorbic acid (AA, >99 %) was received from Alfa Aesar, silver nitrate (AgNO$_3$, >99 %)

was purchased from Fisher Scientific and CTAB (>99 %) was obtained from Fluka. All reactants were used without further purification.

For the electrochemical experiments, the glassware was cleaned according to the methodology described elsewhere [28, 35]. Afterwards, the glassware was rinsed with hot water and kept in aqua regia solution for an hour and finally rinsed with ultrapure water (18.2 MΩ cm at 20 °C).

Nanoparticles preparation

AuNPs of different shapes (spherical, rod, polyhedron and cuboids) were synthesized by wet chemical methods in the presence of CTAB used as capping agent. Such morphology control enables the production of nanoparticles with specific shapes and surface types. It could therefore allow electrocatalytic reactions with great selectivity. The details of the synthesis process were reported elsewhere [5, 14, 16, 28].

*Spherical gold nanoparticle preparation* In a clean 10 mL vial tube, 0.125 mL of 10 mmol L$^{-1}$ HAuCl$_4$ was mixed with 3.75 mL of an aqueous solution of 0.1 mol L$^{-1}$ CTAB. Then, 0.30 mL of an ice-cold 0.05 mol L$^{-1}$ NaBH$_4$ solution was added to the previous solution all at once under stirring for 2 min. Rapidly, the mixture turned light brown indicating the formation of gold seeds. Finally, the solution was kept undisturbed at a constant temperature of 27 °C for 2 h before its use for growing anisotropic nanoparticles. After aging the same solution for 1 day, it was used as spherical gold solution (GNSs).

*Gold nanorod preparation* This procedure requires the preparation of the growth solution as follows. In a clean vial tube, appropriate quantities of reactants were placed, namely 4.75 mL of 0.1 mol L$^{-1}$ CTAB, 0.2 mL of 0.1 mol L$^{-1}$ HAuCl$_4$ and 30 μL of 0.01 mol L$^{-1}$ AgNO$_3$ in this order and mixed gently. The solution appeared bright brown–yellow. Then, 37 μL of freshly prepared 0.1 mol L$^{-1}$ AA was added to the solution which became rapidly colorless due to the reduction of Au$^{III}$ to Au$^{I}$. For the following final reaction step, 10 μL of the seed solution was added, and the mixture was gently mixed for 10 s and left undisturbed for at least 3 h. After ca. 3 min, the colorless solution changed to purple indicating the formation of gold nanorod particles (GNRs).

*Gold nanocuboid preparation* In a typical experiment, 5 mL of a 2.0×10$^{-2}$ mol L$^{-1}$ CTAB solution was mixed with 1.25 mL of 2.0×10$^{-3}$ mol L$^{-1}$ HAuCl$_4$ at 25 °C, into which 12.5 μL of 0.010 mol L$^{-1}$ CuSO$_4$ was added. Then, 0.75 mL of a 0.10 mol L$^{-1}$ freshly prepared AA solution was added. The solution turned from orange to colorless. This solution was used as growth solution. Thereafter, 1.25 μL of the seed solution was added to the growth solution at 27 °C. The color of the solution gradually changed to violet–red within 2–10 min indicating the formation of gold nanocuboid particles (GNCs).

*Gold nanopolyhedron preparation* A procedure similar to that above was used for the polyhedron preparation. Typically, 12.5 μL of $1.0 \times 10^{-2}$ mol $L^{-1}$ silver nitrate and 12.5 μL of $1.0 \times 10^{-2}$ mol $L^{-1}$ $CuSO_4$ were jointly added. Finally, 1.25 μL of the seed solution was added to the growth solution at 25 °C. The color of the solution gradually changed to brownish purple. The combination of silver and copper ions promotes the polyhedron formation.

*Cleaning procedure of AuNPs* The synthesized AuNP solutions were centrifuged in order to remove a large amount of excess surfactant and also to separate the synthesized AuNPs from the other undesired shape particles. Therefore, at a controlled temperature of 27 °C, 5 mL of gold solutions was centrifuged at 14,000 (GNSs), 12,000 (GNRs), and 10,000 rpm (GNCs and GNPs) for 12 min. The supernatant was removed and the solid part was washed with ultrapure water three times.

*UV–vis absorption spectroscopy measurements* Absorption spectra of the GNRs solution were measured using Thermo Fisher Helios Omega UV–vis, near infrared spectrophotometer (NIR). Typical quartz cells of 1 cm optical path length were used for all spectra measurements. Each sample was prepared by dilution of 1 mL of GNRs initial solution with 3 mL of ultrapure water. The UV–visible absorption signal was recorded from 400 to 1,100 nm.

*Dynamic light scattering (DLS) measurements* The effective hydrodynamic size and the zeta potential of synthesized AuNPs were estimated using a Zetasizer Nano-ZS, model ZEN 3600, from Malvern Instruments Ltd., based on multiple light scattering methods; it detects concentration variation in the colloid by scanning the whole height of the sample in transmission and backscattering. Typically, 1.2 mL of 5 μg $mL^{-1}$ of clean AuNPs was introduced in the Zetasizer cell to evaluate the effective hydrodynamic size and the dispersion stability of the AuNPs solution which will be used as a catalytic ink. All sizes reported herein were based on intensity average. The intensity of the average particle size was obtained using a nonnegative least squares analysis method. For each sample, the measurement was performed with a fixed run time of 10 s.

*Transmission electron microscopy analysis* The morphology and the size of the prepared gold nanoparticles were observed with a transmission electron microscope (TEM), a JEOL 2100 UHR (200 kV) electron microscope equipped with $LaB_6$ filament. TEM grids were prepared by placing one drop of the particle solution on a copper grid. This drop was dried in nitrogen atmosphere.

All the electrochemical experiments were carried out at a controlled temperature of 20 °C. The supporting electrolyte was a 0.1 mol $L^{-1}$ NaOH solution. A typical three-electrode Pyrex glass cell was employed. The reference electrode was a reversible hydrogen electrode (RHE). All potentials are referred to this reference electrode. The working electrode was a glassy carbon disk (GC) of 0.071 $cm^2$ mounted in Teflon. The counter electrode was a glassy carbon slab electrically connected with a gold wire. Before any electrochemical measurement, the electrolyte was deaerated by bubbling nitrogen. The nitrogen flow is maintained during the experiment to avoid introducing any oxygen in the electrolyte. Before each experiment, the GC electrode was polished with alumina 0.5 and 0.1 μm, respectively, followed by several ultrasonic cleanings in ultrapure water. Afterwards, a cyclic voltammogram of the GC electrode is recorded in the supporting electrolyte to insure the cleanliness of the cell. After synthesis, the colloidal suspension was collected by centrifugation for subsequent physical and electrochemical characterizations. From this step, the excess of surfactant and the nondesired shape was removed. Gold nanoparticles were first cleaned by centrifugation before their deposition onto a glassy carbon disk, then the working electrode was introduced into the electrolyte under a controlled potential. Voltammetric experiments were performed with a potentiostat (EG&G Princeton Applied Research Model 362) monitored by a computer. The electrochemical cleaning step of AuNPs electrodes was extensively described in the literature [28].

All the current densities in the present work were obtained by dividing the current by the real surface area of each catalytic layer. The real surface area (electrochemically active surface area, EASA) of AuNPs electrodes can be estimated by the charge corresponding to the reduction of the oxide monolayer on the gold surface. This charge also depends on the value of the upper potential limit and equal to 482 μC $cm^{-2}$ in our case [36]. Therefore, the calculated EASA of GNRs, GNCs, GNPs, and GNSs are 0.075, 0.025, 0.054 and 0.012 $cm^2$, respectively.

## Results and discussion

### UV–visible, zeta potential measurements and TEM characterizations

Figure 1 shows the shape-dependent UV–vis spectra of the different AuNPs solutions. For the GNRs, the surface plasmon resonance corresponding to the longitudinal component is observed at 793 nm [4, 14, 37]. However, the other AuNPs present mostly the plasmon resonance excitation at 520, 555, and 560 nm for GNSs, GNCs and GNPs, respectively. In

**Fig. 1** UV–visible absorption
spectra of suspended AuNPs
solutions: GNCs (*dashed–dotted
line*), GNPs (*dotted line*), GNSs
(*dashed line*), and GNRs (*solid
line*). The *inset image* shows the
colors of the AuNPs solutions

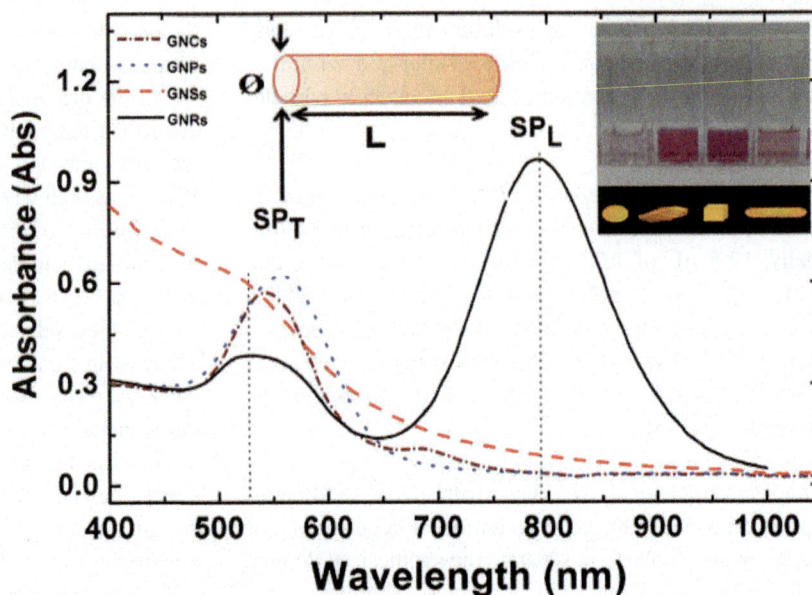

addition, the GNCs show another plasmon resonance excitation at 680 nm which can be assigned to the quadrupole plasmon resonance mode [38] or to the few anisotropic nanoparticles observed during the synthesis. The inset image in Fig. 1 illustrates the different colors of the gold nanoparticle solutions.

After the cleaning step by centrifugation, the AuNPs solutions obtained were optically, spatially homogeneous and can be used for electrochemical investigations. Preventing any nanoparticle aggregation in the colloidal solution after removal of the surfactant can be the way to maintain their properties for various applications. Besides, the size and the shape of the particles and the stability of colloid suspension depend on the environment, i.e. the solvent which contains the particles [39]. Therefore, the study of the free particles' environment becomes essential. The aggregation of the particles leads to a modification of their surface structure, then to an increase of their size. In this study, the measurement of the zeta potential of the samples was performed for understanding the stability of the free gold solution. According to the literature, nanoparticles with zeta potentials greater than +30 mV or less than −30 mV have sufficient electrostatic repulsion to remain stable in solution [40]. Nanoparticles are stabilized either by charge or steric strains. Table 1 shows the hydrodynamic size and the zeta potential values of AuNPs solutions. No sedimentation was observed. For AuNPs preparation, CTAB has been used as surfactant. The zeta potential of 0.1 mol $L^{-1}$ CTAB leads to a positive value *ca.* +59±12 mV. This positive value strongly imposes the positive charge to the nanoparticles synthesized by CTAB. A high value (+75±11 mV) of zeta potential was obtained for spherical AuNPs because of their facility for the Brownian motion. High particle sizes lead to low zeta potential values as in the case of GNCs and GNPs.

The experimental hydrodynamic size obtained for GNRs is 37 nm while their length by TEM is 33.2±6.0 nm.

Figure 2 shows the TEM images of synthesized AuNPs. The AuNPs are quite uniform in shape and size as expected (micrographs *(a)–(d)*). However, a few undesired shapes are observed in the GNRs and GNCs samples. Typically, the average particle sizes are 6.2±1.8, 29.2±2.9 and 36.8± 4.9 nm for spherical, cuboid and polyhedral shapes, respectively. The GNRs have an aspect ratio of 3.3±0.7 for an average length of 33.2±6.0 nm. The percentage of the nonrod-shaped nanoparticles represents about 5 % in this sample. For GNCs, the average particle size is 29±3 nm and this sample also contains around 14 % of noncuboid particles. Typically, the average particle sizes are 6±2 and 37±5 nm for spherical and polyhedral shapes, respectively (see Supplementary Material). In addition, the high-resolution transmission electron microscopy images of GNRs present some dislocation already examined by Wang et al. [41] which can affect their electrochemical behavior.

Electrochemical characterization of the AuNPs surface

Figure 3 shows the cyclic voltammetry of GNSs in 0.1 mol $L^{-1}$ NaOH recorded at 20 mV s$^{-1}$ and 20 °C. This CV represents the typical behavior of spherical gold nanoparticles in alkaline medium. Three important regions can be observed: during the positive scan, the double layer region observed from 0.05 to 0.9 V follows by the oxidation of the GNSs surface from 0.9 to 1.6 V leading to the oxides' formation and the reduction of these oxides during the negative sweep from 1.4 to 0.85 V. In addition, the oxygen evolution begins on these NPs at 1.6 V. In the double layer region which corresponds to the low currents, the chemisorption of OH⁻ species takes place at around 0.8 V.

**Table 1** Summary of physical properties of AuNPs (UV–vis, TEM, DLS data)

| AuNP solution | Plasmon absorption wavelength (nm) | | Mean particle size (nm) | Effective hydrodynamic size $D_H$ (nm) | Zeta potential (mV) |
| --- | --- | --- | --- | --- | --- |
| | SP$_T$ | SP$_L$ | Aspect ratio | | |
| Rod | 535 | 793 | 3.3±0.7 | 37 | 45±19 |
| | SPR | | | | |
| Spherical | 520 | | 6.2±1.8 | 37 | 75±11 |
| Cuboid | 555 | | 29.2±2.9 | 54 | 26±10 |
| Polyhedron | 560 | | 36.8±4.9 | 59 | 28±5 |
| CTAB | – | – | – | | 59±12 |

This adsorption is essential for the surface oxidation of AuNPs. The CV comparison of the different AuNPs synthesized was made and presented in the Supplementary Material.

The surface structure of noble metal nanoparticles can be characterized by electrochemical methods [18, 28]. Indeed, the UPD of lead on gold material is one of the sensitive electrochemical methods, which reveals the crystallographic structure and quantifies the ratio of the existing facets [18, 28]. Figure 4 shows the CVs of the different AuNPs electrodes, GNRs (solid line), GNPs (dashed line), GNCs (dashed dot line), and GNSs (dot line) in 0.1 mol L$^{-1}$ NaOH+1 mmol L$^{-1}$ Pb(NO$_3$)$_2$. During the negative potential sweep from 0.85 to 0.25 V on the AuNPs electrode, two peaks assigned to deposition of lead on (110) and (111) facets were observed at around 0.52 and 0.38 V. The three electrodes (GNSs, GNPs and GNRs) do not show a significant peak corresponding to the (100) facet.

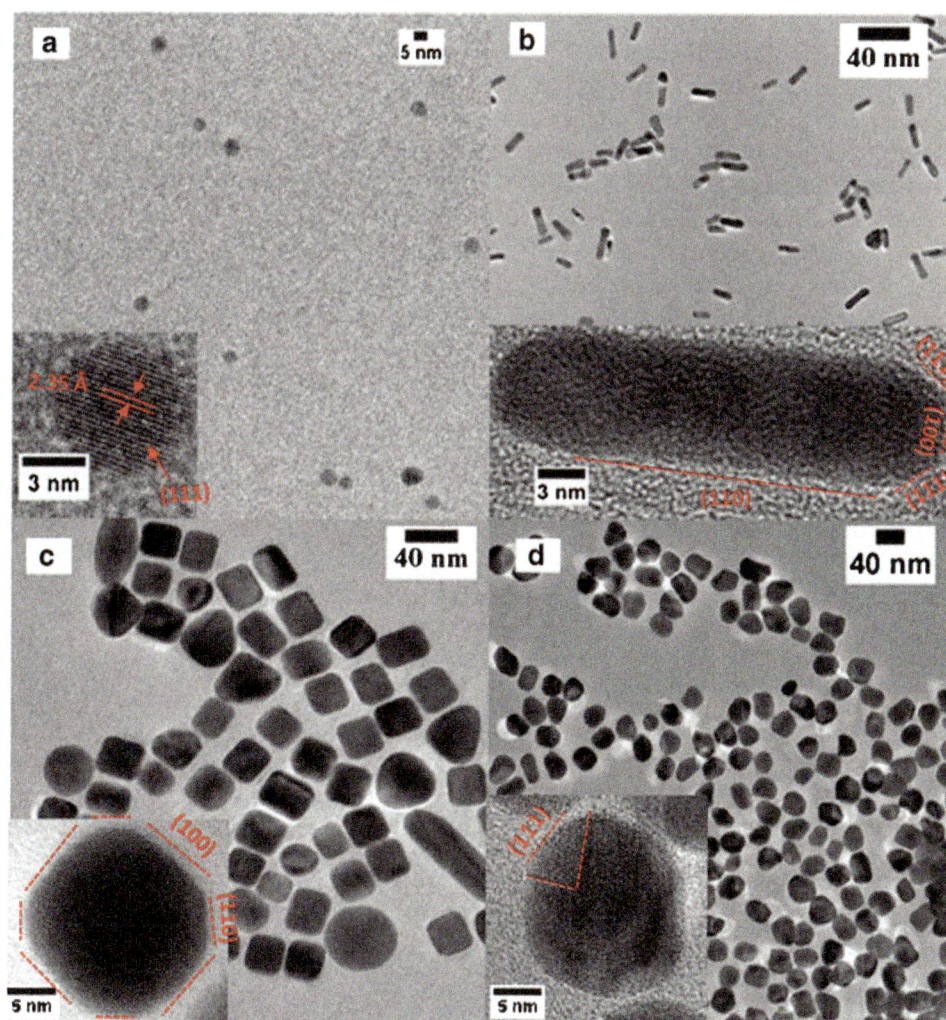

**Fig. 2** Transmission electron micrographs (TEM) and high-resolution transmission electron micrographs (HRTEM) showing the lattice planes and the exposed facets of GNSs, GNRs, GNCs, and GNPs

**Fig. 3** Cyclic voltammogram of GNSs electrode in 0.1 mol L$^{-1}$ NaOH recorded at 20 mV s$^{-1}$ and at controlled temperature of 20 °C

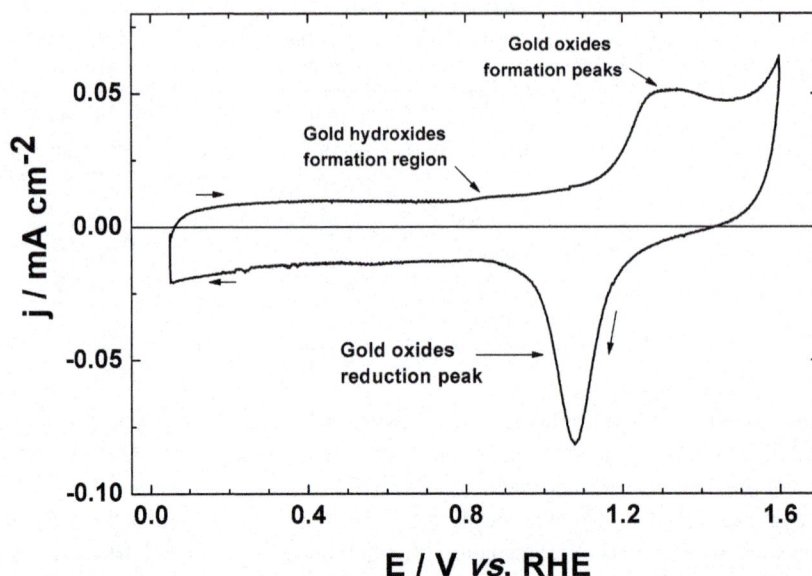

However, this is clearly observed for the GNCs electrode at 0.43 V. During the positive potential sweep, three stripping peaks corresponding to the desorption of lead layer on (111), (100) and (110) facets were observed at around 0.41, 0.47 and 0.57 V, respectively. The desorption peak on the (100) facet is well defined and high in activity for the GNCs compared to GNRs, as expected. No desorption peak of lead on (100) was observed for GNSs and GNPs. The lead desorption on (111) and (110) facets on GNCs is due to their truncated edges. According to Murphy et al. [14], GNRs can exhibit (100) or (111) facets at the end of the rods. In addition, an increase of (111) facets along the GNRs' side can be observed because of the crystallization defects [41].

**Glucose oxidation on gold nanoparticles**

On the low-index single-crystal gold electrodes, it was mentioned that the oxidation of the glucose depends on such surface structure and on the adsorbed intermediates [26, 33]. The formation of (OH)$_{ads}$ on Au is crucial for the electrochemical oxidation of glucose [30]. This oxidation is assumed to occur through the interaction between the adsorbed hemiacetal group and (OH)$_{ads}$. It is generally accepted that the AuOH sites on the Au surface act as the active species for glucose oxidation [42]. Thereby, the oxidation of glucose strongly depends on the number of AuOH sites. The first step of the oxidation of glucose in alkaline media involves the adsorption of glucose on the Au surface through the anomeric carbon leading to the dehydrogenation process via the formation of an

**Fig. 4** Voltammetric UPD profiles of the different AuNPs (GNSs, GNRs, GNCs and GNPs) in 0.1 mol L$^{-1}$ NaOH+ 1 mmol L$^{-1}$ Pb(NO$_3$)$_2$ recorded at 20 mV s$^{-1}$ and at controlled temperature of 20 °C

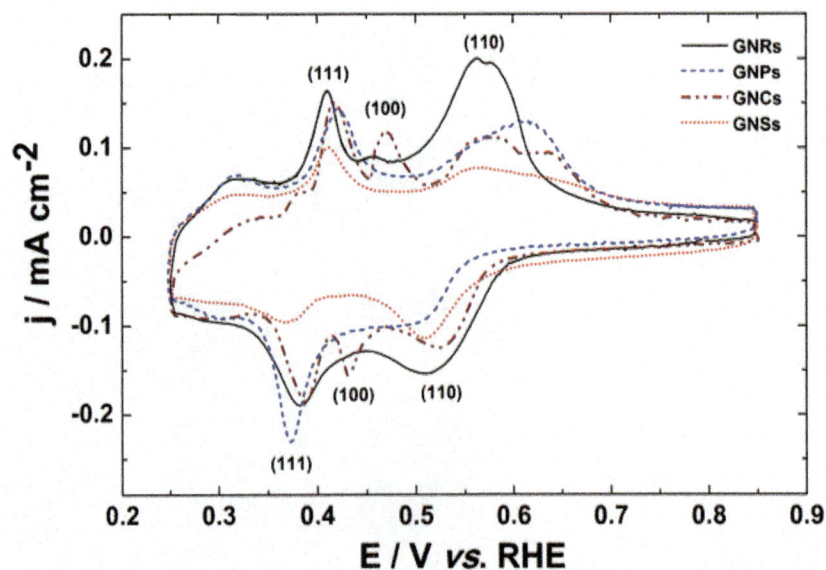

**Fig. 5** Voltammograms of the different AuNP (GNSs, GNRs, GNCs and GNPs) electrodes in 0.1 mol L$^{-1}$ NaOH+10 mmol L$^{-1}$ glucose, recorded at 20 mV s$^{-1}$ and at controlled temperature of 20 °C

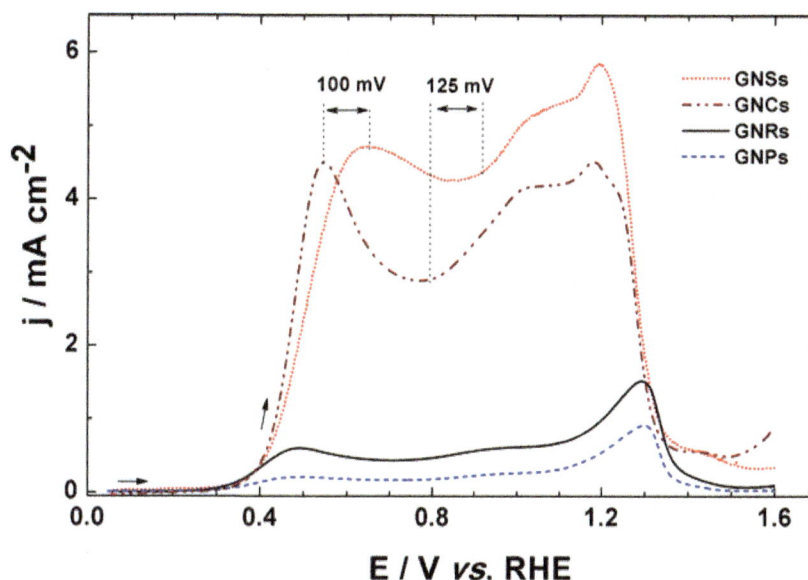

**Fig. 5** Voltammograms of the different AuNP (GNSs, GNRs, GNCs and GNPs) electrodes in 0.1 mol L$^{-1}$ NaOH+10 mmol L$^{-1}$ glucose, recorded at 20 mV s$^{-1}$ and at controlled temperature of 20 °C

adsorbed radical. This process leads to the formation of gluconolactone [29, 33]. Furthermore, gluconolactone can be hydrolyzed in solution to give sodium gluconate [30, 42]. The second step deals with the oxidation of adsorbed intermediates or gluconolactone with the adsorbed OH species. Several products can be obtained as indicated in the literature [29]. Figure 5 presents the positive potential scan of the glucose oxidation reaction on the different AuNPs. The oxidation of glucose begins at ca. 0.3 V for all materials followed by the first oxidation peak which is associated with the formation of gluconolactone species. GNCs (dashed dot line) and GNSs (dot line) present high activities for this process. For GNCs, this oxidation peak is sharp and its maximum appears at 0.54 V (100 mV lower than for GNSs). The lowest activity of the dehydrogenation was observed with GNPs. After this process, the oxidation continues in a large range of potential from 0.8 to 1.3 V. GNRs (solid line) and GNPs (dashed line) show an overlap shoulder followed by a maximum oxidation peak at 1.3 V. However, this peak is observed at 1.2 V for GNSs and GNCs. The high current density observed for GNSs can be related to their small size (around 6±2 nm) and their electronic surface structure. The catalytic effect on GNCs depends on their structure, i.e. the number of sites of low coordination. These NPs show a well-defined structure. GNCs exhibit the mean (100) and (110) facets that promote the dehydrogenation of the molecule of glucose at low potential (0.54 V) compared to the other nanostructured materials. These facets are the most favorable for this reaction [33, 43].

## Conclusion

Different shape-controlled gold nanoparticles were synthesized by wet chemical methods. The UV–vis spectroscopy

measurements and transmission electron microscopy characterizations showed a fine distribution in size and shape of the AuNPs. The zeta potential measurements permitted the confirmation of spherical nanoparticles as more stable than the anisotropic ones.

The electrochemical characterization by UPD showed that the GNRs and GNCs present the three low-index facets, while the GNSs and GNPs do not display the (100) facet. These different surface structures affect strongly the electrocatalytic activity of the gold nanoparticles. The dehydrogenation reaction occurs at low potential on GNCs, whereas the GNSs show the highest current density for this reaction. This study shows clearly that the shape and the structure of the nanoparticles should be combined to understand their electrocatalytic activity.

**Acknowledgments** The authors acknowledge the CNRS and the Region Poitou-Charentes for their financial support.

## References

1. Huang X, El-Sayed MA (2010) Gold nanoparticles: Optical properties and implementations in cancer diagnosis and photothermal therapy. J Adv Res 1(1):13–28
2. Daniel MC, Astruc D (2004) Gold nanoparticles: assembly, supramolecular chemistry, quantum-size-related properties, and applications toward biology, catalysis, and nanotechnology. Chem Rev 104: 293–346
3. Cui C, Gan L, Li HH, Yu SH, Heggen M, Strasser P (2012) Octahedral PtNi Nanoparticle Catalysts: Exceptional Oxygen

Reduction Activity by Tuning the Alloy Particle Surface Composition. Nano Lett 12(11):5885–5889

4. Yu, Chang SS, Lee CL, Wang CRC (1997) Gold nanorods:i electrochemical synthesis and optical properties. J Phys Chem B 101(34): 6661–6664

5. Sau TK, Murphy CJ (2004) Room temperature, high-yield synthesis of multiple shapes of gold nanoparticles in aqueous solution. J Am Chem Soc 126(28):8648–8649

6. Sun Y, Xia Y (2002) Shape-controlled synthesis of gold and silver nanoparticles. Science 298(5601):2176–2179

7. Shankar SS, Rai A, Ankamwar B, Singh A, Ahmad A, Sastry M (2004) Biological synthesis of triangular gold nanoprisms. Nat Mater 3(7):482–488

8. Huang CJ, Chiu PH, Wang YH, Yang CF (2006) Synthesis of the gold nanodumbbells by electrochemical method. J Colloid Interface Sci 303(2):430–436

9. Langille MR, Personick ML, Zhang J, Mirkin CA (2012) defining rules for the shape evolution of gold nanoparticles. J Am Chem Soc 134(35):14542–14554

10. Straney PJ, Andolina CM, Millstone JE (2013) Seedless initiation as an efficient, sustainable route to anisotropic gold nanoparticles. Langmuir 29(13):4396–4403

11. Yong KT, Sahoo Y, Swihart M, Schneeberger P, Prasad P (2008) Templated Synthesis of Gold Nanorods (NRs): The Effects of Cosurfactants and Electrolytes on the Shape and Optical Properties. Top Catal 47(1–2):49–60

12. Zhu YJ, Hu XL (2003) Microwave-polyol preparation of single-crystalline gold nanorods and nanowires. Chem Lett 32(12):1140–1141, 2

13. Simakin AV, Voronov VV, Shafeev GA, Brayner R, Bozon-Verduraz F (2001) Nanodisks of Au and Ag produced by laser ablation in liquid environment. Chem Phys Lett 348(3–4):182–186

14. Sau TK, Murphy CJ (2004) Seeded high yield synthesis of short Au nanorods in aqueous solution. Langmuir 20(15):6414–6420

15. Perez-Juste J, Pastoriza-Santos I, Liz-Marzan LM, Mulvaney P (2005) Gold nanorods: synthesis, characterization and applications. Coord Chem Rev 249(17–18):1870–1901

16. Sun J, Guan M, Shang T, Gao C, Xu Z, Zhu J (2008) Selective synthesis of gold cuboid and decahedral nanoparticles regulated and controlled by Cu2+ Ions. Cryst Growth Des 8(3):906–910

17. Hernandez J, Solla-Gullon J, Herrero E, Aldaz A, Feliu JM (2005) Characterization of the surface structure of gold nanoparticles and nanorods using structure sensitive reactions. J Phys Chem B 109(26): 12651–12654

18. Hernandez J, Solla-Gullon J, Herrero E, Feliu JM, Aldaz A (2009) In Situ surface characterization and oxygen reduction reaction on shape-controlled gold nanoparticles. J Nanosci Nanotechnol 9(4):2256–2273

19. Hvolbæk B, Janssens TVW, Clausen BS, Falsig H, Christensen CH, Nørskov JK (2007) Catalytic activity of Au nanoparticles. Nano Today 2(4):14–18

20. Guo Z, Fan X, Liu L, Bian Z, Gu C, Zhang Y, Gu N, Yang D, Zhang J (2010) Achieving high-purity colloidal gold nanoprisms and their application as biosensing platforms. J Colloid Interface Sci 348(1): 29–36

21. Tominaga M, Shimazoe T, Nagashima M, Kusuda H, Kubo A, Kuwahara Y, Taniguchi I (2006) Electrocatalytic oxidation of glucose at gold-silver alloy, silver and gold nanoparticles in an alkaline solution. J Electroanal Chem 590(1):37–46

22. Tominaga M, Shimazoe T, Nagashima M, Taniguchi I (2005) Electrocatalytic oxidation of glucose at gold nanoparticle-modified carbon electrodes in alkaline and neutral solutions. Electrochem Commun 7(2):189–193

23. Cho S, Shin H, Kang C (2006) Catalytic glucose oxidation on a polycrystalline gold electrode with an amalgamation treatment (TM 05092). Electrochim Acta 51(18):3781–3786

24. Jin CC, Taniguchi I (2007) Electrocatalytic oxidation of glucose on gold nanocomposite electrodes. Chem Eng Technol 30(9):1298–1301

25. Nelson T (2009) Recent progress in electrochemical oxidation of saccharides at gold and copper electrodes in alkaline solutions. Bioelectrochemistry 76(1–2):195–200

26. Wang J, Gong J, Xiong Y, Yang J, Gao Y, Liu Y, Lu X, Tang Z (2011) Shape-dependent electrocatalytic activity of monodispersed gold nanocrystals toward glucose oxidation. Chem Commun 47(24): 6894–6896

27. Tonda-Mikiela P, Napporn TW, Morais C, Servat K, Chen A, Kokoh KB (2012) Synthesis of gold-platinum nanomaterials using bromide anion exchange-synergistic electroactivity toward CO and glucose oxidation. J Electrochem Soc 159(11):H828–H833

28. Hebié S, Cornu L, Napporn TW, Rousseau J, Kokoh BK (2013) Insight on the surface structure effect of free gold nanorods on glucose electrooxidation. J Phys Chem C 117(19):9872–9880, 3

29. Kokoh KB, Léger JM, Beden B, Lamy C (1992) "On line" chromatographic analysis of the products resulting from the electrocatalytic oxidation of d-glucose on Pt. Au and adatoms modified Pt electrodes-Part I. Acid and neutral media. Electrochim Acta 37(8): 1333–1342

30. Hsiao MW, Adžić RR, Yeager EB (1996) Electrochemical oxidation of glucose on single crystal and polycrystalline gold surfaces in phosphate buffer. J Electrochem Soc 143(3):759–767

31. Chen Y, Schuhmann W, Hassel AW (2009) Electrocatalysis on gold nanostructures: is the 110 facet more active than the 111 facet? Electrochem Commun 11(10):2036–2039

32. Vuyyuru KR, Strasser P (2012) Oxidation of biomass derived 5-hydroxymethylfurfural using heterogeneous and electrochemical catalysis. Catal Today 195(1):144–154

33. Adzic RR, Hsiao MW, Yeager EB (1989) Electrochemical oxidation of glucose on single crystal gold surfaces. J Electroanal Chem Interfacial Electrochem 260(2):475–485

34. Martins A, Ferreira V, Queirós A, Aroso I, Silva F, Feliu J (2003) Enantiomeric electrooxidation of d- and l-glucose on chiral gold single crystal surfaces. Electrochem Commun 5(9):741–746

35. Habrioux A, Hebié S, Napporn T, Rousseau J, Servat K, Kokoh KB (2011) One-Step synthesis of clean and size-controlled gold electrocatalysts: modeling by taguchi design of experiments. Electrocatalysis 2(4):279–284

36. Finot MO, Braybrook GD, McDermott MT (1999) Characterization of electrochemically deposited gold nanocrystals on glassy carbon electrodes. J Electroanal Chem 466(2):234–241

37. Eustis S, El-Sayed MA (2006) Determination of the aspect ratio statistical distribution of gold nanorods in solution from a theoretical fit of the observed inhomogeneously broadened longitudinal plasmon resonance absorption spectrum. J Appl Phys 100(4):044324

38. Lee J-H, Kamada K, Enomoto N, Hojo J (2007) Morphology-selective synthesis of polyhedral gold nanoparticles: What factors control the size and morphology of gold nanoparticles in a wet-chemical process. J Colloid Interface Sci 316(2):887–892

39. Pavlin M, Bregar VB (2012) Stability of nanoparticle suspensions in different biologically revelant media. Dig J Nanomater Biostructure 7(4):1389–1400

40. Gauggel S, Derreza-Greeven C, Wimmer J, Wingfield M, Bvd B, Dietrich DR (2012) Characterization of biologically available wood combustion particles in cell culture medium. Altex 29(2):183–200

41. Wang ZL, Gao RP, Nikoobakht B, El-Sayed MA (2000) Surface reconstruction of the unstable 110 Surface in Gold Nanorods. J Phys Chem B 104(23):5417–5420

42. Aoun SB, Dursun Z, Koga T, Bang GS, Sotomura T, Taniguchi I (2004) Effect of metal ad-layers on Au(111) electrodes on electrocatalytic oxidation of glucose in an alkaline solution. J Electroanal Chem 567(2):175–183

43. Beden B, Çetin I, Kahyaoglu A, Takky D, Lamy C (1987) Electrocatalytic oxidation of saturated oxygenated compounds on gold electrodes. J Catal 104(1):37–46

# 28

# Simultaneous determination of uric acid and dopamine using a carbon fiber electrode modified by layer-by-layer assembly of graphene and gold nanoparticles

Jiao Du · Ruirui Yue · Fangfang Ren · Zhangquan Yao · Fengxing Jiang · Ping Yang · Yukou Du

**Abstract** A novel layer-by-layer assembly of graphene sheets and gold nanoparticles modified carbon fiber electrode (GE/Au/GE/CFE) was successfully fabricated and applied to simultaneous determination of dopamine (DA) and uric acid (UA). The structure of GE/Au/GE/CFE was characterized by scanning electron microscopy (SEM). It was observed that the gold nanoparticles were homogeneously assembled between the two layers of GE sheets. Cyclic voltammetry (CV) measurements elucidate that GE/Au/GE/CFE has higher electrocatalytic activity for the oxidation of DA and UA when compared with graphene modified carbon fiber electrode (GE/CFE), and graphene and gold nanoparticles modified carbon fiber electrode (Au/GE/CFE). Simultaneous determination of UA and DA on GE/Au/GE/CFE was conducted with a differential pulse voltammetry technique, and two well-defined and fully resolved anodic oxidation peaks were observed. Anodic oxidation currents of DA and UA are linear with their concentration in the range of 0.59–43.96 µM and 12.6–413.62 µM, respectively. Moreover, the composite electrode displays high reproducibility and selectivity for the determination of UA and DA.

**Keywords** Gold nanoparticles · Graphene · Dopamine · Uric acid

## Introduction

Dopamine (DA), one of the most important catecholamine neurotransmitters, exists in the mammalian central nervous system at very low concentrations and plays a significant role in central nervous, endocrine, and cognitive functions. Brain tissues will be severely impacted when the level of DA fluctuates. Deficiency of DA may result in serious neurological disorders, such as Parkinson's disease and addiction [1–4]. Uric acid (UA), the primary end product of purine metabolism, is present in biological fluids such as blood or urine. Abnormal levels of UA are symptoms of several diseases such as gout and hyperuricemia [5, 6]. Therefore, it is important to determine these species. Since DA and UA are electroactive materials, electrochemical methods appear to be suitable for their quantitative determination. Moreover, electrochemical methods are highly sensitive, easily operable, and cost effective [7–11]. However, a major problem that inhibits electrochemical determination is that UA and DA have similar oxidation potential at most of the conventional solid electrodes. In the actual samples, UA presents in concentrations much higher than DA [12, 13]. Therefore, it is critically important to prepare a sensitive and accurate electrode for simultaneous determination of UA and DA.

Graphene–metal nanoparticles composite has attracted extensive attention in recent years in many fields such as sensor and fuel cells [14, 15]. Gold nanoparticles, which are usually applied to bimolecular immobilization, have many advantages such as excellent catalytic activity, huge surface area, effective mass transport, and hospitable environment. Furthermore, they can serve on electron tunnel to promote electron exchange at electron/protein interface [16, 17], and thereby gold nanoparticles have found many applications in biosensors [18, 19]. Graphene (GE), a one-atom-thick sp$^2$-bonded carbon sheet, has attracted tremendous attention in recent years due to its large amount of edge-plan-like defects and large specific surface area [20–22]. The elusive two-dimensional structure of GE contributes to

J. Du · R. Yue · F. Ren · Z. Yao · F. Jiang · P. Yang · Y. Du (✉)
College of Chemistry, Chemical Engineering and Materials Science, Soochow University, Suzhou 215123, People's Republic of China
e-mail: duyk@suda.edu.cn

numerous unique electronic properties, such as the quantum Hall effect and transport via relativistic Dirac fermions [23, 24], making GE possess high electron conductivity and good biocompatibility. The combination of GE and gold nanoparticles can increase electrode surface and enhance optical, electronic, and catalytic properties. Moreover, GE and gold nanoparticles offer synergistic effects in their composite electrocatalytic applications [14, 24]. Li et al. [25] developed GE and gold nanoparticles modified electrodes which can selectively determine DA with a high detection limit (1.86 μM). To the best of our knowledge, there have been no reports about the fabrication and application of GE and gold nanoparticles layer-by-layer modified electrode for simultaneous detection of DA and UA.

In this work, GE and gold nanoparticles were assembled into a layer-by-layer structure on the carbon fiber electrode, which was endowed with large surface area and high electron transfer ability. The graphene sheets and gold nanoparticles modified carbon fiber electrode (GE/Au/GE/CFE) was used to simultaneously determine UA and DA. For comparison, the electrocatalytic performance of graphene modified carbon fiber electrode (GE/CFE), and graphene and gold nanoparticles modified carbon fiber electrode (Au/GE/CFE) towards UA and DA oxidation was also studied.

## Experimental

### Reagents and materials

Graphite powder, disodium hydrogen phosphate, and sodium dihydrogen phosphate were obtained from Sinopharm Chemicals Reagent Co., Ltd. (China). Uric acid and dopamine were purchased from Acros Organics and used as received. All chemicals were of analytical reagent grade. All solution was prepared with secondary distilled water. Graphene oxide (GO) nanosheets were prepared according to the literature [15, 26].

### Apparatus

The morphology of obtained electrodes was characterized by emission scanning electron microscopy (SEM) (S-4700; Hitachi High Technologies Corporation, Japan). A CHI650D electrochemical workstation (Shanghai Chenhua Instrument Plant, China) was used to take the electrochemical experiments. A standard three-electrode configuration, consisting of obtained sample electrodes as the working electrodes, a platinum wire with a diameter of 0.5 mm as the counter electrode, and a saturated calomel electrode (SCE) as the reference electrode, was used throughout the experiment. Also, 0.1 M phosphate buffer solution (PBS) (pH=7.0) was used as an electrolyte solution. All experiments were carried out at room temperature.

### Electrode preparation

The preparation of 0.5 mg/mL GO nanosheets solution was as follows: 10 mg obtained GO was dispersed into 10 mL of secondly distilled water under ultrasonic bath for 90 min. Impurities were centrifuged from this suspension. The obtained GO ink (0.15 mL) was dropcast onto CFE and then dried in air. The GO on CFE was reduced at a constant potential of −0.9 V in Na–PBS (0.1 M, pH=4.1) solution, and then gold nanoparticles were deposited thereon by reduction of 0.3 mM HAuCl$_4$ solution at a constant potential of −0.2 V with a certain electric quantity of $6.0 \times 10^3$ C calculated as 4.08 μg. Subsequently, the GO ink (0.15 mL) was again dropcast onto the above obtained electrode and dried in air, and then reduced at a constant potential of −0.9 V in Na–PBS (0.1 M, pH=4.1) solution to obtain a GE/Au/GE/CFE composite electrode (shown in Scheme 1). Preparation of GE/CFE or Au/GE/CFE was similar to GE/Au/GE/CFE just using GE or Au and GE as the precursor instead of GE, Au, and GE. The electrode area immersed into the solution is 1 cm×360 μm.

**Scheme 1** Illustration of the fabrication of GE/Au/GE/CFE

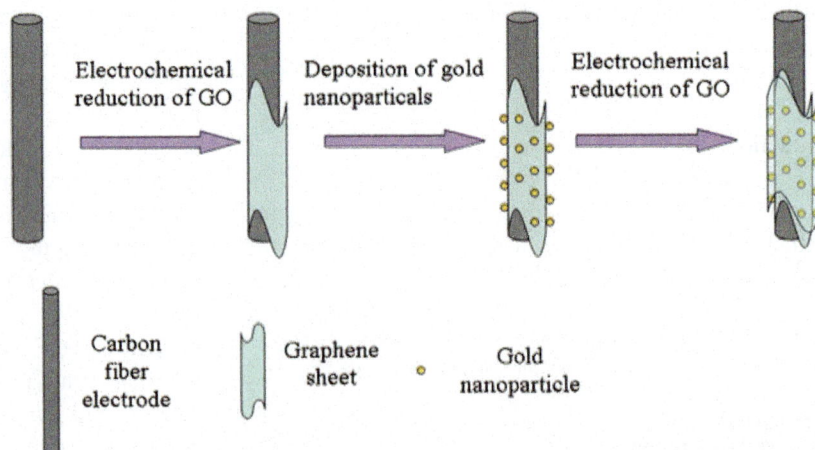

## Results and discussion

### Characterization of obtained electrodes

The morphology of the obtained electrodes was examined by SEM. Figure 1a shows the surface morphology of GE/CFE. It

**Fig. 1** SEM images of **a** GE/CFE, **b** Au/GE/CFE, and **c** GE/Au/GE/CFE electrodes

can be seen that GE uniformly wraps on the surface of CFE. Many wrinkles are observed on the surface of the electrode, indicating the increase in superficial area of the GE/CFE electrode. Figure 1b is the image of Au/GE/CFE. The image shows that gold nanoparticles are well distributed to form a nanocomposite on the surface of GE sheets. The sizes of Au nanoparticles are homogeneous, and the distribution of nanoparticles is very uniform and dense. The SEM image of GE/Au/GE/CFE is shown in Fig. 1c. It can be seen that GE layer likes filmy gauze wholly coated on the surface of gold nanoparticles, and wholly wraps gold nanoparticles. There are no aggregation of GE sheets and free gold nanoparticles. Furthermore, an increase in the surface roughness is observed, which gives rise to much higher active surface area for Au/CFE, and thus enhances its electrochemical activity.

Figure 2 shows the CV responses of the three electrodes in 10 mM $Fe(CN)_6^{3/4}$+0.1 M KCl solution. A pair of poor redox peaks is observed in the CV response of GE/CFE, indicating a poor electron transfer at the interface of GE/CFE. Compared with the CV response of GE/CFE, an obvious increase in the redox peak current and a narrow peak potential difference are observed in the CV response of Au/GE/CFE, suggesting that the introduction of gold nanoparticles can enhance the electrochemical activity of obtained electrode. This may contribute to the huge surface area of gold nanoparticles. A pair of well-defined redox peaks with the smallest peak potential separation of about 100 mV can be observed in the CV response of GE/Au/GE/CFE. The enhanced peak current intensity and the smallest peak potential separation may be due to the high electric transfer kinetics and the huge surface of GE/Au/GE/CFE [1].

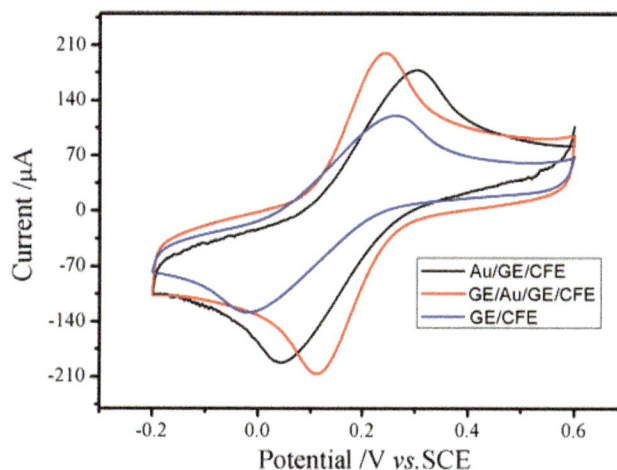

**Fig. 2** CVs of Au/GE/CFE, GE/CFE, and GE/Au/GE/CFE electrodes recorded in 10 mM $Fe(CN)_6^{3-/4-}$+0.1 M KCl solution at scan rate of 50 mV/s

Cyclic voltammetric detection of UA and DA

Figure 3a and b shows the CV responses of DA and UA on GE/CFE, Au/GE/CFE, and GE/Au/GE/CFE, respectively. In Fig. 3a, for 0.5 mM DA solution, the anodic oxidation peak currents are 80 μA, 110 μA, and 150 μA for GE/CFE, Au/GE/CFE, and GE/Au/GE/CFE, respectively. In Fig. 3b, for 1.0 mM UA solution the anodic peak current at GE/CFE is 150 μA, which becomes as low as 25 μA with the Au/GE/CFE and 50 μA with the GE/Au/GE/CFE. The anodic peak potential of UA at GE/CFE is more positive compared with that of the other two electrodes. Different electrodes are used to simultaneously determine UA and DA in 0.1 M PBS (pH=7.0) solution containing 0.25 mM UA and 0.25 mM DA by CV. Figure 3c shows the electrocatalytic oxidation CV responses of UA and DA in the mixture solution on GE/CFE, Au/GE/CFE, and GE/Au/GE/CFE electrodes. As can be seen, two well-defined peaks at 0.3 V and 0.15 V are observed for all the CV responses of different electrodes. Towards UA and DA electrooxidation, the GE/Au/GE/CFE shows the highest oxidation currents of 100 μA and 80 μA, while the oxidation currents are 70 μA and 50 μA, and 50 μA and 40 μA for Au/GE/CFE and

GE/CFE, respectively. The above result indicates that GE/Au/GE/CFE shows higher electrocatalysis response to the oxidation of UA and DA. The reason may be that the higher specific surface area and the larger amount of edge-plan-like defects of GE/Au/GE/CFE can provide more active sites for the oxidation of UA and DA as compared with other electrodes and be conducive to the separation of DA and UA. Additionally, GE is a highly conductive material, and gold nanoparticles can promote electron exchange. The composited GE and gold nanoparticles offer synergistic effects in electric catalytic applications.

Effects of scan rate on the oxidation of UA and DA

The effects of scan rate ($v$) on the electrocatalytic activity of GE/Au/GE/CFE towards oxidation of UA and DA have been investigated by CV (Fig. 4a, b). It can be seen that both the peak potentials ($E_{pa}$) and peak currents ($I_{pa}$) are dependent on the scan rate. When the scan rate increases, small positive shifts in the UA and DA oxidation peaks are observed, suggesting that the adsorption of UA and DA does not occur on GE/Au/GE/CFE in 0.1 M PBS (pH= 7.0) solution. For both UA and DA, the redox peak currents

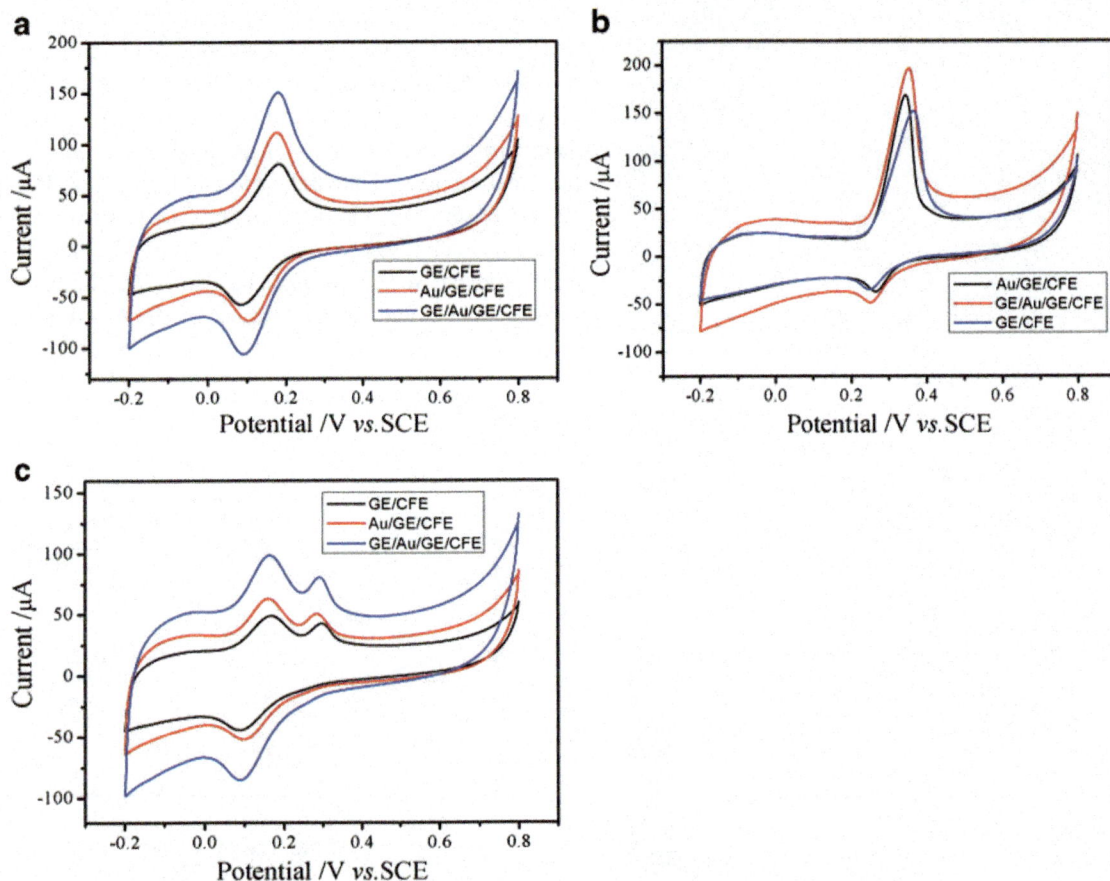

Fig. 3 CV responses of GE/CFE, Au/GE/CFE, and GE/Au/GE/CFE in 0.1 M PBS (pH=7.0) containing a 0.5 mM DA, b 1.0 mM UA, and c 0.25 mM UA and 0.25 mM DA. Scan rate: 50 mV/s

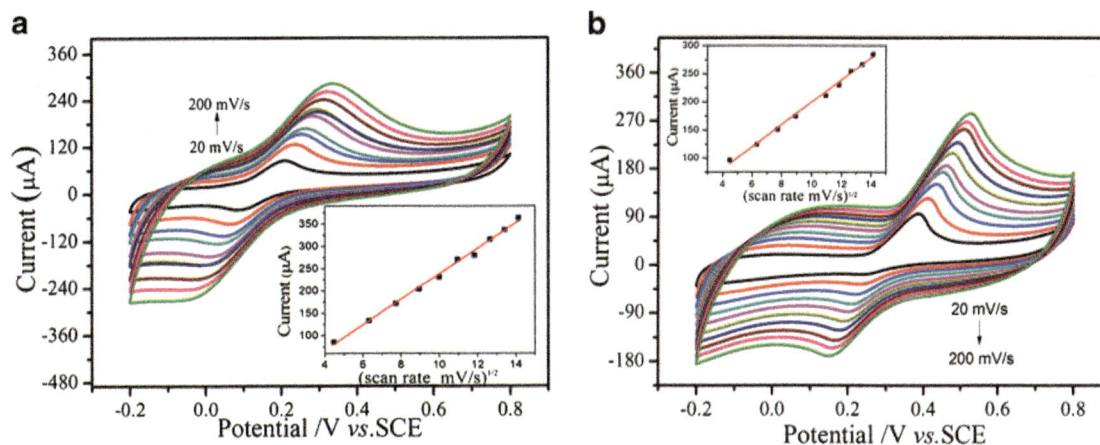

**Fig. 4** CV curves of GE/Au/GE/CFE in 0.1 M PBS (pH=7.0) solution containing 0.25 mM DA (**a**) and 0.5 mM UA (**b**) at different scan rates of 20, 40, 60, 80, 100, 120, 140, 160, 180, and 200 mV/s. *Inset*: plot of peak current versus the square root of scan rate

gradually increase as the scan rate increases, and a linear relation between the anodic currents and the square root of scan rates at the range of 20–200 mV/s can be observed. The linear equations are expressed as follows:

$$I_{pa}(\mu A) = 28.53\upsilon^{1/2}(mV/s) - 47.82 \ (DA); I_{pa}(\mu A)$$

$$= 19.70\upsilon^{1/2}(mV/s) + 1.205(UA),$$

with linear relative coefficients both of 0.995. The above results indicate that the oxidation of UA and DA on GE/Au/GE/CFE is a diffusion-controlled process. In addition, the electrode reactions of DA and UA are quasi-reversible as the redox peaks potentials vary with the scan rates [13].

Effects of pH on the oxidation of UA and DA

The effects of solution pH on the peak currents and peak potentials of UA and DA oxidation are studied by differential pulse voltammetry (DPV). The results are shown in Fig. 5a and b. As shown in Fig. 5a, the oxidation potentials shift negatively with the increase in the

solution pH, and the slopes of −54 and −58 mV/pH for DA and UA in the pH range of 3.0–9.0 are obtained. These slopes are close to a Nernst system, defined as −59 mV/pH at 25 °C, implying that the total number of electrons and protons taking part in the charge transfer is the same, that is, for the oxidation of DA and UA two electrons and two protons are involved, shown as follows [20]:

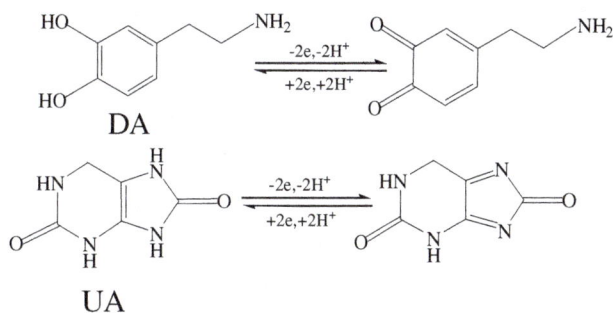

The oxidation peak current of UA increases as the pH value increases up to 6.0 and then decreases quickly with continue increasing of pH. By comparison, the oxidation peak current of DA decreases as the pH is increased. The pH

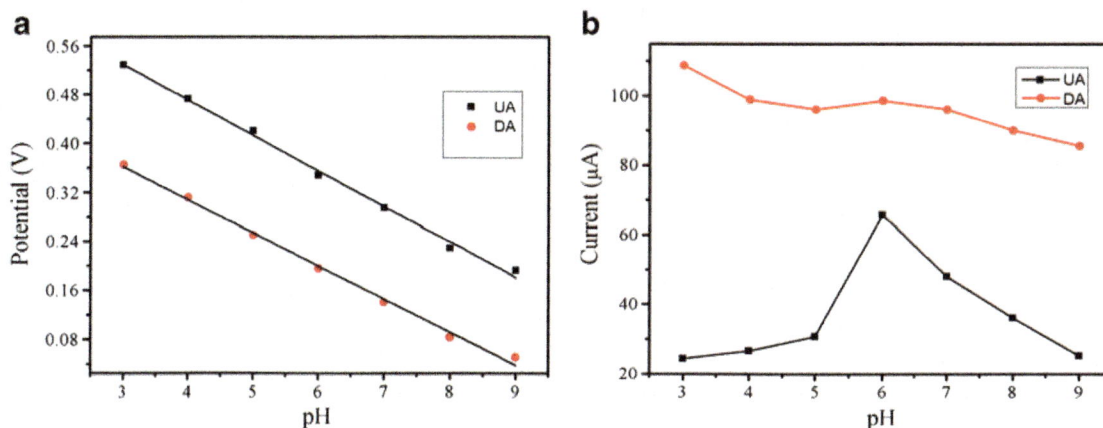

**Fig. 5** Dependence of the **a** anodic peak potentials and **b** anodic currents as a function of the solution pH for 0.5 mM DA and 0.8 mM UA

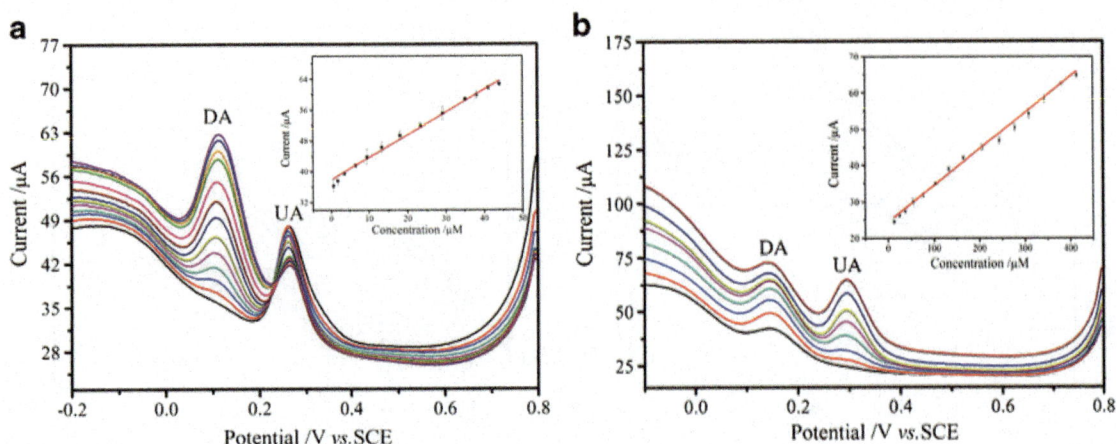

**Fig. 6** DPV curves of **a** 0.15 mM UA with different concentrations of DA and **b** 8 μM DA with different concentrations of UA. *Insets*: plots of oxidation currents versus the concentration of DA and UA, respectively

value of human blood and urine is close to 7.0, thus 0.1 M PBS solution with pH=7.0 is chosen for simultaneous determination of UA and DA.

### Simultaneous determination of UA and DA

As shown by the above results, the difference in the oxidation peak potentials for DA and UA is large enough for separation and simultaneous determination of DA and UA in a mixture. DPV has been used to simultaneously determine UA and DA at GE/Au/GE/CFE. The electrooxidation processes of DA and UA in the mixture are evaluated when the concentration of one species changes and the other one keeps constant. Figure 6a depicts the DPV curves of various DA concentrations at GE/Au/GE/CFE in presence of 0.15 mM UA. As shown in Fig. 6a, the oxidation currents of UA show small change when DA concentration increases; it indicates that the addition of DA does not affect the determination of UA. The inset in Fig. 6a depicts the linear relationship between the oxidation peak currents and DA concentration in the range of 0.59–43.955 μM. The corresponding linear regression equation is defined as $I_{pa}=$ $37.4099+0.5996C_{DA}$ ($R$=0.992), and the lower detection limit for DA is observed to be 0.59 μM (S/N=3), which is lower than previously reported results [27–29]. Meanwhile, a similar experiment was conducted with UA in the presence of 8 μM DA. In Fig. 6b, although the oxidation currents of DA change when UA concentration increases, the determination of UA is not affected. As shown in the inset of Fig. 6b, the linear relationship between the oxidation peak currents of UA and the UA concentration is defined as $I_{pa}=24.262+0.1001C_{UA}$ ($R$=0.997) in the range of 12.6–413.62 μM and the lower detection limit is observed to be 12.6 μM (S/N=3), which is lower than what was previously reported [28, 29]. The above results show that GE/Au/GE/CFE is well response for the determination of DA and UA.

### Reproducibility and interference of GE/Au/GE/CFE

In the reproducibility tests, GE/Au/GE/CFE was used to repeatedly ($n$=6) determine 0.14 mM DA and 0.42 mM UA in the mixture solution. It is found that the relative standard deviations (RSD) for DA and UA are 1.25 % and 2.36 % indicating the excellent reproducibility of the prepared composite electrode. The measured currents of the electrode maintain at 96.6 % for DA and 96.1 % for UA after 7 days. The test about the influence of various interferents that may exist in the real samples on the determination of DA and UA was also carried out. When 10-times concentration of NaCl, KCl, NaNO₃, ascorbic acid, glucose, urea, and NH₄Cl were added into the mixture of 0.25 mM UA and 0.25 mM DA, no obvious interference was observed (shown in Table 1), demonstrating the good anti-interference ability of as-fabricated GE/Au/GE/CFE composite electrode. It is worth to mention that the interference for DA is lower than that of UA, which may be attributed to the π–π interaction between the phenyl structure of DA molecule and the two-dimensional hexagonal carbon structure of GE [30].

**Table 1** Results of interference study on the response of UA and DA

| Interfering substances | Peak current intensity variation (DA) | Peak current intensity variation (UA) |
|---|---|---|
| NaCl | 0.67 % | 2.10 % |
| KCl | 2.18 % | 1.66 % |
| Ascorbic acid | 2.30 % | −1.80 % |
| NaNO₃ | 3.40 % | 3.90 % |
| NH₄Cl | −0.70 % | −5.06 % |
| Glucose | −1.75 % | −3.69 % |
| Urea | 1.87 % | 3.10 % |

## Conclusion

In summary, a novel electrode based on layer-by-layer assembly of GE sheets and gold nanoparticles has been developed by simple electrochemical reduction of GO and potentiostatic electrodeposition of gold nanoparticles onto GE sheets. The prepared GE/Au/GE/CFE composite electrode was used to simultaneously determine UA and DA. It is found that two clear and well-separated oxidation peaks are observed in the determination of UA and DA at GE/Au/GE/CFE as compared with the other two electrodes. The introduction of GE and gold nanoparticles and the structure of GE/Au/GE/CFE increase the effective electroactive surface area, and also enhance the ability of electron transfer. Additionally, the low detection limit and wide detection range evince that the developed electrode is of high selectivity and good reproducibility towards UA and DA determination. Moreover, the fabrication of the developed electrode is low cost and easy workable.

**Acknowledgments**  The authors appreciate the supports of the National Natural Science Foundation of China (grant nos. 20933007 and 51073114), the Priority Academic Program Development of Jiangsu Higher Education Institutions (PAPD), and the Academic Award for Young Graduate Scholar of Soochow University.

## References

1. Ping J, Wu J, Wang Y, Ying Y (2012) Simultaneous determination of ascorbic acid, dopamine and uric acid using high-performance screen-printed graphene electrode. Biosens Bioelectron 34:70–76
2. Ciszewski A, Milczarek G (1999) Polyeugenol-modified platinum electrode for selective detection of dopamine in the presence of ascorbic acid. Anal Chem 71:1055–1061
3. Salimi A, Abdi K, Khayatian GR (2004) Amperometric detection of dopamine in the presence of ascorbic acid using a nafion coated glassy carbon electrode modified with catechin hydrate as a natural antioxidant. Microchim Acta 144:161–169
4. Zhao Y, Gao Y, Zhan D, Liu H, Zhao Q, Kou Y, Shao Y, Li M, Zhuang Q, Zhu Z (2005) Selective detection of dopamine in the presence of ascorbic acid and uric acid by a carbon nanotubes–ionic liquid gel modified electrode. Talanta 66:51–57
5. Ensafi AA, Taei M, Khayamian T, Arabzadeh A (2010) Highly selective determination of ascorbic acid, dopamine, and uric acid by differential pulse voltammetry using poly(sulfonazo III) modified glassy carbon electrode. Sensors Actuators B 147:213–221
6. Mallesha M, Manjunatha R, Nethravathi C, Suresh GS, Rajamathi M, Melo JS, Venkatesha TV (2011) Functionalized-graphene modified graphite electrode for the selective determination of dopamine in presence of uric acid and ascorbic acid. Bioelectrochemistry 81:104–108
7. Wu J, Suls J, Sansen W (2000) Amperometric determination of ascorbic acid on screen-printing ruthenium dioxide electrode. Electrochem Commun 2:90–93
8. Gea S, Yan M, Lua J, Zhang M, Yu F, Yu J, Song X, Yu S (2012) Electrochemical biosensor based on graphene oxide–Au nanoclusters composites for L-cysteine analysis. Biosens Bioelectron 31:49–54
9. Jiang F, Yue R, Du Y, Xu J, Yang P (2013) A one-pot 'green' synthesis of Pd-decorated PEDOT nanospheres for nonenzymatic hydrogen peroxide sensing. Biosens Bioelectron 44:127–131
10. Wu L, Feng L, Ren J, Qu X (2012) Electrochemical detection of dopamine using porphyrin-functionalized graphene. Biosens Bioelectron 34:57–62
11. Wang L, Qin X, Liu S, Luo Y, Asiri A, Al-Youbi A, Sun X (2012) Single-stranded DNA-mediated immobilization of graphene on a gold electrode for sensitive and selective determination of dopamine. ChemPlusChem 77:19–22
12. Mahshid S, Li C, Mahshid SS, Askari M, Dolati A, Yang L, Luo S, Cai Q (2011) Sensitive determination of dopamine in the presence of uric acid and ascorbic acid using $TiO_2$ nanotubes modified with Pd, Pt and Au nanoparticles. Analyst 136:2322–2329
13. Hou S, Kasner ML, Su S, Patel K, Cuellari R (2010) Highly sensitive and selective dopamine biosensor fabricated with silanized graphene. J Phys Chem C 114:14915–14921
14. Guo S, Wen D, Zhai Y, Dong S, Wang E (2010) Platinum nanoparticle ensemble-on-graphene hybrid nanosheet: one-pot, rapid synthesis, and used as new electrode material for electrochemical sensing. ACS Nano 7:3959–3968
15. Yao Z, Zhu M, Jiang F, Du Y, Wang C, Yang P (2012) Highly efficient electrocatalytic performance based on Pt nanoflowers modified reduced graphene oxide/carbon cloth electrode. J Mater Chem 22:13707–13713
16. Li J, Fan C, Xiao F, Yan R, Fan S, Zhao F, Zeng B (2007) Influence of ionic liquids on the direct electrochemistry of glucose oxidase entrapped in nanogold–*N,N*-dimethylformamide–ionic liquid composite film. Electrochim Acta 52:6178–6185
17. Lu J, Liu S, Ge S, Yan M, Yu J, Hu X (2012) Ultrasensitive electrochemical immunosensor based on Au nanoparticles dotted carbon nanotube–graphene composite and functionalized mesoporous materials. Biosens Bioelectron 33:29–35
18. Cui F, Zhang X (2012) Electrochemical sensor for epinephrine based on a glassy carbon electrode modified with graphene/gold nanocomposites. J Electroanal Chem 669:35–41
19. Shanmugasundaram K, Subramanian P, Paramasivam M, Iyengar GA, Lee KP (2011) Preparation of a functional nanofibrous polymer membrane incorporated with poly(2-aminothio phenol) stabilized gold nanoparticles. Gold Bull 44:37–42
20. Yue Y, Hu G, Zheng M, Guo Y, Cao J, Shao S (2012) A mesoporous carbon nanofiber-modified pyrolytic graphite electrode used for the simultaneous determination of dopamine, uric acid, and ascorbic acid. Carbon 50:107–114
21. Kim YR, Bong S, Kang YJ, Yang Y, Mahajan RK, Kim JS, Kim H (2010) Electrochemical detection of dopamine in the presence of ascorbic acid using graphene modified electrodes. Biosens Bioelectron 25:2366–2369
22. Luo J, Jiang S, Zhang H, Jiang J, Liu X (2012) A novel nonenzymatic glucose sensor based on Cu nanoparticle modified graphene sheets electrode. Anal Chim Acta 709:47–53
23. Wang Y, Li Y, Tang L, Lu J, Li J (2009) Application of graphene-modified electrode for selective detection of dopamine. Electrochem Commun 11:889–892
24. Zhou K, Zhu Y, Yang X, Li C (2010) Electrocatalytic oxidation of glucose by the glucose oxidase immobilized in graphene–Au–nafion biocomposite. Electroanalysis 22:259–264
25. Li J, Yang J, Yang Z, Li Y, Yu S, Xu Q, Hu X (2012) Graphene–Au nanoparticles nanocomposite film for selective electrochemical determination of dopamine. Anal Methods 4:1725–1728
26. Hummers WS, Offeman RE (1958) Preparation of graphitic oxide. J Am Chem Soc 80:1339–1339

27. Lin K, Tsai T, Chen S (2010) Performing enzyme-free $H_2O_2$ biosensor and simultaneous determination for AA, DA and UA by MWCNT–PEDOT film. Biosens Bioelectron 26:608–614

28. Thiagarajan S, Chen S (2007) Preparation and characterization of PtAu hybrid film modified electrodes and their use in simultaneous determination of dopamine, ascorbic acid and uric acid. Talanta 74:212–222

29. Brownson D, Foster C, Banks C (2012) The electrochemical performance of graphene modified electrodes: an analytical perspective. Analyst 137:1815–1823

30. Pruneanu S, Pogacean F, Biris AR, Ardelean S, Canpean V, Blanita G, Dervishi E, Biris AS (2011) Novel graphene–gold nanoparticle modified electrodes for the high sensitivity electrochemical spectroscopy detection and analysis of carbamazepine. J Phys Chem C 115:23387–23394

# Permissions

All chapters in this book were first published in GB, by Springer; hereby published with permission under the Creative Commons Attribution License or equivalent. Every chapter published in this book has been scrutinized by our experts. Their significance has been extensively debated. The topics covered herein carry significant findings which will fuel the growth of the discipline. They may even be implemented as practical applications or may be referred to as a beginning point for another development.

The contributors of this book come from diverse backgrounds, making this book a truly international effort. This book will bring forth new frontiers with its revolutionizing research information and detailed analysis of the nascent developments around the world.

We would like to thank all the contributing authors for lending their expertise to make the book truly unique. They have played a crucial role in the development of this book. Without their invaluable contributions this book wouldn't have been possible. They have made vital efforts to compile up to date information on the varied aspects of this subject to make this book a valuable addition to the collection of many professionals and students.

This book was conceptualized with the vision of imparting up-to-date information and advanced data in this field. To ensure the same, a matchless editorial board was set up. Every individual on the board went through rigorous rounds of assessment to prove their worth. After which they invested a large part of their time researching and compiling the most relevant data for our readers.

The editorial board has been involved in producing this book since its inception. They have spent rigorous hours researching and exploring the diverse topics which have resulted in the successful publishing of this book. They have passed on their knowledge of decades through this book. To expedite this challenging task, the publisher supported the team at every step. A small team of assistant editors was also appointed to further simplify the editing procedure and attain best results for the readers.

Apart from the editorial board, the designing team has also invested a significant amount of their time in understanding the subject and creating the most relevant covers. They scrutinized every image to scout for the most suitable representation of the subject and create an appropriate cover for the book.

The publishing team has been an ardent support to the editorial, designing and production team. Their endless efforts to recruit the best for this project, has resulted in the accomplishment of this book. They are a veteran in the field of academics and their pool of knowledge is as vast as their experience in printing. Their expertise and guidance has proved useful at every step. Their uncompromising quality standards have made this book an exceptional effort. Their encouragement from time to time has been an inspiration for everyone.

The publisher and the editorial board hope that this book will prove to be a valuable piece of knowledge for researchers, students, practitioners and scholars across the globe.

# List of Contributors

**Xiaoyun Qin**
Chemical Synthesis and Pollution Control Key Laboratory of Sichuan Province, School of Chemistry and Chemical Industry, China West Normal University, Nanchong 637002 Sichuan, China

**Wenbo Lu**
Chemical Synthesis and Pollution Control Key Laboratory of Sichuan Province, School of Chemistry and Chemical Industry, China West Normal University, Nanchong 637002 Sichuan, China

**Guohui Chang**
Chemical Synthesis and Pollution Control Key Laboratory of Sichuan Province, School of Chemistry and Chemical Industry, China West Normal University, Nanchong 637002 Sichuan, China

**Yonglan Luo**
Chemical Synthesis and Pollution Control Key Laboratory of Sichuan Province, School of Chemistry and Chemical Industry, China West Normal University, Nanchong 637002 Sichuan, China

**Abdullah M. Asiri**
Chemistry Department, Faculty of Science, King Abdulaziz University, Jeddah 21589, Saudi Arabia
Center of Excellence for Advanced Materials Research, King Abdulaziz University, Jeddah 21589, Saudi Arabia

**Abdulrahman O. Al-Youbi**
Chemistry Department, Faculty of Science, King Abdulaziz University, Jeddah 21589, Saudi Arabia
Center of Excellence for Advanced Materials Research, King Abdulaziz University, Jeddah 21589, Saudi Arabia

**Xuping Sun**
Chemistry Department, Faculty of Science, King Abdulaziz University, Jeddah 21589, Saudi Arabia
Chemical Synthesis and Pollution Control Key Laboratory of Sichuan Province, School of Chemistry and Chemical Industry, China West Normal University, Nanchong 637002 Sichuan, China
Center of Excellence for Advanced Materials Research, King Abdulaziz University, Jeddah 21589, Saudi Arabia

**Suman Singh**
Central Scientific Instruments Organization, Chandigarh, India

**Pooja Devi**
Central Scientific Instruments Organization, Chandigarh, India

**Deepak Singh**
Central Scientific Instruments Organization, Chandigarh, India

**D. V. S. Jain**
Indian Institute of Pulses Research, (IIPR), Kanpur, India

**M. L. Singla**
Panjab University, Chandigarh, India

**Wei Wang**
Institute of Colloid and Interface Chemistry, Faculty of Chemistry, Northeast Normal University, Changchun 130024, People's Republic of China
State Key Laboratory on Integrated Optoelectronics, College of Electronic Science and Engineering, Jilin University, Changchun 130012, People's Republic of China

**Yujia Pang**
Institute of Colloid and Interface Chemistry, Faculty of Chemistry, Northeast Normal University, Changchun 130024, People's Republic of China

**Jian Yan**
Institute of Colloid and Interface Chemistry, Faculty of Chemistry, Northeast Normal University, Changchun 130024, People's Republic of China

**Guibao Wang**
State Key Laboratory on Integrated Optoelectronics, College of Electronic Science and Engineering, Jilin University, Changchun 130012, People's Republic of China

**Hui Suo**
State Key Laboratory on Integrated Optoelectronics, College of Electronic Science and Engineering, Jilin University, Changchun 130012, People's Republic of China

**Chun Zhao**
State Key Laboratory on Integrated Optoelectronics, College of Electronic Science and Engineering, Jilin University, Changchun 130012, People's Republic of China

**Shuangxi Xing**
Institute of Colloid and Interface Chemistry, Faculty of Chemistry, Northeast Normal University, Changchun 130024, People's Republic of China

**Guiying Liu**
Zhejiang Key Laboratory for Reactive Chemistry on Solid Surfaces, Institute of Physical Chemistry, Zhejiang Normal University, Jinhua 321004 Zhejiang, People's Republic of China

**Yong Shao**
Zhejiang Key Laboratory for Reactive Chemistry on Solid Surfaces, Institute of Physical Chemistry, Zhejiang Normal University, Jinhua 321004 Zhejiang, People's Republic of China

**Kun Ma**
Zhejiang Key Laboratory for Reactive Chemistry on Solid Surfaces, Institute of Physical Chemistry, Zhejiang Normal University, Jinhua 321004 Zhejiang, People's Republic of China

**Qinghua Cui**
Zhejiang Key Laboratory for Reactive Chemistry on Solid Surfaces, Institute of Physical Chemistry, Zhejiang Normal University, Jinhua 321004 Zhejiang, People's Republic of China

**Fei Wu**
Zhejiang Key Laboratory for Reactive Chemistry on Solid Surfaces, Institute of Physical Chemistry, Zhejiang Normal University, Jinhua 321004 Zhejiang, People's Republic of China

**Shujuan Xu**
Zhejiang Key Laboratory for Reactive Chemistry on Solid Surfaces, Institute of Physical Chemistry, Zhejiang Normal University, Jinhua 321004 Zhejiang, People's Republic of China

**Kara Zabetakis**
Department of Chemistry and Biochemistry, University of Maryland, Baltimore County, Baltimore, MD 21250, USA
School of Civil and Environmental Engineering, University of Maryland, College Park, Baltimore, MD 20742, USA

**William E. Ghann**
Department of Chemistry and Biochemistry, University of Maryland, Baltimore County, Baltimore, MD 21250, USA

**Sanjeev Kumar**
Department of Chemical Engineering, Indian Institute of Science, Bangalore 560012, India

**Marie-Christine Daniel**
Department of Chemistry and Biochemistry, University of Maryland, Baltimore County, Baltimore, MD 21250, USA

**Helin Liu**
NanoScience Technology Center, Department of Chemistry and Burnett School of Biomedical Science, University of Central Florida, Orlando, FL 32826, USA

**Nickisha Pierre-Pierre**
NanoScience Technology Center, Department of Chemistry and Burnett School of Biomedical Science, University of Central Florida, Orlando, FL 32826, USA

**Qun Huo**
NanoScience Technology Center, Department of Chemistry and Burnett School of Biomedical Science, University of Central Florida, Orlando, FL 32826, USA

**Yuan-zhi Song**
Jiangsu Province Key Laboratory for Chemistry of Low-Dimensional Materials, School of Chemistry and Chemical Engineering, Huaiyin Normal University, Huai An 223300, People's Republic of China

**An-feng Zhu**
Jiangsu Province Key Laboratory for Chemistry of Low-Dimensional Materials, School of Chemistry and Chemical Engineering, Huaiyin Normal University, Huai An 223300, People's Republic of China

**Yang Song**
College of Materials Science and Engineering, Beijing University of Chemical Technology, Peking 100029, People's Republic of China

**Zhi-peng Cheng**
Jiangsu Province Key Laboratory for Chemistry of Low-Dimensional Materials, School of Chemistry and Chemical Engineering, Huaiyin Normal University, Huai An 223300, People's Republic of China

**Jian Xu**
College of Materials Science and Engineering, Beijing University of Chemical Technology, Peking 100029, People's Republic of China

**Jian-feng Zhou**
Jiangsu Province Key Laboratory for Chemistry of Low-Dimensional Materials, School of Chemistry and Chemical Engineering, Huaiyin Normal University, Huai An 223300, People's Republic of China

**Ali Ayati**
Laboratory of Green Chemistry, Lappeenranta University of Technology, Jääkärinkatu 31, 50100 Mikkeli, Finland
Department of Chemical Engineering, Ferdowsi University of Mashhad, Mashhad, Iran

**Ali Ahmadpour**
Department of Chemical Engineering, Ferdowsi University of Mashhad, Mashhad, Iran

**Fatemeh F. Bamoharram**
Department of Chemistry, Mashhad Branch, Islamic Azad University, Mashhad, Iran

**Majid M. Heravi**
Department of Chemistry, School of Sciences, Alzahra University, Tehran, Iran

**Mika Sillanpää**
Laboratory of Green Chemistry, Lappeenranta University of Technology, Jääkärinkatu 31, 50100 Mikkeli, Finland

**Se Yeon Park**
Department of Chemical Engineering, Hanyang University, Seoul 133-791, South Korea

**So Min Lee**
Department of Bio Engineering, Hanyang University, Seoul 133-791, South Korea

**Gae Baik Kim**
Department of Life Science, Hanyang University, Seoul 133-791, South Korea
Research Institute for Natural Sciences, Hanyang University, Seoul 133-791, South Korea

**Young-Pil Kim**
Department of Life Science, Hanyang University, Seoul 133-791, South Korea
Research Institute for Natural Sciences, Hanyang University, Seoul 133-791, South Korea

**DaJian Wu**
School of Physics, Nanjing University, Nanjing 210093, China
Faculty of Science, Jiangsu University, henjiang 212013, China

**ShuMin Jiang**
Faculty of Science, Jiangsu University, henjiang 212013, China

**Ying Cheng**
School of Physics, Nanjing University, Nanjing 210093, China

**XiaoJun Liu**
School of Physics, Nanjing University, Nanjing 210093, China

**Kazu Okumura**
Department of Chemistry and Biotechnology, Graduate School of Engineering, Tottori University, 4-101 Koyama-cho Minami, Tottori 680-8552, Japan

**Chika Murakami**
Department of Chemistry and Biotechnology, Graduate School of Engineering, Tottori University, 4-101 Koyama-cho Minami, Tottori 680-8552, Japan

**Tetsuya Oyama**
Department of Chemistry and Biotechnology, Graduate School of Engineering, Tottori University, 4-101 Koyama-cho Minami, Tottori 680-8552, Japan

**Takashi Sanada**
Department of Chemistry and Biotechnology, Graduate School of Engineering, Tottori University, 4-101 Koyama-cho Minami, Tottori 680-8552, Japan
Research Department, NISSAN ARC, LTD., Yokosuka 237-0061, Japan

**Ayano Isoda**
Research Department, NISSAN ARC, LTD., Yokosuka 237-0061, Japan

**Naonobu Katada**
Department of Chemistry and Biotechnology, Graduate School of Engineering, Tottori University, 4-101 Koyama-cho Minami, Tottori 680-8552, Japan

**Ana L. Daniel-da-Silva**
Department of Chemistry, CICECO, Aveiro Institute of Nanotechnology, University of Aveiro, 3810-193 Aveiro, Portugal

**Ana M. Salgueiro**
Department of Chemistry, CICECO, Aveiro Institute of Nanotechnology, University of Aveiro, 3810-193 Aveiro, Portugal

**Tito Trindade**
Department of Chemistry, CICECO, Aveiro Institute of Nanotechnology, University of Aveiro, 3810-193 Aveiro, Portugal

**N. Alissawi**
Institute for Materials Science, Multicomponent Materials, Faculty of Engineering, Christian-Albrechts-University (CAU) Kiel, Kiel, Germany

**V. Zaporojtchenko**
Institute for Materials Science, Multicomponent Materials, Faculty of Engineering, Christian-Albrechts-University (CAU) Kiel, Kiel, Germany

**T. Strunskus**
Institute for Materials Science, Multicomponent Materials, Faculty of Engineering, Christian-Albrechts-University (CAU) Kiel, Kiel, Germany

**I. Kocabas**
Institute for Materials Science, Multicomponent Materials, Faculty of Engineering, Christian-Albrechts-University (CAU) Kiel, Kiel, Germany

**V. S. K. Chakravadhanula**
Institute for Materials Science, Synthesis and Real Structure, Faculty of Engineering, Christian-Albrechts-University (CAU) Kiel, Kaiser Str. 2, 24143 Kiel, Germany Helmholtz Institute Ulm (HIU) Electrochemical energy storage, Karlsruhe Institute of Technology (KIT), Albert-Einstein-Allee 11, 89081, Ulm, Germany

**L. Kienle**
Institute for Materials Science, Synthesis and Real Structure, Faculty of Engineering, Christian-Albrechts-University (CAU) Kiel, Kaiser Str. 2, 24143 Kiel, Germany

**D. Garbe-Schönberg**
Institute of Geosciences/ICP-MS Lab, Christian-Albrechts-University (CAU) Kiel, Ludewig-Meyn-Str. 10, 24118 Kiel, Germany

**F. Faupel**
Institute for Materials Science, Multicomponent Materials, Faculty of Engineering, Christian-Albrechts-University (CAU) Kiel, Kiel, Germany

**Sónia Carabineiro**
Laboratory of Catalysis and Materials Department of Chemical Engineering, Faculty of Engineering, University of Porto, Rua Dr. Roberto Frias, s/n, 4200-465 Porto, Portugal

**Thomas Borrmann**
School of Chemical and Physical Sciences, Victoria University of Wellington, PO Box 600, 6140 Wellington, New Zealand
Othbergstr.10, 37632 Eschershausen, Germany

**Teck Hock Lim**
School of Chemical and Physical Sciences, Victoria University of Wellington, PO Box 600, 6140 Wellington, New Zealand

**Hannah Cope**
School of Chemical and Physical Sciences, Victoria University of Wellington, PO Box 600, 6140 Wellington, New Zealand

**Kerstin Lucas**
School of Chemical and Physical Sciences, Victoria University of Wellington, PO Box 600, 6140 Wellington, New Zealand

**Michael Lorden**
School of Chemical and Physical Sciences, Victoria University of Wellington, PO Box 600, 6140 Wellington, New Zealand

**Zeinab Esmail Nazari**
Institute of Physics and Nanotechnology, Aalborg University, Aalborg, Denmark

**Maryam Banoee**
Azad University of North Tehran Branch, Tehran, Iran

**Abbas Akhavan Sepahi**
Azad University of North Tehran Branch, Tehran, Iran

**Fatemeh Rafii**
Division of Microbiology, National Center for Toxicological Research, Jefferson, AR, USA

**Ahmad Reza Shahverdi**
Department of Pharmaceutical Biotechnology and Biotechnology Research Center, Faculty of Pharmacy, Tehran University of Medical Sciences, Tehran, Iran

**Yolaine Joubert**
Université de Nantes, MMS EA 2160, LUNAM Université, 9, Rue Bias, 44035 Nantes, France

**Jin-Fen Pan**
Université de Nantes, MMS EA 2160, LUNAM Université, 9, Rue Bias, 44035 Nantes, France
College of Environmental Science and Engineering, Ocean University of China, Qingdao 266100 China

**Pierre-Emmanuel Buffet**
Université de Nantes, MMS EA 2160, LUNAM Université, 9, Rue Bias, 44035 Nantes, France
Université Catholique de l'Ouest, MMS EA 2160, LUNAM Université, 3, Place André Leroy, 49000 Angers, France

**Paul Pilet**
School of dental Surgery, INSERM U791, 1 Place Alexis Ricordeau, 44042 Nantes Cedex 1, France

**Douglas Gilliland**
Institute For Health and Consumer Protection European Commission—DG JRC, Via E. Fermi, 19 I-21027 Ispra, VA, Italy

**Eugenia Valsami-Jones**
School of Geography, Earth and Environmental Sciences, University of Birmingham, Edgbaston, Birmingham B15 2TT, UK
Department of Earth Sciences, Natural History Museum London, Cromwell Road, London SW7 5BD, UK

**Catherine Mouneyrac**
Université de Nantes, MMS EA 2160, LUNAM Université, 9, Rue Bias, 44035 Nantes, France

Université Catholique de l'Ouest, MMS EA 2160, LUNAM Université, 3, Place André Leroy, 49000 Angers, France

**Claude Amiard-Triquet**
Université de Nantes, MMS EA 2160, LUNAM Université, 9, Rue Bias, 44035 Nantes, France

**Gérrard Eddy Jai Poinern**
Murdoch Applied Nanotechnology Research Group, Department of Physics, Energy Studies, and Nanotechnology, School of Engineering and Energy, Murdoch University, Murdoch, WA 6150, Australia

**Peter Chapman**
Department of Chemistry, Curtin University of Technology, Bentley, WA 6102, Australia

**Xuan Le**
Murdoch Applied Nanotechnology Research Group, Department of Physics, Energy Studies, and Nanotechnology, School of Engineering and Energy, Murdoch University, Murdoch, WA 6150, Australia

**Derek Fawcett**
Murdoch Applied Nanotechnology Research Group, Department of Physics, Energy Studies, and Nanotechnology, School of Engineering and Energy, Murdoch University, Murdoch, WA 6150, Australia

**Néné Thioune**
Laboratoire CSPBAT UMR CNRS 7244 UFR SMBH, Université Paris 13, Sorbonne Paris Cité, 74 rueMarcel Cachin, 93017 Bobigny, France
Département d'Odontologie de Dakar, Université Cheikh Anta Diop (UCAD) BP 5005, Dakar, Senegal

**Nathalie Lidgi-Guigui**
Laboratoire CSPBAT UMR CNRS 7244 UFR SMBH, Université Paris 13, Sorbonne Paris Cité, 74 rueMarcel Cachin, 93017 Bobigny, France

**Maximilien Cottat**
Laboratoire CSPBAT UMR CNRS 7244 UFR SMBH, Université Paris 13, Sorbonne Paris Cité, 74 rueMarcel Cachin, 93017 Bobigny, France

**Ana-Maria Gabudean**
Nanobiophotonics and Laser Microspectroscopy Center, Interdisciplinary Research Institute on Bio-Nano-Sciences, Faculty of Physics, Babes-Bolyai University, T. Laurian 42, 400271 Cluj-Napoca, Romania

**Monica Focsan**
Nanobiophotonics and Laser Microspectroscopy Center, Interdisciplinary Research Institute on Bio-Nano-Sciences, Faculty of Physics, Babes-Bolyai University, T. Laurian 42, 400271 Cluj-Napoca, Romania

**Henri-Michel Benoist**
Département d'Odontologie de Dakar, Université Cheikh Anta Diop (UCAD) BP 5005, Dakar, Senegal

**Simion Astilean**
Nanobiophotonics and Laser Microspectroscopy Center, Interdisciplinary Research Institute on Bio-Nano-Sciences, Faculty of Physics, Babes-Bolyai University, T. Laurian 42, 400271 Cluj-Napoca, Romania

**Marc Lamy de la Chapelle**
Laboratoire CSPBAT UMR CNRS 7244 UFR SMBH, Université Paris 13, Sorbonne Paris Cité, 74 rueMarcel Cachin, 93017 Bobigny, France

**Yukiko Yasukawa**
Department of Computer Science and Engineering, Faculty of Engineering, Shinshu University, 4-17-1 Wakasato, Nagano 380-8553, Japan

**Xiaoxi Liu**
Department of Computer Science and Engineering, Faculty of Engineering, Shinshu University, 4-17-1 Wakasato, Nagano 380-8553, Japan

**Akimitsu Morisako**
Department of Computer Science and Engineering, Faculty of Engineering, Shinshu University, 4-17-1 Wakasato, Nagano 380-8553, Japan

**Mariana Pannerec-Varna**
Université Paris Diderot, Sorbonne Paris Cité, UMR-S 728, 75010 Paris, France
INSERM, U728, 75010 Paris, France

**Philippe Ratajczak**
Université Paris Diderot, Sorbonne Paris Cité, UMR-S 728, 75010 Paris, France
INSERM, U728, 75010 Paris, France

**Guilhem Bousquet**
Université Paris Diderot, Sorbonne Paris Cité, UMR-S 728, 75010 Paris, France
INSERM, U728, 75010 Paris, France

**Irmine Ferreira**
Université Paris Diderot, Sorbonne Paris Cité, UMR-S 728, 75010 Paris, France
INSERM, U728, 75010 Paris, France

**Christophe Leboeuf**
Université Paris Diderot, Sorbonne Paris Cité, UMR-S 728, 75010 Paris, France
INSERM, U728, 75010 Paris, France

**Raphaël Boisgard**
Laboratoire de Photonique Quantique etMoléculaire, Ecole Centrale Paris, Grande Voie des Vignes, 92295 Châtenay-Malabry Cedex, France

**Guillaume Gapihan**
Université Paris Diderot, Sorbonne Paris Cité, UMR-S 728, 75010 Paris, France
INSERM, U728, 75010 Paris, France

**Jérôme Verine**
Université Paris Diderot, Sorbonne Paris Cité, UMR-S 728, 75010 Paris, France
INSERM, U728, 75010 Paris, France

**Bruno Palpant**
Commissariat d'Energie Atomique et aux Energies Alternatives, SHFJ, INSERM, U1023, 91401 Orsay, France

**Emmanuel Bossy**
Université Paris Diderot, Sorbonne Paris Cité, UMR-S 728, 75010 Paris, France
INSERM, U728, 75010 Paris, France

**Eric Doris**
Université Paris Diderot, Sorbonne Paris Cité, UMR-S 728, 75010 Paris, France
INSERM, U728, 75010 Paris, France

**Joel Poupon**
Université Paris Diderot, Sorbonne Paris Cité, UMR-S 728, 75010 Paris, France
INSERM, U728, 75010 Paris, France

**Emmanuel Fort**
Université Paris Diderot, Sorbonne Paris Cité, UMR-S 728, 75010 Paris, France
INSERM, U728, 75010 Paris, France

**Anne Janin**
Université Paris Diderot, Sorbonne Paris Cité, UMR-S 728, 75010 Paris, France
INSERM, U728, 75010 Paris, France

**Brett W. Boote**
Department of Chemistry, Illinois State University, Normal, IL 61790-4160, USA

**Hongsik Byun**
Department of Chemical System Engineering, Keimyung University, Daegu 704-701, South Korea

**Jun-Hyun Kim**
Department of Chemistry, Illinois State University, Normal, IL 61790-4160, USA

**Maroua Ben Haddada**
UPMCUniv Paris 6,UMRCNRS 7197, Laboratoire de Réactivité de Surface, 75005 Paris, France
Laboratoire de Réactivité de Surface, UMR CNRS 7197, Université Pierre et Marie Curie-Paris VI, 4 Place Jussieu, 75252 Paris cedex 05, France

**Juliette Blanchard**
UPMCUniv Paris 6,UMRCNRS 7197, Laboratoire de Réactivité de Surface, 75005 Paris, France
Laboratoire de Réactivité de Surface, UMR CNRS 7197, Université Pierre et Marie Curie-Paris VI, 4 Place Jussieu, 75252 Paris cedex 05, France

**Sandra Casale**
UPMCUniv Paris 6,UMRCNRS 7197, Laboratoire de Réactivité de Surface, 75005 Paris, France
Laboratoire de Réactivité de Surface, UMR CNRS 7197, Université Pierre et Marie Curie-Paris VI, 4 Place Jussieu, 75252 Paris cedex 05, France

**Jean-Marc Krafft**
UPMCUniv Paris 6,UMRCNRS 7197, Laboratoire de Réactivité de Surface, 75005 Paris, France
Laboratoire de Réactivité de Surface, UMR CNRS 7197, Université Pierre et Marie Curie-Paris VI, 4 Place Jussieu, 75252 Paris cedex 05, France

**Anne Vallée**
UPMCUniv Paris 6,UMRCNRS 7197, Laboratoire de Réactivité de Surface, 75005 Paris, France
Laboratoire de Réactivité de Surface, UMR CNRS 7197, Université Pierre et Marie Curie-Paris VI, 4 Place Jussieu, 75252 Paris cedex 05, France

**Christophe Méthivier**
UPMCUniv Paris 6,UMRCNRS 7197, Laboratoire de Réactivité de Surface, 75005 Paris, France
Laboratoire de Réactivité de Surface, UMR CNRS 7197, Université Pierre et Marie Curie-Paris VI, 4 Place Jussieu, 75252 Paris cedex 05, France

**Souhir Boujday**
UPMCUniv Paris 6,UMRCNRS 7197, Laboratoire de Réactivité de Surface, 75005 Paris, France
Laboratoire de Réactivité de Surface, UMR CNRS 7197, Université Pierre et Marie Curie-Paris VI, 4 Place Jussieu, 75252 Paris cedex 05, France

**Damien Canneson**
Groupe d' ´Etude de la Matière Condensée, Université de Versailles-Saint-Quentin-en-Yvelines, CNRS UMR8635, 45 avenue des ´Etats-Unis, 78035 Versailles, France

**Stéphanie Buil**
Groupe d' ´Etude de la Matière Condensée, Université de Versailles-Saint-Quentin-en-Yvelines, CNRS UMR8635, 45 avenue des ´Etats-Unis, 78035 Versailles, France

**Xavier Quélin**
Groupe d' ´Etude de la Matière Condensée, Université de Versailles-Saint-Quentin-en-Yvelines, CNRS UMR8635, 45 avenue des ´Etats-Unis, 78035 Versailles, France

**Clémentine Javaux**
Laboratoire de Physique et d' ´ Etude des Matériaux, CNRS UMR8213, ESPCI, 10 rue Vauquelin, 75005 Paris, France

**Benoît Dubertret**
Institut Universitaire de France, 103 boulevard Saint-Michel, 75005 Paris, France

**Jean-Pierre Hermier**
Groupe d' ´ Etude de la Matière Condensée, Université de Versailles-Saint-Quentin-en-Yvelines, CNRS UMR8635, 45 avenue des ´ Etats-Unis, 78035 Versailles, France

**Stéphane Lemonier**
Laboratoire IMRCP, CNRS UMR 5623, Université de Toulouse, 118, route de Narbonne, 31062 Toulouse, France

**Pierre Moutet**
LPCNO, INSA-UPS-CNRS, Université de Toulouse, 135 avenue de Rangueil, 31077 Toulouse Cedex 4, France

**Wissam Moussa**
Laboratoire IMRCP, CNRS UMR 5623, Université de Toulouse, 118, route de Narbonne, 31062 Toulouse, France

**Mathias Destarac**
Laboratoire HFA, CNRS UMR 5069, Université de Toulouse, 118, route de Narbonne, 31062 Toulouse, France

**Laurence Ressier**
LPCNO, INSA-UPS-CNRS, Université de Toulouse, 135 avenue de Rangueil, 31077 Toulouse Cedex 4, France

**Jean-Daniel Marty**
Laboratoire IMRCP, CNRS UMR 5623, Université de Toulouse, 118, route de Narbonne, 31062 Toulouse, France

**N. Pellerin**
CNRS, UPR3079 CEMHTI, 1D avenue de la recherche scientifique, 45071, Orléans Cedex 2, France
Université d'Orléans, Faculté des Sciences, Avenue du parc floral, BP 6749, 45067, Orléans Cedex 2, France

**J-P. Blondeau**
CNRS, UPR3079 CEMHTI, 1D avenue de la recherche scientifique, 45071, Orléans Cedex 2, France
Université d'Orléans, Faculté des Sciences, Avenue du parc floral, BP 6749, 45067, Orléans Cedex 2, France

**S. Noui**
CNRS, UPR3079 CEMHTI, 1D avenue de la recherche scientifique, 45071, Orléans Cedex 2, France
Université d'Orléans, Faculté des Sciences, Avenue du parc floral, BP 6749, 45067, Orléans Cedex 2, France

**M. Allix**
CNRS, UPR3079 CEMHTI, 1D avenue de la recherche scientifique, 45071, Orléans Cedex 2, France
Université d'Orléans, Faculté des Sciences, Avenue du parc floral, BP 6749, 45067, Orléans Cedex 2, France

**S. Ory**
CNRS, UPR3079 CEMHTI, 1D avenue de la recherche scientifique, 45071, Orléans Cedex 2, France
Université d'Orléans, Faculté des Sciences, Avenue du parc floral, BP 6749, 45067, Orléans Cedex 2, France

**O. Veron**
CNRS, UPR3079 CEMHTI, 1D avenue de la recherche scientifique, 45071, Orléans Cedex 2, France
Université d'Orléans, Faculté des Sciences, Avenue du parc floral, BP 6749, 45067, Orléans Cedex 2, France

**D. De Sousa Meneses**
CNRS, UPR3079 CEMHTI, 1D avenue de la recherche scientifique, 45071, Orléans Cedex 2, France
Institut de Chimie, CNRS, 3 rue Michel-Ange, 75794, Paris cedex 16, France

**D. Massiot**
CNRS, UPR3079 CEMHTI, 1D avenue de la recherche scientifique, 45071, Orléans Cedex 2, France
Université d'Orléans, Faculté des Sciences, Avenue du parc floral, BP 6749, 45067, Orléans Cedex 2, France

**Seydou Hebié**
IC2MP, UMR 7285 CNRS, Université de Poitiers, 4, rue Michel Brunet B-27, BP 633, 86022 Poitiers Cedex, France

**K. Boniface Kokoh**
IC2MP, UMR 7285 CNRS, Université de Poitiers, 4, rue Michel Brunet B-27, BP 633, 86022 Poitiers Cedex, France

**Karine Servat**
IC2MP, UMR 7285 CNRS, Université de Poitiers, 4, rue Michel Brunet B-27, BP 633, 86022 Poitiers Cedex, France

**Teko W. Napporn**
IC2MP, UMR 7285 CNRS, Université de Poitiers, 4, rue Michel Brunet B-27, BP 633, 86022 Poitiers Cedex, France

CPSIA information can be obtained
at www.ICGtesting.com
Printed in the USA
BVOW10*0421050716
454361BV00018B/142/P